MONOCHROM

AND

COLOUR TELEVI

MONOCHROME
AND
COLOUR TELEVISION

R R Gulati

**NEW AGE
TECHNO
PRESS**

An Imprint of

NEW AGE INTERNATIONAL (P) LIMITED, PUBLISHERS
New Delhi • Bangalore • Chennai • Cochin • Guwahati
Hyderabad • Kolkata • Lucknow • Mumbai
Visit us at **www.newagepublishers.com**

BRANCHES

- **Bangalore** 37/10, 8th Cross (Near Hanuman Temple), Azad Nagar, Chamarajpet, Bangalore-560 018
  Tel.: (080) 26756823, **Telefax:** 26756820, **E-mail:** bangalore@newagepublishers.com

- **Chennai** 26, Damodaran Street, T. Nagar, Chennai-600 017
  Tel.: (044) 24353401, **Telefax:** 24351463, **E-mail:** chennai@newagepublishers.com

- **Cochin** CC-39/1016, Carrier Station Road, Ernakulam South, Cochin-682 016
  Tel.: (0484) 2377004, **Telefax:** 4051303, **E-mail:** cochin@newagepublishers.com

- **Guwahati** Hemsen Complex, Mohd. Shah Road, Paltan Bazar, Near Starline Hotel, Guwahati-781 008
  Tel.: (0361) 2513881, **Telefax:** 2543669, **E-mail:** guwahati@newagepublishers.com

- **Hyderabad** 105, 1st Floor, Madhiray Kaveri Tower, 3-2-19, Azam Jahi Road, Nimboliadda, Hyderabad-500 027
  Tel.: (040) 24652456, **Telefax:** 24652457, **E-mail:** hyderabad@newagepublishers.com

- **Kolkata** RDB Chambers (Formerly Lotus Cinema) 106A, 1st Floor, S.N. Banerjee Road, Kolkata-700 014
  Tel.: (033) 22273773, **Telefax:** 22275247, **E-mail:** kolkata@newagepublishers.com

- **Lucknow** 16-A, Jopling Road, Lucknow-226 001
  Tel.: (0522) 2209578, 4045297, **Telefax:** 2204098, **E-mail:** lucknow@newagepublishers.com

- **Mumbai** 142C, Victor House, Ground Floor, N.M. Joshi Marg, Lower Parel, Mumbai-400 013
  Tel.: (022) 24927869, **Telefax:** 24915415, **E-mail:** mumbai@newagepublishers.com

- **New Delhi** 22, Golden House, Daryaganj, New Delhi-110 002
  Tel.: (011) 23262368, 23262370, **Telefax:** 43551305, **E-mail:** sales@newagepublishers.com

---

ISBN: 978-81-224-3606-8

₹ 399.00

C-13-06-6959

Printed in India at Rakesh Press, Noida (UP)
Typeset at In-house, Delhi.

PUBLISHING FOR ONE WORLD

NEW AGE INTERNATIONAL (P) LIMITED, PUBLISHERS
7/30 A, Daryaganj, New Delhi-110002
Visit us at **www.newagepublishers.com**

To

My Wife

PARKASH

PREFACE TO THE THIRD EDITION

Out of the various television systems in use in different countries, India adopted the 625-B monochrome (black and white) and the compatible PAL-B colour systems. Most European and many other countries are also using these standards. This book is therefore based on these systems and presents an integrated approach with equal emphasis on both monochrome and colour television. It thus meets the requirements of a complete text on Television Engineering. Comprehensive design criteria for various sections of the receiver have been given in each chapter of the book without going into rigorous mathematical details. Due emphasis has also been laid on TV receiver servicing and servicing equipment. Detailed charts for locating faults and trouble shooting together with alignment procedures for various sections of the receiver have also been included.

Early TV receivers manufactured in India and other countries used vacuum tube circuitry. However, with rapid advances in technology, hybrid circuitry soon came into use and transistors replaced most vacuum tubes. With the widespread development of integrated circuits, special ICs are now available replacing discrete circuitry employing transistors. Since these developments have been very fast, sets employing tubes, only transistors and ICs are in use simultaneously. In view of this fact, discussion of circuits using tubes, hybrid circuitry and ICs has been included in the chapters devoted to various sections of the TV receiver. The stress, however, is more on solid state receiver circuits and design.

Because of the importance of colour transmission and reception, two comprehensive chapters have been exclusively devoted to the techniques of colour television and various colour television systems. The American NTSC and French SECAM television systems have also been accorded due coverage while presenting the adopted PAL colour system.

Over the years there has been rapid advances in television technology which has enabled Satellite Television, Plasma and LCD panel TV receivers, Projection TV and 3D televisions. Thus separate chapters on these topics have been added in this edition to introduce these developments.

For the convenience of engineering students, tables of conversion factors and prefixes, transient response and wave shaping and frequency bands of TV broadcast channels have been added as appendices. A set of revision questions have also been included at the end of each chapter.

R R Gulati

CONTENTS

Introduction

Introduction

Development of Television

Television* means 'to see from a distance'. The desire in man to do so has been there for ages. In the early years of the twentieth century many scientists experimented with the idea of using selenium photosensitive cells for converting light from pictures into electrical signals and transmitting them through wires.

The first demonstration of actual television was given by J.L. Baird in UK and C.F. Jenkins in USA around 1927 by using the technique of mechanical scanning employing rotating discs. However, the real breakthrough occurred with the invention of the cathode ray tube and the success of V.K. Zworykin of the USA in perfecting the first camera tube (the iconoscope) based on the storage principle. By 1930 electromagnetic scanning of both camera and picture tubes and other ancillary circuits such as for beam deflection, video amplification, etc. were developed. Though television broadcast started in 1935, world political developments and the second world war slowed down the progress of television. With the end of the war, television rapidly grew into a popular medium for dispersion of news and mass entertainment.

Television Systems

At the outset, in the absence of any international standards, three monochrome (*i.e.* black and white) systems grew independently. These are the 525 line American, the 625 line European and the 819 line French systems. This naturally prevents direct exchange of programme between countries using different television standards. Later, efforts by the all world committee on radio and television (CCIR) for changing to a common 625 line system by all concerned proved ineffective and thus all the three systems have apparently come to stay. The inability to change over to a common system is mainly due to the high cost of replacing both the transmitting equipment and the millions of receivers already in use. However the UK, where initially a 415 line monochrome system was in use, has changed to the 625 line system with some modification in the channel bandwidth. In India, where television transmission started in 1959, the 625-B monochrome system has been adopted.

The three different standards of black and white television have resulted in the development of three different systems of colour television, respectively compatible with the three monochrome systems. The original colour system was that adopted by the USA in 1953 on the recommendations of its National Television Systems Committee and hence called the NTSC system. The other two

*From the Greek *tele* (= far) and the Latin *visionis* (from *videre* = to see).

colour systems–PAL and SECAM are later modifications of the NTSC system, with minor improvements, to conform to the other two monochrome standards.

Regular colour transmission started in the USA in 1954. In 1960, Japan adopted the NTSC system, followed by Canada and several other countries. The PAL colour system which is compatible with the 625 line monochrome European system, and is a variant of the NTSC system, was developed at the Telefunken Laboratories in the Federal Republic of Germany (FRG). This system incorporates certain features that tend to reduce colour display errors that occur in the NTSC system during transmission. The PAL system was adopted by FRG and UK in 1967. Subsequently Australia, Spain, Iran and several other countries in West and South Asia have opted for the PAL system. Since this system is compatible with the 625-B monochrome system, India also decided to adopt the PAL system. The third colour TV system in use is the SECAM system. This was initially developed and adopted in France in 1967. Later versions, known as SECAM IV and SECAM V were developed at the Russian National Institute of Research (NIR) and are sometimes referred to as the NIR-SECAM systems. This system has been adopted by the USSR, German Democratic Republic, Hungary, some other East European countries and Algeria. When both the quality of reproduction and the cost of equipment are taken into account, it is difficult to definitely establish the superiority of any one of these systems over the other two. All three systems have found acceptance in their respective countries. The deciding factor for adoption was compatibility with the already existing monochrome system.

Applications of Television

Impact of television is far and wide, and has opened new avenues in diverse fields like public entertainment, social education, mass communication, newscasts, weather reports, political organization and campaigns, announcements and guidance at public places like airport terminals, sales promotion and many others. Though the capital cost and operational expenses in the production and broadcasting of TV programmes are high compared to other media, its importance for mass communication and propagation of social objectives like education are well recognized and TV broadcasts are widely used for such purposes.

Closed Circuit Television (CCTV) is a special application in which the camera signals are made available over cable circuits only to specified destinations. This has important applications where viewers need to see an area to which they may not go for reasons of safety or convenience. Group demonstrations of surgical operations or scientific experiments, inspection of noxious or dangerous industrial or scientific processes (*e.g.* nuclear fuel processing) or of underwater operations and surveillance of areas for security purposes are some typical examples.

A special type of CCTV is what might be called wired community TV. Small communities that fall in the 'shadow' of tall geographical features like hills can jointly put up an antenna at a suitable altitude and distribute the programme to the subscribers' premises through cable circuits. Another potential use of CCTV that can become popular and is already technically feasible is a video-telephone or 'visiphone'.

Equipment

Television broadcasting requires a collection of sophisticated equipment, instruments and components that require well trained personnel. Television studios employ extensive lighting facilities, cameras, microphones, and control equipment. Transmitting equipment for modulation, amplification and radiation of the signals at the high frequencies used for television broadcast are complex and expensive. A wide variety of support equipment essential in broadcast studios, control rooms and outside includes video tape recorders, telecine machines, special effects equipment plus all the apparatus for high quality sound broadcast.

Coverage

Most programmes are produced live in the studio but recorded on video tape at a convenient time to be broadcast later. Of course, provision for live broadcast also has to be there for VIP interviews, sports events and the like. For remote pick-ups the signal is relayed by cable or RF link to the studio for broadcasting in the assigned channel. Each television broadcast station is assigned a channel bandwidth of 7 MHz (6 MHz in the American, 8 MHz in the British and 14 MHz in the French systems). In the earlier days TV broadcast was confined to assigned VHF bands of 41 to 68 MHz and 174 to 230 MHz. Later additional channel allocations have been made in the UHF band between 470 and 890 MHz. Because of the use of VHF-UHF frequencies for television broadcast, reception of TV signals is limited to roughly the line of sight distance. This usually varies between 75 and 140 km depending on the topography and radiated power. Area of TV broadcast coverage can be extended by means of relay stations that rebroadcast signals received via microwave links or coaxial cables. A matrix of such relay stations can be used to provide complete national coverage. With the rapid strides made in the technology of space and satellite communication it has now become possible to have global coverage by linking national TV systems through satellites. Besides their use for international TV networks, large countries can use satellites for distributing national programmes over the whole area. One method for such national coverage is to set up a network of sensitive ground stations for receiving signals relayed by a satellite and retelecasting them to the surrounding area. Another method is to employ somewhat higher transmitter power on the satellite and receive the down transmissions directly through larger dish antenna on conventional television receivers fitted with an extra front-end converter. A combination of both the methods was successfully tested in India where NASA's ATS-6 satellite was used for the SITE programme trials in 1975-76. This resulted in the launching of INSAT 1-A in April 1982.

Recent Trends

In the last decade, transistors and integrated circuits have greatly improved the quality of performance of TV broadcasting and reception. Modern camera tubes like vidicon and plumbicon have made TV broadcast of even dimly lit scenes possible. Special camera tubes are now used for different specific applications. The most sensitive camera tubes available today can produce usable signals even from the scenes where the human eye sees total darkness. With rapid advances in solid state technology, rugged solid state image scanners may conceivably replace the fragile camera tubes in the not-too-distant future. Experimental solid state cameras are already in use for some special applications. Solid state 'picture-plates' for use in receivers are under active development. Before long the highly vulnerable high vacuum glass envelope of the picture tube may be a thing of the past. Since solid state charge coupled devices are scanned by digital addressing, the camera scanner and picture plate can work in exact synchronism with no non-linear distortions of the reconstructed picture.

An important recent technological advance is the use of pseudo-random scan. The signal so generated requires much less bandwidth than the one for conventional method of scanning. Besides all this, wider use of composite devices, made by integrated solid state technology, for television studio and transmitter equipment as well as for receivers will result in higher quality of reproduction, lower costs and power consumptions with increased reliability and compactness. Special mention may be made of the surface acoustic wave filter to replace the clumsy and expensive IF transformer. Further, large screen TV reception systems based on projection techniques now under development will make it possible to show TV programmes to large audience as in a theatre.

With the rapid development of large scale integrated (LSI) electronics in the last decade, digital communication by pulse code modulation (PCM) has made immense progress. The advantage gained is, that virtual freedom from all noise and interference is obtained by using a somewhat larger bandwidth and a specially coded signal. Even if the final transmission in TV is retained in its present form, so that all previous receivers remain usable, the processing of pictures from the camera to the transmitter input is likely to change over to PCM techniques. Unlike the case of monochrome TV standards, the International Telecommunication Union (ITU), a UN special agency, has already adopted a single set of standards accepted by all member countries for the production and processing of picture signals by digital methods. Digital TV has become all the more attractive since solid state cameras compatible with digital signal processing and deflection circuitry have also been developed and are at present in the field testing stage.

Advances in Television Technology

While fundamentals and basic techniques do not change with time, technological advances do continue to take place. During the past two decades or so research efforts enabled the development of analog and digital electronic devices and systems that have resulted in improved performance of television and other user facilities.

(i) Broadcasting: Till recently the main drawback of terrestrial broadcasting was the limiting range because of the earth's curvature which eventually breaks the signal path thus preventing reception over long distances. This problem has been solved with geo-stationary satellites that orbit the globe at the same speed as the rotation of earth. The curvature of earth thus no longer presents any problem and communication over long distances is carried out through satellites. Now we have a large number of geo-synchronous communication satellites orbiting the earth enabling not only the national but also international television transmissions around the world.

(ii) Digitisation of Data Transmission: The transmission of data—radio, audio or video by digital techniques was another step forward in the recent past. As we know, any analog signal like the output of a microphone can have any amplitude (within a range) for different frequency components. In contrast, in a digital system information is represented in discrete or digitized form rather than continuous as in analog. In digital, a binary code has been chosen which means that for any signal only **two** discrete states are possible which are normally denoted **as** 0 and 1. Such a usage of only two discrete signal levels results in a very significant noise immunity advantage for digital circuits. Also, the fixed (0,1) signal designations reduces distortion in signal components, both during transmission and reception.

However, while it is very advantageous to choose signal processing in digital form, the real world is analog. Therefore, analog signals must first be converted to digital form before processing can be done and similarly the digital output must be converted back to analog form for human viewing. This needs high speed analog to digital (A/D) converters at the sending and digital to analog (D/A) converters at the receiving end. With the rapid strides in VLSI technology such devices are now available. This has impacted television transmission, reception and signal processing with the result that we get clear sound and well defined pictures in the television screens. Television receivers have D/A circuitry in its input to enable channel selection and conversion of signals to analog form for further processing in video and audio circuits of the receivers.

Television Viewing Screens: During the period after around the year 1990, another equally important success was the development of large viewing screens for television and allied applications. From the time television became available TV receivers have been built around a cathode ray tube (CRT) for display of pictures. In a color receiver, three guns (R,G & B) in the CRT fire beams of electrons towards the screen to excite corresponding color phosphorous coated on its inner surface. The phosphor atoms on excitation emit R,G & B color lights which on scanning enable a picture on the CRT screen. Though the pictures are quite bright and crisp, such receivers are bulky, mainly due to the weight of the cathode ray tube. However, over the years, demand for bigger screen receivers has been growing for better viewing of television programs. While efforts wave made to increase the CRT screen size but because of limitations in glass tube technology it has not been possible to go beyond around 40″ screens. As such research work has been going on and the first success came in the development of Liquid Crystal Displays (LCD). In a LCD display, the screen consists of a liquid crystal solution in-between two clear glass panels. An electric current passed through the solution causes the crystals to act like a shutter, either blocking the incident light or allowing it to pass through. This phenomenon is used to cause light and dark areas on the LCD screen which when regulated results in pictures. Another alternative that became available is the plasma display, a better choice for TV screens. In it, a solution is coated on the inner side of the glass panels. This solution has millions of phosphor coated miniature glass bubbles containing plasma, which is a gas made up of free flowing ions and electrons. An electric current flow through the solution causes certain plasma containing bubbles to emit ultraviolet (UV) rays which trigger the phosphors to produce color lights. These when combined and controlled, result in color pictures on the screen.

Several other developments that gained importance are the projection display systems, television home theaters and three dimensional (3D) Televisions. All this has further enhanced the gamut of television application.

Introduction

Elements of a Television System

The fundamental aim of a television system is to extend the sense of sight beyond its natural limits, along with the sound associated with the scene being televised. Essentially then, a TV system is an extension of the science of radio communication with the additional complexity that besides sound the picture details are also to be transmitted.

In most television systems, as also in the C.C.I.R. 625 line monochrome system adopted by India, the picture signal is amplitude modulated and sound signal frequency modulated before transmission. The carrier frequencies are suitably spaced and the modulated outputs radiated through a common antenna. Thus each broadcasting station can have its own carrier frequency and the receiver can then be tuned to select any desired station. Figure 1.1 shows a simplified block representation of a TV transmitter and receiver.

1.1 PICTURE TRANSMISSION

The picture information is optical in character and may be thought of as an assemblage of a large number of bright and dark areas representing picture details. These elementary areas into which the picture details may be broken up are known as 'picture elements', which when viewed together, represent the visual information of the scene. Thus the problem of picture transimission is fundamentally much more complex, because, at any instant there are almost an infinite number of pieces of information, existing simultaneously, each representing the level of brightness of the scene to the reproduced. In other words the information is a function of two variables, time and space. Ideally then, it would need an infinite number of channels to transmit optical information corresponding to all the picture elements simultaneously. Presently the practical difficulties of transmitting all the information simultaneously and decoding it at the receiving end seem insurmountable and so a method known as scanning is used instcad. Here the conversion of optical information to electrical form and its transmission are carried out element by element, one at a time and in a sequential manner to cover the entire scene which is to be televised. Scanning of the elements is done at a very fast rate and this process is repeated a large number of times per second to create an illusion of simultaneous pick-up and transmission of picture details.

A TV camera, the heart of which is a camera tube, is used to convert the optical information into a corresponding electrical signal, the amplitude of which varies in accordance with the variations of brightness. Fig. 1.2 (a) shows very elementary details of one type of camera tube (vidicon) to illustrate this principle. An optical image of the scene to be transmitted is focused by a lens assembly on the rectangular glass face-plate of the camera tube. The inner side of the glass face-plate has a transparent conductive coating on which is laid a very thin layer of photoconductive material. The photolayer has a very high resistance when no light falls on it, but decreases

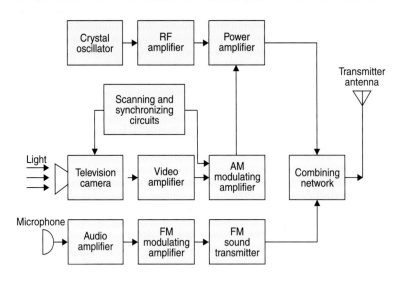

Fig. 1.1 (a) *Basic monochrome television transmitter.*

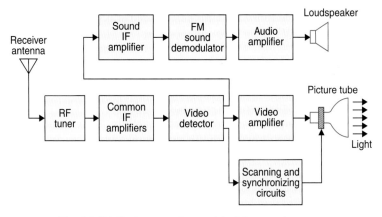

Fig. 1.1 (b) *Basic monochrome television receiver.*

Fig. 1.1 *Simplified block diagram of a monochrome television broadcasting system.*

depending on the intensity of light falling on it. Thus depending on the light intensity variations in the focused optical image, the conductivity of each element of the photolayer changes accordingly. An electron beam is used to pick-up the picture information now available on the target plate in terms of varying resistance at each point. The beam is formed by an electron gun in the TV camera tube. On its way to the inner side of the glass face-plate it is deflected by a pair of deflecting coils mounted on the glass envelope and kept mutually perpendicular to each other to achieve scanning of the entire target area. Scanning is done in the same way as one reads a written page to cover all the words in one line and all the lines on the page (see Fig. 1.2 (*b*)). To achieve this the deflecting coils are fed separately from two sweep oscillators which continuously generate saw-tooth waveforms, each operating at a different desired frequency. The magnetic deflection caused by the current in one coil gives horizontal motion to the beam from left to right at a uniform rate and then brings it quickly to the left side to commence the trace of next line. The other coil is used to deflect the beam from top to bottom at a uniform rate and for its quick retrace back to the top of the plate to start this process all over again. Two simultaneous motions are thus given to the beam, one from left to right across the target plate and the other from top to bottom thereby covering the entire area on which the electrical image of the picture is available. As the beam moves from element to element, it encounters a different resistance across the target-plate, depending on the resistance of the photoconductive coating. The result is a flow of current which varies in magnitude as the elements are scanned. This current passes through a load resistance R_L, connected to the conductive coating on one side and to a dc supply source on the other. Depending on the magnitude of the current a varying voltage appears across the resistance R_L and this corresponds to the optical information of the picture.

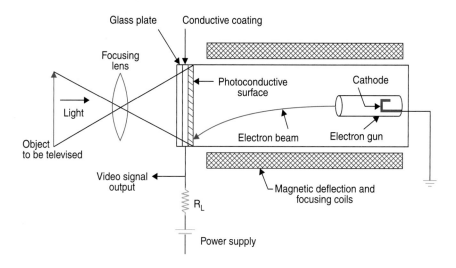

Fig. 1.2 (a) *Simplified cross-sectional view of a Vidicon TV camera tube.*

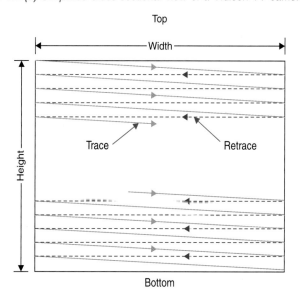

Fig. 1.2 (b) *Path of scanning beam in covering picture area.*

If the scanning beam moves at such a rate that any portion of the scene content does not have time to move perceptibly in the time required for one complete scan of the image, the resultant electrical signal contains the true information existing in the picture during the time of the scan. The desired information is now in the form of a signal varying with time and scanning may thus be identified as a particular process which permits the conversion of information existing in space and time coordinates into time variations only. The electrical information obtained from the TV camera tube is generally referred to as video signal (video is Latin for 'see'). This signal is amplified and then amplitude modulated with the channel picture carrier frequency. The modulated output is fed to the transmitter antenna for radiation along with the sound signal.

1.2 SOUND TRANSMISSION

The microphone converts the sound associated with the picture being televised into proportionate electrical signal, which is normally a voltage. This electrical output, regardless of the complexity of its waveform, is a single valued function of time and so needs a single channel for its transmission. The audio signal from the microphone after amplification is frequency modulated, employing the assigned carrier frequency. In FM, the amplitude of the carrier signal is held constant, whereas its frequency is varied in accordance with amplitude variations of the modulating signal. As shown in Fig. 1.1 (a), output of the sound FM transmitter is finally combined with the AM picture transmitter output, through a combining network, and fed to a common antenna for radiation of energy in the form of electromagnetic waves.

1.3 PICTURE RECEPTION

The receiving antenna intercepts the radiated picture and sound carrier signals and feeds them to the RF tuner (see Fig. 1.1 (b)). The receiver is of the heterodyne type and employs two or three stages of intermediate frequency (IF) amplification. The output from the last IF stage

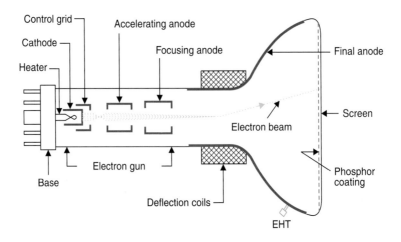

Fig. 1.3 Elements of a picture tube.

is demodulated to recover the video signal. This signal that carries the picture information is amplified and coupled to the picture tube which converts the electrical signal back into picture elements of the same degree of black and white. The picture tube shown in Fig. 1.3 is very similar to the cathode-ray tube used in an oscilloscope. The glass envelope contains an electron-gun structure that produces a beam of electrons aimed at the fluorescent screen. When the electron beam strikes the screen, light is emitted. The beam is deflected by a pair of deflecting coils mounted on the neck of the picture tube in the same way and rate as the beam scans the target in the camera tube. The amplitudes of the currents in the horizontal and vertical deflecting coils are so adjusted that the entire screen, called raster, gets illuminated because of the fast rate of scanning.

The video signal is fed to the grid or cathode of the picture tube. When the varying signal voltage makes the control grid less negative, the beam current is increased, making the spot of light on the screen brighter. More negative grid voltage reduces the brightness. if the grid voltages is negative enough to cut-off the electron beam current at the picture tube there will be no light. This state corresponds to black. Thus the video signal illuminates the fluorescent screen from white to black through various shades of grey depending on its amplitude at any instant. This corresponds to the brightness changes encountered by the electron beam of the camera tube while scanning the picture details element by element. The rate at which the spot of light moves is so fast that the eye is unable to follow it and so a complete picture is seen because of the storage capability of the human eye.

1.4 SOUND RECEPTION

The path of the sound signal is common with the picture signal from antenna to the video detector section of the receiver. Here the two signals are separated and fed to their respective channels. The frequency modulated audio signal is demodulated after at least one stage of amplification. The audio output from the FM detector is given due amplification before feeding it to the loudspeaker.

1.5 SYNCHRONIZATION

It is essential that the same coordinates be scanned at any instant both at the camera tube target plate and at the raster of the picture tube, otherwise, the picture details would split and get distorted. To ensure perfect synchronization between the scene being televised and the picture produced on the raster, synchronizing pulses are transmitted during the retrace, *i.e.*, fly-back intervals of horizontal and vertical motions of the camera scanning beam. Thus, in addition to carrying picture detail, the radiated signal at the transmitter also contains synchronizing pulses. These pulses which are distinct for horizontal and vertical motion control, are processed at the receiver and fed to the picture tube sweep circuitry thus ensuring that the receiver picture tube beam is in step with the transmitter camera tube beam.

1.6 RECEIVER CONTROLS

The front view of a typical monochrome TV receiver, having various controls is shown in Fig. 1.4.

The channel selector switch is used for selecting the desired channel. The fine tuning control is provided for obtaining best picture details in the selected channel. The hold control is used to get a steady picture in case it rolls up or down. The brightness control varies the beam intensity of the picture tube and is set for optimum average brightness of the picture. The contrast control is actually the gain control of the video amplifier. This can be varied to obtain the desired contrast between the white and black contents of the reproduced picture. The volume and tone controls form part of the audio amplifier in the sound section, and are used for setting the volume and tonal quality of the sound output from the loudspeaker.

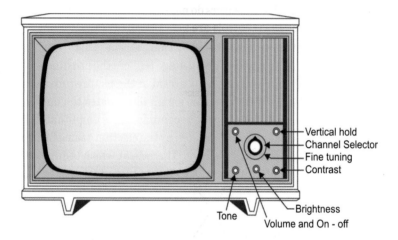

Fig. 1.4 *Television receiver controls*

1.7   COLOUR TELEVISION

Colour television is based on the theory of additive colour mixing, where all colours including white can be created by mixing red, green, and blue lights. The colour camera provides video signals for the red, green, and blue information. These are combined and transmitted along with the brightness (monochrome) signal.

Each colour TV system* is compatible with the corresponding monochrome system. Compatibility means that colour broadcasts can be received as black and white on monochrome receivers. Conversely colour receivers are able to receive black and white TV broadcasts. This is illustrated in Fig. 1.5 where the transmission paths from the colour and monochrome cameras are shown to both colour and monochrome receivers.

At the receiver, the three colour signals are separated and fed to the three electron guns of colour picture tube. The screen of the picture tube has red, green, and blue phosphors arranged in alternate dots. Each gun produces an electron beam to illuminate the three colour phosphors separately on the fluorescent screen. The eye then integrates the red, green and blue colour information and their luminance to perceive the actual colour and brightness of the picture being televised.

---

* The three compatible colour television systems are NTSC, PAL and SECAM.

Colour Receiver Controls

NTSC colour television receivers have two additional controls, known as Colour and Hue controls. These are provided at the front panel along with other controls. The colour or saturation control varies the intensity or amount of colour in the reproduced picture. For example, this control determines whether the leaves of a tree in the picture are dark green or light green, and whether the sky in the picture is dark blue or light blue. The tint or hue control selects the correct colour to be displayed. This is primarily used to set the correct skin colour, since when flesh tones are correct, all other colours are correctly reproduced.

It may be noted that PAL colour receivers do not need any tint control while in SECAM colour receivers, both tint and saturation controls are not necessary. The reasons for such differences are explained in chapters exclusively devoted to colour television.

Chapter 1

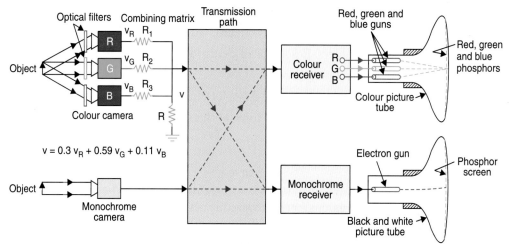

Fig. 1.5. Signal transmission paths illustrating compatibility between colour and monochrome TV systems. R, G and B represent three camera tubes which develop video signals corresponding to the red, green and blue contents of the scene being televised.

REVIEW QUESTIONS

1. Why is scanning necessary in TV transmission ? Why is it carried out at a fast rate ?
2. What is the basic principle of operation of a television camera tube ?
3. What is a raster and how is it produced on the picture tube screen ?
4. Why are synchronizing pulses transmitted along with the picture signal ?
5. Why is FM preferred to AM for sound signal transmission ?
6. Describe briefly the functions of various controls provided on the front panel of a TV receiver.
7. Describe the basic principle of colour television transmission and reception.
8. Describe the function of saturation and hue controls in a NTSC colour TV receiver.

2

Analysis and Synthesis of Television Pictures

The basic factors with which the television system must deal for successful transmission and reception of pictures are:

(a) *Gross Structure:* Geometric form and aspect ratio of the picture.

(b) *Image Continuity:* Scanning and its sequence.

(c) *Number of Scanning Lines:* Resolution of picture details.

(d) *Flicker:* Interlaced scanning.

(e) *Fine Structure:* Vertical and horizontal resolution.

(f) *Tonal Gradation:* Picture brightness transfer characteristics of the system.

2.1 GROSS STRUCTURE

The frame adopted in all television systems is rectangular with width/height ratio, *i.e.*, aspect ratio = 4/3. There are many reasons for this choice. In human affairs most of the motion occurs in the horizontal plane and so a larger width is desirable. The eyes can view with more ease and comfort when the width of a picture is more than its height. The usage of rectangular frame in motion pictures with a width/height ratio of 4/3 is another important reason for adopting this shape and aspect ratio. This enables direct television transmission of film programmes without wastage of any film area.

It is not necessary that the size of the picture produced on the receiver screen be same as that being televised but it is essential that the aspect ratio of the two be same, otherwise the scene details would look too thin or too wide. This is achieved by setting the magnitudes of the current in the deflection coils to correct values, both at the TV camera and receiving picture tube. Another important requirement is that the same coordinates should be scanned at any instant both by the camera tube beam and the picture tube beam in the receiver. Synchronizing pulses are transmitted along with the picture information to achieve exact congruence between transmitter and receiver scanning systems.

2.2 IMAGE CONTINUITY

While televising picture elements of the frame by means of the scanning process, it is necessary to present the picture to the eye in such a way that an illusion of continuity is created and any motion in the scene appears on the picture tube screen as a smooth and continuous change. To achieve this, advantage is taken of 'persistence of vision' or storage characteristics of the human eye. This arises from the fact that the sensation produced when nerves of the eye's retina are stimulated by incident light does not cease immediately after the light is removed but persists for about 1/16th of a second. Thus if the scanning rate per second is made greater than sixteen, or the number of pictures shown per second is more than sixteen, the eye is able to integrate the changing levels of brightness in the scene. So when the picture elements are scanned rapidly enough, they appear to the eye as a complete picture unit, with none of the individual elements visible separately.

In present day motion pictures twenty-four still pictures of the scene are taken per second and later projected on the screen at the same rate. Each picture or frame is projected individually as a still picture, but they are shown one after the other in rapid succession to produce the illusion of continuous motion of the scene being shown. A shutter in the projector rotates in front of the light source and allows the film to be projected on the screen when the film frame is still, but blanks out any light from the screen during the time when the next film frame is being moved into position. As a result, a rapid succession of still-film frames is seen on the screen. With all light removed during the change from one frame to the next, the eye sees a rapid sequence of still pictures that provides the illusion of continuous motion.

Scanning. A similar process is carried out in the television system. The scene is scanned rapidly both in the horizontal and vertical directions simultaneously to provide sufficient number of complete pictures or frames per second to give the illusion of continuous motion. Instead of the 24 as in commercial motion picture practice, the frame repetition rate is 25 per second in most television systems.

Horizontal scanning. Fig. 2.1 (*a*) shows the trace and retrace of several horizontal lines. The linear rise of current in the horizontal deflection coils (Fig. 2.1 (*b*)) deflects the beam across the screen with a continuous, uniform motion for the trace from left to right. At the peak of the rise, the sawtooth wave reverses direction and decreases rapidly to its initial value. This fast reversal produces the retrace or flyback. The start of the horizontal trace is at the left

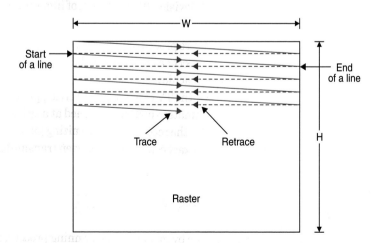

Fig. 2.1 (a) Path of scanning beam in covering picture area (Raster).

edge of raster. The finish is at the right edge, where the flyback produces retrace back to the left edge.

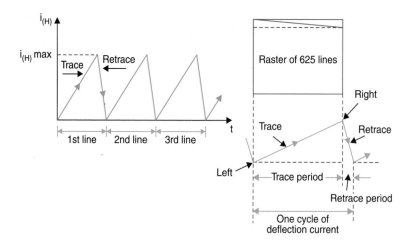

Fig. 2.1 (b) *Waveform of current in the horizontal deflection coils producing linear (constant velocity) scanning in the horizontal direction.*

Note, that 'up' on the sawtooth wave corresponds to horizontal deflection to the right. The heavy lines in Fig. 2.1 (*a*) indicate the useful scanning time and the dashed lines correspond to the retrace time.

Vertical scanning. The sawtooth current in the vertical deflection coils (see Fig. 2.2) moves the electron beam from top to bottom of the raster at a uniform speed while the electron beam is being deflected horizontally. Thus the beem produces complete horizontal lines one below the other while moving from top to bottom.

As shown in Fig. 2.2 (*c*), the trace part of the sawtooth wave for vertical scanning deflects the beam to the bottom of the raster. Then the rapid vertical retrace returns the beam to the top. Note that the maximum amplitude of the vertical sweep current brings the beam to the bottom of the raster. As shown in Fig. 2.2 (*b*) during vertical retrace the horizontal scanning continues and several lines get scanned during this period. Because of motion in the scene being televised, the information or brightness at the top of the target plate or picture tube screen normally changes by the time the beam returns to the top to recommence the whole process. This information is picked up during the next scanning cycle and the whole process is repeated 25 times to cause an illusion of continuity. The actual scanning sequence is however a little more complex than that just described and is explained in a later section of this chapter. It must however be noted, that both during horizontal retrace and vertical retrace intervals the scanning beams at the camera tube and picture tube are blanked and no picture information is either picked up or reproduced. Instead, on a time division basis, these short retrace intervals are utilized for transmitting distinct narrow pulses to keep the sweep oscillators of the picture tube deflection circuits of the receiver in synchronism with those of the camera at the transmitter. This ensures exact correspondence in scanning at the two ends and results in distortionless reproduction of the picture details.

Chapter 2

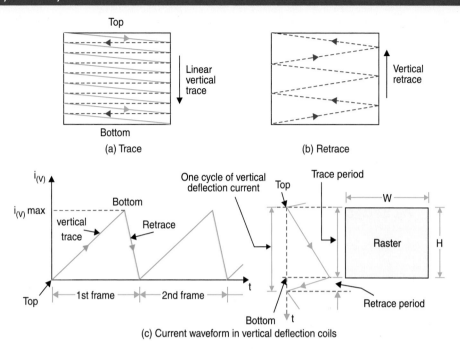

Fig. 2.2 Vertical deflection and deflection current waveform.

2.3 NUMBER OF SCANNING LINES

Most scenes have brightness gradations in the vertical direction. The ability of the scanning beam to allow reproduction of electrical signals according to these variations and the capability of the human eye to resolve these distinctly, while viewing the reproduced picture, depends on the total number of lines employed for scanning.

It is possible to arrive at some estimates of the number of lines necessary by considering the bar pattern shown in Fig. 2.3 (*a*), where alternate lines are black and white. If the thickness of the scanning beam is equal to the width of each white and black bar, and the number of scanning lines is chosen equal to the number of bars, the electrical information corresponding to the brightness of each bar will be correctly reproduced during the scanning process. Obviously the greater the number of lines into which the picture is divided in the vertical plane, the better will be the resolution.However, the total number of lines that need be employed is limited by the resolving capability of the human eye at the minimum viewing distance.

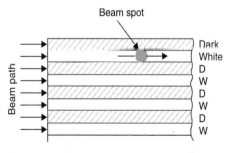

Fig. 2.3 (a) Scanning spot perfectly aligned with black and white lines.

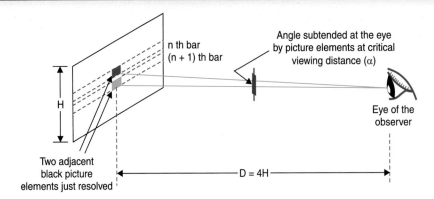

Fig. 2.3 (b) *Critical viewing distance as determined by the ability of the eye to resolve two separate picture elements.*

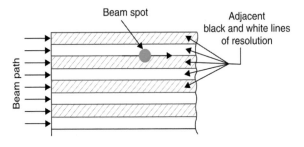

Fig. 2.3 (c) *Scanning beam focused on the junction of black and white lines.*

The maximum number of alternate light and dark elements (lines) which can be resolved by the eye is given by

$$N_v = \frac{1}{\alpha \rho}$$

where N_v = total number of lines (elements) to be resolved in the vertical direction, α = minimum resolving angle of the eye expressed in radians, and $\rho = D/H$ = viewing-distance/picture height.

For the eye this resolution is determined by the structure of the retina, and the brightness level of the picture. It has been determined experimently that with reasonable brightness variations and a minimum viewing distance of four times the picture height ($D/H = 4$), the angle that any two adjacent elements must subtend at the eye for distinct resolution is approximately one minute (1/60 degree). This is illustrated in Fig. 2.3 (*b*). Substituting these values of α and ρ, we get

$$N_v = \frac{1}{(\pi / 180 \times 1/ 60) \times 4} \approx 860$$

Thus if the total number of scanning lines is chosen close to 860 and the scanning beam as illustrated in Fig. 2.3 (*a*) just passes over each bar (line) separately while scanning all the lines from top to bottom of the picture frame, a distinct pick up of the picture information results and this is the best that can be expected from the system. This perhaps explains the use of 819 lines in the original French TV system.

In practice however, the picture elements are not arranged as equally spaced segments but have random distribution of black, grey and white depending on the nature of the picture details or the scene under consideration. Statistical analysis and subjective tests carried out to determine the average number of effective lines suggest that about 70 per cent of the total lines or segments get separately scanned in the vertical direction and the remaining 30 per cent get merged with other elements due to the beam spot falling equally on two consecutive lines. This is illustrated in Fig. 2.3 (c). Thus the effective number of lines distinctly resolved, i.e., $N_r = N_v \times k$, where k is the resolution factor whose value lies between 0.65 to 0.75. Assuming the value of $k = 0.7$ we get, $N_r = N_v \times k = 860 \times 0.7 = 602$.

However, there are other factors which also influence the choice of total number of lines in a TV system. Tests conducted with many observers have shown that though the eye can detect the effective sharpness provided by about 800 scanning lines, but the improvement is not very significant with line numbers greater than 500 while viewing pictures having motion. Also the channel bandwidth increases with increase in number of lines and this not only adds to the cost of the system but also reduces the number of television channels that can be provided in a given VHF or UHF transmission band. Thus as a compromise between quality and cost, the total number of lines inclusive of those lost during vertical retrace has been chosen to be 625 in the 625-B monochrome TV system. In the 525 line American system, the total number of lines has been fixed at 525 because of a somewhat higher scanning rate employed in this system.

2.4 FLICKER

Although the rate of 24 pictures per second in motion pictures and that of scanning 25 frames per second in television pictures is enough to cause an illusion of continuity, they are not rapid enough to allow the brightness of one picture or frame to blend smoothly into the next through the time when the screen is blanked between successive frames. This results in a definite flicker of light that is very annoying to the observer when the screen is made alternately bright and dark.

This problem is solved in motion pictures by showing each picture twice, so that 48 views of the scene are shown per second although there are still the same 24 picture frames per second. As a result of the increased blanking rate, flicker is eliminated.

Interlaced scanning. In television pictures an effective rate of 50 vertical scans per second is utilized to reduce flicker. This is accomplished by increasing the downward rate of travel of the scanning electron beam, so that every alternate line gets scanned instead of every successive line. Then, when the beam reaches the bottom of the picture frame, it quickly returns to the top to scan those lines that were missed in the previous scanning. Thus the total number of lines are divided into two groups called 'fields'. Each field is scanned alternately. This method of scanning is known as interlaced scanning and is illustrated in Fig. 2.4. It reduces flicker to an acceptable level since the area of the screen is covered at twice the rate. This is like reading alternate lines of a page from top to bottom once and then going back to read the remaining lines down to the bottom.

Chapter 2

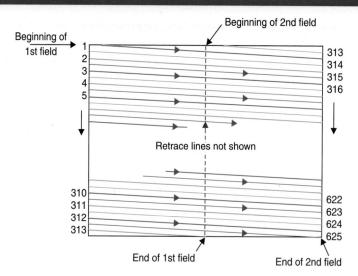

Fig. 2.4 *Principle of interlaced scanning. Note that the vertical retrace time has been assumed to be zero.*

In the 625 line monochrome system, for successful interlaced scanning, the 625 lines of each frame or picture are divided into sets of 312.5 lines and each set is scanned alternately to cover the entire picture area. To achieve this the horizontal sweep oscillator is made to work at a frequency of 15625 Hz (312.5 × 50 = 15625) to scan the same number of lines per frame (15625/25 = 625 lines), but the vertical sweep circuit is run at a frequency of 50 instead of 25 Hz. Note that since the beam is now deflected from top to bottom in half the time and the horizontal oscillator is still operating at 15625 Hz, only half the total lines, *i.e.*, 312.5 (625/2 = 312.5) get scanned during each vertical sweep. Since the first field ends in a half line and the second field commences at middle of the line on the top of the target plate or screen (see Fig. 2.4), the beam is able to scan the remaining 312.5 alternate lines during its downward journey. In all then, the beam scans 625 lines (312.5 × 2 = 625) per frame at the same rate of 15625 lines (312.5 × 50 = 15625) per second. Therefore, with interlaced scanning the flicker effect is eliminated without increasing the speed of scanning, which in turn does not need any increase in the channel bandwidth.

It may be noted that the frame repetition rate of 25 (rather than 24 as used in motion pictures) was chosen to make the field frequency equal to the power line frequency of 50 Hz. This helps in reducing the undesired effects of hum due to pickup from the mains, because then such effects in the picture stay still, instead of drifting up or down on the screen. In the American TV system, a field frequency of 60 was adopted because the supply frequency is 60 Hz in USA. This brings the total number of lines scanned per second ((525/2) × 60 = 15750) lines to practically the same as in the 625 line system.

Scanning periods. The waveshapes of both horizontal and vertical sweep currents are shown in Fig. 2.5. As shown there the retrace times involved (both horizontal and vertical) are due to physical limitations of practical scanning systems and are not utilized for transmitting or receiving any video signal. The nominal duration of the horizontal line as shown in Fig. 2.5 (*a*) is 64 μs (10^6/15625 = 64 μs), out of which the active line period is 52 μs and the remaining 12 μs is the line blanking period. The beam returns during this short interval to the extreme left side of the frame to start tracing the next line.

Similarly with the field frequency set at 50 Hz, the nominal duration of the vertical trace (see Fig. 2.5(*b*)) is 20 ms (1/50 = 20 ms). Out of this period of 20 ms, 18.720 ms are spent in bringing the beam from top to bottom and the remaining 1.280 ms is taken by the beam to return back to the top to commence the next cycle. Since the horizontal and vertical sweep oscillators operate continuously to achieve the fast sequence of interlaced scanning, 20 horizontal lines

$$\left(\frac{1280\ \mu s}{64\ \mu s} = 20\ \text{lines} \right) \text{get traced during each vertical retrace interval. Thus 40 scanning lines are}$$

lost per frame, as blanked lines during the retrace interval of two fields. This leaves the active number of lines, N_a, for scanning the picture details equal to 625 – 40 = 585, instead of the 625 lines actually scanned per frame.

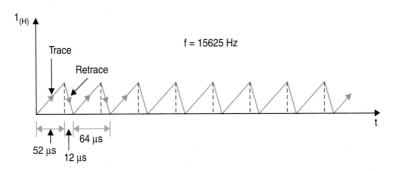

Fig. 2.5 (a) *Horizontal deflection current.*

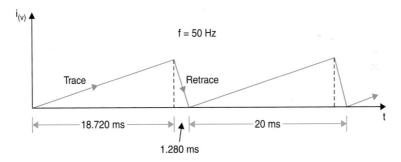

Fig. 2.5 (b) *Vertical deflection current.*

Scanning sequence. The complete geometry of the standard interlaced scanning pattern is illustrated in Fig. 2.6. Note that the lines are numbered in the sequence in which these are actually scanned. During the first vertical trace actually 292.5 lines are scanned. The beam starts at A, and sweeps across the frame with uniform velocity to cover all the picture elements in one horizontal line. At the end of this trace the beam then retraces rapidly to the left side of the frame as shown by the dashed line in the illustration to begin the next horizontal line. Note that the horizontal lines slope downwards in the direction of scanning because the vertical deflecting current simultaneously produces a vertical scanning motion, which is very slow compared with horizontal scanning. The slope of the horizontal trace from left to right is greater than during retrace from right to left. The reason is that the faster retrace does not allow the beam so much time to be deflected vertically. After line one, the beam is at the left side ready to scan line 3, omitting the second line. However, as mentioned earlier it is convenient to number the lines as they are scanned and so the next scanned line skipping one line, is numbered two and not three. This process continues till the last line gets scanned half when the vertical motion reaches the bottom of the raster or frame. As explained earlier skipping of lines is accomplished by doubling the vertical scanning frequency from the frame or picture repetition rate of 25 to the field frequency of 50 Hz. With the field frequency of 50 Hz the height of the raster is so set that 292.5 lines get

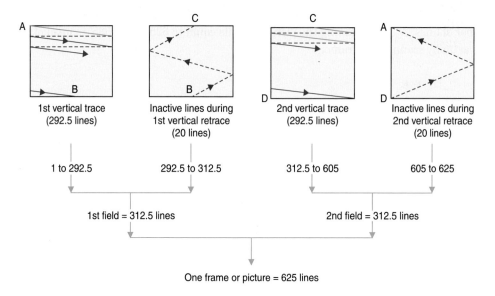

Fig. 2.6 *Odd line interlaced scanning procedure.*

scanned as the beam travels from top to bottom and reaches point B. Now the retrace starts and takes a period equal to 20 horizontal line periods to reach the top marked C. These 20 lines are known as inactive lines, as the scanning beam is cut-off during this period. Thus the second field starts at the middle of the raster and the first line scanned is the 2nd half of line number 313. The scanning of second field, starting at the middle of the raster automatically enables the beam to scan the alternative lines left unscanned during the first field. The vertical scanning motion otherwise is exactly the same as in the previous field giving all the horizontal lines the same slope downwards in the direction of scanning. As a result 292.5 lines again get scanned and the beam reaches the bottom of the frame when it has completed full scanning of line number 605. The inactive vertical retrace again begins and brings the beam back to the top at point A in a period during which 20 blanked horizontal lines (605 to 625) get scanned. Back at point A, the scanning beam has just completed two fields or one frame and is ready to start the third field covering the same area (no. of lines) as scanned during the first field. This process (of scanning fields) is continued at a fast rate of 50 times a second, which not only creates an illusion of continuity but also solves the problem of flicker satisfactorily.

2.5 FINE STRUCTURE

The ability of the image reproducing system to represent the fine structure of an object is known as its resolving power or resolution. It is necessary to consider this aspect separately in the vertical and horizontal planes of the picture.

Vertical resolution. The extent to which the scanning system is capable of resolving picture details in the vertical direction is referred to as its vertical resolution. It has already been explained that the vertical resolution is a function of the scanning lines into which the picture is divided in the vertical plane. Based on that discussion the vertical resolution in the 625 lines system can then be expressed as

$$V_r = N_a \times k$$

where V_r is the vertical resolution expressed in number of lines, N_a is the active number of lines and k is the resolution factor (also known as Kell factor).

Assuming a reasonable value of $k = 0.69$,

$$V_r = 585 \times 0.69 = 400 \text{ lines}$$

It is of interest to note that the corresponding resolution of 35 mm motion pictures is about 515 lines and thus produces greater details as compared to television pictures.

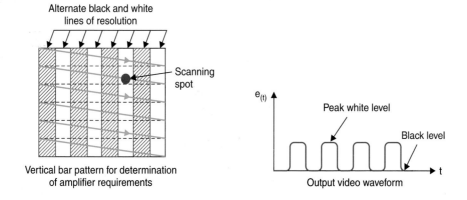

Fig. 2.7 (a) Determination of horizontal resolution.

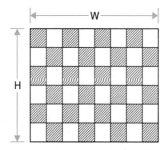

Fig. 2.7(b) Chess-board pattern for studying vertical and horizontal resolution.

Chapter 2

Horizontal resolution. The capability of the system to resolve maximum number of picture elements along the scanning lines determines horizontal resolution. This can be evaluated by considering a vertical bar pattern as shown in Fig. 2.7(a). It would be realistic to aim at equal vertical and horizontal resolution and as such the number of alternate black and white bars that should be considered is equal to

$$N_a \times \text{aspect ratio} = 585 \times 4/3 = 780$$

Before proceeding further it must be recognised that as all lines in the vertical plane are not fully effective, in a similar way all parts of an individual line are not fully effective all the time. As explained earlier, it ultimately depends on the random distribution of black and white areas in the picture. Thus for equal vertical and horizontal resolution, the same resolution factor may be used while determining the effective number of distinct picture elements in a horizontal line. Therefore, the effective number of alternate black and white segments in one horizontal line for equal vertical and horizontal resolution are :

$$N = N_a \times \text{aspect ratio} \times k = 585 \times 4/3 \times 0.69 = 533$$

To resolve these 533 squares or picture elements the scanning spot must develop a video signal of square wave nature switching continuously along the line between voltage levels corresponding to black and peak white. This is shown along the bar pattern drawn in Fig. 2.7(a). Since along one line there are $533/2 \approx 267$ complete cyclic changes, 267 complete square wave cycles get generated during the time the beam takes to travel along the width of the pattern. Thus the time duration t_h of one square wave cycle is equal to

$$t_h = \frac{\text{active period of each horizontal line}}{\text{number of cycles}}$$

$$= \frac{52 \times 10^{-6}}{267} \text{ seconds}$$

∴   the frequency of the periodic wave

$$f_h = \frac{1}{t_h} = \frac{267 \times 10^6}{52} = 5 \text{ MHz}$$

Since the consideration of both vertical and horizontal resolutions is based on identical black and white bars in the horizontal and vertical planes of the picture frame, it amounts to considering a chessboard pattern as the most stringent case and is illustrated in Fig. 2.7(b). Here each alternate black and white square element takes the place of bars for determining the capability of the scanning system to reproduce the fine structure of the object being televised. The actual size of each square element in the chess pattern is very small and is equal to thickness of the scanning beam. It would be instructive to know as an illustration that the size of such a square element on the screen of a 51 cm picture tube is about 0.5 mm^2 only.

Since the spacing of these small elements in the above consideration corresponds to the limiting resolution of the eye, it will distinguish only the alternate light and dark areas but not the shape of the variations along the scanning line. Thus the eye will fail to distinguish the difference between a square wave of brightness variation and a sine wave of brightness variation in the reproduced picture. Therefore, if the amplifier for the square-wave signal is capable of reproducing a sine-wave of frequency equal to the repetition frequency of the rectangular wave, it is satisfactory for the purpose of TV signals. It may be mentioned that even otherwise to handle a 5 MHz square wave would necessitate reproduction up to 11th harmonic of a periodic sinusoidal wave of 5 MHz by the associated electronic circuitry. This would mean a bandwidth of atleast up to $5 \times 11 = 55$ MHz which is excessive and almost impossible to provide in practice. Another justification for restricting the bandwidth up to 5 MHz is that in practice it is rare when alternate picture elements are black and white throughout the picture width and height, and a bandwidth up to 5 MHz has been found to be quite adequate to produce most details of the scene being televised.

Therefore, the highest approximate modulating frequency 'f_h' that the 625 line television system must be capable of handling for successful transmission and reception of picture details is

$$f_h = \frac{\text{No. of active lines} \times \text{aspect ratio} \times \text{resolution factor}}{2 \times \text{time duration of one active line}}$$

$$= \frac{585 \times 4/3 \times 0.69}{2 \times 52 \times 10^{-6}}$$

$$\approx 5 \text{ MHz}$$

In the second (525 line) widely used television system, where the active number of lines is 485 and the duration of one active line is 57 μs, the highest modulating frequency $f_h \approx 4$ MHz.

This explains the allocation of 6 MHz as the channel bandwidth in USA and other countries employing the 525 line system in comparison to a channel bandwidth allocation of 7 MHz in countries that have adopted the 625 line system. Similarly in the French 819 TV system where the highest modulating frequency comes to 10.4 MHz a channel bandwidth of 14 MHz is allowed.

Colour resolution and bandwidth. As explained above a bandwidth of 5 MHz (4 MHz in the American system) is needed for transmission of maximum horizontal detail in monochrome. However, this bandwidth is not necessary for the colour video signals. The reason is that the human eye's colour response changes with the size of the object. For very small objects the eye can perceive only the brightness rather than the colours in the scene. Perception of colours by the eye is limited to objects which result in a video frequency output up to about 1.5 MHz. Thus the colour information needs much less bandwidth than monochrome details and can be easily accommodated in the channel bandwidth allotted for monochrome transmission.

Low-frequency requirements. The analysis of the signals produced by the bar pattern gives no information regarding the low-frequency requirement of a video amplifier used to handle such signals. This requirement may be determined from consideration of a pattern shown in Fig. 2.8(a). The signal output during vertical excursions of the beam would be a square wave (see Fig. 2.8(b)) at vertical field frequency. It is apparent then, that any amplifier capable of reproducing this waveform would be required to have good square-wave response at 50 Hz. Any degradation in response as shown in Fig. 2.8(c) would result in brightness distortion. In order to have satisfactory square-wave response at field frequency, an amplifier must have good sine-wave response with negligible phase distortion down to a much lower frequency than the field frequency. In addition, to correct phase and amplitude response at the field frequency, it is necessary to preserve the dc component of the brightness signal. Thus a good frequency response from dc to about 5 MHz becomes necessary for true reproduction of the brightness variations and find details of any scene.

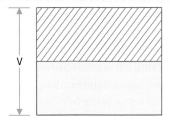

Fig. 2.8(a) *Single bar pattern.*

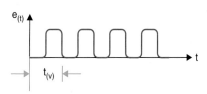

Fig. 2.8(b) *Ideal response to scanning of single bar pattern.*

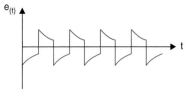

Fig. 2.8(c) *Distorted response due to poor low frequency response of the system.*

Influence of number of lines on bandwidth. As the number of lines employed in a television picture is increased, the bandwidth necessary for a given quality of definition also increases. This is due to the fact that increasing the number of lines per picture decreases the time duration of each line. This means that the spot travels across the screen at a higher velocity and results in increase of the highest modulating frequency. For example doubling the number of lines per frame would very much improve the vertical resolution, infact it would get doubled but would need increasing the bandwidth in the same ratio. If now, it is required to increase the horizontal resolution so that it again equals the vertical resolution it would be necessary to scan double the number of alternate black and white signal elements in a line, and this would necessitate multiplying the original highest video frequency by a factor of four. The conclusion is that, if the number of lines employed in a television system is increased, it is necessary to increase the video frequency bandwidth in direct proportion to the increase in number of lines to maintain the same degree of vertical definition (as before), and in order to increase horizontal definition in the same proportion as the increase in vertical resolution the video frequency bandwidth must increase as the square of the increase in number of lines.

Effect of interlaced scanning on bandwidth. As already explained, interlaced scanning reduces flicker. However, scanning 50 complete frames of 625 lines in a progressive manner would also eliminate flicker in the picture but this would need double the scanning speed which in turn would double the video frequencies corresponding to the picture elements in a line. This would necessitate double the channel bandwidth of that required with interlaced scanning. It should be noted that by employing interlaced scanning, the basic concept of interchangeability of time and bandwidth is not violated, because more time in allowed for transmission and this results in decrease of bandwidth needed for each TV channel. Thus interlaced scanning reduces flicker and conserves bandwidth.

Effect of field frequency on bandwidth. With increase in field frequency the time available for each field decreases and this results in a proportionate decrease of the active line period. Hence, bandwidth increases in direct proportion to the increase in the field frequency.

Chapter 2

Bandwidth requirement for transmission of synchronizing pulses. The equalizing pulses to be discussed later have a pulse width of 2.3 μs with an allowed rise time of 0.2 μs. The highest sinusoidal frequency which must lie in the pass band of the system for effective transmission of these pulses is given by the expression :

$$\text{Highest necessary frequency} = \frac{1}{2 \times \text{allowed rise time}} = \frac{10^6}{2 \times 0.2} = 2.5 \text{ MHz}$$

It is then clear that all sync pulses are safely preserved in the video circuitry where, as has been shown, a frequency bandwidth considerably in excess of this figure has to be maintained in order to preserve the required picture definition.

Interlace error. As explained earlier interlaced scanning provides a means of decreasing the effect of flicker in the TV picture without increasing the system bandwidth. The selection of 2 : 1 as the interlace ratio is the simplest with least circuit complications. Here, by selecting an odd number of lines, the symmetry in frame blanking pulses is achieved and this enables perfect interlaced scanning. Any error in scanning timings and sequence would leave a large number of picture elements unresolved and thus the quality of the reproduced picture gets impaired. Fig. 2.9 shows various cases of interlace error. For convenience of explanation the retrace time has been assumed to be zero. Interlace error occurs due to the time difference in starting the second field. For perfect interlace the second field should start from point '*b*' (see Fig. 2.9 (*a*)), *i.e.*, 32 μs away from '*a*', the starting point of the first field. If it starts early or late interlace error will be there. For a 16 μs delay in the start of the second field (Fig. 2.9 (*b*)), starting points of the two fields will be 48 μs apart instead of the desired 32 μs. Then the percentage interlace error

$$= \frac{48 - 32}{32} \times 100 = 50\%$$

if the second field starts 16 μs early even then the error would be 50%. For a delay of 32 μs the two fields will overlap (Fig. 2.9 (*c*)) and the interlace error would be 100%, *i.e.*, half the picture area will go unscanned.

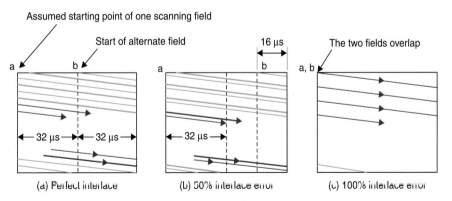

(a) Perfect interlace   (b) 50% interlace error   (c) 100% interlace error

Fig. 2.9 *Examples of interlace error.*

The above examples demonstrate that incorrect start of any field produces vertical displacement between the lines of the two fields. This brings these lines closer leaving gaps between the pairs thus formed. The result is a deterioration of the picture's vertical resolution because certain areas do not get scanned at all.

For perfect interlaced scanning it is essential that the starting points at the top of the frame is separated exactly one half line between first and second fields. To achieve this it is necessary to feed two regularly spaced synchronising pulses to the field time base during each frame period. One of these pulses must arrive in the middle of a line and the next at the end of a line. This is shown in Fig. 2.10. Thus the vertical time base must be triggered 50 times per second in the manner explained above. For half line separation between the two fields only the topmost and the extreme bottom lines are then half lines whereas the remaining lines are all full lines. If there are x number of full lines per field, where x may be even or odd, the total number of full lines per frame is then $2x$, an even number. To this, when the two half lines get added the total number of lines per frame becomes odd. Thus for interlaced scanning the total number of lines in any TV system must be odd. With an even number of lines the two fields are bound to fall on each other and interlaced scanning would not take place.

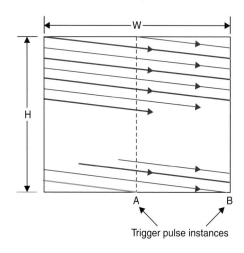

Fig. 2.10 *Vertical trigger pulse instances.*
A—after 1st field, B—after 2nd field.

Further for correct interlacing it becomes necessary that at the transmitter automatic frequency control must be utilized to maintain a horizontal scanning frequency that is exactly 312.5 times as great as the field frequency, *i.e.*, 50 Hz. This is accomplished by generating a stable frequency at 15625 Hz by crystal controlled oscillator circuits. A frequency doubling circuit produces a frequency of 31250 Hz, which is utilized to control the correct generation of equalizing and vertical sync pulses. Four frequency division circuits each with a ratio of 5 : 1 are employed to derive 50 Hz, the vertical scanning frequency ($31250 = 5 \times 5 \times 5 \times 5 \times 50$). Thus all the required frequencies are derived from a common stable source and they automatically remain interlocked in the correct ratios. To achieve this, *i.e.*, frequency division, the total number of lines per frame must be a product of small whole numbers. The frame frequency of 625 satisfies all the above requirements. Similarly 525 lines in the American system and 819 lines in the French system also meet these requirements.

Comparison of various TV systems. Picture and sound signal standards for the principal monochrome television systems are given at the end of chapter 4. The CCIR 625-B monochrome system used in most parts of Europe and adopted by India has a video bandwidth of 5 MHz, whereas the British 625 line system has a video bandwidth of 5.5 MHz. Obviously, here 0.73 has been used as the resolution factor instead of the 0.69 used in our system. So the British system is marginally better than the European system. The French TV system employs 819 lines with a video bandwidth of 10.4 MHz. This system therefore has both much improved vertical resolution and a better horizontal resolution.

The American 525 line system employs a frame frequency of 30 as compared to 25 in the CCIR 625-B monochrome system. Thus, the line frequency in this system is 15750, which compares very closely to our system where the line frequency is 15625. However, the American system employs a bandwidth of 4 MHz which suggests that the horizontal resolution of this system is less than all other systems in use. It must be noted that the number of lines employed by a given TV system is not in itself, a guide to the quality of resolution available from the system. It is true that greater the number of lines the better the vertical resolution, but an assessment of the horizontal resolution, *i.e.*, the bandwidth employed by the system is a better overall guide to the quality of definition.

2.6   TONAL GRADATION

In addition to proper bandwidth required to produce the details allowed by the scanning system at the transmitting end and the picture tube at the receiving end, the signal-transmission system should have proper transfer characteristics to preserve same brightness gradation as the eye would perceive when viewing the scene directly. Any non-linearity in the pick-up and picture tube should also be corrected by providing inverse nonlinearities in the channel circuitry to obtain overall linear characteristics. Note that the sensation in the eye to detect changes or brightness is logarithmic in nature and this must be taken into account while designing the overall channel.

Various other factors that influence the tonal quality of the reproduced picture are :

(a) Contrast. This is the difference in intensity between black and white parts of the picture over and above the brightness level.

(b) Contrast ratio. The ratio of maximum to minimum brightness relative to the original picture is called contrast ratio. In broad daylight the variations in brightness are very wide with ratio as high as 10000 : 1, whereas the picture tube, because of certain limitations, cannot produce a contrast with variations more than 50 : 1 or atmost 100 : 1. Ratio of brightness variations in the reproduced picture on the screen of the picture tube, to the brightness variations in the original scene is known as Gamma of the picture. Its value is close to 0.5. In studios, under controlled conditions of light, the variations are less wide than outside and so the brightness variations that can be reproduced by the picture tube are not very much different than that of the scene. Realism is still maintained because the viewer does not actually see the scene being televised. Another factor which makes stringent demands from the system unnecessary is the fact that our eye can accommodate not more than 10 : 1 variations of light intensity at any time. Too bright a representation of the bright areas in a picture would make grey areas appear as dark in comparison. This is true at all levels of light intensity with brightness variations in relative ratios of 10 : 1.

Chapter 2

When a TV receiver is off, there is no beam impinging on the fluorescent screen of the picture tube and no light gets emitted. Then with normal light in the room the screen appears as dull white. But when the receiver is no, and a TV programme is being received the bright portions of the scene appear quite bright because the corresponding amplitude of the video signal makes the control-grid of the picture tube much less negative and the consequent increased beam current causes more light on the screen. However, for a very dark portion of the scene the corresponding video signal makes the grid highly negative with respect to the cathode and thus cuts-off the beam current and no light is emitted on the corresponding portions on the screen. These areas appear to the eye as dark in comparison with the high light areas of the screen, whereas the same area in the absence of beam current when the set was off appeared close to a white shade. This as explained earlier is due to the logarithmic response of the human eye and its inability to accommodate light intensity variations greater than 10 : 1.

(c) *Viewing distance.* The viewing distance from the screen of the TV receiver should not be so large that the eye cannot resolve details of the picture. The distance should also not be so small that picture elements become separately visible. The above conditions are met when the vertical picture size subtends an angle of approximately 15° at the eye. The distance also depends on habit, varies from person to person, and lies between 3 to 8 times the picture height. Most people prefer a distance close to five times the picture height. While viewing TV, a small light should be kept on in the room to reduce contrast. This does not strain the eyes and there is less fatigue.

REVIEW QUESTIONS

1. Justify the choice of rectangular frame with width to height ratio = 4/3 for television transmission and reception.

2. How is the illusion of continuity created in television pictures ? Why has the frame reception rate been chosen to be 25 and not 24 as in motion pictures ?

3. What do you understand by interlaced scanning ? Show that it reduces flicker and conserve bandwidth.

4. What do you understand by active and blanking periods in horizontal and vertical scanning ? Give the periods of nominal, active and retrace intervals of horizontal and vertical scanning as used in the 625 line system.

5. How many horizontal lines get traced during each vertical retrace ? What is the active number of lines that are actually used for picture information pick up and reception ?

6. Draw a picture frame chart showing the total number of active and inactive lines during each field and establish the need for terminating the first field in a half line and the beginning the second at the middle of a line at the top.

7. Justify the choice of 625 lines for TV transmission. Why is the total number of lines kept odd in all television systems ? What is the significance of choosing the number of lines as 625 and not 623 or 627 ?

8. What do you understand by resolution or Kell-factor ? How does it affect the vertical resolution of a television picture ? Show that the vertical resolution increases with increase in number of scanning lines.

9. What is meant by equal vertical and horizontal 'resolution' ? Derive an expression for the highest modulating frequency in a television system and show that it is nearly 5 MHz. in the 625-B monochrome system.

Chapter 2

10. Show that if the number of lines employed in a TV system is increased then the highest video frequency must increase as the square of the increase in number of lines for equal improvement in vertical and horizontal resolution.

11. Show that the 625-B TV system is only marginally superior to the 525 line American system.

12. What do you understand by interlace error and how does it affect the quality of the picture ? Calculate the percentage interlace error when the second field is delayed by 8 μs. Retrace time may be assumed to be negligible.

13. In the British 625 lines system the resolution factor employed is 0.73 instead of 0.69 as used in the 625-B monochrome system. All other scanning details remaining the same, calculate the highest modulating frequency used in the British system.

14. Explain the need for providing very good low frequency response and phase characteristics in amplifiers used in any TV link, for proper reproduction of brightness variations.

15. The relevant data for a closed circuit TV system is given below. Calculate the highest modulating frequency that will be generated while scanning the most stringent case of alternate black and white dots for equal vertical and horizontal resolution.

No. of lines	= 250
Interlace ratio	= 1 : 1
Picture repetition rate	= 50/sec
Aspect ratio	= 4/3
Vertical retrace time	= 10% of the picture frame time
Horizontal retrace time	= 20% of the total line time
Assume resolution factor	= 0.8
Ans	≈ 2 MHz

16. Explain the meaning of terms-tonal gradation, contrast, contrast ratio and gamma of the picture.

When a TV receiver is off, no electron beam strikes the picture tube screen and the screen face looks a dull white. With the set on and a black and white picture showing on the screen, no electron beam impinges on the darker area of the reproduced picture. But these areas now appear quite black instead of the dull white of the switched-off set. Explain the reason for this difference in appearance.

Chapter 2

Composite Video Signal

Composite video signal consists of a camera signal corresponding to the desired picture information, blanking pulses to make the retrace invisible, and synchronizing pulses to synchronize the transmitter and receiver scanning. A horizontal synchronizing pulse is needed at the end of each active line period whereas a vertical sync pulse is required after each field is scanned. The amplitude of both horizontal and vertical sync pulses is kept the same to obtain higher efficiency of picture signal transmission but their duration (width) is chosen to be different for separating them at the receiver. Since sync pulses are needed consecutively and not simultaneously with the picture signal, these are sent on a time division basis and thus form a part of the composite video signal.

3.1 VIDEO SIGNAL DIMENSIONS

Figure 3.1 shows the composite video signal details of three different lines each corresponding to a different brightness level of the scene. As illustrated there, the video signal is constrained to vary between certain amplitude limits. The level of the video signal when the picture detail being transmitted corresponds to the maximum whiteness to be handled, is referred to as peak-white level. This is fixed at 10 to 12.5 percent of the maximum value of the signal while the black level corresponds to approximately 72 percent. The sync pulses are added at 75 percent level called the blanking level. The difference between the black level and blanking level is known as the 'Pedestal'. However, in actual practice, these two levels, being very close, tend to merge with each other as shown in the figure. Thus the picture information may vary between 10 percent to about 75 percent of the composite video signal depending on the relative brightness of the picture at any instant. The darker the picture the higher will be the voltage within those limits.

Note that the lowest 10 percent of the voltage range (whiter than white range) is not used to minimize noise effects. This also ensures enough margin for excessive bright spots to be accommodated without causing amplitude distortion at the modulator.

At the receiver the picture tube is biased to ensure that a received video voltage corresponding to about 10 percent modulation yields complete whiteness at that particular point on the screen, and an analogous arrangement is made for the black level. Besides this, the television receivers are provided with 'brightness' and 'contrast' controls to enable the viewer to make final adjustments as he thinks fit.

D.C. component of the video signal. In addition to continuous amplitude variations for individual picture elements, the video signal has an average value or dc component corresponding to the average brightness of the scene. In the absence of dc component the receiver cannot follow changes in brightness, as the ac camera signal, say for grey picture elements on a black background will then be the same as a signal for white area on a grey back-ground. In Fig. 3.1, dc components of the signal for three lines have been identified, each representing a different level of average brightness in the scene. It may be noted that the break shown in the illustration after each line signal is to emphasize that dc component of the video signal is the average value for complete frames rather than lines since the background information of the picture indicates the brightness of the scene. Thus Fig. 3.1 illustrates the concept of change in the average brightness of the scene with the help of three lines in separate frames because the average brightness can change only from frame to frame and not from line to line.

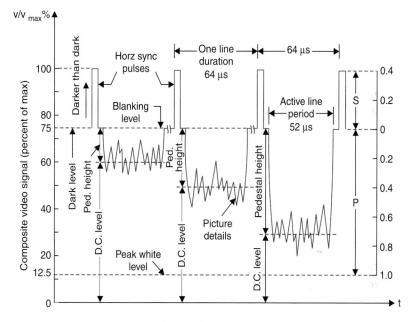

Fig. 3.1 *Arbitrary picture signal details of three scanning lines with different average brightness levels. Note that picture to sync ratio P/S = 10/4.*

Pedestal height. As noted in Fig. 3.1 the pedestal height is the distance between the pedestal level and the average value (dc level) axis of the video signal. This indicates average brightness since it measures how much the average value differs from the black level. Even when the signal loses its dc value when passed through a capacitor-coupled circuit the distance between the pedestal and the dc level stays the same and thus it is convenient to use the pedestal level as the reference level to indicate average brightness of the scene.

Setting the pedestal level. The output signal from the TV camera is of very small amplitude and is passed through several stages of ac coupled high gain amplifiers before being coupled to a control amplifier. Here sync pulses and blanking pulses are added and then clipped at the correct level to form the pedestals. Since the pedestal height determines the average brightness of the scene, any smaller value than the correct one will make the scene darker while a larger pedestal height will result in higher average brightness. The video control operator who observes the scene at the studio sets the level for the desired brightness in the reproduced picture which he is viewing on a monitor receiver. This is known as dc insertion because this amounts to adding a dc component to the ac signal. Once the dc insertion has been acomplished the pedestal level becomes the black reference and the pedestal height indicates correct relative brightness for the reproduced picture. However, the dc level inserted in the control amplifier is usually lost in succeeding stages because of capacitive coupling, but still the correct dc component can be reinserted when necessary because the pedestal height remains the same.

The blanking pulses. The composite video signal contains blanking pulses to make the retrace lines invisible by raising the signal amplitude slightly above the black level (75 per cent) during the time the scanning circuits produce retraces. As illustrated in Fig. 3.2, the composite video signal contains horizontal and vertical blanking pulses to blank the corresponding retrace intervals. The repetition rate of horizontal blanking pulses is therefore equal to the line scanning frequency of 15625 Hz. Similarly the frequency of the vertical blanking pulses is equal to the field-scanning frequency of 50 Hz. It may be noted that though the level of the blanking pulses is distinctly above the picture signal information, these are not used as sync pulses. The reason is that any occasional signal corresponding to any extreme black portion in the picture may rise above the blanking level and might conceivably interfere with the synchronization of the scanning generators. Therefore, the sync pulses, specially designed for triggering the sweep oscillators are placed in the upper 25 per cent (75 per cent to 100 per cent of the carrier amplitude) of the video signal, and are transmitted along with the picture signal.

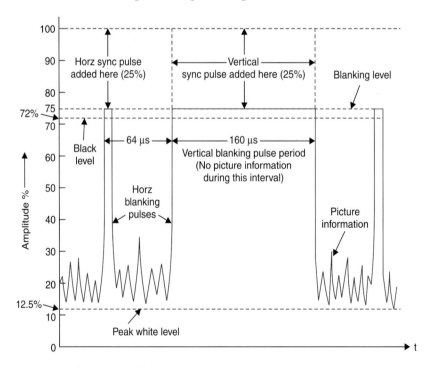

Fig. 3.2 *Horizontal and vertical blanking pulses in video signal. Sync pulses are added above the blanking level and occupy upper 25% of the composite video signal amplitude.*

Sync pulse and video signal amplitude ratio. The overall arrangement of combining the picture signal and sync pulses may be thought of as a kind of voltage division multiplexing where about 65 per cent of the carrier amplitude is occupied by the video signal and the upper 25 per cent by the sync pulses. Thus, as shown in Fig. 3.1, the final radiated signal has a picture to sync signal ratio (P/S) equal to 10/4. This ratio has been found most satisfactory because if the picture signal amplitude is increased at the expense of sync pulses, then when the signal to noise ratio of the received signal falls, a point is reached when the sync pulse amplitude becomes insufficient to keep the picture locked even though the picture voltage is still of adequate amplitude to yield an acceptable picture. On the other hand if sync pulse height is increased at the expense of the picture detail, then under similar conditions the raster remains locked but the picture content is of too low an amplitude to set up a worthwhile picture. A ratio of P/S = 10/4, or thereabout, results in a situation such that when the signal to noise ratio reaches a certain low level, the sync amplitude becomes insufficient, *i.e.,* the sync fails at the same time as the picture ceases to be of entertainment value. This represents the most efficient use of the television system.

3.2 HORIZONTAL SYNC DETAILS

The horizontal blanking period and sync pulse details are illustrated in Fig. 3.3. The interval between horizontal scanning lines is indicated by *H*. As explained earlier, out of a total line

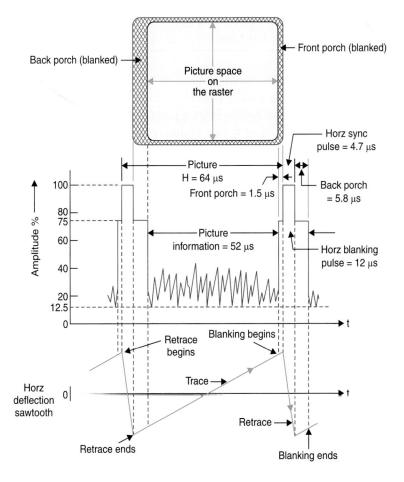

Fig. 3.3 *Horz line and sync details compared to horizontal deflection sawtooth and picture space on the raster.*

period of 64 µs, the line blanking period is 12 µs. During this interval a line synchronizing pulse is inserted. The pulses corresponding to the differentiated leading edges of the sync pulses are actually used to synchronize the horizontal scanning oscillator. This is the reason why in Fig. 3.3 and other figures to follow, all time intervals are shown between sync pulse leading edges.

The line blanking period is divided into three sections. These are the 'front porch', the 'line sync' pulse and the 'back porch'. The time intervals allowed to each part are summarized below and their location and effect on the raster is illustrated in Fig. 3.3.

Details of Horizontal Scanning

Period	Time (µs)
Total line (H)	64
Horz blanking	12 ± .3
Horz sync pulse	4.7 ± 0.2
Front porch	1.5 ± .3
Back porch	5.8 ± .3
Visible line time	52

Front porch. This is a brief cushioning period of 1.5 µs inserted between the end of the picture detail for that line and the leading edge of the line sync pulse. This interval allows the receiver video circuit to settle down from whatever picture voltage level exists at the end of the picture line to the blanking level before the sync pulse occurs. Thus sync circuits at the receiver are isolated from the influence of end of the line picture details. The most stringent demand is made on the video circuits when peak white detail occurs at the end of a line. Despite the existence of the front porch when the line ends in an extreme white detail, and the signal amplitude touches almost zero level, the video voltage level fails to decay to the blanking level before the leading-edge of the line sync pulse occurs. This results in late triggering of the time base circuit thus upsetting the 'horz' line sync circuit. As a result the spot (beam) is late in arriving at the left of the screen and picture information on the next line is displaced to the left. This effect is known as 'pulling-on-whites'.

Line sync pulse. After the front proch of blanking, horizontal retrace is produced when the sync pulse starts. The flyback is definitely blanked out because the sync level is blacker than black. Line sync pulses are separated at the receiver and utilized to keep the receiver line time base in precise synchronism with the distant transmitter. The nominal time duration for the line sync pulses is 4.7 µs. During this period the beam on the raster almost completes its back stroke (retrace) and arrives at the extreme left end of the raster.

Back porch. This period of 5.8 µs at the blanking level allows plenty of time for line flyback to be completed. It also permits time for the horizontal time-base circuit to reverse direction of current for the initiation of the scanning of next line. Infact, the relative timings are so set that small black bars (see Fig. 3.3) are formed at both the ends of the raster in the horizontal plane. These blanked bars at the sides have no effect on the picture details reproduced during the active line period.

The back porch* also provides the necessary amplitude equal to the blanking level (reference level) and enables to preserve the dc content of the picture information at the transmitter. At the receiver this level which is independent of the picture details is utilized in the AGC (automatic gain control) circuits to develop true AGC voltage proportional to the signal strength picked up at the antenna.

3.3 VERTICAL SYNC DETAILS

The vertical sync pulse train added after each field is somewhat complex in nature. The reason for this stems from the fact that it has to meet several exacting requirements. Therefore, in order to fully appreciate the various constituents of the pulse train, the vertical sync details are explored step by step while explaining the need for its various components.

The basic vertical sync added at the end of both even and odd fields is shown in Fig. 3.4. Its width has to be kept much larger than the horizontal sync pulse, in order to derive a suitable field sync pulse at the receiver to trigger the field sweep oscillator.

The standards specify that the vertical sync period should be 2.5 to 3 times the horizontal line period. If the width is less than this, it becomes difficult to distinguish between horizontal and vertical pulses at the receiver.

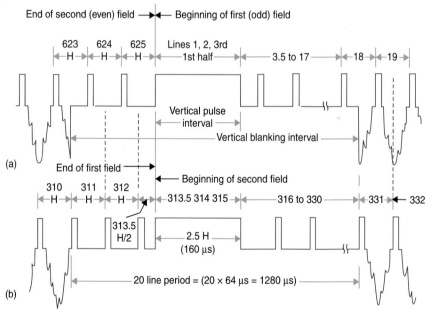

Fig. 3.4 *Composite video waveforms showing horizontal and basic vertical sync pulses at the end of (a) second (even) field, (b) first (odd) field. Note, the widths of horizontal blanking intervals and sync pulses are exaggerated.*

If the width is greater than this, the transmitter must operate at peak power for an unnecessarily long interval of time. In the 625 line system 2.5 line period (2.5 × 64 = 160 μs) has

---

*In colour TV transmission a short sample (8 to 10 cycles) of the colour subcarrier oscillator output is sent to the receiver for proper detection of colour signal sidebands. This is known as colour burst and is located at the back porch of the horizontal blanking pedestal.

been allotted for the vertical sync pulses. Thus a vertical sync pulse commences at the end of 1st half of 313th line (end of first field) and terminates at the end fo 315th line. Similarly after an exact interval of 20 ms (one field period) the next sync pulse occupies line numbers—1st, 2nd and 1st half of third, just after the second field is over. Note that the beginning of these pulses has been aligned in the figure to signify that these must occur after the end of vertical stroke of the beam in each field, *i.e.*, after each 1/50th of a second. This alignment of vertical sync pulses, one at the end of a half-line period and the other after a full line period (see Fig. 3.4), results in a relative misalignment of the horizontal sync pulses and they do not appear one above the other but occur at half-line intervals with respect to each other. However, a detailed examination of the pulse trains in the two fields would show that horizontal sync pulses continue to occur exactly at 64 μs intervals (except during the vertical sync pulse periods) throughout the scanning period from frame to frame and the apparent shift of 32 μs is only due to the alignment of vertical sync instances in the figure.

As already mentioned the horizontal sync information is extracted from the sync pulse train by differentiation, *i.e.*, by passing the pulse train through a high-pass filter. Indeed pulses corresponding to the differentiated leading edges of sync pulses are used to synchronise the horizontal scanning oscillator. The process of deriving these pulses is illustrated in Fig. 3.5. Furthermore, receivers often use monostable multivibrators to generate horizontal scan, and so a pulse is required to initiate each and every cycle of the horizontal oscillator in the receiver.

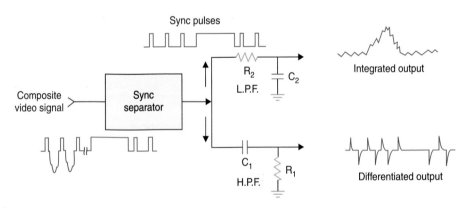

Fig. 3.5 Sync pulse separation and generation of vertical and horizontal sync pulses.

This brings out the first and most obvious shortcoming of the waveforms shown in Fig. 3.4. The horizontal sync pulses are available both during the active and blanked line periods but there are no sync pulses (leading edges) available during the 2.5 line vertical sync period. Thus the horizontal sweep oscillator that operates at 15625 Hz, would tend to step out of synchronism during each vertical sync period. The situation after an odd field is even worse. As shown in Fig. 3.4, the vertical blanking period at the end of an odd field begins midway through a horizontal line. Consequently, looking further along this waveform, we see that the leading edge of the vertical sync pulse comes at the wrong time to provide synchronization for the horizontal oscillator. Therefore, it becomes necessary to cut slots in the vertical sync pulse at half-line-intervals to provide horizontal sync pulses at the correct instances both after even and odd fields. The technique is to take the video signal amplitude back to the blanking level 4.7 μs before the line pulses are needed. The waveform is then returned back to the maximum level at the moment the line sweep circuit needs synchronization. Thus five narrow slots of 4.7 μs width get formed in each vertical sync pulse at intervals of 32 μs. The trailing but rising edges of these pulses are actually used to trigger the horizontal oscillator. The resulting waveforms together with line numbers and the differentiated output of both the field trains is illustrated in Fig. 3.6. This insertion of short pulses is known as notching or serration of the broad field pulses.

Note that though the vertical pulse has been broken to yield horizontal sync pulses, the effect on the vertical pulse is substantially unchanged. It still remains above the blanking voltage level all of the time it is acting. The pulse width is still much wider than the horizontal pulse width and thus can be easily separated at the receiver. Returning to Fig. 3.6 it is seen that each horizontal sync pulse yields a positive spiked output from its leading edge and a negative spiked pulse from its trailing edge. Time-constant of the differentiating circuit is so chosen, that by the time a trailing edge arrives, the pulse due to the leading edge has just about decayed. The negative-going triggering pulses may be removed with a diode since only the positive going pulses are effective in locking the horizontal oscillator.

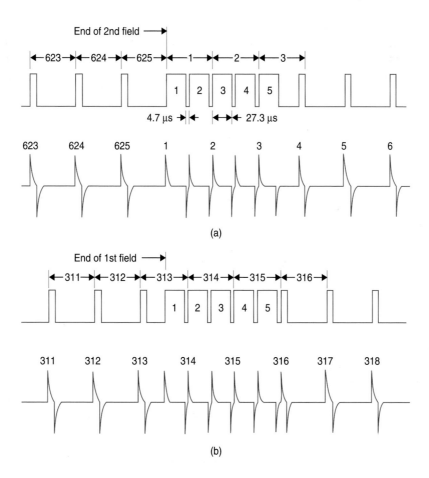

Fig. 3.6 Differentiating waveforms (a) pulses at the end of even (2nd) field and the corresponding output of the differentiator (H.P.F.) (b) pulses at the end of odd (1st) field and the corresponding output of the differentiator (H.P.F.) Note, the differentiated pulses bearing line numbers are the only ones needed at the end of each field.

Chapter 3

However, the pulses actually utilized are the ones that occur sequentially at 64 µs intervals. Such pulses are marked with line numbers for both the fields. Note that during the intervals of serrated vertical pulse trains, alternate vertical spikes are utilized. The pulses not used in one field are the ones utilized during the second field. This happens because of the half-line difference at the commencement of each field and the fact that notched vertical sync pulses occur at intervals of 32 µs and not 64 µs as required by the horizontal sweep oscillator. The pulses that come at a time when they cannot trigger the oscillator are ignored. Thus the requirement of keeping the horizontal sweep circuit locked despite insertion of vertical sync pulses is realized.

Now we turn to the second shortcoming of the waveform of Fig. 3.4. First it must be mentioned that synchronization of the vertical sweep oscillator in the receiver is obtained from vertical sync pulses by integration. This is illustrated in Fig. 3.5 where the time-constant R_2C_2 is chosen to be large compared to the duration of horizontal pulses but not with respect to width of the vertical sync pulses. The integrating circuit may equally be looked upon as a low-pass filter, with a cuit-off frequency such that the horizontal sync pulses produce very little output, while the vertical pulses have a frequency that falls in the pass-band of the filter. The voltage built across the capacitor of the low-pass filter (integrating circuit) corresponding to the sync pulse trains of both the fields is shown in Fig. 3.7. Note that each horizontal pulse causes a slight rise in voltage across the capacitor but this is reduced to zero by the time the next pulse arrives. This is so, because the charging period for the capacitor is only 4.7 µs and the voltage at the input to the integrator remains at zero for the rest of the period of 59.3 µs. Hence there is no residual voltage across the vertical filter (L.P. filter) due to horizontal sync-pulses. Once the broad serrated vertical pulse arrives the voltage across the output of the filter starts increasing. However, the built up voltage differs for each field. The reason is not difficult to find. At the beginning of the first field (odd field) the last horz sync pulse corresponding to the beginning of 625th line is separated from the 1st vertical pulse by full one line and any voltage developed across the filter will have enough time to return to zero before the arrival of the first vertical pulse, and thus the filter output voltage builds up from zero in response to the five successive broad vertical sync pulses. The voltage builds up because the capacitor has more time to charge and only 4.7 µs to discharge. The situation, however, is not the same for the beginning of the 2nd (even) field. Here the last horizontal pulse corresponding to the beginning of 313th line is separated from the first vertical pulse by only half-a-line. The voltage developed across the vertical filter will thus not have enough time to reach zero before the arrival of the first vertical pulse, which means that the voltage build-up does not start from zero, as in the case of the 1st field. The residual voltage on account of the half line discrepancy gets added to the voltage developed on account of the broad vertical pulses and thus the voltage developed across the output filter is some what higher at each instant as compared to the voltage developed at the beginning of the first-field. This is shown in dotted chain line in Fig. 3.7.

The vertical oscillator trigger potential level marked as trigger level in the diagram (Fig. 3.7) intersects the two filter output profiles at different points which indicates that in the case of second field the oscillator will get triggered a fraction of a second too soon as compared to the first field. Note that this inequality in potential levels for the two fields continues during the period of discharge of the capacitor once the vertical sync pulses are over and the horizontal sync pulses take-over. Though the actual time difference is quite short it does prove sufficient to upset the desired interlacing sequence.

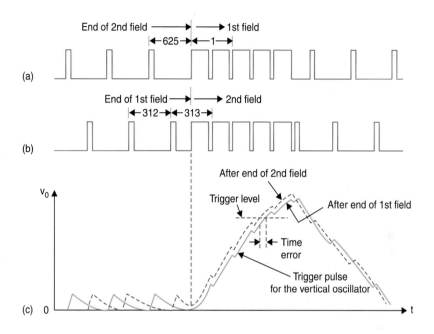

Fig. 3.7 *Integrating waveforms (a) pulses at the end of 2nd (even) field (b) pulses at the end of 1st (odd) field (c) integrator output. Note the above sync pulses have purposely been drawn without equalizing pulses.*

Equalizing pulses. To take care of this drawback which occurs on account of the half-line discrepancy five narrow pulses are added on either side of the vertical sync pulses. These are known as pre-equalizing and post-equalizing pulses. Each set consists of five narrow pulses occupying 2.5 lines period on either side of the vertical sync pulses. Pre-equalizing and post-equalizing pulse details with line numbers occupied by them in each field are given in Fig. 3.8. The effect of these pulses is to shift the half-line discrepancy away both from the beginning and end of vertical sync pulses. Pre-equalizing pulses being of 2.3 μs duration result in the discharge of the capacitor to essentially zero voltage in both the fields, despite the half-line discrepancy before the voltage build-up starts with the arrival of vertical sync pulses. This is illustrated in Fig. 3.9. Post-equalizing pulses are necessary for a fast discharge of the capacitor to ensure triggering of the vertical oscillator at proper time. If the decay of voltage across the capacitor is slow as would happen in the absence of post-equalizing pulses, the oscillator may trigger at the trailing edge which may be far-away from the leading edge and this could lead to an error in triggering.

Thus with the insertion of narrow pre and post equalizing pulses, the voltage rise and fall profile is essentially the same for both the field sequences (see Fig. 3.9) and the vertical oscillator is triggered at the proper instants, *i.e.,* exactly at an interval of 1/50th of a second. This problem

Chapter 3

could possibly also be solved by using an integrating circuit with a much larger time constant, to ensure that the capacitor remains virtually uncharged by the horizontal pulses. However, this would have the effect of significantly reducing the integrator output for vertical pulses so that a vertical sync amplifier would have to be used. In a broadcasting situation, there are thousands of receivers for every transmitter. Consequently it is much more efficient and economical to cure this problem in one transmitter than in thousands of receivers. This, as explained above, is achieved by the use of pre and post equalizing pulses. The complete pulse trains for both the fields incorporating equalizing pulses are shown in Fig. 3.10.

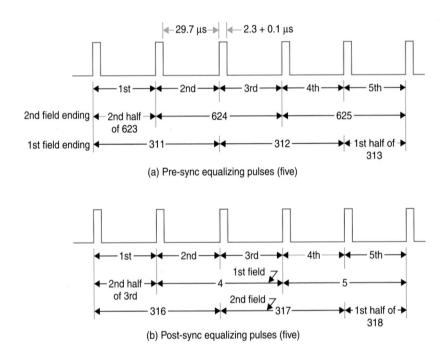

(a) Pre-sync equalizing pulses (five)

(b) Post-sync equalizing pulses (five)

Fig. 3.8 *Pre-sync equalizing and Post-sync equalizing pulses.*

From the comparison of the horizontal and vertical output pulse forms shown in Figs. 3.7 and 3.9 it appears that the vertical trigger pulse (output of the low-pass filter) is not very sharp but actually it is not so. The scale chosen exaggerates the extent of the vertical pulses. The voltage build-up period is only 160 μs and so far as the vertical synchronizing oscillator is concerned this pulse occurs rapidly and represents a sudden change in voltage which decays very fast.

The polarity of the pulses as obtained at the outputs of their respective fields may not be suitable for direct application in the controlled synchronizing oscillator and might need inversion depending on the type of oscillator used. This aspect will be fully developed in the chapter devoted to vertical and horizontal oscillators.

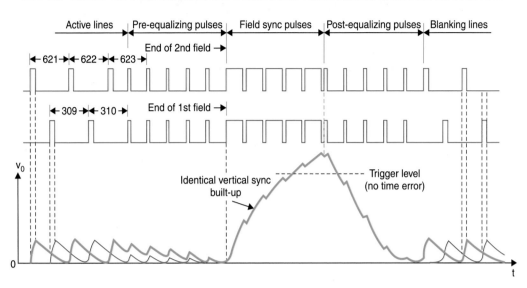

Fig. 3.9 *Identical vertical sync voltage built-up across the integrating capacitor.*

3.4 SCANNING SEQUENCE DETAILS

A complete chart giving line numbers and pulse designations for both the fields (corresponding to Fig. 3.10) is given below :

First Field (odd field)

Line numbers : one to 1st-half of 313th line (312.5 lines)

1, 2 and 3rd 1st-half, lines	2.5 lines—Vertical sync pulses
3rd 2nd-half, 4, and 5	2.5 lines—Post-vertical sync equalizing pulses.
6 to 17, and 18th 1st-half	12.5 lines—Blanking retrace pulses
18th 2nd-half to 310	292.5 lines—Picture details
311, 312, and 313th 1st-half	2.5 lines—Pre-vertical sync equalizing pulses for the 2nd field.

Total number of lines = 312.5

Second Field (even field)

Line numbers : 313th 2nd-half to 625 (312.5 lines)

313th 2nd-half, 314, 315	2.5 lines—Vertical sync pulses
316, 317, 318th 1st-half	2.5 lines—Post-vertical sync equalizing pulses
318th 2nd-half-to 330	12.5 lines—Blanking retrace pulses
331 to 1st-half of 623rd	292.5 line—Picture details
623 2nd-half, 624 and 625	2.5 lines—Pre-vertical sync equalizing pulses for the 1st field

Total number of lines = 312.5

Total Number of Lines per Frame = 625

Chapter 3

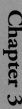

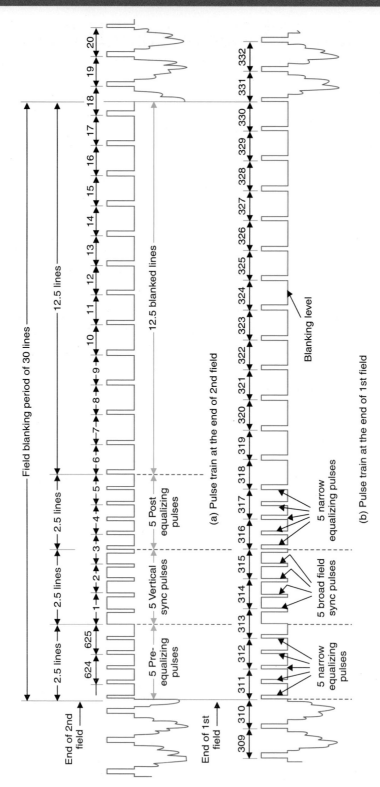

Fig. 3.10 Field synchronizing pulse trains of the 625 lines TV system.

Approximate location of line numbers. The serrated vertical sync pulse forces the vertical deflection circuitry to start the flyback. However, the flyback generally does not begin with the start of vertical sync because the sync pulse must build up a minimum voltage across the capacitor to trigger the scanning oscillator. If it is assumed that vertical flyback starts with the leading edge of the fourth serration, a time of 1.5 lines passes during vertical sync before vertical flyback starts. Also five equalizing pulses occur before vertical sync pulse train starts. Then four lines (2.5 + 1.5 = 4) are blanked at the bottom of the picture before vertical retrace begins. A typical vertical retrace time is five lines. Thus the remaining eleven (20 – (4 + 5) = 11) lines are blanked at the top of the raster. These lines provide the sweep oscillator enough time to adjust to a linear rise for uniform pick-up and reproduction of the picture.

3.5 FUNCTIONS OF VERTICAL PULSE TRAIN

By serrating the vertical sync pulses and the providing pre- and post-equalizing pulses the following basic requirements necessary for successful interlaced scanning are ensured.

(*a*) A suitable field sync pulse is derived for triggering the field oscillator.

(*b*) The line oscillator continues to receive triggering pulses at correct intervals while the process of initiation and completion of the field time-base stroke is going on.

(*c*) It becomes possible to insert vertical sync pulses at the end of a line after the 2nd field and at the middle of a line at the end of the 1st field without causing any interlace error.

(*d*) The vertical sync build up at the receiver has precisely the same shape and timing on odd and even fields.

3.6 SYNC DETAILS OF THE 525 LINE SYSTEM

In the 525 line American TV system where the total number of lines scanned per second is 15750, the sync pulse details are as under :

Details of Horz Blanking

Period	Time (μs)
Field line (H)	63.5
Horz blanking	9.5 to 11.5
Horz sync pulse	4.75 ± 0.5
Front porch	1.26 (minimum)
Back porch	3.81 (minimum)
Visible line	52 to 54

Details of Vertical Blanking

Period	Time
Total field (*V*) period	= 1/60 sec. = 16.7 ms
Visible field time	= 15 to 16 ms
Vertical blanking	= 0.8 to 1.3 ms

Chapter 3

Total duration of six (serrated)
vertical sync pulses $= 3H = 190.5 \, \mu s$

Each serrated pulse $= H/2 = 31.75 \, \mu s$

Each equalizing pulse
(Six pre- and six post-equailzing pulses
are provided at $H/2$ intervals) $= 0.04 \, H = 2.54 \, \mu s$

REVIEW QUESTIONS

1. Sketch composite video signal waveform for at least three successive lines and indicate :
    (i) extreme white level, (ii) blanking level, (ii) pedestal height and (iv) sync pulse level. Justify
    the choice of P/S ratio = 10/4 in the composite signal. Why is the combining of picture signal and
    sync pulses called a voltage division multiplex ?

2. Sketch composite video signal waveforms for the picture information shown in Fig. P 3.1.

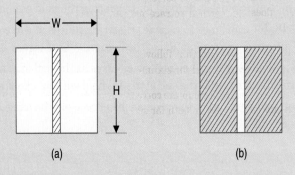

(a)   (b)

Fig. P3.1

3. Show picture information on a raster for the video signals drawn in Fig. P3.2.

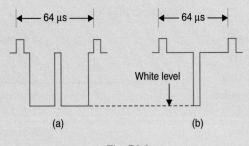

(a)   (b)

Fig. P3.2

4. Sketch the details of horizontal blanking and sync pulses. Label on it (i) front porch, (ii) horizon-
    tal sync pulse, (iii) back porch and (iv) active line periods. Why are the front porch and back
    porch intervals provided before and after the horizontal sync pulse ? Explain why the blanking
    pulses are not used as sync pulses.

5. Enumerate the basic requirements that must be satisfied by the pulse train added after each field. Why is it necessary to serrate the broad vertical sync pulse ?

6. Sketch the pulse trains that follow after the second and first field of active scanning. Why are the vertical sync pulses notched at 32 μs interval and not at 64 μs interval to provide horizontal sync pulses ?

7. Explain how the horizontal and vertical sync pulses are separated and shaped at the receiver. For a time constant of 5 μs for the differentiating circuit, and 100 μs for the integrating circuit, plot the output waveforms from both the circuits for the entire vertical period. Calculate the error in timing for successive vertical fields in the absence of equalizing pulses.

8. Sketch the complete pulse trains that follow at the end of both odd and even fields. Fully label them and explain how the half line discrepancy is removed by insertion of pre-equalizing pulses.

9. Justify the need for pre and post equalizing pulses. Why it is necessary to keep their duration equal to the half-line period ?

10. Justify the need for a blanking period corresponding to 20 complete lines after each active field of scanning. Why does the vertical retrace not begin with the incoming of the first serrated vertical sync pulse ?

11. Sketch the complete pulse trains that follow at the end of odd and even fields in the 525 line television system. Justify the need for six instead of five pre and post equalizing pulses.

12. Show by any suitable means approximate correspondence between line numbers and the location of the electron beam on the screen, both for odd and even fields.

Chapter 3

4

Signal Transmission and Channel Bandwidth

In most television systems as also in the C.C.I.R 625 line, the picture signal is amplitude modulated and sound signal frequency modulated before transmission. The channel bandwidth is determined by the highest video frequency required for proper picture reception and the maximum sound carrier frequency deviation permitted in a TV system.

Need for modulation. The need for modulation stems from the fact that it is impossible to transmit a signal by itself. The greatest difficulty in the use of unmodulated wave is the need for long antennas for efficient radiation and reception. For example, a quarter-wavelength antenna for the transmitting frequency of 15 kHz would be 5000 meters long. A vertical antenna of this size is unthinkable and in fact impracticable.

Another important reason for not transmitting signal frequencies directly is that both picture and sound signals from different stations are concentrated within the same range of frequencies. Therefore, radiation from different stations would be hopelessly and inextricably mixed up and it would be impossible to separate one from the other at the receiving end. Thus in order to be able to separate the intelligence from different stations, it is necessary to translate them all to different portions of the electromagnetic spectrum depending on the carrier frequency assigned to each station. This also overcomes the difficulties of poor radiation at low frequencies. Once signals are translated before transmission, a tuned circuit provided in the RF section of the receiver can be used to select the desired station.

4.1 AMPLITUDE MODULATION

In amplitude modulation the intelligence to be conveyed is used to vary the amplitude of the carrier wave. As an illustration, an amplitude modulated signal is shown in Fig. 4.1 (a) where

$$e_c = E_c \cos \omega_c t \text{ is the carrier wave and}$$

$$e_m = E_m \cos \omega_m t \text{ is the modulating signal.}$$

Note that the camera signal is actually complex in nature but a single modulating frequency has been chosen for convenience of analysis.

The equation of the modulated wave is :

$$e = A \cos \omega_c t$$

where $A = (E_c + kE_m \cos \omega_m t)$ when k is a constant of the modulator.

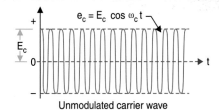

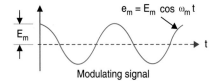

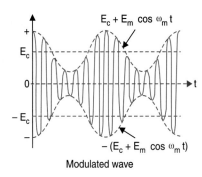

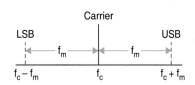

Fig. 4.1(a) *Modulation of R.F. carrier with a signal frequency.*

Fig. 4.1(b) *Frequency spectrum of AM wave.*

On substituting the value of A we get :

$$e = (E_c + kE_m \cos \omega_m t) \cos \omega_c t = E_c (1 + m \cos \omega_m t) \cos \omega_c t \qquad ...(4.1)$$

where $m = \dfrac{kE_m}{E_c}$ is the modulation index.

It may be noted that at $kE_m = E_c$, $m = 1$ and the corresponding depth of modulation is then termed as 100%.

Equation (4.1) may be expanded by the use of trigonometrical identities and expressed as :

$$e = E_c \cos \omega_{ct} + \frac{mE_c}{2} \cos (\omega_c - \omega_m) \, t - \frac{mE_c}{2} \cos (\omega_c + \omega_m)t \qquad ...(4.2)$$

This result shows that if a carrier wave having frequency equal to f_c is amplitude modulated with a single frequency f_m, the resultant wave consists of the carrier (f_c) and the sum and difference components ($f_c \pm f_m$) of the carrier frequency and the modulating frequency. However, if the modulating signal consists of more than a single frequency, as it would be for a video signal, the equation can be extended to include the sum and difference of the carrier and all frequency components of the modulating signal. This is illustrated in Fig. 4.1 (b) where f_m has been shown to be the highest modulating frequency. The region between f_c and ($f_c + f_m$) is called the upper sideband (USB) and that between f_c and ($f_c - f_m$) the lower sideband (LSB). Therefore if the modulated

wave is to be transmitted without distortion by this method, the transmission channel must be atleast of width $2f_m$ centred on f_c.

4.2 CHANNEL BANDWIDTH

In the 625 line TV system where the frequency components present in the video signal extend from dc (zero Hz) to 5MHz, a double sideband AM transmission would occupy a total bandwidth of 10 MHz. The actual band space allocated to the television channel would have to be still greater, because with practical filter characteristics it is not possible to terminate the bandwidth of a signal abruptly at the edges of the sidebands. Therefore, an attenuation slope of 0.5 MHz is provided at each edge of the two sidebands. This adds 1 MHz to the required total band space. In addition to this, each television channel has its associated FM (frequency modulated) sound signal, the carrier frequency of which is situated just outside the upper limit of 5.5 MHz of the picture signal. This, together with a small guard band, adds another 0.25 MHz to the channel width, so that a practical figure for the channel bandwidth would be 11.25 MHz. This is illustrated in Fig. 4.2.

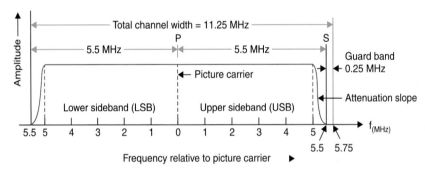

Fig. 4.2 *Total channel bandwidth using double sideband picture signal. P is picture carrier and S is sound carrier.*

Such a bandwidth is too large, and if used, would limit the number of channels in a given high frequency spectrum allocated for TV transmission. Therefore, to ensure spectrum conservation, some saving in the bandwidth allotted to each channel is desirable.

Single sideband transmission (SSB). A careful look at eqn. (4.2) reveals that the carrier component conveys no information because its amplitude and frequency remain constant no matter what the amplitude of the modulating voltage is. However, the presence of the carrier frequency is necessary at the receiver for recovering the modulating frequency f_m, from the upper sideband by taking $(f_c + f_m) - f_c$ or from the lower sideband by taking $f_c - (f_c - f_m)$. Therefore, though superfluous from the point of view of transmission of intelligence, the carrier frequency is radiated along with the sideband components in all radio-broadcast and TV systems. Such an arrangement results in simpler transmitting equipment and needs a very simple and inexpensive diode detector at the receiver for recovering the modulation components without undue distortion.

From eqn. (4.2) it is also obvious that the two sidebands are images of each other, since each is equally affected by changes in the modulating voltage amplitude via the component $\dfrac{mE_c}{2}$.

Also, any change in the frequency of the modulating signal results in identical changes in the band spread of the two sidebands. It is seen, therefore, that all the information can be conveyed by the use of one sideband only and this results in a saving of 5 MHz per channel. It may, however, be noted that the magnitude of the detected signal in the receiver will be just half of that obtained when both the sidebands are transmitted. This is no serious drawback because the IF (intermediate frequency) amplifier stages of the receiver provide enough gain to develop reasonable amplitude of the video signal at the output of video detector.

4.3 VESTIGIAL SIDEBAND TRANSMISSION

In the video signal very low frequency modulating components exist along with the rest of the signal. These components give rise to sidebands very close to the carrier frequency which are difficult to remove by physically realizable filters. Thus it is not possible to go to the extreme and fully suppress one complete sideband in the case of television signals. The low video frequencies contain the most important information of the picture and any effort to completely suppress the lower sideband would result in objectionable phase distortion at these frequencies. This distortion will be seen by the eye as 'smear' in the reproduced picture. Therefore, as a compromise, only a part of the lower sideband, is suppressed, and the radiated signal then consists of a full upper sideband together with the carrier, and the vestige (remaining part) of the partially suppressed lower sideband. This pattern of transmission of the modulated signal is known as vestigial sideband or A5C transmission. In the 625 line system, frquencies up to 0.75 MHz in the lower sideband are fully radiated. The net result is a normal double sideband transmission for the lower video frequencies corresponding to the main body of picture information.

As stated earlier, because of fillter design difficulties it is not possible to terminate the bandwidth of a signal abruptly at the edges of the sidebands. Therefore, an attenuation slope covering approximately 0.5 MHz is allowed at either end. Any distortion at the higher frequency end, if attenuation slope were not allowed, would mean a serious loss in horizontal detail, since the high frequency components of the video modulation determine the amount of horizontal detail in the picture. Fig. 4.3 illustrates the saving of band space which results from vestigial sideband transmission. The picture signal is seen to occupy a bandwidth of 6.75 MHz instead to 11 MHz.

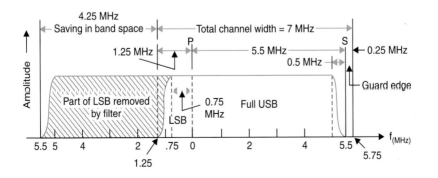

Fig. 4.3 Total channel bandwidth using vestigial' lower sideband.

Chapter 4

4.4 TRANSMISSION EFFICIENCY

Though the total power that is developed and radiated at the transmitter has no direct bearing on bandwidth requirements, the saving in power that can be effected by suppressing the carrier and one of the sidebands cannot be totally ignored. This can be demonstrated by considering the power relations in the modulated wave. Based on eqn. (4.2) the total power P_t, in the modulated wave is the sum of the carrier power P_c, and the power in the two sidebands. This can be expressed as

$$P_t = P_c + P_{USB} + P_{LSB} = \frac{E_c^2}{2R} + \frac{m^2 E_c^2}{8R} + \frac{m^2 E_c^2}{8R} \qquad ...(4.3)$$

where $\frac{E_c}{\sqrt{2}}$ is the r.m.s. value of the sinusoidal carrier wave, and R is the resistance in which the power is dissipated. Equation (4.3) can be simplified to read as

$$P_t = P_c + \frac{m^2}{4} P_c + \frac{m^2}{4} P_c = P_c \left(1 + \frac{m^2}{2} \right) \qquad ...(4.4)$$

Note from the above expression that P_c remains constant but P_t depends on the value of the modulation index m. Also note that when several frequency components of different amplitudes modulate the carrier wave, which in fact is the rule rather than an exception, the carrier power P_t is unaffected but the total sideband power gets distributed in the individual sideband component powers. This is so because the total modulating voltage is equal to the square root of the sum of the squares of individual modulating voltages.

It can be seen from eqn. (4.4), that at 100% modulation ($m = 1$) the transmitted power attains its maximum possible value. $P_{t(max)} = 1.5\,P_c$, where the power contained in the two sidebands has a maximum value of 50% of the carrier power. It is clear then, that the carrier component that is redundent, so far as the transmission of intelligence is concerned, constitutes about 72% of the total power that is radiated in the double sideband, full carrier (better known as A3 modulation) AM system. Therefore, a lot of economy can be effected if the carrier power is suppressed and not transmitted. Furthermore, suppression of one sideband results in more economy and also halves the bandwidth requirements for transmission as compared to A3. In practice SSB is used to save power and bandwidth in mobile communication systems, telemetry, radio navigation, military and several other such applications. However, such a system needs the generation of carrier frequency at the receiver for detection and this necessitates the transmission of a low level pilot carrier along with either of the two sidebands. In addition to this, a single sideband with suppressed carrier requires excellent frequency stability on the part of both transmitter and receiver. Any deviation in frequency and phase of the generated carrier at the receiver would severely impair the quality of the picture when used for television signal transmission. Such difficulties are not unsurmountable, but this tends to make the receiver circuitry more complicated, which in turn adds to the cost of the receiver. In point to point communication systems, where only one receiver is necessary, the additional expense is justifiable and infact SSB is now the accepted mode of communication for such applications. However, in television and radio broadcast systems, where a very large number of receivers simultaneously receive programme from one transmitter, additional cost of receivers is not justified and as such SSB cannot be recommended. Therefore, as stated earlier, in all TV systems, full carrier is radiated and vestigial sideband transmission is used. In radio broadcast where the channel bandwidth is only 100 kHz, both the sidebands are transmitted along with full carrier.

4.5 COMPLETE CHANNEL BANDWIDTH

The sound carrier is always positioned at the extremity of the fully radiated upper sideband and hence is 5.5 MHz away from the picture carrier. This is its logical place since it makes for minimum interference between the two signals. The FM sound signal occupies a frequency spectrum of about ± 75 KHz around the sound carrier. However, a guard band of 0.25 MHz is allowed on the sound carrier side of the television channel to allow for adequate inter-channel separation. The total channel bandwidth thus occupies 7 MHz and this represents a bandspace saving of 4.25 MHz per channel, when compared with the 11.25 MHz space, which would be required by the corresponding double sideband signal. Figure 4.4 show the complete channel. The frequency axis is scaled ralative to the picture carrier, which is marked as 0 MHz. This makes the diagram very informative, since details such as the widths of the upper and lower sidebands and the relative position of the sound carrier are easily read off.

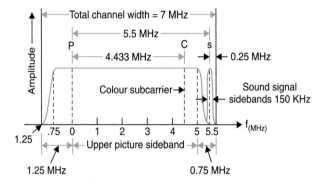

Fig. 4.4 *C.C.I.R. (Indian and European) TV channel sideband spectrum.*
C is colour subcarrier frequency.

Fig. 4.5 (*a*) show television channel details of the British 625 line system, where the highest modulating frequency employed is 5.5 MHz and the lower sideband up to 1.25 MHz

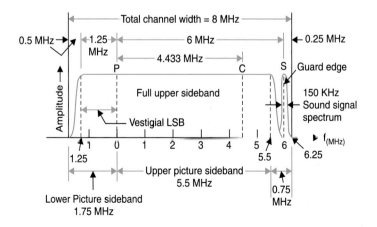

Fig. 4.5(a) *U.K. TV channel standards.*

Chapter 4

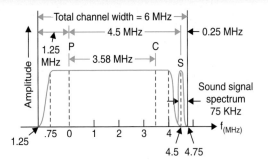

Fig. 4.5(b) American TV channel standards.

is allowed to be radiated. The total bandwidth per channel is 8 MHz. Fig. 4.5 (b) illustrates channel details of 525 line American system, where the highest allowed modulating frequency is 4 MHz with a total bandwidth of 6 MHz. In the French 819 line system where the highest modulating frequency is 10.4 MHz a channel bandwidth equal to 14 MHz is allowed. The diagram in Fig. 4.6 shows how two adjacent C.C.I.R. 625 line channels in the VHF Band-I are disposed one after the other.

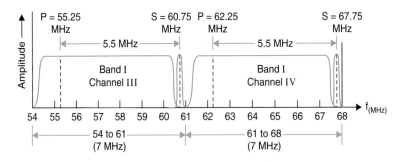

Fig. 4.6 Sideband spectrum of two adjacent channels of the lower VHF band of television station allocations.

4.6 RECEPTION OF VESTIGIAL SIDEBAND SIGNALS

In principle an SSB signal with carrier cannot be demodulated by an envelope detector. Either synchronous demodulation or a square law device to produce effective multiplication of the carrier with the sideband is required. However, it can be shown that if the sideband amplitude is small compared to the carrier, then the envelope of the SSB with carrier signal nearly corresponds to the modulating signal. In that case, envelope detection can be used and is the normal practice in television receivers. With vestigial sideband however, the relative amplitude of the frequencies for which both sidebands exist is double that of the true SSB component at the envelope detector output. In the video signal it would be so for the low frequency content of the picture signal, and in effect, amounts to distortion in terms of relative amplitudes for different frequencies and needs correction at the receiver.

This when expressed in another way means that if the picture carrier were successively modulated to an equal depth by a series of frequencies throughout the video frequency range employed by the system, and the resulting voltage output from the detector recorded, the output voltage against input frequency characteristic obtained would have the form shown in Fig. 4.7. The vestigial sideband extends to 0.75 MHz below the carrier and thereafter this sideband is linearly attenuated down to zero at 1.25 MHz. The detector output voltage would thus be twice as great between 0 Hz and 0.75 MHz than between 1.25 MHz to 5 MHz.

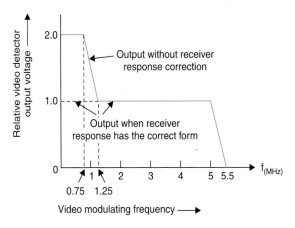

Fig. 4.7 *Receiver video detector output vs modulating frequency characteristics illustrating the need for specially shaped receiver IF response curves.*

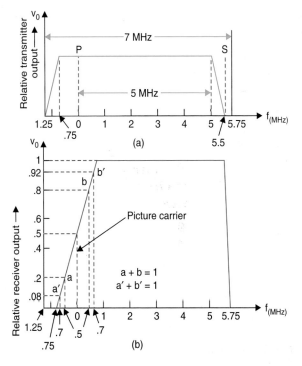

Fig. 4.8 *Ideal characteristics of a TV transmitter and receiver. (a) transmitter output characteristics for vestigial sideband signals. (b) desired receiver characteristics for correct reproduction of video signals. Note that the picture carrier is positioned half-way down the response curve.*

Between 0.75 MHz and 1.25 MHz the output voltage would fall linearly following the sideband attenuation slope of the transmitter. To correct this discrepancy, it is necessary to so shape the receiver response curve, that the frequencies present 'twice' are afforded less amplification than those occurring in one sideband only. The desired response is shown in Fig. 4.8. The response curve is shaped to place the picture carrier half-way down the side corresponding to the suppressed sideband. The width of the sloping edge on which the carrier is positioned is twice the width of the vestigial sideband. To understand how this achieves the desired result, refer to Fig. 4.8 and consider the treatment afforded to various frequencies within the video bandwidth. Frequencies between 5 MHz and 0.75 MHz *i.e.*, those present in the upper sideband only, are seen to give unit output. Next, consider a frequency component at 0.5 MHz. This is present in both the sidebands. The total detector output is again unity. The component in the lower sideband gives rise to an output of a volts, while that in the upper sideband gives rise to b volts. From the geometry of the figure we see that $(a + b) = 1$. As a further example consider the response at 0.7 MHz. This component in the vestigial sideband gives rise to an output = 0.08 V, whilst in the upper sideband, it gives rise to 0.92 V. Again the sum of the two is unity, so that the same output is achieved for frequencies between 0.75 MHz and 5 MHz. Note that at 0.75 MHz the output in the vestigial sideband is zero, and that in the upper sideband it is equal to one. The necessary correction detailed above is carried out at the Intermediate Frequency (IF) amplifier stages of the television receiver by suitably shaping the passband characteristics of the tuned amplifiers. This matter is fully dealt with in Chapter 8.

Demerits of Vestigial Sideband Transmission

(*a*) A small portion of the transmitter power is wasted in the vestigial sideband filters which remove the remaining lower sideband.

(*b*) The attenuation slope of the receiver to correct the boost at lower video frequencies places the carrier at 50 per cent output voltage which amounts to introducing a loss of about 6 dB in the signal to noise voltage ratio relative to what be available if double sideband transmission is used.

(*c*) Some phase and amplitude distortion of the picture signal occurs despite careful filter design at the transmitter. Also, it is very difficult to tune IF stages of the receiver to correspond exactly with the ideal desired response as shown in Fig. 4.8 and this too introduces some phase and amplitude distortion.

(*d*) More critical tuning at the receiver becomes necessary because for a given amount of local oscillator mismatch or drift after initial tuning, the degeneration of picture quality is less with wider lower sideband than with narrow lower sideband. In this respect the British 625 line system is superior because it allows 1.25 MHz unattenuated lower sideband transmission as compared to 0.75 MHz in most other systems.

Despite these demerits of vestigial sideband transmission it is used in all television systems because of the large saving it effects in the bandwidth required for each channel.

4.7   FREQUENCY MODULATION

The sound signal is frequency modulated because of its inherent merits of interference-free reception. Here the amplitude of the modulated carrier remains constant, whereas its frequency is varied in accordance with variations in the modulating signal. The variation in carrier frequency is made proportional to the instantaneous value of the modulating voltage. The rate at which this frequency variation takes place is equal to the modulating frequency. It is assumed that the phase relations of a complex modulating signal will be preserved. However, for simplicity, it is again assumed that the modulating signal is sinusoidal. The situation is illustrated in Fig. 4.9 which shown the modulating voltage, and the resulting frequency modulated wave. Fig. 4.9 also shows the frequency variation with time, which is seen to be identical to the variations with time of the modulating voltage.

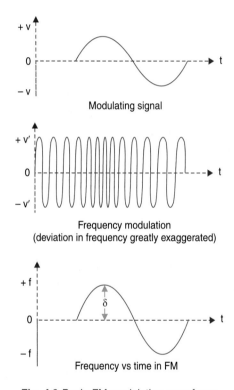

Fig. 4.9 *Basic FM modulation waveforms.*

Analysis of Frequency-Modulated (FM) Wave. In order to understand clearly the meaning of instantaneous frequency f_i and the associated instantaneous angular velocity $\omega_i = 2\pi f \omega_i$, the equation of an ac wave in the generalized form may first be written as :

$$e = A \sin \phi(t)$$

where e = instantaneous amplitude
  A = peak amplitude
  $\phi(t)$ = total angular displacement at time t.

The instantaneous angular velocity ω_t is, by definition, the instantaneous rate of change $\dfrac{d\phi(t)}{dt}$ of angular displacement $\phi(t)$.

Thus $\omega_t = \dfrac{d\phi(t)}{dt}$ ...(4.5)

Chapter 4

A sinusoidal wave of constant frequency say $f_c(\omega_c = 2\pi f_c)$ is a special case of eqn. (4.5) and then $\phi(t)$ = $\omega_{ct} + \theta$ where θ is the angular position at $t = 0$. Application of eqn. (4.5) yields the result

$$\omega_t = \frac{d\phi(t)}{dt} = \omega_c$$

A frequency modulated wave with sinusoidal modulation can now be expressed as:

$$\omega_i = \omega_c + 2\pi\Delta f \cos \omega_m t \qquad \qquad ...(4.6)$$

where ω_i = instantaneous angular velocity

ω_c = angular velocity of carrier wave

(average angular velocity).

ω_m = 2π times the modulating frequency f_m.

Δf = maximum deviation of instantaneous frequency from the average value.

It may be emphasized that the frequency deviation Δf is proportional to the peak amplitude $(\cos \omega_m t = \pm 1)$ of the modulating signal and is independent of the modulating frequency.

The equation of the FM wave can now be obtained by combining eqn. (4.5) and (4.6) to give the value of $\phi(t)$. The steps involved are as follows :

$$\omega_i = \frac{d\phi(t)}{dt} = \omega_c + 2\pi\Delta f \cos \omega_m t$$

Integration gives :

$$\phi(t) = \omega_c t + \left(\frac{2\pi\Delta f}{\omega_m}\right) \sin \omega_m t + \theta$$

where the constant of integration θ defines the angular position at time $t = 0$.

Substituting the above value of $\phi(t)$ into the generalized form $e = A \sin \phi(t)$ yields :

$$e = A \sin \left(\omega_c t + \frac{2\pi\Delta f}{\omega_m} \sin \omega_m t \right) \qquad \qquad ...(4.7)$$

where for the sake of simplicity angle θ has been assumed to be equal to zero.

Equation (4.7) is commonly written in the form

$$e = A \sin (\omega_c t + m_f \sin \omega_m t) \qquad \qquad ...(4.8)$$

where m_f is termed the 'modulation index' of the FM wave and is defined as :

$$m_f = \text{modulation index} = \frac{\text{frequency deviation}}{\text{modulating frequency}} = \frac{\Delta f}{f_m}$$

It may be noted that for a given frequency deviation, the modulation index varies inversely as the modulating frequency. Also m_f is defined only for sinusoidal modulation unlike m of AM which is defined for any modulating signal.

Frequency Spectrum of the FM Wave. The eqn. (4.8) is of the form, sine of a sine, and can be expressed as :

$$e = A [(\sin \omega_c t \cos (m_f \sin \omega_m t) + \cos \omega_c t \sin (m_f \sin \omega_m t)] \qquad ...(4.9)$$

The term $\cos(m_f \sin \omega_m t)$ can be expanded into

$$J_0(m_f) + 2J_2(m_f) \cos 2\omega_m t + 2J_4(m_f) \cos 4\omega_m t + \ldots\ldots$$

and $\sin(m_f \sin \omega_m t)$ into :

$$2J_1(m_f) \sin \omega_m t + 2J_3(m_f) \sin 3\omega_m t + \ldots\ldots$$

Substitution of these results into eqn. (4.9) and some manipulation yields :

$$\begin{aligned}
e = A\{ &J_0(m_f) \sin \omega_c t \\
&+ J_1(m_f)[\sin(\omega_c + \omega_m)t - \sin(\omega_c - \omega_m)t] \\
&+ J_2(m_f)[\sin(\omega_c + 2\omega_m)t - \sin(\omega_c - 2\omega_m)t] \\
&+ J_3(m_f)[\sin(\omega_c + 3\omega_m)t - \sin(\omega_c - 3\omega_m)t] + \ldots\ldots \\
&+ J_n(m_f) \ldots\ldots\ldots\} + \ldots\ldots
\end{aligned}$$

$$\ldots(4.10)$$

where $J_n^*(m_f)$ are Bessel functions of the first kind and nth order with argument m_f.

The final expression obtained in eqn. (4.10) yields the following information :

(a) FM has infinite number of sidebands besides the carrier. The sidebands are separated from the carrier by integer multiples of f_m.

(b) For a given m_f, J_n coefficients eventually decrease to negligible values as n increases and the values for different n may be positive, negative or zero.

(c) The sidebands on either side of the carrier at equal distance from f_c have equal amplitudes so that the sideband distribution is symmetrical about the carrier frequency.

(d) For a given modulating frequency, increase in the amplitude of the modulating signal results in an increase of Δf and therefore of m_f causing larger number of sidebands to acquire significant amplitudes. Thus higher amplitude signals would need more sidebands for transmission without distortion. However, the total transmitted power stays constant.

(e) The way the number of significant J coefficients increase with m_f is illustrated in the table below.

Table 1. Bessel Functions of the 1st Kind

X	n or order								
(m_f)	J_0	J_1	J_2	J_3	J_4	J_5	J_6	J_7	J_8
0.00	1.0	—	—	—	—	—	—	—	—
0.25	0.98	0.12	—	—	—	—	—	—	—
0.5	0.94	0.24	0.03	—	—	—	—	—	—
1.0	0.77	0.44	0.11	0.02	—	—	—	—	—
2.0	0.22	0.58	0.35	0.13	0.03	—	—	—	—
3.0	− 0.26	0.34	0.49	0.31	0.13	0.04			
4.0	− 0.40	− 0.07	0.36	0.43	0.28	0.13	0.05	—	—
5.0	− 0.18	− 0.33	0.05	0.36	0.39	0.26	0.13	0.05	—
6.0	0.15	− 0.28	− 0.24	0.11	0.36	0.36	0.25	0.13	0.06
7.0	0.30	0.00	− 0.30	− 0.17	0.16	0.55	0.34	0.23	0.13

*Theory of Bessel's functions is not necessary for us. Tabulated values of Bessel's function are widely available.

Chapter 4

(f) As seen in the table when the modulation index (m_f) is less than 0.5, i.e., when the frequency deviation is less than half the modulating frequency, the second and higher-order sideband components are relatively small and the frequency band required to acommodate the essential part of the signal is the same as in amplitude modulation. On the other hand when m_f exceeds unity, there are important higher-order sideband components contained in the wave and this results in increased bandwidth requirements.

(g) The modulation index actually depends on both the amplitude and frequency of the modulating tone. It is higher in FM systems that permit large frequency deviation for a maximum amplitude of the modulating tone. This is turn results in higher order significant J coefficients and a larger bandwidth is required for reasonably distortion free transmission.

(h) Since a lot of the higher sidebands have insignificant relative amplitudes, their exclusion will not distort the modulated wave unduly, and while calculating channel bandwidth J coefficients having values less than 0.05 for a calculated value of m_f can be neglected.

4.8   FM CHANNEL BANDWIDTH

Based on the above discussion the channel bandwidth

$$\text{BW} = 2nf_m \qquad \qquad ...(4.11)$$

where f_m is the frequency of the modulating wave and n is the number of the significant side-frequency components. The value of n is determined from the modulation index.

Though the higher frequencies in speech or music have much less amplitude as compared to lower audio frequencies, we shall estimate the channel bandwidth for the worst case where even the highest frequency to be transmitted causes maximum permitted frequency deviation.

The maximum frequency deviation of commercial FM is limited to 75 kHz, and the modulating frequencies typically cover 25 Hz to 15 kHz.

If a 15 kHz tone has unit amplitude, i.e., equal to the maximum allowed amplitude, then

$m_f = \dfrac{75}{15} = 5$. From the Bessel function table, for $m_f = 5$, the significant (0.05) value of $J_n = 7$, i.e., $n = 7$. Therefore

$$\text{BW} = 2 \times 7 \times 15 = 210 \text{ kHz}$$

Had the amplitude been less, the maximum frequency deviation would not be developed, and the bandwidth would be smaller. This brings out an interesting observation that in frequency modulation (with fixed Δf) the bandwidth depends on the tone amplitude, whereas in amplitude modulation, the bandwidth depends on the tone frequency.

Similarly in the 625-B television system where the standards specify that the maximum deviation (Δf) should not exceed $\pm$ 50 kHz for the highest modulating frequency of 15 kHz, m_f

$$= \frac{50}{15} \approx 3$$

This gives a value of $n = 5$ as seen in the given chart.

$$\therefore \qquad \qquad \text{BW} = 2 \times 5 \times 15 = 150 \text{ kHz.}$$

The bandwidth can also be estimated from 'Carson's Rule' which states that to a good approximation, the bandwidth required to pass an FM wave is equal to twice the sum of the deviation and the highest modulating frequency. Thus, for the standard FM transmission the required bandwidth $= 2(75 + 15) = 180$ kHz. This nearly checks with the value of *210 kHz estimated earlier.

Similarly for the 625 line system the Carson's Rule yields a bandwidth requirement of $2(50 + 15) = 130$ kHz and this is close to the value calculated earlier. The resultant deviation of ± 75 kHz around the sound carrier is very much within the guard-band edge and reasonably away from any significant video sideband components.

It may be noted that in the American television system where the maximum permissible deviation is ± 25 kHz around the sound carrier, a bandwidth of about 100 kHz is enough for sound signal transmission.

4.9 CHANNEL BANDWIDTH FOR COLOUR TRANSMISSION

As explained in the chapter devoted to the analysis and synthesis of TV pictures the colour video signal does not extend beyond about 1.5 MHz. Therefore, the colour information can be transmitted with a restricted bandwidth much less than 5 MHz. This feature allows the narrow band chrominance (colour) signal to be multiplexed with the wideband luminance (brightness) signal in the standard 7 MHz television channel. This is achieved by modulating the colour signal with a carrier frequency which lies within the normal channel bandwidth. This is called colour subcarrier frequency and is located towards the upper edge of the video frequencies to avoid interference with the monochrome signal.

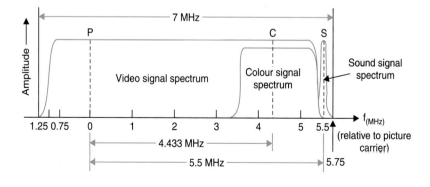

Fig. 4.10 *C.C.I.R. 625 lines monochrome and the compatible PAL Colour channel bandwidth details.*

*In commercial FM broadcast a frequency spectrum of 200 kHz is allotted for each channel.

Chapter 4

In the PAL colour system which is compatible with the C.C.I.R. 625 line monochrome system the colour subcarrier frequency is located 4.433 MHz way from the picture carrier. The bandwidth of colour signals is restricted to about ± 1.2 MHz around the subcarrier. Fig. 4.10 gives necessary details of the location of monochrome (picture), colour and sound signal spectrums all within the same channel bandwidth of 7 MHz. It may be noted that in the American television system where the channel bandwidth is 6 MHz, the colour subcarrier is located 3.58 MHz away from the picture carrier.

4.10   ALLOCATION OF FREQUENCY BANDS FOR TELEVISION SIGNAL TRANSMISSION

For effective amplitude modulation and better selectivity at the RF and IF tuned amplifiers in the receiver, it is essential that the carrier frequency be chosen about ten times that of the highest modulating frequency. Since the highest modulating frequency for picture signal transmission is 5 MHz, the minimum carrier frequency that can be employed, cannot be much less than 40 MHz. As an illustration consider a carrier frequency f_c = 10 MHz. With the highest video modulating frequency = 5 MHz, a deviation of 50 per cent from the centre frequency would be necessary in any tuned circuit to accommodate the lower and upper sideband frequencies. However, if the carrier frequency is fixed at, say 50 MHz, the percentage deviation required to pass the upper and lower sideband frequencies for the same modulating frequency would be only 10 per cent. It is obvious from these observations that selectivity is bound to be poor at the receiver tuned amplifiers with a carrier frequency of 10 MHz. The 3 db down points with a carrier frequency of 50 MHz are within 5 per cent deviation from the carrier frequency and thus the selectivity is bound to be much better. Further, each television channel occupies about 7 MHz. In order to accommodate several TV channels, the carrier frequencies have to be in the region of the spectrum above about 40 MHz. This explains why television transmission has to be carried out at very high frequencies in the VHF and UHF bands. In radio broadcast where the highest modulating frequency is only 5 kHz, lower carrier frequencies can be used, and accordingly transmission is carried out in the medium wave band (550 kHz to 1600 kHz) and short wave bands extending up to about 30 MHz. Transmission at very high frequencies has its own problems and limitations for long distance transmission and these are discussed in another chapter.

4.11   TELEVISION STANDARDS

After having learnt about various aspects of television transmission and reception, it would be instructive to review, in detail, the picture and sound signal standards as specified by the International Radio Consulative Committee (C.C.I.R) for the 625-B monochrome system and also to compare its main characteristics with that of other principal television system. This is detailed in Tables 2 and 3.

Chapter 4

Table 2. Television Signal Standards

Vision and Sound Signal Standards for the 625-B Monochrome System Adopted by India as Recommended by the International Radio Consultative Committee (C.C.I.R.)

Characteristics of the 625-B Monochrome TV System

No. of lines per picture (frame)	625
Field frequency (Fields/second)	50
Interlace ratio, *i.e.*, No. of fields/picture	2/1
Picture (frame) frequency, *i.e.*, Pictures/second	25
Line frequency and tolerance in lines/second, (when operated non-synchronously)	$15625 \pm 0.1\%$
Aspect Ratio (width/height)	4/3
Scanning sequence	(*i*) Line : Left to right
	(*ii*) Field : Top to bottom
System capable of operating independently of power supply frequency	YES
Approximate gamma of picture signal	0.5
Nominal video bandwidth, *i.e.*, highest video modulating frequency (MHz)	5
Nominal Radio frequency bandwidth, *i.e.*, channel bandwidth (MHz)	7
Sound carrier relative to vision carrier (MHz)	+ 5.5
Sound carrier relative to nearest edge of channel (MHz)	– 0.25
Nearest edge of channel relative to picture carrier (MHz)	– 1.25
Fully radiated sideband	Upper
Nominal width of main sideband (upper) (MHz)	5
Width of end-slope of full (Main) sideband (MHz)	0.5
Nominal width of vestigial sideband (MHz)	0.75
Vestigial (attenuated) sideband	Lower
Min : attenuation of vestigial sideband in db, (below the ideal demodulated curve)	(at 1.25 MHz) 20 db
	(at 4.43 MHz) 30 db
Width of end-slope of attenuated (vestigial) sideband (MHz)	0.5
Type and polarity of vision modulation	(A5C) Negative
Synchronizing level as a percentage of peak carrier	100
Blanking level as percentage of peak carrier	72.5 to 77.5
Difference between black and blanking level as a percentage of peak carrier	0 to 7
Peak white level as a percentage of peak carrier	10 to 12.5
Type of sound modulation	FM, $\pm$ 50 KHz
Pre-emphasis	50 µs
Resolution	400 max
Ratio of effective radiated powers of vision and sound	5/1 to 10/1

Chapter 4

The values to be considered are : (*i*) the r.m.s. value of the carrier at the peak of modulation envelope for the vision signal. (*ii*) the r.m.s. value of the unmodulated carrier for amplitude modulated and frequency modulated sound transmissions.

Details of Line-Blanking Intervals

Nominal duration of a horizontal line	$= 64\ \mu s = H$
Line blanking interval	$= 12 \pm 0.3\ \mu s$
Front porch	$= 1.5 \pm 0.3\ \mu s$
Sync pulse width	$= 4.7 \pm 0.2\ \mu s$
Back porch	$= 5.8 \pm 0.3\ \mu s$
Build up time (10% to 90%) of line blanking edges	$= 0.3 \pm 0.1\ \mu s$
Interval between datum level and black edge of line blanking signal (average calculated time for information)	$= 10.5\ \mu s$

Details of Field-Blanking Intervals

Field blanking period $= 20\ H$ (20 lines)	$= 1280\ \mu s$
Pre sync equalizing pulses, 5 pulses of duration $1/2\ H$, *i.e.*, 32 μs, total time	$= 160\ \mu s$
Equalizing pulses are narrow pulses with pulse width	$= 2.35 \pm 0.1\ \mu s$
Field sync pulses at $\frac{1}{2} H$ intervals, 5 such pulse, each pulsewidth	$= 27.3\ \mu s$
Interval between field sync pulses	$= 4.7 \pm 0.2\ \mu s$
Interval between equalizing pulses	$= 29.65\ \mu s$
Post-sync equalizing pulses, 5 pulses	same as for pre-sync. eq. pulses
Build up time of field blanking edges.	$= 0.3 \pm 0.1\ \mu s$
Build up time for field sync pulses	$= 0.2 \pm 0.1\ \mu s$

Table 3. Principal Television System

Particulars	Western Europe, Middle East, India and most Asian countries	North and South America including US, Canada, Mexico and Japan	England	USSR	France
Lines per frame	625	525	625	625	625
Frames per second	25	30	25	25	25
Field frequency (Hz)	50	60	50	50	50
Line frequency (Hz)	15,625	15,750	15,625	15,625	15,625
Video bandwidth (MHz)	5 ot 6	4.2	5.5	6	6

(*Contd.*)...

Channel bandwidth (MHz)	7 or 8	6	8	8	8
Video modulation	Negative	Negative	Negative	Negative	Positive
Picture modulation	AM	AM	AM	AM	FM
Sound signal modulation	FM	FM	FM	FM	AM
Colour system	PAL	NTSC	PAL	SECAM	SECAM

(*i*) England earlier used 405 line system in the 5 MHz channel.

(*ii*) France earlier used 819 line system with a channel bandwidth of 14 MHz.

REVIEW QUESTIONS

1. Why is it necessary to modulate the picture and sound signals before transmission ? Why is TV transmission carried out in the UHF and VHF bands ?

2. Show that in the 625-B system, a total channel bandwidth of 11.25 MHz would be necessary if both the sidebands of the amplitude modulated picture signal are fully radiated along with the frequency modulated picture signal.

3. Why is an attenuation slope of 0.5 MHz allowed at both the edges of the AM picture signal sidebands ? Why is a guard band provided at the sound signal edge of the television channel ?

4. Why is it necessary to affect economy in channel bandwidth ? Why SSB is not used for picture signal transmission ?

5. What is vestigial sideband transmission and why it is used for transmission of TV picture signals ?

6. Why is a portion of the lower sideband of the AM picture signal transmitted along with the carrier and full USB ? Does it need any correction somewhere in the television link ? If so where is it carried out ?

7. Sketch and fully label the desired response of a TV receiver that includes necessary correction on account of the discrepancy caused by VSB transmission. Comment on the response curve drawn by you.

8. Show that a total channel bandwidth of 7 MHz is necessary for successful transmission of both picture and sound signals in the 625 line TV system. Sketch frequency distribution of the channel and mark the location of picture and sound signal carrier frequencies. Why is the sound carrier located 5.5 MHz away from the picture carrier ?

9. Justify the allocation of 8 MHz in the British TV system and 6 MHz in the American system for each TV channel. What is the separation between picture and sound carriers in each of these systems ?

10. What is 'modulation index' in FM transmission and how does it affect the bandwidth required for each FM channel ?

11. Explain how you would proceed to determine the channel bandwidth for transmission of sound signals (highest modulating frequency = 15 kHz) by frequency modulation. How does the permitted maximum deviation affect the bandwidth requirements ?

12. Show that in the 625-B system where the maximum allowed frequency deviation is ± 50 kHz, a bandwidth of 150 kHz is necessary for almost distortion free transmission by frequency modulation, the highest modulating frequency being 15 kHz, Repeat this for the American system where the maximum allowed deviation is ± 25 kHz. Verify the results by 'Carson's Rule' of determining channel bandwidth.

Chapter 4

5

The Picture Tube

The picture tube or 'kinescope' that serves as the screen for a television receiver is a specialized form of cathode-ray tube. It consists of an evacuated glass bulb or envelope, inside the neck of which is rigidly supported an electron gun that supplies the electron beam. A luminescent phosphor coating provided on the inner surface of its face plate produces light when hit by the electrons of the fast moving beam.

A monochrome picture tube has one electron gun and a continuous phosphor coating that produces a picture in black and white. For colour picture tubes the screen is formed of three different phosphors and there are three electron beams, one for each colour phosphor. The three colours—red, green and blue produced by three phosphors combine to produce different colours. More details of colour picture tubes are given in chapters devoted to colour television.

5.1 MONOCHROME PICTURE TUBE

Modern monochrome picture tubes employ electrostatic focussing and electromagnetic deflection. A typical black and white picture tube is shown in Fig. 5.1. The deflection coils are

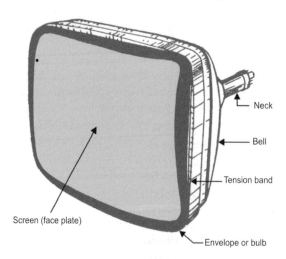

Fig. 5.1. A rectangular picture tube.

mounted externally in a specially designed yoke that is fixed close to the neck of the tube. The coils when fed simultaneously with vertical and horizontal scanning currents deflect the beam at a fast rate to produce the raster. The composite video signal that is injected either at the grid or cathode of the tube, modulates the electron beam to produce brightness variations of the tube, modulates the electron beam to produce brightness variations on the screen. This results in reconstruction of the picture on the raster, bit by bit, as a function of time. However, the information thus obtained on the screen is perceived by the eye as a complete and continuous scene because of the rapid rate of scanning.

Electron Gun

The various electrodes that constitute the electron gun are shown in Fig. 5.2. The cathode is indirectly heated and consists of a cylinder of nickel that is coated at its end with thoriated tungsten or barium and strontium oxides. These emitting materials have low work-function

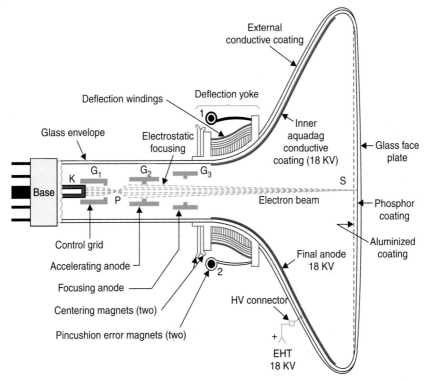

Fig. 5.2. Elements of a picture tube employing low voltage electrostatic focusing and magnetic deflection.

and when heated permit release of sufficient electrons to form the necessary stream of electrons within the tube. The control grid (Grid No. 1) is maintained at a negative potential with respect to cathode and controls the flow of electrons from the cathode. However, instead of a wiremesh structure, as in a conventional amplifier tube, it is a cylinder with a small circular opening to confine the electron stream to a small area. The grids that follow the control grid are the accelerating or screen grid (Grid No. 2) and the focusing grid (Grid No. 3). These are maintained at different positive potentials with respect to the cathode that vary between + 200 V to + 600 V. All the elements of the electron gun are connected to the base pins and receive their rated voltages from the tube socket that is wired to the various sections of the receiver.

Electrostatic Focussing

The electric field due to the positive potential at the accelerating grid (also known as 1st anode) extends through the opening of the control grid right to the cathode surface. The orientation of this field is such that besides accelerating the electrons down the tube, it also brings all the electrons in the stream into a tiny spot called the crossover. This is known as the first electrostatic lens action. The resultant convergence of the beam is shown in Fig. 5.2. The second lens system that consists of the screen grid and focus electrode draws electrons from the crossover point and brings them to a focus at the viewing screen. The focus anode is larger in diameter and is operated at a higher potential than the first anode. The resulting field configuration between the two anodes is such that the electrons leaving the crossover point at various angles are subjected to both convergent and divergent forces as they more along the axis of the tube. This in turn alters the path of the electrons in such a way that they meet at another point on the axis. The electrode voltages are so chosen or the electric field is so varied that the second point where all the electrons get focused is the screen of the picture tube. Electrostatic focusing is preferred over magnetic focusing because it is not affected very much by changes in the line voltage and needs no ion-spot correction.

Beam Velocity

In order to give the electron stream sufficient velocity to reach the screen material with proper energy to cause it to fluoresce, a second anode is included within the tube. This is a conductive coating with colloidal graphite on the inside of the wide bell of the tube. This coating, called aquadag, usually extends from almost half-way into the narrow neck to within 3 cm of the fluorescent screen as shown in Fig. 5.2. It is connected through a specially provided pin at the top or side of the glass bell to a very high potential of over 15 kV. The exact voltage depends on the tube size and is about 18 kV for a 48 cm monochrome tube. The electrons that get accelerated under the influence of the high voltage anode area, attain very high velocities before they hit the screen. Most of these electrons go straight and are not collected by the positive coating because its circular structure provides a symmetrical accelerating field around all sides of the beam. The kinetic energy gained by the electrons while in motion is delivered to the atoms of the phosphor coating when the beam hits the screen. This energy is actually gained by the outer valence electrons of the atoms and they move to higher energy levels. While recturning to their original levels they give out energy in the form of electromagnetic radiation, the frequency of which lies in the spectral region and is thus perceived by the eye as spots of light of varying intensity depending on the strength of the electron beam bombarding the screen.

Because of very high velocities of the electrons which hit the screen, secondary emission takes place. If these secondary emitted electrons are not collected, a negative space charge gets formed near the screen which prevents the primary beam from arriving at the screen. The conductive coating being at a very high positive potential collects the secondary emitted electrons and thus serves the dual purpose of increasing the beam velocity and removing unwanted secondary electrons. The path of the electron current flow is thus from cathode to screen, to the conductive coating through the secondary emitted electrons and back to the cathode through the high voltage supply. A typical value of beam current is about 0.6 mA with 20 kV applied at the aquadag coating.

5.2 BEAM DEFLECTION

Both electric and magnetic fields can be employed for deflecting the electron beam. However, in television picture tubes electromagnetic deflection is preferred for the following reasons :

(*a*) As already stated the electron beam must attain a very high velocity to deliver enough energy to the atoms of the phosphor coating. Because of this the electrons of the beam remain under the influence of the deflecting field for a very short time. This necessitates application of high deflecting fields to achieve the desired deflection. For example with an anode voltage of about 1 kV, as would be the case in most oscilloscopes, some 10 V would be necessary for 1 cm deflection of the beam on the screen, whereas in a picture tube with 15 kV at the final anode, about 7500 V would be necessary to get full deflection on a 50 cm screen. It is very difficult to generate such high voltages at the deflection frequencies. On the other hand with magnetic deflection it is a large current that would be necessary to achieve the same deflection. Since it is more convenient to generate large currents than high voltages, all picture tubes employ electromagnetic deflection.

(*b*) With electrostatic deflection the beam electrons gain energy. Thus larger deflection angles tend to defocus the beam. Further, the deflection plates need to be placed further apart as the deflection angle is made larger, thus requiring higher voltages to produce the same deflection field. Magnetic deflection is free from both these shortcomings and much larger deflection angles can be achieved without defocusing or nonlinearities with consequent saving in tube length and cabinet size.

(*c*) For electrostatic deflection two delicate pairs of deflecting plates, are needed inside the picture tube, whereas for magnetic deflection two pairs of deflecting coils are mounted outside and close to the neck of the tube. Such a provision is economical and somewhat more rugged.

Deflection Yoke

The physical placement of the two pairs of coils around the neck of the picture tube is illustrated in Fig. 5.3 and the orientation of the magnetic fields produced by them is shown in Fig. 5.4. In combination, the vertical and horizontal deflection coils are called the 'Yoke'. This yoke is fixed outside and close to the neck of the tube just before it begins to flare out (see Fig. 5.2).

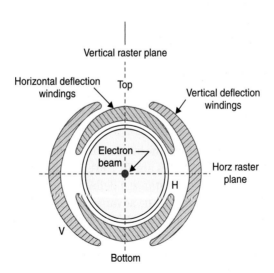

Fig. 5.3. Cross-sectional view of a yoke showing location of vertical and horizontal deflection windings about the neck of the picture tube.

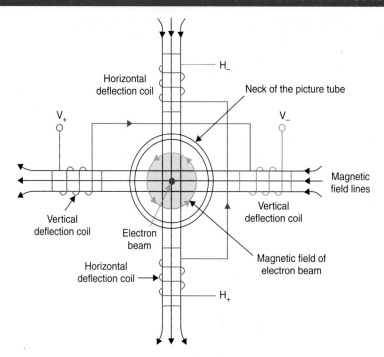

Fig. 5.4. *Horizontal and Vertical deflecting coils (pairs) around the neck of the picture tube. Note that the location of the beam on the picture tube screen will depend on the strength and direction of currents in the two pairs of coils. For the directions of current shown the beam will be deflected upwards and to the left.*

The magnetic field of the coils reacts with the electron beam to cause its deflection. The horizontal deflection coil which sweeps the beam across the face of the tube from left to right is split into two sections and mounted above and below the beam axis. The vertical deflection coil is also split into two sections and placed left and right on the neck in order to pull the beam gradually downward as the horizontal coils sweep the beam across the tube face. Each coil gets its respective sweep input from the associated sweep circuits, and together they form the raster upon which the picture information is traced. It may be noted that a perpendicular displacement results because the magnetic field due to each coil reacts with the magnetic field of the electron beam to produce a force that deflects the electrons at right angles to both the beam axis and the deflection field.

Deflection Angle

This is the maximum angle through which the beam can be deflected without striking the side of the bulb. Typical values of deflection angles are 70°, 90°, 110° and 114°. As shown in Fig. 5.5, it is the total angle that is specified. For instance a deflection angle of 110° means the electron beam can be deflected 55° from the centre. The advantage of a large deflection angle is that for equal picture size the picture tube is shorter and can be installed in a smaller cabinet. However, a large deflection angle requires more power from the deflection circuits. For this reason the tubes are made with a narrow neck to put the deflection yoke closer to the electron beam. A 110° yoke has a smaller hole diameter (about 3 cm) compared with neck diameters for tubes with lesser deflection angles. Different screen sizes can be filled with the same deflection angle, because bigger tubes have larger axial lengths.

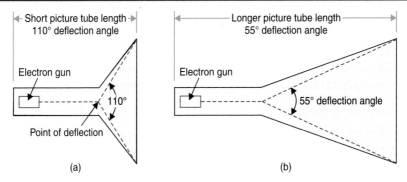

Fig. 5.5. *Effect of deflection angle on picture tube length for the same face plate size.*
(a) Picture tube rated for 110° deflection angle, (b) Picture tube rated for 55° deflection angle
Note : *The nominal deflection angle that is listed for picture tubes*
is usually the diagonal deflection angle.

Cosine Winding

With increased deflection angles it becomes necessary to use a special type of winding to generate uniform magnetic fields for linear deflection. In this arrangement the thickness of the deflection winding varies as the cosine of the angle from a central reference line. Such a winding is known as 'Cosine winding' and its appearance in a deflection yoke is shown in Fig. 5.3. Nearly all present day yokes are wound in this manner to ensure linear deflection.

5.3 SCREEN PHOSPHOR

The phosphor chemicals are generally light metals such as zinc and cadmium in the form of sulphide, sulphate, and phosphate compounds. This material is processed to produce very fine particles which are then applied on the inside of the glass plate. As already explained the high velocity electrons of the beam on hitting the phosphor excite its atoms with the result that the corresponding spot fluoresces and emits light. The phosphorescent characteristics of the chemicals used are such that an afterglow remains on the screen for a short time after the beam moves away from any screen spot. This afterglow is known as persistence. Medium persistence is desirable to increase the average brightness and to reduce flicker. However, the persistence must be less than 1/25 second for picture tube screens so that one frame does not persist into the next and cause blurring of objects in motion. The decay time of picture tube phosphors is approximately 5 ms, and its persistence is referred to as P_4 by the industry.

5.4 FACE PLATE

A rectangular image on a circular screen is wasteful of screen area. Therefore, all present day picture tubes have rectangular face plates, with a breadth to height ratio of 4 : 3. A rectangular tube with 54 cm screen means that the distance between the two diagonal points is 54 centimeters. Approximately 1.5 cm thickness provides the strength required for the large face plate to withstand the air pressure on the evacuated glass envelope. In older receivers special glass or plastic shields were placed in the cabinet in front of the picture tube to prevent any glass from hitting the viewer in case of an implosion. Modern picture tubes incorporate integral implosion protection. There are a number of different systems in use. In one arrangement, known as kimcode, a metal rim band (see Fig. 5.1) is held around the tube by a tension strap or with a layer of epoxy cement. In another system called Panoply, a special faceplate is held in front of the tube by epoxy cement. In all cases it is essential to check for implosion proofing while replacing any picture tube.

Yoke and Centering Magnets

The yoke on all black and white tubes is positioned right up against the flare of the tube in order to achieve complete coverge of the full screen area. If the yoke is not moved as far forward as possible, the electron beam will strike the neck of the picture tube and cause a shadow near the corners of the face plate. The mounting system permits positioning of the yoke against the tube funnel and allows rotation of the yoke to ensure that horizontal lines run parallel to the natural horizontal axis.

Electrical centering of the beam can be accomplished by supplying direct current through the horizontal and vertical deflection coils. However, this method is not used now because of the added current drain on the low voltage power supply. Modern tubes have a pair of permanent magnets (see Fig. 5.2) for centering, in the form of rings usually mounted on the yoke cover. Poles of both the magnets can be suitably shifted with a pair of projecting tabs provided on the magnetic rings. When the two tabs (one from each ring) coincide with each other, the strongest field is achieved; that is, the beam will be pushed furthest off centre. When the two tabs are 180° apart (on opposite sides) the field is minimum and so is the decentering. The two rings are rotated together to change the direction in which decentering occurs. This is illustrated in Fig. 5.6.

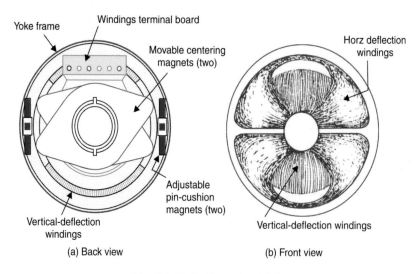

Fig. 5.6. Deflection yoke details.

The edge of the yoke linear (see Fig. 5.2) is used to hold small permanent magnets. As shown in Fig. 5.6 these are positioned to correct any 'pincushion error'.

Screen Brightness

It is estimated that about 50 per cent of the light emitted at the screen, when the electron beam strikes it, travels back into the tube. Another 20 percent or so is lost in the glass of the tube because of internal reflections and only about 20 percent of it reaches the viewer. Image contrast is also impaired because of interference caused by the light which is returned to the screen after reflection from some other points.

Also any ions is the beam, which do exist despite best precautions while degassing, damage the phosphor material on hitting it and thus cause a dark brownish patch on the screen. This area usually centers around the middle of the screen because the greater mass of the ions prevents any appreciable deflection during their transit, with the result, that they arrive almost at the centre of the screen.

To overcome these serious drawbacks practically all modern picture tubes employ a very thin coating of aluminium on the back surface of the screen phosphor. The aluminized coating is very thin and with a final anode voltages of 10 kV or more, the electrons of the beam have enough velocity to penetrate this coating and excite the phosphor. Thus most of the light that would normally travel back and get lost in the tube is now reflected back to the screen by the metal backing and this results in a much improved brilliancy. The aluminized coating is connected to the high voltage anode coating and thus helps in draining off the secondary emitted electrons at the screen. This further improves the brightness.

Ion-trap

In older picture tubes a magnetic beam, bender commonly known as 'ion-trap' was employed to deflect the heavy ions away from the screen. In present day picture tubes having a thin metal coating on the screen, it is no longer necessary to provide an ion-trap. This is because the ions on account of their greater mass fail to penetrate the metal backing and do not reach the phosphor screen.

Thus an aluminized coating when provided on the phosphor screen, not only improves screen brightness and contrast but also makes the use of 'ion-traps' unnecessary.

High Voltage Filter Capacitor

A grounded coating is provided on the outer surface of the picture tube. This provides shielding from stray fields and also acts as one plate of the capacitor, the other plate being the inner anode coating with the glass bulb serving as the insulator between the two. The capacitor thus formed (see Fig. 5.2) serves as a filter capacitor for the high voltage supply. This capacitor can hold charge for a long time after the anode voltage is switched off and so before handling the picture tube the capacitor must be discharged by shorting the anode button to the grounded wall coating.

Spark-gap Protection

On account of close spacing between the various-electrodes and the use of very high voltages, arcing of flashover can occur in the electron gum especially at the control grid. This arcing causes voltage surges, which result in damage to the associated circuit components. Therefore for protection of the receiver circuit, due to any arcing, metallic spark-gaps are provided as shunt paths for the surge currents. In some designs neon bulbs are used as spark gaps. The gas in the neon tube ionizes when the potential exceeds a certain limit and thus provides a shunt path for the high voltage arc current.

5.5 | PICTURE TUBE CHARACTERISTICS

As shown in Fig. 5.7 the transfer characteristics of picture tubes are similar to the grid-plate characteristics of vacuum tubes. The grid of the picture tube has a fixed bias that is set with the brightness control for optimum average brightness of the picture on the screen. The video signal that finally controls the brightness variations on the screen may be applied either at the grid or cathode of the picture tube. Each method has its own merits and demerits and are discussed in another chapter. This method of varying the beam current to control the instantaneous screen brightness is called intensity or 'Z' axis modulation. The peak-to-peak amplitude of the ac video signal determines the contrast in the picture, between peak white with maximum beam current and black at cut-off. The contrast control is in the video amplifier, which controls the peak-to-peak amplitude of the video signal applied to the picture tube.

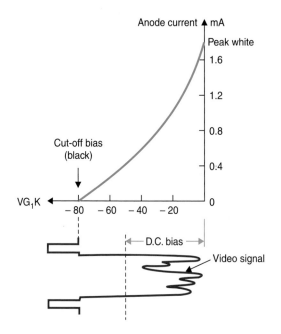

Fig. 5.7. *Transfer characteristics of a picture tube. Note the alignment of blanking level with cut-off bias.*

At cut-off the grid voltage is negative enough to reduce the beam current to a value low enough to extinguish the beam, and this corresponds to the black level in the picture. The parts of the screen without any luminescence look black in comparison with the adjacent white areas.

5.6 | PICTURE TUBE CIRCUIT CONTROLS

Manufacturers usually recommend a sufficiently high voltage to the second anode of the picture tube to produce adequate screen brilliancy for normal viewing. This voltage is always obtained from the output of the horizontal deflection circuit. The dc voltages to the screen grid and focus grid are also taken from the horizontal stage and adjusted to suitable values by resistive potential divider networks. This is shown in Fig. 5.8.

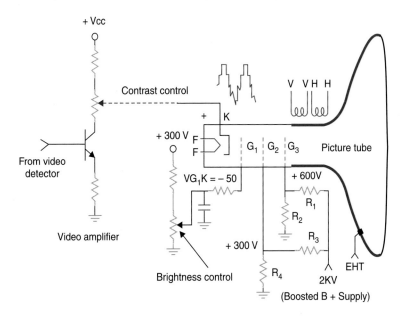

Fig. 5.8. *Picture tube circuit and associated controls.*

A variable bias control either in the cathode circuit or control grid lead is provided to control the electron density, which in turn controls the brightness on the screen. This control, known as the 'brightness control', is brought out at the front panel of the receiver to enable the viewer to adjust brightness.

As discussed earlier most modern picture tubes do not require critical focus adjustment. Therefore no focus control is normally provided and instead dc voltage at the focus electrode is carefully set as explained above.

The contrast control through not strictly a part of the picture tube circuit forms part of cathode or control grid circuit. This control is also provided at the front panel of the receiver and its variation enables adjustment of contrast in the reproduced picture.

Picture Tube Handling

The very high vacuum in a modern picture tube means that there is a danger of implosion if the tube is struck with a hard object or if it is made to rest on its neck. Because of the large volume of the tube, there is a very high pressure on the glass shell.

In case it breaks the resulting implosion will often cause tube fragments to fly in all directions at high speed. This may cause severe injury to the persons hit by the tube fragments. Manufacturers recommend the use of protective goggles and gloves whenever picture tubes are handled and such precautions should be observed. The tube neck is particularly fragile and must be handled with care.

REVIEW QUESTIONS

1. Sketch the sectional view of a picture tube that employs electrostatic focusing and electromagnetic deflection and label all the electrodes.

2. Explain briefly, how the electron beam is focused on the tube screen. What is meant by crossover point in the electron gun ?

3. What type of phosphor is employed for picture tube screens ? Why is a medium persistence phosphor preferred ?

4. What is the function of aquadag coating on the inner side of the tube bell ? Why is a grounded coating provided on the outer surface of the picture tube ?

5. Why is an aluminized coating provided on the phosphor screen ? How are any stary ions prevented from hitting the screen ?

6. What do you understand by a 54 cm picture tube ? Why is it necessary to employ implosion protection in picture tubes ?

7. What precautions must be observed while handling a picture tube ? Why is it necessary to provide spark-gap protection between the various electrodes ?

8. Discuss the merits of electromagnetic deflection over electrostatic deflection in television picture tubes. Why is 'cosine winding' used for deflection coils ?

9. What is meant by the deflection angle of a picture tube ? What is the advantage of providing a large deflection angle yoke ?

10. Explain how the yoke is mounted on the tube neck. Describe how the centering of the electron beam is accomplished with the help of centering magnets. Why are small permanent magnets provided at the edges of the yoke liner ?

11. Show with a circuit diagram how dc potentials are supplied to the various electrodes of the picture tube.

12. What are the functions of 'brightness' and 'contrast' controls ? Explain their action with suitable circuit diagrams.

6

Television Camera Tubes

A TV camera tube may be called the eye of a TV system. For such an analogy to be correct the tube must possess characteristic that are similar to its human counterpart. Some of the more important functions must be (*i*) sensitivity to visible light, (*ii*) wide dynamic range with respect to light intensity, and (*iii*) ability to resolve details while viewing a multielement scene.

During the development of television, the limiting factor on the ultimate performance had always been the optical-electrical conversion device, *i.e.,* the pick-up tube. Most types developed have suffered to a greater or lesser extent from (*i*) poor sensitivity, (*ii*) poor resolution, (*iii*) high noise level, (*iv*) undesirable spectral response, (*v*) instability, (*vi*) poor contrast range and (*vii*) difficulties of processing.

However, development work during the past fifty years or so, has enabled scientists and engineers to develop image pick-up tubes, which not only meet the desired requirements but infact excel the human eye in certain respects. Such sensitive tubes have now been developed which deliver output even where our eyes see complete darkness. Spectral response has been so perfected, that pick-up outside the visible range (in infra-red and ultraviolet regions) has become possible. Infact, now there is a tube available for any special application.

6.1 BASIC PRINCIPLE

When minute details of a picture are taken into account, any picture appears to be composed of small elementary areas of light or shade, which are known as picture elements. The elements thus contain the visual image of the scene. The purpose of a TV pick-up tube is to sense each element independently and develop a signal in electrical form proportional to the brightness of each element. As already explained in Chapter 1, light from the scene is focused on a photosensitive surface known as the image plate, and the optical image thus formed with a lens system represents light intensity variations of the scene. The photoelectric proportion of the image plate then convert different light intensities into corresponding electrical variations. In addition to this photoelectric conversion whereby the optical information is transduced to electrical charge distribution on the photosensitive image plate, it is necessary to pick-up this information as fast as possible. Since simultaneous pick-up is not possible, scanning by an electron beam is resorted to. The electron beam moves across the image plate line by line, and field by field to provide signal variations in a successive order. This scanning process divides the image into its basic picture elements. Through the entire image plate is photoelectric, its construction isolates the picture elements so that each discrete small area can produce its own signal variations.

Photoelectric Effects

The two photoelectric effects used for converting variations of light intensity into electrical variations are (i) photoemission and (ii) photoconductivity. Certain metals emit electrons when light falls on their surface. These emitted electrons are called photoelectrons and the emitting surface a photocathode. Light consists of small bundles of energy called photons. When light is made incident on a photocathode, the photons give away their energy to the outer valence electrons to allow them to overcome the potential-energy barrier at the surface. The number of electrons which can overcome the potential barrier and get emitted, depends on the light intensity. Alkali metals are used as photocathode because they have very low work-function. Cesium-silver or bismuth-silver-cesium oxides are preferred as photoemissive surfaces because they are sensitive to incandescent light and have spectral response very close to the human eye.

The second method of producing an electrical image is by photoconduction, where the conductivity or resistivity of the photosensitive surface varies in proportion to the intensity of light focused on it. In general the semiconductor metals including selnium, tellurium and lead with their oxides have this property known as photoconductivity. The variations in resistance at each point across the surface of the material is utilized to develop a varying signal by scanning it uniformly with an electron beam.

Image Storage Principle

Television cameras developed during the initial stages of development were of the non-storage type, where the signal output from the camera for the light on each picture element is produced only at the instant it is scanned. Most of the illumination is wasted. Since the effect of light on the image plate cannot be stored, any instantaneous pick-up has low sensitivity. Image disector and flying-spot camera are examples of non-storage type of tubes. These are no longer in use and will not be discussed. High camera sensitivity is necessary to televise scenes at low light levels and to achieve this, storage type tubes have been developed. In storage type camera tubes the effect of illumination on every picture element is allowed to accumulate between the times it is scanned in successive frames. With light storage tubes the amount of photoelectric signal can be increased 10,000 times approximately compared with the earlier non-storage type.

The Electron Scanning Beam

As in the case of picture tubes an electron gun produces a narrow beam of electrons for scanning. In camera tubes magnetic focusing is normally employed. The electrons must be focused to a very narrow and thin beam because this is what determines the resolving capability of the camera. The diameter of the beam determines the size of the smallest picture element and hence the finest detail of the scene to which it can be resolved. Any movement of electric charge is a flow of current and thus the electron beam constitutes a very small current which leaves the cathode in the electron gun and scans the target plate. The scanning is done by deflecting the beam with the help of magnetic fields produced by horizontal and vertical coils in the deflection yoke put around the tubes. The beam scans 312.5 lines per field and 50 such fields are scanned per second.

Video Signal

In tubes employing photoemissive target plates the electron beam deposits some charge on the target plate, which is proportional to the light intensity variations in the scene being televised. The beam motion is so controlled by electric and magnetic fields, that it is decelerated before it reaches the target and lands on it with almost zero velocity to avoid any secondary emission. Because of the negative acceleration the beam is made to move back from the target and on its return journey, which is very accurately controlled by the focusing and deflection coils, it strikes an electrode which is located very close to the cathode from where it started. The number of electrons in the returning beam will thus vary in accordance with the charge deposited on the target plate. This in turn implies that the current which enters the collecting electrode varies in amplitude and represents brightness variations of the picture. This current is finally made to flow through a resistance and the varying voltage developed across this resistance constitutes the video signal. Figure 6.1 (a) illustrates the essentials of this technique of developing video signal.

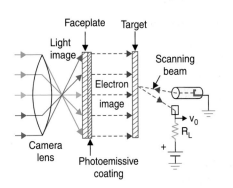

Fig. 6.1(a). Production of video signal by photoemission.

Fig. 6.1(b). Production of video signal by photoconduction.

In camera tubes employing photoconductive cathodes the scanning electron beam causes a flow of current through the photoconductive material. The amplitude of this current varies in accordance with the resistance offered by the surface at different points. Since the conductivity of the material varies in accordance with the light falling on it, the magnitude of the current represents the brightness variations of the scene. This varying current completes its path under the influence of an applied dc voltage through a load resistance connected in series with path of the current. The instantaneous voltage developed across the load resistance is the video signal which, after due amplification and processing is amplitude modulated and transmitted. Figure 6.1 (b) shows a simplified illustration of this method of developing video signal.

Electron Multiplier

When the surface of a metal is bombarded by incident electrons having high velocities, secondary emission takes place. Aluminium, as an example, can release several secondary electrons for each incident primary electron. Camera tubes often include an electron multiplier structure, making use of the secondary emission effect to amplify the small amount of photoelectric current that is later employed to develop video signal. The electron multiplier is a series of cold anode-cathode electrodes called dynodes mounted internally, with each at a progressively higher positive potential as illustrated in Fig. 6.2. The few electrons emitted by the photocathode are accelerated to a more positive dynode. The primary electrons can then force the ejection of secondary emission electrons when the velocity of the incident electrons is large enough. The secondary emission ratio is normally three or four, depending on the surface and the potential applied. The number of electrons available is multiplied each time the secondary electrons strike the emitting surface of the next more positive dynode. The current amplification thus obtained is noise free because the electron multiplier does not have any active device or resistors. Since the signal amplitude is very low any conventional amplifier, if used instead of the electron multiplier, woul cause serious *S/N* ratio problems.

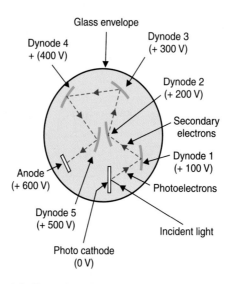

Fig. 6.2. *Illustration of an electron-multiplier structure.*

Types of Camera Tubes

The first developed storage type of camera tube was 'Iconoscope' which has now been replaced by image-orthicon because of its high light sensitivity, stability and high quality picture capabilities. The light sensitivity is the ratio of the signal output to the incident illumination. Next to be developed was the *vidicon* and is much simpler in operation. Similar to the vidicon is another tube known as *plumbicon.* The latest device in use for image scanning is the solid state image scanner.

6.2   IMAGE ORTHICON

This tube makes use of the high photoemissive sensitivity obtainable from photocathodes, image multiplication at the target caused by secondary emission and an electron multiplier. A sectional view of an image orthicon is shown in Fig. 6.3. It has three main sections: image section, scanning section and electron gun-cum-multiplier section.

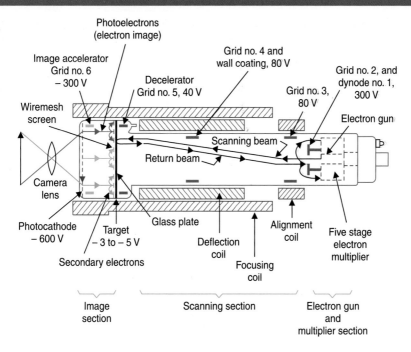

Fig. 6.3. *Principle of operation of Image Orthicon (non-field mesh type).*

(i) Image Section

The inside of the glass face plate at the front is coated with a silverantimony coating sensitized with cesium, to serve as photocathode. Light from the scene to be televised is focused on the photocathode surface by a lens system and the optical image thus formed results in the release of electrons from each point on the photocathode in proportion to the incident light intensity. Photocathode surface is semitransparent and the light rays penetrate it to reach its inner surface from where electron emission takes place. Since the number of electrons emitted at any point in the photocathode has a distribution corresponding to the brightness of the optical image, an electron image of the scene or picture gets formed on the target side of the photocoating and extends towards it. Through the convertion efficiency of the photocathode is quite high, it cannot store charge being a conductor. For this reason, the electron image produced at the photocathode is made to move towards the target plate located at a short distance from it. The target plate is made of a very thin sheet of glass and can store the charge received by it. This is maintained at about 400 volts more positive with respect to the photocathode, and the resultant electric field gives the desired acceleration and motion to the emitted electrons towards it. The electrons, while in motion, have a tendency to repel each other and this can result in distortion of the information now available as charge image. To prevent this divergence effect an axial magnetic field, generated in this region by the 'long focus coil' is employed. This magnetic field imparts helical motion of increasing pitch and focuses the emitted electrons on the target into a well defined electron image of the original optical image. The image side of the target has a very small deposit of cesium and thus has a high secondary emission ratio. Because of the high velocity attained by the electrons while in motion from photocathode to the target plate, secondary emission results, as the electrons bombard the target surface. These secondary electrons are collected by a wire-mesh screen, which is located in front of the target on the image side and is maintained at a slightly higher potential with respect to the target. The wire-mesh screen has about 300 meshes per cm^2 with an open area of 50 to 75 per cent, so that the screen wires do not interfere with the electron image. The secondary electrons leave behind on the target plate surface, a positive charge distribution, corresponding to the light intensity distribution on the original photocathode.

For storage action this charge on the target plate should not spread laterally over its surface, during the storage time, since this would destory the resolution of the device. To achieve this the target is made out of extremely thin sheet of glass. The positive charge distribution builds up during the frame storage time (40 ms) and thus enhances the sensitivity of the tube. It should be clearly understood, that the light from the scene being televised continuously falls on the photocathode, and the resultant emitted electrons on reaching the target plate cause continuous secondary emission. This continuous release of electrons results in the building up of positive charge on the target plate.

Because of the high secondary emission ratio, the intensity of the positive charge distribution is four to five times more as compared to the charge liberated by the photocathode. This increase in charge density relative to the charge liberated at the photocathode is known as 'image multiplication' and contributes to the increased sensitivity of image orthicon. As shown in Fig. 6.3, the two-sided target has the charge image on one side while an electron beam scans the opposite side. Thus, while the target plate must have high resistivity laterally for storage action, it must have low resistivity along its thickness, to enable the positive charge to conduct to the other side which is scanned. It is for this reason that the target plate is very thin, with thickness close to 0.004 mm. Thus, whatever charge distribution builds up on one side of the target plate due to the focused image, appears on the other side, which is scanned, and it is from here that the video signal is obtained.

(ii) Scanning Section

The electron gun structure produces a beam of electrons that is accelerated towards the target. As indicated in the figure, positive accelerating potentials of 80 to 330 volts are applied to grid 2, grid 3, and grid 4 which is connected internally to the metalized conductive coating on the inside wall of the tube. The electron beam is focused at the target by magnetic field of the external focus coil and by voltage supplied to grid 4. The alignment coil provides magnetic field that can be varied to adjust the scanning beam's position, if necessary, for correct location. Deflection of electron beam's to scan the entire target plate is accomplished by magnetic fields of vertical and horizontal deflecting coils mounted on yoke external to the tube. These coils are fed from two oscillators, one working at 15625 Hz, for horizontal deflection, and the other operating at 50 Hz, for vertical deflection.

The target plate is close to zero potential and therefore electrons in the scanning beam can be made to stop their forward motion at its surface and then return towards the gun structure. The grid 4 voltage is adjusted to produce uniform deceleration of electrons for the entire target area. As a result, electrons in the scanning beam are slowed down near the target. This eliminates any possibility of secondary emission from this side of the target plate. If a certain element area on the target plate reaches a potential of, say, 2 volts during the storage time, then as a result of its thinness the scanning beam 'sees' the charge deposited on it, part of which gets diffused to the scanned side and deposits an equal number of negative charges on the opposite side. Thus out of the total electrons in the beam, some get deposited on the target plate, while the remaining stop at its surface and turn back to go towards the first electrode of the electron multiplier. Because of low resistivity across the two sides of the target, the deposited negative charge neutralizes the existing positive charge in less than a frame time. The target can again become charged as a result of the incident picture information, to be scanned during the successive frames. As the target is scanned element by element, if there are no positive charges at certain points, all the electrons in the beam return towards the electron gun and none gets deposited on the target plate. The number of electrons, leaving cathode of the gun, is practically constant, and out of this, some get deposited and remaining electrons, which travel backwards provide signal current that varies in amplitude in accordance with the picture information. Obviously then, the signal current is maximum for black areas on the picture, because absence of light from black areas on the picture does not result in any emission on the photocathode, and there is no secondary emission at the corresponding points on the target, and no electrons are needed from the beam to neutralize them. On the contrary for high light areas, on the picture, there is maximum loss of electrons from the target plate, due to secondary emission, and this results in large deposits of electrons from the beam and this reduces the amplitude of the returning beam current. The resultant beam current that turns away from the target, is thus, maximum for black areas and minimum for bright areas on the picture. High intensity light causes large charge imbalance on the glass target plate. The scanning beam is not able to completely neutralize it in one scan. Therefore the earlier impression persists for several scans.

Image Resolution. It may be mentioned at this stage that since the beam is of low velocity type, being reduced to near zero velocity in the region of the target it is subjected to stray electric fields in its vicinity, which can cause defocusing and thus loss of resolution. Also on contact with the target, the electrons would normally glide along its surface tangentially for a short distance and the point of contact becomes ill defined. The beam must strike the target at right angle at all points of the target, for better resolution. These difficulties are overcome in the image-orthicon by the combined action of electrostatic field because of potential on grid 4, and magnetic field of the long focusing coil. The interaction of two fields gives rise to cycloidal motion to the beam in the vicinity of target, which then hits it at right angle no matter which point is being scanned. This very much improves the resolving capability of the picture tube.

(iii) Electron Multiplier

The returning stream of electrons arrive at the gun close to the aperture from which electron beam emerged. The aperture is a part of a metal disc covering the gun electrode. When the returning electrons strike the disc which is at a positive potential of about 300 volts, with respect to the target, they produce secondary emission. The disc serves as first stage of the electron multiplier. Successive stages of the electron multiplier are arranged symmetrically around and back of the first stage. Therefore secondary electrons are attracted to the dynodes at progressively higher positive potentials. Five stages of multiplication are used, details of which are shown in Fig. 6.4. Each multiplier stage provides a gain of approximately 4 and thus a total gain of $(4)^5 \approx 1000$ is obtained at the electron multiplier. This is known as signal multiplication. The multiplication so obtained maintains a high signal to noise ratio. The secondary electrons are finally collected by the anode, which is connected to the highest supply voltage of + 1500 volts in series with a load resistance R_L. The anode current through R_L has the same variations that are present in the return beam from the target and amplified by the electron multiplier. Therefore

Chapter 6

voltage across R_L is the desired video signal; the amplitude of which varies in accordance with light intensity variations of scene being televised. The output across R_L is capacitively coupled to the camera signal amplifier. With R_L = 20 K-ohms and typical dark and high light currents of magnitudes 30 μA and 5 μA respectively, the camera output signal will have an amplitude of 500 mV peak-to-peak.

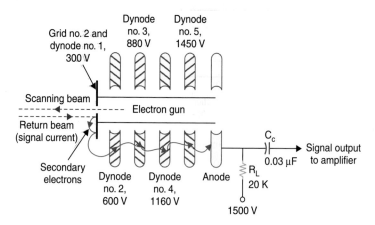

Fig. 6.4. Electron-multiplier section of the Image Orthicon.

Field Mesh Image Orthicon. The tube described above is a non-field mesh image orthicon. In some designs an additional pancake-shaped magnetic coil is provided in front of the face plate. This is connected in series with the main focusing coil. The location of the coil results in a graded magnetic field such that the optically focused photocathode image is magnified by about 1.5 times. Thus the charge image produced on the target plate is bigger in size and this results in improved resolution and better overall performance. Such a camera tube is known as a field mesh Image Orthicon.

Light Transfer Characteristics and Applications—During the evolution of image orthicon tubes, two separate types were developed, one with a very close target-mesh spacing (less than 0.001 cm) and the other with somewhat wider spacing. The tube, with very close target mesh spacing, has very high signal to noise ratio but this is obtained at the expense of sensitivity and contrast ratio. This is a worthwhile exchange where lighting conditions can be controlled and picture quality is of primary importance. This is generally used for live shows in the studios. The other type with wider target-mesh spacing has high sensitivity and contrast ratio with more desirable spectral response. This tube has wider application for outdoor or other remote pickups where a wide range of lighting conditions have to be accommodated. More recent tubes with improved photocathodes have sensitivities several times those of previous tubes and much improved spectral response. Overall transfer characteristics of such tubes are drawn in Fig. 6.5. Tube 'A' is intended primarily for outdoor pick-ups where as tube 'B' is much suited for studio use and requires strong illumination. The knee of the transfer characteristics is reached when the illumination causes the target to be fully charged with respect to the mesh between successive scans by the electron beam. The tube is sometimes operated slightly above the knee, to obtain the black border effect (also known as Halo effect) around the high light areas of the target.

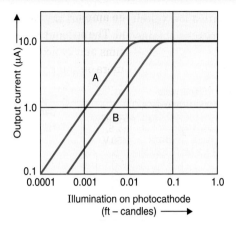

Fig. 6.5. *Light transfer characteristics of two different Image Orthicons.*

6.3 VIDICON

The Vidicon came into general use in the early 50's and gained immediate popularity because of its small size and ease of operation. It functions on the principle of photoconductivity, where the resistance of the target material shows a marked decrease when exposed to light. Fig. 6.6

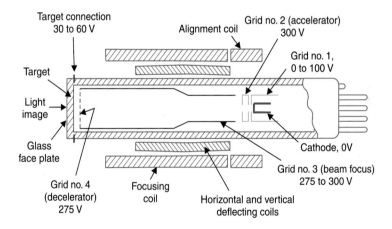

Fig. 6.6. *Vidicon camera tube cross-section.*

illustrates the structural configuration of a typical vidicon, and Fig. 6.7 shows the circuit arrangement for developing camera signal output. As shown there, the target consists of a thin photo conductive layer of either selenium or anti-mony compounds. This is deposited on a transparent conducting film, coated on the inner surface of the face plate. This conductive coating is known as signal electrode or plate. Image side of the photolayer, which is in contact with the signal electrode, is connected to DC supply through the load resistance R_L. The beam that emerges from the electron gun is focused on surface of the photo conductive layer by combined action of uniform magnetic field of an external coil and electrostatic field of grid No 3. Grid No. 4 provides a uniform decelerating field between itself, and the photo conductive layer, so that the electron beam approaches the layer with a low velocity to prevent any secondary emission. Deflection of the beam, for scanning the target, is obtained by vertical and horizontal deflecting coils, placed around the tube.

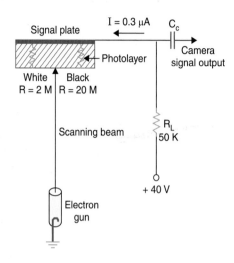

Fig. 6.7. Circuit for output signal from a Vidicon camera tube.

Charge Image

The photolayer has a thickness of about 0.0001 cm, and behaves like an insulator with a resistance of approximately 20 MΩ when in dark. With light focused on it, the photon energy enables more electrons to go to the conduction band and this reduces its resistivity. When bright light falls on any area of the photoconductive coating, resistance across the thickness of that portion gets reduces to about 2 MΩ. Thus, with an image on the target, each point on the gun side of the photolayer assumes a certain potential with respect to the DC supply, depending on its resistance to the signal plate. For example, with a B + source of 40 V (see Fig. 6.7), an area with high illumination may attain a potential of about + 39 V on the beam side. Similarly dark areas, on account of high resistance of the photolayer may rise to only about + 35 volts. Thus, a pattern of positive potentials appears, on the gun side of the photolayer, producing a charge image, that corresponds to the incident optical image.

Storage Action

Though light from the scene falls continuously on the target, each element of the photocoating is scanned at intervals equal to the frame time. This results in storage action and the net change in resistance, at any point or element on the photoconductive layer, depends on the time, which elapses between two successive scannings and the intensity of incident light. Since storage time for all points on the target plate is same, the net change in resistance of all elementary areas is proportional to light intensity variations in the scene being televised.

Signal Current

As the beam scans the target plate, it encounters different positive potentials on the side of the photolayer that faces the gun. Sufficient number of electrons from the beam are then deposited on the photolayer surface to reduce the potential of each element towards the zero cathode potential. The remaining electrons, not deposited on the target, return back and are not utilized in the vidicon. However, the sudden change in potential on each element while the beam scans, causes a current flow in the signal electrode circuit producing a varying voltage across the load resistance R_L. Obviously, the amplitude of current and the consequent output voltage across R_L are directly proportional to the light intensity variations on the scene. Note that, since, a large current would cause a higher voltage drop across R_L, the output voltage is most negative for white areas. The video output voltage, that thus develops across the load resistance (50 K-ohms) is adequate and does not need any image or signal multiplication as in an image orthicon. The output signal is further amplified by conventional amplifiers before it leaves the camera unit. This makes the vidicon a much simpler picture tube.

Leaky Capacitor Concept

Another way of explaining the development of 'charge image' on the photolayer is to consider it as an array of individual target elements, each consisting of a capacitor paralleled with a light dependent resistor. A number of such representations are shown in Fig. 6.8. As seen there, one end of these target elements is connected to the signal electrode and the other end is unterminated facing the beam.

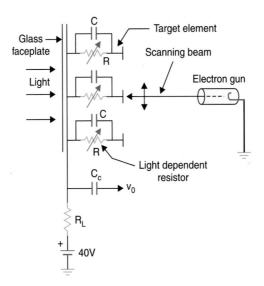

Fig. 6.8. Schematic representation of a Vidicon target area.

In the absence of any light image, the capacitors attain a charge almost equal to the B + (40 V) voltage in due course of time. However, when an image is focused on the target the resistors in parallel with the capacitors change in value depending on the intensity of light on each unit element. For a high light element, the resistance across the capacitor drops to a fairly low value, and this permits lot of charge from the capacitor to leak away. At the time of scanning, more electrons are deposited, on the unterminated end of this capacitor to recharge it to the full supply voltage of + 40 V. The consequent flow of current that completes its path through R_L develops a signal voltage across it. Similarly for black areas of the picture, the resistance across the capacitors remains fairly high, and not much charge is allowed to leak from the corresponding capacitors. This in turn needs fewer number of electrons from the beam to recharge the capacitors. The resultant small current that flows, develops a lower voltage across the load resistance.

The electron beam thus 'sees' the charge on each capacitor, while scanning the target, and delivers more or less number of electrons to recharge them to the supply voltage. This process is repeated every 40 ms to provide the necessary video signal corresponding to the picture details at the upper end of the load resistor. The video signal is fed through a blocking capacitor to an amplifier for necessary amplification.

Light Transfer Characteristics

Vidicon output characteristics are shown in Fig. 6.9. Each curve is for a specific value of 'dark' current, which is the output with no light. The 'dark' current is set by adjusting the target voltage. Sensitivity and dark current both increase as the target voltage is increased. Typical output for the vidicon is 0.4 µA for bright light with a dark current of 0.02 µA. The photoconductive layer has a time lag, which can cause smear with a trail following fast moving objects. The photoconductive lag increases at high target voltages, where the vidicon has its highest sensitivity.

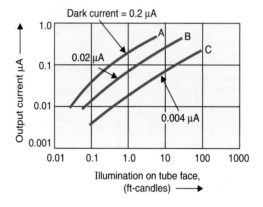

Fig. 6.9. *Light transfer characteristics of Vidicon.*

Applications

Earlier types of vidicons were used only where there was no fast movement, because of inherent lag. These applications included slides, pictures, closed circuit TV etc. The present day improved vidicon finds wide applications in education, medicine, industry, aerospace and oceanography. It is, perhaps, the most popular tube in the television industry. Vidicon is a short tube with a length of 12 to 20 cm and diameter between 1.5 and 4 cm. Its life is estimated to be between 5000 and 20,000 hours.

6.4 THE PLUMBICON

This picture tube has overcome many of the less favourable features of standard vidicon. It has fast response and produces high quality pictures at low light levels. Its smaller size and light weight, together with low-power operating characteristics, makes it an ideal tube for transistorized television cameras.

Except for the target, plumbicon is very similar to the standard vidicon. Focus and deflection are both obtained magnetically. Its target operates effectively as a P–I–N semi-conductor diode. The inner surface of the faceplate is coated with a thin transparent conductive layer of tin oxide

(SnO$_2$). This forms a strong N type (N$_+$) layer and serves as the signal plate of the target. On the scanning side of this layer is deposited a photoconductive layer of pure lead monoxide (PbO) which is intrinsic or 'I' type. Finally the pure PbO is doped to form a P type semiconductor on which the scanning beam lands. The details of the target are shown in Fig. 6.10 (a). The overall thickness of the target is 15 × 10^{-6} m. Figure 6.10 (b) shows necessary circuit details for developing the video signal. The photoconductive target of the plumbicon functions similar to the photoconductive target in the vidicon, except for the method of discharging each storage element. In the standard vidicon, each element acts as a leaky capacitor, with the leakage resistance decreasing with increasing light intensity. In the plumbicon, however, each element serves as a capacitor in series with a reverse biased light controlled diode. In the signal circuit, the conductive film of tin oxide (SnO$_2$), is connected to the target supply of 40 volts through an external load resistance R_L to develop the camera output signal voltage. Light from the scene being televised is focussed through the transparent layer of tin-oxide on the photoconductive lead monoxide. Without light the target prevents any conduction because of absence of any charge carriers and so there is little or no output current. A typical value of dark current is around 4 nA (4 × 10^{-9} Amp). The incidence of light on the target results in photoexcitation of semiconductor junction between the pure PbO and doped layer. The resultant decrease in resistance causes signal current flow which is proportional to the incident light on each photo element. The overall thickness of the target is 10 to 20 μm.

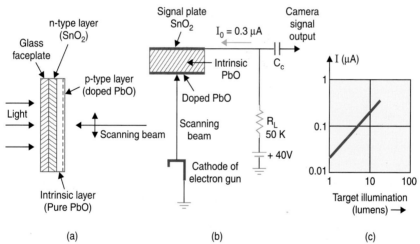

Fig. 6.10. Plumbicon camera tube (a) target details
(b) output signal current and (c) characteristics.

Light Transfer Characteristics

The current output versus target illumination response of a plumbicon is shown in Fig. 6.1 (c). It is a straight line with a higher slope as compared to the response curve of a vidicon. The higher value of current output, *i.e.*, higher sensitivity, is due to much reduced recombination of photogenerated electrons and holes in the intrinsic layer which contains very few discontinuities. For target voltages higher than about 20 volts, all the generated carriers are swept quickly across the target without much recombinations and thus the tube operates in a photosaturated mode. The spectral response of the plumbicon is closer to that of the human eye except in the red colour region.

6.5 SILICON DIODE ARRAY VIDICON

This is another variation of vidicon where the target is prepared from a thin n-type silicon wafer instead of deposited layers on the glass faceplate. The final result is an array of silicon photodiodes for the target plate. Figure 6.11 shows constructional details of such a target. As shown there, one side of the substrate (n-type silicon) is oxidized to form a film of silicon dioxide (SiO_2) which is an insulator. Then by photomasking and etching processes, an array of fine openings is made in the oxide layer. These openings are used as a diffusion mask for producing corresponding number of individual photodiodes. Boron, as a dopent is vapourized through the array of holes, forming islands of p-type silicon on one side of the n-type silicon substrate. Finally a very thin layer of gold is deposited on each p-type opening to form contacts for signal output. The other side of the substrate is given an antiflection coating. The resulting p-n photodiodes are about 8 μm in diameter. The silicon target plate thus formed is typically 0.003 cm thick, 1.5 cm square having an array of 540 × 540 photodiodes. This target plate is mounted in a vidicon type of camera tube.

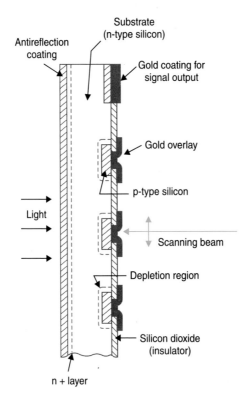

Fig. 6.11. Constructional details (enlarged) of a silicon diode array target plate.

Scanning and Operation

The photodiodes are reverse biased by applying +10 V or so to the n + layer on the substrate. This side is illuminated by the light focused on to it from the image. The incidence of light generates electron-hole pairs in the substrate. Under influence of the applied electric field, holes are swept over to the 'p' side of the depletion region thus reducing reverse bias on the diodes. This process continues to produce storage action till the scanning beam of electron gun scans the photodiode side of the substrate. The scanning beam deposits electrons on the p-side thus returning the diodes to their original reverse bias. The consequent sudden increase in current across each diode caused by the scanning beam represents the video signal. The current flows through a load resistance in the battery circuit and develops a video signal proportional to the intensity of light falling on the array of photodiodes. A typical value of peak signal current is 7 µA for bright white light.

The vidicon employing such a multidiode silicon target is less susceptible to damage or burns due to excessive high lights. It also has low lag time and high sensitivity to visible light which can be extended to the infrared region. A particular make of such a vidicon has the trade name of 'Epicon'. Such camera tubes have wide applications in industrial, educational and CCTV (closed circuit television) services.

6.6   SOLID STATE IMAGE SCANNERS*

The operation of solid state image scanners is based on the functioning of charge coupled devices (CCDs) which is a new concept in metal-oxide-semiconductor (MOS) circuitry. The CCD may be thought of to be a shift register formed by a string of very closely spaced MOS capacitors. It can store and transfer analog charge signals—either electrons or holes—that may be introduced electrically or optically.

The constructional details and the manner in which storing and transferring of charge occurs is illustrated in Fig. 6.12. The chip consists of a p-type substrate, the one side of which is oxidized to form a film of silicon dioxide, which is an insulator. Then by photolithographic processes, similar to those used in miniature integrated circuits an array of metal electrodes, known as gates, are deposited on the insulator film. This results in the creation of a very large number of tiny MOS capacitors on the entire surface of the chip.

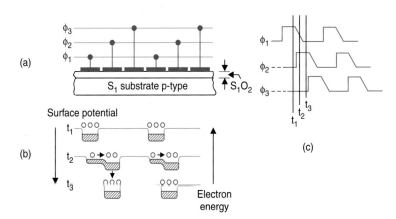

Fig. 6.12. *A three phase n-channel MOS charge coupled device. (a) construction (b) transfer of electrons between potential wells (c) different phases of clocking voltage waveform.*

* For more details on Solid State Image Scanners refer to IEEE Trans on 'Charge Coupled Devices— Technology and Applications' Edited by Roger Melen and Dennis Buss.

The application of small positive potentials to the gate electrodes results in the development of depletion regions just below them. These are called potential wells. The depth of each well (depletion region) varies with the magnitude of the applied potential. As shown in Fig. 6.12 (a), the gate electrodes operate in groups of three, with every third electrode connected to a common conductor. The spots under them serve as light sensitive elements. When any image is focused onto the silicon chip, electrons are generated within it, but very close to the surface. The number of electrons depends on the intensity of incident light. Once produced they collect in the nearby potential wells. As a result the pattern of collected charges represents the optical image.

Charge Transfer

The charge of one element is transferred along the surface of the silicon chip by applying a more positive voltage to the adjacent electrode or gate, while reducing the voltage on it. The minority carriers (electrons in this case) while accumulating in the so called wells reduce their depths much like the way a fluid fills up in a container. The acumulation of charge carries under the first potential wells of two consecutive trios is shown in Fig. 6.12 (b) where at instant t_1 a potential ϕ_1 exists at the corresponding gate electrodes. In practice the charge transfer is effected by multiphase clock voltage pulses (see Fig. 6.12 (c)) which are applied to the gates in a suitable sequence. The manner in which the transition takes place from potential wells under ϕ_1 to those under ϕ_2 is illustrated in Fig. 6.12 (b). A similar transfer moves charges from ϕ_2 to ϕ_3 and then from ϕ_3 to ϕ_1 under the influence of continuing clock pulses. Thus, after one complete clock cycle, the charge pattern moves one stage (three gates) to the right. The clocking sequence continues and the charge finally reaches the end of the array where it is collected to form the signal current.

Scanning of Television Pictures

A large number of CCD arrays are packed together to form the image plate. It does not need an electron gun, scanning beam, high voltage or vacuum envelope of a conventional camera tube. The potential required to move the charge is only 5 to 10 volt. The spot under each trio serves as the resolution cell. When light image is focused on the chip, electrons are generated in proportion to the intensity of light falling on each cell.

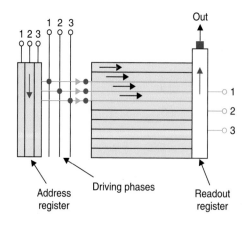

Fig. 6.13. *Basic organization of line addressed charge transfer area imaging devices.*

The principle of one-dimensional charge transfer as explained above can be integrated in various ways to render a solid-state area image device. The straightforward approach consists of arranging a set of linear imaging structures so that each one corresponds to a scan line in the display. The lines are then independently addressed and read into a common output diode by application of driving pulses through a set of switches controlled by an address register as shown in Fig. 6.13. To reduce capacitance, the output can be simply a small diffused diode in one corner of the array. The charge packets emerging from any line are carried to this diode by an additional vertical output register. In such a line addressed structure (Fig. 6.13) where the sequence of addressing the lines is determined by the driving circuitry, interlacing can be accomplished in a natural way.

Cameras Employing Solid-State Scanners

CCDs have a bright future in the field of solid state imaging. Full TV line-scan arrays have already been constructed for TV cameras. However, the quality of such sensors is not yet suitable for normal TV studio use. RCA SID 51232 is one such 24 lead dual-in-line image senser. It is a self-scanned senser intended primarily for use in generating standard interlaced 525 line television pictures. The device contains 512×320 elements and is constructed with a 3 phase n-channel, vertical frame transfer organization using a sealed silicon gate structure. Its block diagrams is shown in Fig. 6.14 (a). The image scanner's overall picture performance is comparable to that 2/3 inch vidicon camera tubes but undesirable characteristics such as lag and microphonics are eliminated.

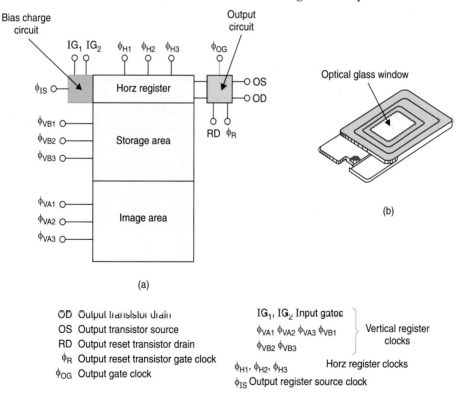

(a)

OD	Output transistor drain	IG$_1$, IG$_2$ Input gates
OS	Output transistor source	ϕ_{VA1} ϕ_{VA2} ϕ_{VA3} ϕ_{VB1} Vertical register
RD	Output reset transistor drain	ϕ_{VB2} ϕ_{VB3} clocks
ϕ_R	Output reset transistor gate clock	
ϕ_{OG}	Output gate clock	ϕ_{H1}, ϕ_{H2}, ϕ_{H3} Horz register clocks
		ϕ_{IS} Output register source clock

Fig. 6.14. *A 512 × 320 element senser (RCA SID 51232) for very compact TV cameras, (a) chip's block diagram, (b) view with optical glass window.*

The SID 51232 is supplied in a hermetic, edge contacted, 24-connection ceramic dual-in-line package. The package contains an optical glass window (see Fig. 6.14 (b)) which allows an image to be focussed into senser's 12.2 mm image diagonal.

REVIEW QUESTIONS

1. What is the basic principle of a camera pick-up tube ? Describe the two photoelectric effects used for converting variations of light intensity into electrical signals.

2. What do you understand by image storage capability of a modern television pick-up tube ? Explain why storage type tubes have must higher sensitivity as compared to the earlier non-storage type.

3. Draw cross-sectional view of an image orthicon camera tube and explain how it develops video signal when light from any scene is focused on its face plate.

4. What do you understand by image multiplication and signal multiplication in an image orthicon camera tube ? Why is an electron multiplier preferred over conventional amplifiers for amplifying the video signal at the output of the camera tube ?

5. In an image orthicon, what is the function of the wire-mesh screen and why is it located very close to the target plate ? Explain with the help of transfer characteristics the effect of target-mesh spacing on the overall performance of the tube.

6. In an image orthicon :
    (a) Why is the electron beam given a cycloidal motion before it hits the target plate ?
    (b) Why is the electron beam velocity brought close to zero on reaching the target plate ?
    (c) What is the function of the decelerator grid ?

7. Explain with the help of suitable sketches, how video signal is developed in a vidicon camera tube ? How is the vidicon different from an image-orthicon and what are its special applications?

8. What do you understand by 'dark current' in a vidicon ? Explain how the inherent smear effect in a vidicon is overcome in a Plumbicon. Explain with a suitable sketch the mechanism by which the video siganl is developed from the P-I-N structure of its target.

9. Give constructional details of the vidicon target prepared from a thin n-type silicon wafer which operates as an array of photodiodes. Explain how the signal voltage is developed from such a target.

10. Explain with suitable sketches the basic principle of a solid state image scanner. Describe briefly the manner in which the CCD array is scanned to provide interlaced scanning.

7

Basic Television Broadcasting

The composite video signal generated by camera and associated circuitry is processed in the control room before routing it to the transmitter. At transmitter, picture carrier frequency assigned to the station is generated, amplified and later amplitude modulated with the incoming video signal. The sound output associated with the scene is simultaneously processed and frequency modulated with channel sound carrier frequency. The two outputs, one from picture signal transmitter and the other from sound signal transmitter are combined in a suitable network and then fed to a common antenna network for transmission. As is obvious, the picture and sound signals, though generated and processed simultaneously pass through two independent transmitters at the broadcasting station. Thus, it is logical, that the two transmitting arrangements be studied separately. The first part of this chapter deals with television studio setup and picture signal transmission, while the later part is devoted to frequency modulation and sound signal transmission.

7.1 TELEVISION STUDIO

A TV studio is an acoustically treated compact anechoic room. It is suitably furnished and equipped with flood lights for proper light effects. The use of dimmerstats with flood lights enables suitable illumination level of any particular area of the studio depending on the scene to be televised. Several cameras are used to telecast the scene from different angles. Similarly a large number of microphones are provided at different locations to pick up sound associated with the programme.

The camera and microphone outputs are fed into the control room by coaxial cables. The control room has several monitors to view pictures picked up by different cameras. A monitor is a TV receiver that contains no provision for receiving broadcast signals but operates on a direct input of unmodulated signal. A large number of such monitors are used to keep a check on the content and quality of pictures being telecast. Similarly, headphones are used to monitor and regulate sound output received from different microphones through audio mixers.

In addition to a live studio, video tape recording and telecine machine rooms are located close to the control room. In most cases, programmes as enacted in the studio are recorded on a video tape recorder (VTR) through the control room. These are later broadcast with the VTR output passing through the same control room. Figure 7.1 illustrates a typical layout of a television studio setup. As shown, the telecine machines together with a slide scanner are installed next to the control room. Such a facility enables telecasting of cinematograph films and advertisement slides. All the rooms are interconnected by coaxial cables and shielded wires. In large establishments, there are several such studio units with their outputs feeding

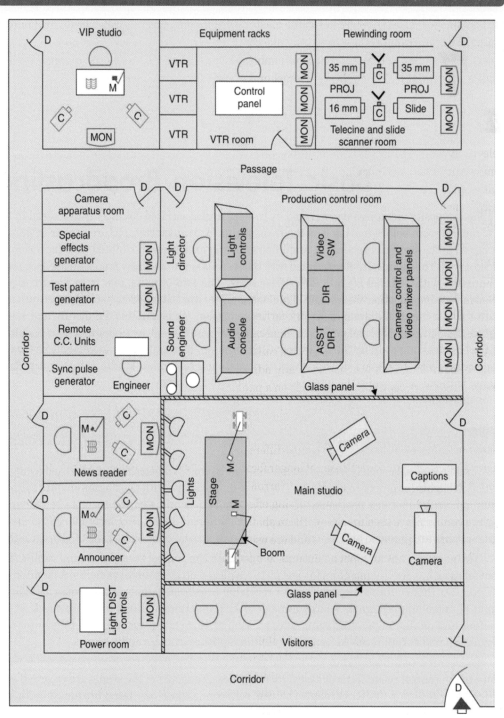

Fig. 7.1 *Plan of a typical television studio (Abbreviations used are :*
MON-Monitor, C-Camera, D-Door, PROJ-Projector, ASST DIR- Assistant Director,
DIR-Director, SW-Switcher, C.C. Units Camera Control Units, M-Microphone,
VTR–Video Tape Recorder, DIST-Distribution).

the transmitter through a switcher in the master control room, which selects one programme at a time. Even in studio set-ups with only one control room, there are several studio rooms, all connected to the same control room. This enables preparation of different programmes in other studios while a programme is being telecast or recorded from one studio.

7.2 TELEVISION CAMERAS

Television cameras may assume different physical and electrical configurations. However, in general they may be divided into two basic groups—self contained cameras and two-unit systems that employ separate camera heads driven by remote camera control equipment located in the central apparatus room (see Fig. 7.1). A self contained camera has all the elements necessary to view a scene and generate a complete television signal. Such units are employed for outdoor locations and normally have a VTR and baby flood lights as an integral part of the televising setup. The remote camera head usually contains only photosensitive pick-up tube, its associated deflection circuitry, video preamplifier and a video monitor. Thus the bulk of the circuitry is contained in the camera control unit, which is connected to the camera head by means of a multiconductor cable. This cable not only carries video, deflection and sync signals but also feeds high voltage supplies necessary for the camera tube. The remote camera control unit contains most of the electrical operating and set-up controls. For this reason, it is usually located near a viewing monitor so that the results of any adjustments may be easily viewed on the monitor screen. All camera controls are available on a panel in the production control room.

Camera Lenses

Television cameras can produce images to different scales depending on the focal length (viewing angle) of the lens employed. Lenses of longer focal length are narrow angle lenses while those of shorter focal length are wide angle lenses. Narrow angle lenses (below 20°) are suitable for closeups of distant objects because of the magnifying effect due to their longer focal length. Lenses with angles over 60° are most suited for location shots which cover large areas. Medium angle lenses (20 to 60°) are called universal lenses and are used for televising normal scenes. All lenses consist of a combination of simple lens elements to minimize spherical aberration and other optical distortions.

Lens Turret

A judicious choice of lens can considerably improve the quality of image, depth of field and the impact which is intended to be created on the viewer. Accordingly a number of lenses with different viewing angles are provided. Their focal lengths are slightly adjustable by movement of the front element of the lens located on the barrel of the lens assembly. This lens compliment of the TV camera is mounted on a turret. The lens turret is screwed in the front of the camera and rotation of the turret brings the desired lens in front of the camera tube. An image orthicon turret assembly holds four lenses of focal lengths 35 mm, 50 mm, 150 mm and a zoom lens of 40 to 400 mm. Figure 7.2 shows such a lens turret mounted in front of a television camera.

Chapter 7

Fig. 7.2. A television camera with lens turret and view finder.

Zoom Lens

A zoom lens has a variable focal length with a range of 10 : 1 or more. In this lens the viewing angle and field view can be varied without loss of focus. This enables dramatic close-up control. The smooth and gradual change of focal length by the cameraman while televising a scene appears to the viewer as it he is approaching or receding from the scene.

The variable focal length is obtained by moving individual lens elements of a compound lens assembly. A zoom lens can in principle simulate any fixed lens which has a focal length within the zoom range. It may, however, be noted that the zoom lens is not a fast lens. The speed of a lens is determined by the amount of light it allows to pass through it. Thus under poor lighting conditions, faster fixed focal length lenses mounted on the turret are preferred.

In many camera units only a zoom lens is provided instead of the turret lens assembly. This alone enables the camera operator to have close-ups, wide coverage of the scene and distant shots without loss of focus. This is particularly so in colour TV cameras where the scene is often well defined and suitably illuminated for proper reproduction of colour details.

Camera Mountings

As shown in Fig. 7.2, studio cameras are mounted on light weight tripod stands with rubber wheels to enable the operator to shift the camera as and when required. It is often necessary to be able to move the camera up and down and around its central axis to pick-up different sections of the scene. In such cases, pan-tilt units may be used which typically provide a 360° rotational capability and allow tilting action of plus or minus 90°. In many applications, primarily closed circuit systems, where it is desirable to be able to remotely move the camera both horizontally and vertically, small servo motors are provided as part of the camera mount. Small motors are also used for remote focusing of the lens unit. In exceptional cases when an overview of a scene is necessary, a remotely controlled camera is hung from the ceiling.

View Finder

To permit the camera operator to frame the scene and maintain proper focus, an electronic view-finder is provided with most TV cameras. This view-finder is essentially a monitor which reproduces the scene on a small picture tube. It receives video signals from the control room stabilizing amplifier. The view-finder has its own deflection circuitry as in any other monitor, to produce the raster. The view-finder also has a built-in dc restorer for maintaining average brightness of the scene being televised.

Studio Lightings

In a television studio it is necessary to illuminate each area of action separately besides providing an average level of brightness over the entire scene. Lighting scheme is so designed that shadows are prevented. As many as 50 to 100 light fittings of different types are often provided in most studios. The light fixtures used include spot lights, broads and flood lights of 0.5 KW to 5 KW ratings. A number of such fittings are suspended from the top so that these can be shifted unseen by the viewer. In big studios catwalks (passages close to the ceiling) are built for ease of changing location of the suspended light fixtures.

The brightness level in different locations of the studio is controlled by varying effective current flow through the corresponding lamps. For a smooth current control, dimmerstats (autotransformers) are used for low rating lamps are silicon controlled rectifiers (SCRs) for higher power lights. The power to all the lines is fed through automatic voltage stabilizers in order to maintain a steady voltage supply. The mains distribution boards and switches are located in a separate room close to the studios. The dimmerstats and other light control equipment is mounted on a separate panel in the programme control room.

Audio Pick-up

The location and placement of microphones depends on the type of programme. For panel discussions, news-reading and musical programmes the microphones may be visible to the viewer and so can be put on a desk or mounted on floor stands. However, for plays and many other similar programmes the microphones must be kept out of view. For such applications these are either hidden suitably or mounted on booms. A microphone boom is an adjustable extended rod from a stand which is mounted on a movable platform. The booms carry microphones close to the area of pick-up but keep them high enough to be out of the camera range. Boom operators manipulate boom arms for distinct sound pick-up yet keeping the microphones out of camera view.

7.3 PROGRAMME CONTROL ROOM

As explained earlier all video and audio outputs are routed through a common control room. This is necessary for a smooth flow and effective control of the programme material. This room is called the Programme Control Room (PCR). It is manned by the programme director, his assistant, a camera control unit engineer, a video mixer expert, a sound engineer and a lighting director. The programme director with the help of this staff effects overall control of the programme while it is telecast live or recorded on a VTR. The camera and sound outputs from the announcer's booth and VIP studios are also routed through the programme control room.

The video and audio outputs from different studios and other ancillary sources are terminated on separate panels in the control room. One panel contains the camera control unit and video

Chapter 7

mixer. In front of this panel are located a number of monitors for editing and previewing all incoming and outgoing programmes. Similarly another panel (see Fig. 7.1) houses microphone controls and switch-in controls of other allied equipment. This panel is under control of the sound engineer who in consulation with the programme director selects and controls all available sound outputs.

Fig. 7.3. A typical programme control room.

The producer and the programme assistant have in front of them a talk-back control panel for giving instructions to the cameramen, boom operator, audio engineer and floor manager. The producer can also talk over the intercom system to the VTR and telecine machine operators. The lighting is controlled by switches and faders from a dimmer console which is also located in the programme control room. Figure 7.3 shows a view of a typical programme control room.

Camera Control Unit (C.C.U.)

The camera control unit has provision to control zoom lens action and pan-tilt movement besides beam focus and brightness control of camera tubes. The C.C.U. engineer manipulates various controls under directions from the producer. In broadcast stations, the video signal must be maintained within very close tolerances of amplitudes. The C.C.U. engineer has the necessary facilities to adjust parameters such as video gain, camera sensitivity, blanking level video polarity etc. For live broadcast of programmes televised far away from the studios, microwave links are used. The modulated composite video signal received over the microwave link is demodulated and processed in the usual manner by the C.C.U. engineer for transmission on the channel allocated to the station.

7.4 VIDEO SWITCHER

A video switcher is a multicontact crossbar switch matrix with provision for selecting any one or more out of a large number of inputs and switching them on to outgoing circuits. The input sources include cameras, VTRs and telecine machine outputs, besides test signals and special effects generators. Thus at this point the programme producer with the assistance of video switcher may select the output of any camera, or mix the output of two or more cameras. Similarly various effects such as fades, wipes, dissolves, supers and so on may be introduced and controlled with a mixer. The results obtained from these switching procedures are quite familiar to any one who has watched commercial television programmes. Through switching, a rather restricted two-dimensional picture presented by one camera can be given additional perspective by changing the display to another camera that views the same scene from a somewhat different angle. Also information being viewed by a number of cameras at various locations can be presented on a single monitor. The ultimate destination of the outputs from the video switcher may be transmitter or a VTR. It could as well as be a string of monitors in a closed circuit television system.

Types of Video Switchers. Broadcast switchers incorporate some method of vertical blanking interval controlled switching. Switching in this manner, during the vertical blanking period, eliminates any visible evidence of switching that might be observed as a disruption during the normal vertical scan. There are three types of video switchers :

(i) *Mechanical Pushbutton Switcher.* In this type the signals are terminated on the actual switch contacts. The bank of switches is interlocked to prevent simultaneous operation. This type of switcher is used primarily for portable field units or in CCTV systems because switching is not frequent and momentary disturbances in the picture during switching can be tolerated.

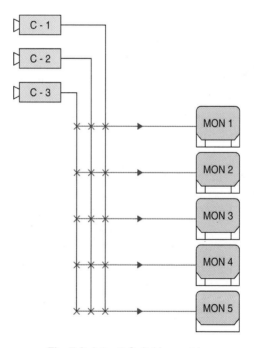

Fig. 7.4. *A 3 × 5 Switching matrix.*

(ii) *Relay Switcher.* The relay switcher or relay cross-bar is an electromechanical switcher. Here magnetically activated read switch contacts are used to effect switching. The relays can be operated by remote control lines. Reed relays have fast operate time (around 1 ms) and so can be used to enable switching during the vertical blanking interval. Figure 7.4 shows a 3 × 5 switching matrix employing a reed relay switcher. This is an example of a distribution-type switcher system where all inputs are available to all monitors. The X's indicate the possibility of combinations that may be achieved. If the crosspoint indicated by the 'X' on the intersection of the lines from camera 3 and monitor 2 were selected, the scene viewed by camera 3 would appear an monitor 2. Similarly any or all of the remaining monitors may be selected to view any camera that is desired. Isolation amplifiers, though not shown, are used in-between the cameras and monitors.

(iii) *Electronic Switcher.* These are all electronic switchers and use solid state devices that provide transition times of the order of a few micro-seconds. Their size is generally very small and due to inherent reliability need much less maintenance. Almost all present day switchers employed in broadcasting are electronic switchers.

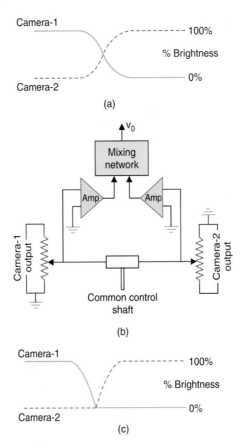

Fig. 7.5. Two types of switching transitions (a) lap-dissolve
(b) circuit for affecting lap-dissolve transition, (c) fade out-fade in transition.

Types of Switching Transitions. The actual switching transition is either carried out by a lap-dissolve operation or a fade out-fade in form of switching. Both methods are illustrated in Fig. 7.5. The lap-dissolve switching (Fig. 7.5(a)) may be accomplished by two potentiometers connected to the two signals that are to be switched. Signal amplitude from say camera number 1 is slowly reduced, while that from camera number 2 is increased at the same rate. This, as shown in Fig. 7.5(b) may be done through variable resistors connected at the inputs of two amplifiers. The values of these potentiometers may be changed as described above to control the gain of amplifiers whose outputs are then combined, Fig. 7.5(c) illustrates the fade out-fade in method of video transfer. It can also be carried out by two potentiometers employed in the lap-dissolve method. However, in this case, by separate actuation of the two potentiometers, signal number 1 is slowly reduced in amplitude until zero signal level is obtained, and then signal No. 2 is slowly raised from 0 to 100 per cent by the second potentiometer. The amplifiers used are commonly called mixers or faders.

Electronic Switcher Configuration

Figure 7.6 is a functional block diagram of very simple broadcast switcher-mixer. It has five inputs out of which any two may be selected to drive the two buffer amplifiers. These, in turn feed into a mixer amplifier.

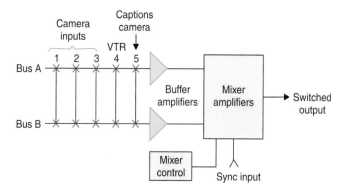

Fig. 7.6. A simple switcher for mixing outputs from two buses.

The mixer transfers video signals by fade out-fade in method. The potentiometers at the remote mixer amplifier can be positioned to select 100 per cent output from either A or B bus. Assume A and B inputs were at 100 per cent and 0 per cent levels respectively. If camera No. 2 is selected on the A bus, it would appear at the output. Similarly, if camera No. 3 is selected on the B bus, it will not appear at the mixer output. However, when the levers that control the potentiometers are moved through their full travel, the output from the mixer amplifier would transfer from A bus to B bus at a relatively slow rate providing a transition from camera No. 2 to camera No. 3. Similarly more complex switchers can be designed to provide different switching matrices.

Special Effects Generator

A special effects generator is normally located along with the camera control units in the camera apparatus room. It is programmed to generate video signals for providing special effects. Its output is available at a panel in the production control room. The special effects signals include curtain moving effects, both horizontal and vertical. These are inserted while changing from one scene to another. Similarly many other patterns are available which can be interposed in-between any two programmes. Infact several options are available and can be selected while ordering the equipment.

7.5 SYNCHRONIZING SYSTEM

To generate a meaningful picture on the raster of a monitor or receiver, some means are needed to synchronize the scanning systems of both the camera and the monitor. In a multicamera system, as is often the case in broadcasting, it is necessary to have them all synchronized by a single sync pulse generator. Accordingly a common sync drive circuitry is provided which controls scanning sequence, insertion and timing of sync pulses in all the cameras. With such a control, when the scene shifts from one camera to another, the synchronizing waveforms are in phase so that the monitor or home receiver is not interrupted in its scanning process. In the absence of such a provision, while switching from one camera to another, the monitor or receiver would have to read just its scanning procedure for the incoming camera and the picture might roll momentarily. Figure 7.7(a) shows one method of driving multiple cameras from a single sync generator. The sync line is terminated in a 75 ohm resistor because the output of most TV camera equipment is designed to work into a 75 ohm load.

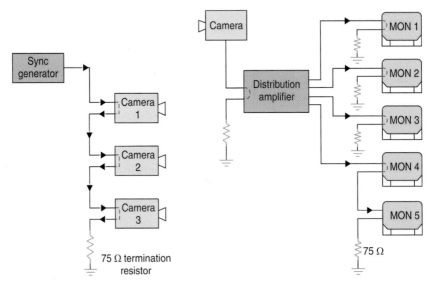

Fig. 7.7(a) A sync generator driving several camera units

Fig. 7.7(b) Connections to several monitors for displaying the output of a single camera.

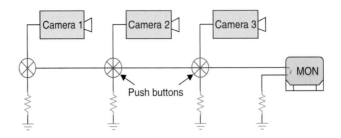

Fig. 7.7(c). Switcher for selecting any camera output to one monitor.

Distribution of Camera Outputs to Monitors

Monitors usually have well designed video amplifiers with bandwidth as large as 30 MHz. This enables excellent reproduction of pictures. Any defects are clearly seen. This is useful while testing and adjusting studio and other allied equipment. Several monitors may be used to display the scene viewed by one camera. When the number of monitors is large and they are located at a considerable distance from each other, a distribution amplifier (see Fig. 7.7(b)) is used to route the video signal to all of them. Similarly it is often desirable to provide means for viewing the output from different cameras on a monitor. This is simply done by selecting the monitor inputs from one camera or another by push button switches as shown in Fig. 7.7(c). The switches not depressed connect terminating resistors to the appropriate cameras. In operation all switches are interlocked so that only one camera can be connected to the monitor at any time. Depressing one switch releases all other.

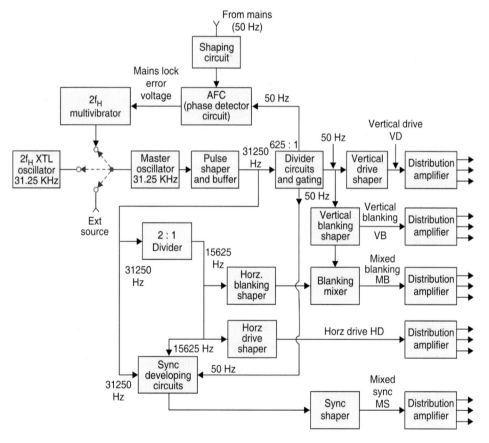

Fig. 7.8. Block diagram of a drive and sync pulse generator (SPG).

Sync Pulse Generation (SPG) Circuitry

Figure 7.8 is a block representation of a drive and sync pulse generator which would provide all the pulse waveforms necessary to meet C.C.I.R. standards. The generator contains (i) a crystal controlled or mains locked timing system, (ii) pulse shapers which generate required pulse trains for blanking, synchronizing and deflection drives and (iii) distribution amplifiers.

A synchronized master oscillator generates an output at $2f_H$, viz., $2 \times 15625 = 31250$ Hz which drives other circuits. As shown in the figure, this oscillator can be synchronized from (i) a crystal controlled oscillator operating at exact $2f_H$, (ii) an external source, or (iii) 50 Hz mains supply. In the mains frequency lock mode, the $2f_H$ multivibrator is controlled by a phase detector circuit which compares the 50 Hz square wave derived from the master oscillator through a divider chain with the 50 Hz square wave derived from mains supply. The error signal which develops at output of the phase detector corrects the frequency of the multivibrator and locks it with the mains frequency. While broadcasting programme received from another TV station, its sync pulses are processed and fed at the 'external source' input to slave the sync circuitry of the station to that of the incoming station.

The master oscillator frequency is twice the horizontal frequency, and is coincident with the frequency of the equalizing and serration pulses which are driven from its output. The buffer amplifier isolates the master oscillator from the rest of the circuitry. The divider and gating block serves two functions. The most important function is to accurately divide the master oscillator frequency by a factor of 625, thus deriving a 50 Hz output that is phase locked with the original 31.25 kHz source. The output is correctly shaped in a shaper circuit before using it to initiate vertical drive and vertical blanking waveforms. The sync developing circuit provides equalizing and serration pulses.

The output from the master oscillator (31.25 kHz) is also fed to a $2:1$ divider to derive output at 15625 Hz, the horizontal frequency. This is used to derive horizontal blanking, horizontal drive and horizontal sync signals in appropriate circuits.

The basic building block necessary for generating and shaping the various sync and drive sources include frequency dividers, pulse shapers or stretchers, delay circuits, adders and logic gates. The circuitry of a modern SPG (sync pulse generator) employs ICs and transistors. This results in a compact, accurate and reliable unit.

The various outputs from the SPG are derived through distribution amplifiers which develop the necessary power and act as buffers between the generation and distribution points.

7.6 MASTER CONTROL ROOM (MCR)

In small broadcasting houses the PCR has a master switcher for routing the composite video signal and allied audio output directly to the transmitter. The ancillary equipment is mostly located in the Central apparatus room. However, in bigger establishments which have a large number of studios and production control rooms, all outputs from various sources are routed through the master control room. This room houses centralized video equipment like sync pulse generators, special effects generator, test equipment, video and audio monitors besides a master routing switcher.

Picture Signal Transmission

At the production control room video signal amplitude as received from the camera is very low and direct coupled amplifiers are used to preserve dc component of the signal. Further on, ac coupling is provided because it is often technically easier and less expensive to use such a coupling. This involves loss of dc component which, however, is reinserted at the transmitter before modulation. This is carried out by a dc restorer circuit often called a blanking level clamp.

In the master control room the composite video signal is raised to about one volt P-P level before feeding it to the cable that connects the control room to the transmitter. Though the transmitter is located close to the studios, often in the same building, matching networks are provided at both ends of the connecting cable to avoid unnecessary attenuation and frequency distortion.

7.7 GENERATION OF AMPLITUDE MODULATION

In AM transmitters where efficiency is the prime requirement, amplitude modulation is effected by making the output current of a class C amplifier proportional to the modulating voltage. This amounts to applying a series of current pulses at the frequency of the carrier to the output tuned (tank) circuit where the amplitude of each pulse follows the variations of the modulating signal. The resonant frequency of the tuned circuit is set equal to the carrier frequency. In the tank circuit each current pulse causes a complete sine wave at the resonant frequency whose amplitude is proportional to the applied current pulse.

The accumulative effect of this flywheel action of the resonant circuit is generation of a continuous sine wave voltage at the output of tank circuit. The frequency of this voltage is equal to carrier frequency having amplitude variations proportional to magnitude of the modulating signal.

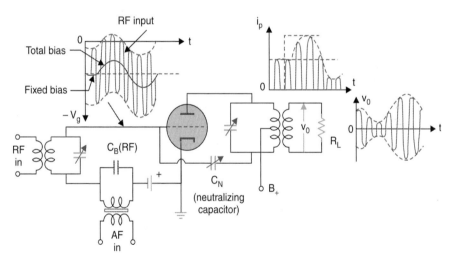

Fig. 7.9. Grid modulated class C amplifier.

In practice AM may be generated by applying the modulating voltage source in series with any of the DC supplies of the class C amplifier. Thus grid (or base), plate (or collector) and cathode (or emitter) modulation are all possible. As an illustration Fig. 7.9 shows the basic circuit and corresponding waveforms of a grid modulated class C amplifier. The modulating voltage is in series with the fixed negative grid bias and so the amplitude of the total bias is proportional to the amplitude of the modulating signal, and varies at the rate of the modulating frequency. Since the carrier RF source is also in series with the bias and modulating voltage, these get superimposed and the total bias appears as shown in Fig. 7.9. The resulting plate current flows in pulses and the amplitude of each pulse is proportional to the instantaneous bias and therefore to the instantaneous modulating voltage. As explained earlier the application of these pulses to the tank circuit then yields amplitude modulation.

In an AM transmitter, amplitude modulation can be generated at any point after the crystal oscillator. If the output stage in the transmitter is plate modulated the method is called high level modulation. If modulation is applied at any other point, including some other electrode of the output amplifier then so-called low-level modulation is produced. The end product, however, of both the system is the same.

It is not practicable to use plate modulation at the output stage in a television transmitter, because of the difficulty of generating high video powers at the large bandwidths required. Accordingly, grid modulation of the output stage is the highest level of modulation employed in TV transmitters. It is called 'high-level' modulation in TV broadcasting and anything else is then called 'low-level' modulation. In recent television transmitter designs, it has become a standard practice to affect modulation in two stages. A carrier frequency of 40 MHz is employed in a balanced modulator configuration and a band ranging from 35 to 45 MHz is obtained as its output. In some designs vestigial sideband correction is also carried out in the modulator output circuit. Its output is then mixed with an appropriate high frequency to get the desired channel carrier frequency and its sidebands as the difference product. This process is illustrated for channel 4 (Band-1) in Fig. 7.10.

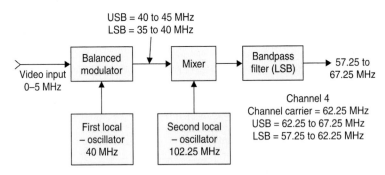

Fig. 7.10. *Modulation and frequency translation.*

7.8　TELEVISION TRANSMITTER

A simplified functional block diagram of a television transmitter is shown in Fig. 7.11. Necessary details of video signal modulation with picture carrier of allotted channel are shown in picture transmitter section of the diagram. Note the inclusion of a dc restorer circuit (DC clamp) before the modulator. Also note that because of modulation at a relatively low power level, an amplifier is used after the modulated RF amplifier to raise the power level. Accordingly this amplifier must be a class-B push-pull linear RF amplifier. Both the modulator and power amplifier sections of the transmitter employ specially designed VHF triodes for VHF channels and klystrons in transmitters that operate in UHF channels.

Vestigial Sideband Filter

The modulated output is fed to a filter designed to filter out part of the lower sideband frequencies. As already explained this results in saving of band space.

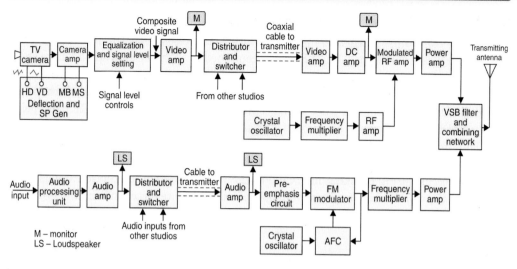

Fig. 7.11. *Simplified block diagram of a television transmitter.*

Antenna

The filter output feeds into a combining network where the output from the FM sound transmitter is added to it. This network is designed in such a way that while combining, either signal does not interfere with the working of the other transmitter.

A coaxial cable connects the combined output to the antenna system mounted on a high tower situated close to the transmitter. A turnstile antenna array is used to radiate equal power in all directions. The antenna is mounted horizontally for better signal to noise ratio.

7.9   POSITIVE AND NEGATIVE MODULATION

When the intensity of picture brightness causes increase in amplitude of the modulated envelope, it is called 'positive' modulation. When the polarity of modulating video signal is so chosen that sync tips lie at the 100 per cent level of carrier amplitude and increasing brightness produces decrease in the modulation envelope, it is called 'negative modulation'. The two polarities of modulation are illustrated in Fig. 7.12.

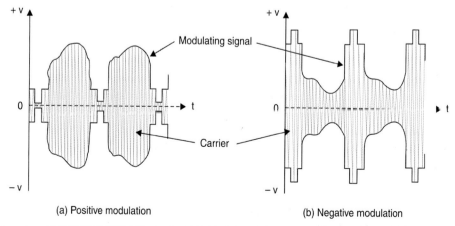

Fig. 7.12. *RF waveforms of an amplitude modulated composite video signal.*

Chapter 7

Comparison of Positive and Negative Modulation

(*a*) *Effect of Noise Interference on Picture Signal.* Noise pulses created by automobile ignition systems are most troublesome. The RF energy contained in such pulses is spread more or less uniformly over a wide frequency range and has a random distribution of phase and amplitude. When such RF pulses are added to sidebands of the desired signal, and sum of signal and noise is demodulated, the demodulated video signal contains pulses corresponding to RF noise peaks, which extend principally in the direction of increasing envelope amplitude. This is shown in Fig. 7.13. Thus in negative system of modulation, noise pulse extends in black direction of the signal when they occur during the active scanning intervals. They extend in the direction of sync pulses when they occur during blanking intervals. In the positive system, the noise extends in the direction of the white during active scanning, *i.e.,* in the opposite direction from the sync pulse during blanking.

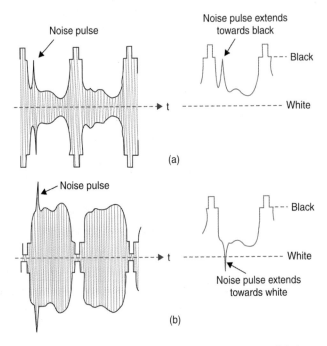

Fig. 7.13. *Effect of noise pulses (a) with negative modulation,*
(b) with positive modulation.

Obviously the effect of noise on the picture itself is less pronounced when negative modulation is used. With positive modulation noise pulses will produce white blobs on the screen whereas in negative modulation the noise pulses would tend to produce black spots which are less noticeable against a grey background. This merit of lesser noise interference on picture information with negative modulation has led to its use in most TV systems.

(*b*) *Effect of Noise Interference on Synchronization.* Sync pulses with positive modulation being at a lesser level of the modulated carrier envelope are not much affected by noise pulses. However, in the case of negatively modulated signal, it is sync pulses which exist at maximum carrier amplitude, and the effect of interference is both to mutilate some of these, and to produce lot of spurious random pulses. This can completely upset the synchronization of the receiver time bases unless something is done about it. Because of almost universal use of negative modulation, special horizontal stabilizing circuits have been developed for use in receivers to overcome the adverse effect of noise on synchronization.

(c) *Peak Power Available from the Transmitter.* With positive modulation, signal corresponding to white has maximum carrier amplitude. The RF modulator cannot be driven harder to extract more power because the non-linear distortion thus introduced would affect the amplitude scale of the picture signal and introduce brightness distortion in very bright areas of the picture. In negative modulation, the transmitter may be over-modulated during the sync pulses without adverse effects, since the non-linear distortion thereby introduced, does not very much affect the shape of sync pulses. Consequently, the negative polarity of modulation permits a large increase in peak power output and for a given setup in the final transmitter stage the output increases by about 40%.

(d) *Use of AGC (Automatic Gain Control) Circuits in the Receiver.* Most AGC circuits in receivers measure the peak level of the incoming carrier signal and adjust the gain of the RF and IF amplifiers accordingly. To perform this measurement simply, a stable reference level must be available in the signal. In negative system of modulation, such a level is the peak of sync pulses which remains fixed at 100 per cent of signal amplitude and is not affected by picture details. This level may be selected simply by passing the composite video signal through a peak detector. In the positive system of modulation the corresponding stable level is zero amplitude at the carrier and obviously zero is no reference, and it has no relation to the signal strength. The maximum carrier amplitude in this case depends not only on the strength of the signal but also on the nature of picture modulation and hence cannot be utilized to develop true AGC voltage. Accordingly AGC circuits for positive modulation must select some other level (blanking level) and this being at low amplitude needs elaborate circuitry in the receiver. Thus negative modulation has a definite advantage over positive modulation in this respect.

The merits of negative modulation over positive modulation, so far as picture signal distortion and AGC voltage source are concerned, have led to the use of negative modulation in almost all TV systems now in use.

7.10 SOUND SIGNAL TRANSMISSION

The outputs of all the microphones are terminated in sockets on the sound panel in the production control room. The audio signal is accorded enough amplification before feeding it to switchers and mixers for selecting and mixing outputs from different microphones. The sound engineer in the control room does so in consultation with the programme director. Some pre-recorded music and special sound effects are also available on tapes and are mixed with sound output from the studio at the discretion of programme director. All this needs prior planning and a lot of rehearsing otherwise the desired effects cannot be produced. As in the case of picture transmission, audio monitors are provided at several stages along the audio channel to keep a check over the quality and volume of sound.

Chapter 7

Preference of FM over AM for Sound Transmission

Contrary to popular belief both FM and AM are capable of giving the same fidelity if the desired bandwidth is allotted. Because of crowding in the medium and short wave bands in radio transmission, the highest modulating audio frequency used in 5 kHz and not the full audio range which extends up to about 15 kHz. This limit of the highest modulating frequency results in channel bandwidth saving and only a bandwidth of 10 kHz is needed per channel. Thus, it becomes possible to accommodate a large number of radio broadcast stations in the limited broadcast band. Since most of the sound signal energy is limited to lower audio frequencies, the sound reproduction is quite satisfactory.

Frequency modulation, that is capable of providing almost noise free and high fidelity output needs a wider swing in frequency on either side of the carrier. This can be easily allowed in a TV channel, where, because of very high video frequencies a channel bandwidth of 7 MHz is allotted. In FM, where highest audio frequency allowed is 15 kHz, the sideband frequencies do not extend too far and can be easily accommodated around the sound carrier that lies 5.5 MHz away from the picture carrier. The bandwidth assigned to the FM sound signal is about 200 kHz of which not more than 100 kHz is occupied by sidebands of significant amplitude. The latter figure is only 1.4 per cent of the total channel bandwidth of 7 MHz. Thus, without encroaching much, in a relative sense, on the available band space for television transmission all the advantages of FM can be availed.

7.11   MERITS OF FREQUENCY MODULATION

Frequency modulation has the following advantages over amplitude modulation.

(a) Noise Reduction

The greatest advantage of FM is its ability to eliminate noise interference and thus increase the signal to noise ratio. This important advantage stems from the fact that in FM, amplitude variations of the modulating signal cause frequency deviations and not a change in the amplitude of the carrier. Noise interference results in amplitude variations of the carrier and thus can be easily removed by the use of amplitude limiters.

It is also possible to reduce noise in FM by increasing frequency deviation. This deviation can be made as large as required without increasing the transmitter power. Higher audio frequencies are mostly harmonics of the lower audio range. They have low amplitudes and so cause a small deviation of the carrier frequency. Noise power interference is also generally low in amplitude and so results in frequency deviation similar to that caused by higher audio frequencies. Thus higher audio frequencies are most susceptible to noise effects. If these frequencies were artificially boosted in amplitude at the transmitter and correspondingly reduced at the receiver, improvement in noise immunity could be expected. This in fact is the standard practice in all FM transmission and reception. In AM on the other hand, the signal modulation can be increased relative to noise modulation only by increasing the transmitter output power. A 20 db improvement in signal-to-noise voltage ratio requires ten-times increase in frequency deviation in FM but an increase of 100 times in AM power output. Evidently an AM system in this respect reaches an economical limit long before the FM system, provided additional bandwidth is available for FM transmission.

In an FM receiver, if two signals are being received simultaneously, the weaker signal will be eliminated almost entirely if it possesses less than half the amplitude of the other stronger signal.

However, in AM the interfering signal or station can be heard or received even when a 100 : 1 relationship exists between their amplitudes.

Pre-emphasis and De-emphasis. The boosting of higher audio modulating frequencies, in accordance with a prearranged response curve is termed pre-emphasis, and the compensation at the receiver is called de-emphasis. Examples of circuits used for each function are shown in Fig. 7.14. As is obvious from these configurations, the pre-emphasis and de-emphasis networks are high-pass and low-pass filters respectively. The time constant of the filter for pre-emphasis at transmitter and later de-emphasis at receiver has been standardized at 50 μs in all the CCIR systems. However, in systems employing American FM and TV standards, networks having a time constant of 75 μs are used. A 50 μs (= RC) de-emphasis corresponds to a frequency response curve that is 3 db down at the frequency given by $\dfrac{1}{2\pi RC}$, which comes to 3180 Hz. Figure 7.15 shows pre-emphasis and de-emphasis curves corresponding to a time constant of 50 μs.

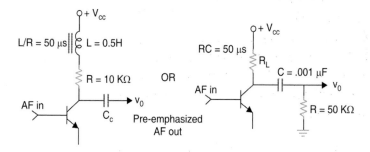

Fig. 7.14(a). Pre-emphasis circuits.

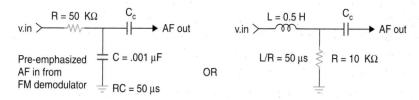

Fig. 7.14(b). De-emphasis circuits.

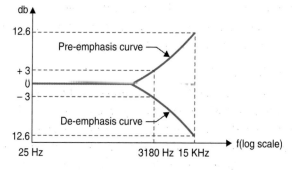

Fig. 7.15. 50 ms emphasis curves.

(b) Transmitter Efficiency

The amplitude of the FM wave is independent of the depth of modulation, whereas in AM it is dependent on this parameter. This means that low level modulation can be used in FM and all succeeding amplifiers can be class 'C' which are more efficient. Thus, unlike AM, all amplifiers handle constant power and this results in more economical FM transmitters.

(c) Adjacent Channel Interference

Because of the provision of a guard band in between any two TV channels, there is less interference than in conventional AM broadcasts.

(d) Co-channel Interference

The amplitude limiter in the FM section of the receiver works on the principle of passing the stronger signal and eliminating the weaker. In this manner, a relatively weak interfering signal or any pick-up from a co-channel station (a station operating at the same carrier frequency) gets eliminated in a FM system.

It may be noted that from general broadcast point of view FM needs much wider bandwidth than AM. It is 7 to 15 times as large as that needed for AM. Besides, FM transmitters and receivers tend to be more complex and hence are expensive. However, in TV transmission and reception, where handling of the picture signal is equally complex, FM sound does not add very much to the cost of equipment.

7.12 GENERATION OF FREQUENCY MODULATION

The primary requirement of an FM generator is a variable output frequency, where the variations are proportional to the instantaneous value of the modulating voltage. In one method of FM generation, either the inductance or capacitance of the tank circuit of an LC oscillator is varied to change the frequency. If this variation can be made directly proportional to the amplitude of the modulating voltage, true FM will be obtained.

A voltage-variable reactance is generally placed across the oscillator tank circuit. The oscillator is tuned to deliver the assigned carrier frequency with the average reactance of the variable element present in parallel with its own tank circuit. The capacitance (or inductance) of the reactance element changes on application of modulating voltage to cause frequency deviations in the oscillator frequency. Larger the departure of modulating voltage from zero, greater is the reactance variation, and in turn higher is the frequency deviation.

Basic Reactance Modulator

An FET, tube or transistor when suitably biased can be used as a variable reactance element. Similarly a varactor diode can also be used for this purpose. The basic circuit arrangements of a reactance modulator either with an FET or with a vacuum tube are shown in Fig. 7.16. Provided certain simple conditions are met, the impedance Z as seen at the terminals marked A – A' in the figures, is almost entirely reactive. The conditions which must be met are, (i) the current i_g should be negligible compared to the plate (or drain) current, and (ii) the impedance $XC >> R$, preferably by more than 5 : 1.

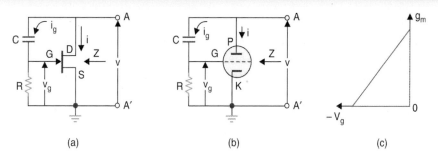

Fig. 7.16. *Basic reactance modulator circuits (a) employing an FET (b) employing a VT (c) relation between bias voltage and transconductance of the device.*

With these assumptions, the following analysis, which is valid for both the circuits can be made : With a voltage v applied across the terminals $A - A'$, currents i and i_g will flow in the plate (or drain) circuit and bias circuit respectively. The current i_g will develop a voltage $v_g = i_g \times R =$

$\dfrac{V \times R}{R - jX_C}$. The current i because of tube or FET action $= g_m v_g = \dfrac{g_m \times R \times V}{R - jX_C}$, where g_m is the

mutual conductance of the device.

$\therefore$
$$Z = \frac{v}{i} = \frac{R - jX_C}{g_m \times R} = \frac{1}{g_m}\left(1 - j\frac{X_C}{R}\right).$$

If $X_C \gg R$, the above equation will reduce to ;

$$Z = -j\frac{X_C}{g_m \times R}$$

This impedance is clearly a capacitive reactance, which may be written as :

$$X_{eq} = \frac{X_C}{g_m R} = \frac{1}{2\pi f g_m RC} = \frac{1}{2\pi f\, C_{eq}}$$

$\therefore$
$$C_{eq} = g_m RC$$

The following conclusions follow from this result :

(*i*) The equivalent capacitance depends on the device transconductance and can therefore be varied with bias voltage. The approximate relation between the bias voltage and g_m is illustrated in Fig. 7.16 (*c*).

(*ii*) The capacitance can be originally adjusted to any value (within reasonable limits), by variation of the components R and C.

(*iii*) $g_m RC$ has the dimensions of capacitance.

(*iv*) From the circuits and analysis made, it is clear that if R and C are interchanged, the

impedance across $A - A'$ will become inductive with $L_{eq} = \dfrac{RC}{g_m}$.

(*v*) Similarly it can be shown that by using L and R instead of C and R in the biasing circuit, both capacitive and inductive reactances can be obtained across the terminals $A - A'$.

Chapter 7

Transistor Reactance Modulator

Figure 7.17 shows a circuit of an LC oscillator that is frequency modulated by a capacitive reactance (RC) transistor modulator. As shown, the audio frequency voltage is applied at the base of the transistor Q_1. The amplitude variations of this driving voltage vary the forward bias to change the transistor collector current. This changes β of the transistor which results in a proportionate change in the equivalent capacitance across the oscillator tank circuit. Note the use of RF chokes in the circuit. They are used to isolate various points of the circuit for ac current while providing a dc path.

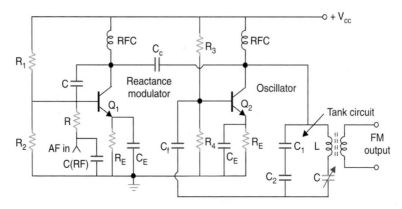

Fig. 7.17. *Reactance modulator circuit.*

Varactor Diode Modulator

The circuit of Fig. 7.18 shows such a modulator. It is seen that the varactor diode has been back-biased to provide the junction capacitance effect. The bias is varied by the modulating voltage which is injected in series with the dc bias source through transformer T_1. The instantaneous changes in the bias voltage cause corresponding changes in the junction capacitance, which in turn vary the oscillator frequency accordingly. It is often used for automatic frequency control and remote tuning.

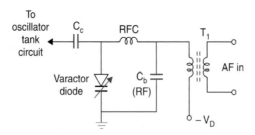

Fig. 7.18. *Varactor diode modulator.*

7.13 STABILIZED REACTANCE MODULATOR

Although the oscillator on which the reactance modulator operates cannot be crystal controlled, it must nevertheless have the stability of a crystal if it is to be a part of a commercial transmitter.

This suggests that frequency stabilization of the reactance modulator is required. The block diagram of Fig. 7.19 shows a typical AFC (automatic frequency control) system used with the reactance controlled FM transmitter. Here a fraction of the output is taken from the limiter and fed to a mixer, which also receives the signal from a crystal oscillator. The resulting difference signal at a frequency which is much less than that of the master oscillator, is amplified and fed to a phase* discriminator. The output of the discriminator which is connected to the reactance modulator, provides a dc control voltage to counteract rapidly any drift in the average frequency of the master oscillator. The time constant of the discriminator load is quite large (of the order of 100 ms). Hence the discriminator will react to slow changes in the incoming frequency, but not to normal frequency changes due to frequency modulation, which are too fast. Note that the discriminator must be connected to give a positive output if the input frequency is higher than the discriminator tuned frequency and a negative output if it is lower. Thus any drift in the oscillator frequency towards the higher side will produce a positive control voltage and this fed in series with the input of the reactance modulator, will increase its trans-conductance. This increases its output capacitance ($C_{eq} = g_m RC$) and thus lowers the oscillator's centre frequency. Analogous sequence of operations will take place to raise the oscillator frequency when it drifts on the lower side of its centre frequency. In the transmitter block diagram of Fig. 7.11 a reactance modulator with such a frequency control has been incorporated.

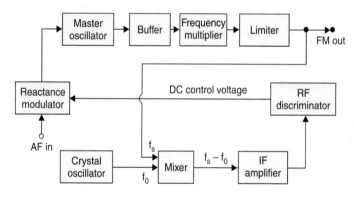

Fig. 7.19. *A typical transmitter AFC system.*

7.14 GENERATION OF FM FROM PM

The direct modulators have the disadvantage of being based on an oscillator which is not stable enough for communication or broadcast purposes. As explained above, it needs stabilization which adds to circuit complexity. Note that the use of a crystal oscillator is not possible because it cannot be successfully frequency modulated. It is possible, however, to generate FM via Phase Modulation (PM) where a crystal oscillator can be used. It has been shown in Chapter 4 that the instantaneous angular frequency for an FM signal $\omega_i = \dfrac{d\phi(t)}{dt} = \omega_c + 2\pi\Delta f \cos \omega_m t$ (eqn. 4.6) for a sinusoidal modulating signal. Conversely an FM signal of amplitude

---

*Phase discriminator circuits are discussed in Chapter 21, which is devoted to FM sound detection.

A is given by $v_{FM(t)} = A \cos \left[\int_{-\infty}^{t} \omega_i \, dt \right]$

$$= A \cos \left[\omega_c t + 2\pi\Delta f \int_{-\infty}^{t} \cos \omega_m t \right]$$

$$= A \cos \left[\omega_c t + \frac{\Delta f}{f_m} \sin \omega_m t \right]$$

In contrast a phase modulated (FM) signal

$$v_{PM(t)} = A \cos \left[\omega_c t + \Delta\phi \cos \omega_m t \right]$$

where both Δf and $\Delta\phi$ are independent of f_m and depend only on the modulating signal amplitude and system constants. Hence the only difference between FM and PM is that if the modulating signal is integrated before performing PM then we get FM. An integrator is a low-pass (bass-boost) circuit. A simple R-L integrator is shown in Fig. 7.20 (a). Here $R/(2\pi L)$ is set at about 30 Hz, so that in the audio range the response falls with frequency at 6 db/decade.

Armstrong FM System

Figure 7.20 (b) shows the functional block diagram of an Armstrong FM system. As explained above the audio voltage enters the modulator, which is essentially a phase modulator, after bass-boosting through an equalizer. The carrier frequency from the crystal oscillator after a phase shift of 90° is fed to a balanced modulator which also receives equalized audio signal. The two sidebands obtained from the balanced modulator are added to the unmodulated carrier in the combining amplifier. The amplitude of carrier voltage obtained from the crystal oscillator through the buffer stage is kept quite large in comparison with that of the sidebands. This as usual, is essential for effective modulation.

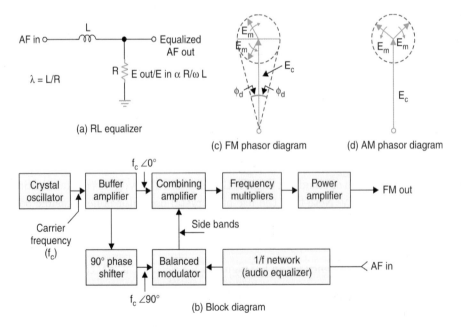

(a) RL equalizer

(c) FM phasor diagram   (d) AM phasor diagram

(b) Block diagram

Fig. 7.20. *The Armstrong frequency-modulation system.*

As shown in the phasor diagram of Fig. 7.20 (c), the two sideband phasors rotate in unison but in opposite direction, and the sum of the carrier and sidebands causes the carrier to change phase by the angle ϕ_d. It will be noted that the phase relation of carrier and side frequencies is at 90° with respect to the AM situation as shown in Fig. 7.20 (d) where only the amplitude of the carrier is varied and not the phase. It may also be noted that the resultant phasor (see Fig. 7.20 (c)) in the case of phase modulation varies in amplitude besides phase deviation and thus a little amplitude modulation is also present in the output. However, the change in amplitude is made very small, though at a price, by making the phase deviation very small. A typical value is $\Delta f =$ 50 Hz at a carrier frequency of 1 MHz corresponding to a maximum phase deviation of 5×10^{-5} radians. For such a small phase deviation, incidental amplitude variation is negligible.

The most convenient crystal oscillator frequency is around 1 MHz, while TV broadcast carrier frequencies are in the region of 50 to 100 MHz. Frequency multipliers are used to raise the carrier frequency. It may be noted that in frequency multiplication, Δf and the carrier frequency get multiplied by the same factor. Usually to achieve the final Δf of ± 75 kHz, a larger multiplying factor is required than that needed to raise the carrier frequency to the required value. This is accomplished by shifting the carrier frequency down by a hetrodyning process at some point within the multiplier chain.

7.15   FM SOUND SIGNAL

As explained in an earlier section, audio signals from different microphones are received at the sound panel in the production control room. After due amplification all the outputs are fed into a switcher, where if necessary they are mixed and the desired output is selected. The final output goes to a distributor in the master control room, where both picture and sound signals from different studios are received. This distributor is switched to select corresponding picture and sound signals from the desired studio. As in case of video signals the audio signals are also routed to the sound transmitter through a cable (see Fig. 7.11) with matching networks on either side.

At transmitter the audio signal is frequency modulated and transferred to assigned channel sound carrier frequency by the use of multipliers. It is later amplified through several stages of power amplifiers to raise the power output to desired level. Audio monitors are provided at various points to keep a check on the sound quality. It is finally fed to the common antenna array through a combining network for radiation along with the modulated picture signal.

Chapter 7

REVIEW QUESTIONS

1. Draw the layout of a typical television studio and explain how the picture and sound signals are processed in the control room. What is the role of a special effects generator ?
2. What is the difference between a self-contained and a two-unit camera system ? What is the function of view finder which is provided at the hood of camera ?
3. Explain how in a multicamera system, synchronization is maintained between the cameras and control monitor. Explain with a functional block diagram how sync and equalizing pulses are generated and kept phase-locked with a common master oscillator.

4. Draw a 3 × 5 switching matrix designed to select output from different cameras on any of the five monitors. Explain how an electronic video switcher accomplishes a smooth change-over from one output to another through a mixer amplifier.

5. Draw a block diagram to show how the video signal is modulated and processed at the picture transmitter. Why is high level modulation not used in a TV transmitter ?

6. Discuss the merits and demerits of positive and negative amplitude modulation and justify the choise of negative modulation in most TV systems.

7. Why is FM chosen for transmission of sound signal in TV systems ? Why are pre-emphasis and de-emphasis circuits provided at the FM transmitter and receiver respectively ?

8. How is frequency modulation produced ? Draw the circuit of a basic reactance modulator and prove that its output reactance varies with changes in the amplitude of the drive voltage.

9. Draw the block diagram of an AFC circuit that forms part of a reactance FM modulator to stabilize the centre frequency of the master oscillator. Explain its control action.

10. Explain briefly how a phase modulator can be used to generate frequency modulation. Draw the block schematic diagram of an Armstrong modulator and explain how an FM output is obtained from it. What is the main merit of this modulator ?

8

Television Receiver

The preceding chapters were devoted to the complexities and essential requirements of generation and transmission of composite video and sound signals associated with the televised scene. As is logical, we should now turn our attention to the receiver. In effect, a television receiver is a combination of an AM receiver for the picture signal and an FM receiver for the associated sound. In addition, the receiver also provides suitable scanning and synchronizing circuitry for reproduction of image on the screen of picture tube. We shall confine our discussion to monochrome (black and white) receivers. The basic principle and essential details of colour receivers are described in Chapters 25 and 26. However, it may be noted that all the circuits for a black-and-white picture are also needed in a colour receiver. The colour television picture is just a monochrome picture with colour added in the main areas of picture information.

8.1 TYPES OF TELEVISION RECEIVERS

The receiver may use tubes for all stages, have all solid-state devices-transistors and integrated circuits, or combine tubes and transistors as a hybrid receiver. A typical chassis of a monochrome receiver is shown in Fig. 8.1.

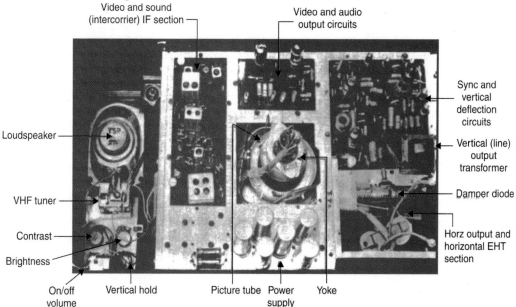

Fig. 8.1. *Rear view of a black and white receiver with the back removed.*

121

(a) *All Tube Receivers.* This type mainly applies to earlier monochrome receivers. All the functions are provided by about 12 tubes including several multipurpose tubes with two or three stages in one glass envelope. The dc supply for tubes is between 140 to 280 volts.

(b) *Solid-State Receivers.* In this type all states except the picture tube use semiconductor diodes, transistors and integrated circuits. The dc supply is between 12 to 100 volts for various stages. The heater power to the picture tube is supplied through a separate filament transformer.

(c) *Hybrid Receivers.* In this type the deflection circuits generally use power tubes, while the signal circuits use transistor and integrated circuits. These receivers usually have a line connected power supply, with series heaters for the tubes. Two dc sources, one for semiconductor devices and the other for tubes are provided.

8.2   RECEIVER SECTIONS

It is desirable to have a general idea of the organization of the receiver before going into circuit details. Figure 8.2 shows block schematic diagram of a typical monochrome TV receiver. As shown there, the receiver has been divided into several main sections depending on their functions and are discussed below.

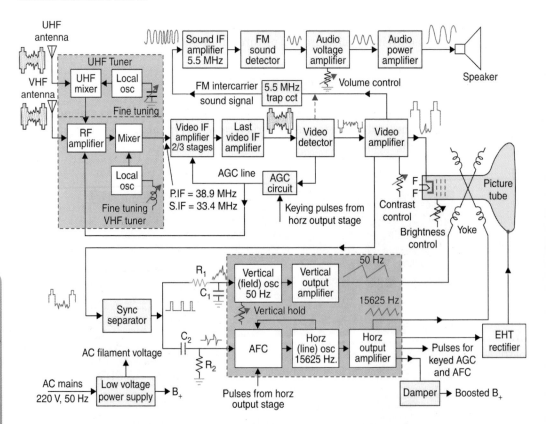

Fig. 8.2. Block diagram of a monochrome television receiver.

Antenna System

Strongest signal is induced in the antenna if it has same polarization as the transmitting antenna. All TV antennas are mounted in horizontal position for better reception and favourable signal to noise ratio. The need for good signal strength has led to the use of tuned antennas. For channels located in the VHF band, a half wave-length antenna is most widely used. Such antennas behave like low 'Q' tuned circuits and a single antenna tuned to the middle frequency of various channels of interest can serve the purpose. Various antennas in use are of dipole type with reflectors and directors. A folded dipole with a reflector is used because its response is more uniform over a band of frequencies. A Yagi antenna, *i.e.*, a dipole with one reflector and two or more directors, is a compact high gain directional array, and is often used in fringe areas. In areas where signal strength is very low, booster amplifiers with suitable matching network are used. On the other hand, in areas situated close to a transmitting antenna, where signal strength is quite high, various types of indoor antennas are frequently employed.

Since it is not possible for one dipole antenna to cover both upper and lower VHF band channels effectively; high and low band dipoles are mounted together and connected to a common transmission line. For channels in the UHF bound, where the attenuation is very high and the signals reaching the antenna are weaker, special antennas like fan dipole, rhombic and parabolic reflector type are often used.

A transmission line connects antenna to the receiver input terminals for the RF tuner. A twin-lead is generally used. This type is an unshielded balanced line with characteristic impedance equal to 300 ohms. When there is a problem of interference, a shielded coaxial cable is used. This cable has high attenuation, especially at UHF channel frequencies. It has a characteristic impedance of 75 ohms.

The current practice is to design input circuit of the TV receiver for a 300 ohm transmission line. It has been found that a 300 ohm transmission line used with a half-wave dipole produces a broad frequency response without too large a loss due to mismatching. A folded dipole has an impedance close to 300 ohms at its resonant frequency, and a much uniform response is obtained with this antenna. Receivers designed to receive UHF channels have two inputs; one to match a 300 ohm transmission line and the other for a 75 ohm coaxial cable. A signal strength of the order of 500 µV to 1 mV and a signal to noise ratio of 30 : 1 are considered adequate for satisfactory reception of both picture details and sound output.

RF Section

This section consists of RF amplifier, mixer and local oscillator and is normally mounted on a separate sub-chassis, called the 'Front End' or 'RF Tuner'. Either tubes or transistors can be used. With tubes, local oscillator and mixer functions are usually combined in one stage called the 'frequency converter'. The purpose of the tuner unit is to amplify both sound and picture signals picked up by the antenna and to convert the carrier frequencies and their associated bands into the intermediate frequencies and their sidebands . The receiver uses superhetrodyne principle as used in radio receivers. The signal voltage or information from various stations modulated over different carrier frequencies is hetrodyned in the mixer with the output from a local oscillator to transfer original information on a common fixed carrier frequency called the intermediate frequency (IF). The setting of the local oscillator frequency enables selection of desired station. The standard intermediate frequencies for the 625-B system are-Picture IF = 38.9 MHz, Sound IF = 33.4 MHz.

Chapter 8

In principle an RF amplifier is not necessary and signal could be fed directly to the tuned input circuit of the mixer. However, the problems of a relatively weak input signal with low signal to noise ratio, local oscillator radiation and image rejection are such, that a stage of amplification ahead of the mixer is desirable. The tuning for different channels is carried out with a channel selector switch which changes resonant frequencies of the associated tuned circuits by varying either inductance or capacitance of these circuits. The RF section is shown separately in Fig. 8.3 (*a*), where the channel selector switch has been set for channel 4 (Band I),

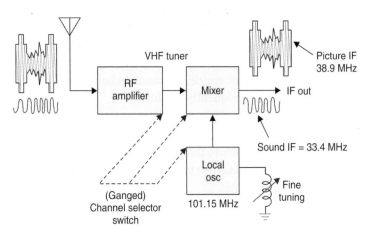

Fig. 8.3 (a). *Block diagram of a VHF tuner. Selector switch set for channel 4 band I (61-68 MHz).*

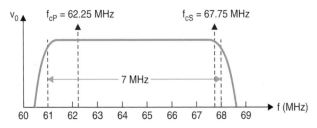

Fig. 8.3 (b). *Ideal response curve of the RF amplifier when set for channel 4.*

i.e., 61 to 68 MHz. The picture carrier frequency in this channel is 62.25 MHz and the sound carrier 67.75 MHz. The RF amplifier must have sufficient bandwidth to accept both the picture and sound signals. This is illustrated in Fig. 8.3 (*b*). The local oscillator frequency is set at 101.15 MHz. In the mixer, both sum and difference (sideband) frequencies are generated. The output circuit of the mixer is however, tuned to deliver difference frequencies *i.e.*, the intermediate frequencies and their sidebands. The required IFs are then produced as here: (Local oscillator frequency of 101.15 MHz)–(Picture carrier frequency of 62.25 MHz) = Picture IF of 38.9 MHz, (Local oscillator frequency of 101.15 MHz)–(Sound carrier frequency of 67.75 MHz) = 33.4 MHz. The desired output response from the mixer is shown in Fig. 8.4. Notice that frequency changing process reverses the relative positions of the sound and picture signals. This is obvious, since the oscillator works above the signal frequencies, and 'difference' frequencies produced, when the picture and sound frequencies are substracted must give a higher IF from the lower frequency picture signal. It may be noted that picture and sound signals would remain in the same relative position, *i.e.*, with sound carrier frequency higher than picture carrier frequency if local oscillator frequency is set below, instead of above the carriers. The local oscillator frequency is kept higher because of ease of oscillator design and several other merits. The ratio of highest to lowest radio frequency that the local oscillator must generate, when the oscillator frequency is chosen to be higher than the incoming carrier frequency, is much less than when the local oscillator frequency is kept below the incoming channel frequency. It is much easier to design an oscillator that maintains almost a constant output amplitude and a sinusoidal waveshape when its overall frequency range is less. This justifies the choice of higher local oscillator frequency.

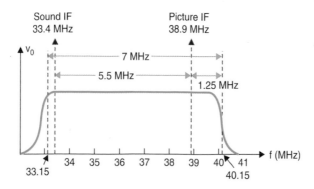

Fig. 8.4. *Location of sound and picture IF frequencies at the output of mixer.*

The tuning of RF and oscillator tuned circuits is pre-set for switching in different channels. Despite the fact that modern tuner units are remarkably stable, most receiver manufacturers provide a fine tuning control for small adjustments of local oscillator frequencies. The control is varied to obtain best picture results on the screen.

IF Amplifier Section

A short length of coaxial cable feeds tuner output to the first IF amplifier. This section is also called video IF amplifier since composite video signal is the envelope of the modulated picture IF signal. Practically all the gain and selectivity of the receiver is provided by the IF section. With tubes, 2 or 3 IF stages are used. With transistors, 3 to 4 IF stages are needed. In integrated circuits, one IC chip contains all the IF amplifier stages.

Essential Functions of the IF Section

The main function of this sections is to amplify modulated IF signal over its entire bandwidth with an input of about 0.5 mV signal from the mixer to deliver about 4 V into the video detector. This needs an overall gain of about 8000. This gain should be adjustable, by automatic gain control, over a wide range to accommodate input signal variations at the antenna from 50 μV to 0.5 V, to deliver about 4 V peak-to-peak signal at the input of the video detector. To achieve desired gain, atleast three stages of tuned amplifiers are cascaded and to obtain desired bandwidth the resonant frequencies of these stages are staggered. Such an arrangement provides desired gain and selectivity.

Chapter 8

8.3 VESTIGIAL SIDEBAND CORRECTION

Another important function assigned to IF section is to equalize the amplitudes of side-band components, because of vestigial sideband transmission. The need for this was fully explained in Chapter 4, and a reference to Fig. 4.8 will show, that for vestigial side-band correction the picture carrier frequency gain must be 6 db down on the IF frequency response curve. It is also necessary to shape the IF response curve around the picture IF frequency in such a way that all lower video frequencies, which got a boost on account of partial lower side-band transmission (besides the full upper side band), are duly attenuated and get restored to their actual level. This is achieved by suitable tuning and shaping the response of the IF stages. This is fully illustrated in Fig. 8.5, which shows ideal overall response of the IF section.

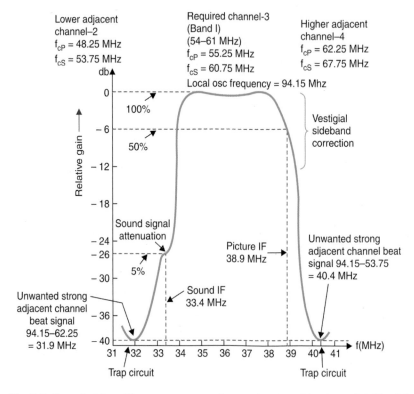

Fig. 8.5. Overall picture IF response curve of a receiver tuned to channel 3-(Band I).
The diagram shows disposition of IF frequencies, vestigial sideband correction,
sound signal attenuation and locations of unwanted adjacent channel
interfering beat frequencies.

In the IF amplifier circuitry, provision must be made for rejection of signals from adjacent channels. For this purpose special tuned circuits, called trap circuits, are connected in the signal path in such a way that the offending frequencies are removed. These trap circuits are disposed at convenient places in the IF amplifiers. Their position will vary from receiver to receiver, but generally they are placed in the input circuit of the first IF amplifier. The way in which these unwanted adjacent channel IF signals appear is illustrated in Fig. 8.5. As an example, suppose that the receiver is switched to channel 3 on Band I. The local oscillator frequency for channel 3 is (55.25 + 38.9) 91.15 MHz. This would beat with the channel picture and sound carrier frequencies to give the desired picture and sound IF frequencies. Besides these, the sound carrier of channel 2, which is close to the beginning of channel 3, will beat with the local oscillator to give unwanted difference frequency of 40.4 MHz (94.15 – 53.75), which would lie close to the upper skirt of desired IF response. Similarly, the picture carrier of upper adjacent channel 4, will also beat with the local oscillator to produce another unwanted difference frequency signal of 31.9 MHz (94.15 – 62.25). This is close to the lower skirt of IF response. The trap circuits are designed to attenuate these two adjacent channel interfering frequencies by about 40 db as shown in the figure. It is understood that such interference would occur only if transmitters operating at channels 2 and 4 are located close to the transmitter operating at channel 3.

8.4 CHOICE OF INTERMEDIATE FREQUENCIES

Since the picture and sound carriers in any channel are spaced by 5.5 MHz, it is natural that the corresponding IF frequencies are also located at the same difference. Accordingly, if the picture IF is fixed at a certain frequency the sound IF automatically gets fixed at a frequency 5.5 MHz less than the picture IF frequency. Therefore we shall refer mostly to picture IF frequency.

The factors which influence the choice of intermediate frequencies in TV receivers are:

(i) Image Rejection Ratio

For a desired input signal at 100 MHz the local oscillator frequency is set at 110 MHz if the IF frequency is fixed at 10 MHz. However, for the same input signal frequency, if the IF frequency is chosen to be 40 MHz, the local oscillator must be set to give an output at 140 MHz. This is shown in Fig. 8.6 (a) and (b). In the first case, if another station is operating at 120 MHz, it will

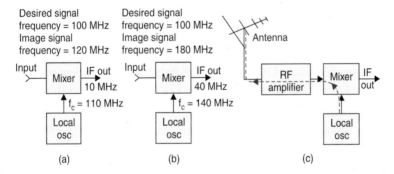

Fig. 8.6(a) and (b). Illustration of image signal interference, (c) Local oscillator signal radiation.

also be received with equal strength because the incoming signal will beat with the local oscillator frequency of 110 MHz to develop output at 10 MHz. Similarly in the second case, a station operating at 180 MHz will be received equally well because the output circuit of mixer is tuned to deliver output at 40 MHz. Note that in each case the undesired signal which gets received is spaced at a gap of twice the IF frequency, and is known as 'Image Signal'. The image rejection ratio is defined as the output due to desired station divided by output due to image signal. Without the use of an RF amplifier prior to the mixer, there is nothing that can stop the reception of image signal if that is present. With RF amplifier the output due to image signal can be very much reduced or completely eliminated. With lower IF frequency, say 10 MHz, the image frequency at 120 MHz is not very far away from desired frequency of 100 MHz and might pass through the pass-band of RF amplifier through somewhat attenuated. But if the IF frequency is kept high, as shown in Fig. 8.6 (b), image signal frequency is 80 MHz away from the desired signal and has no chance of passing through the RF amplifier. Thus the use of RF amplifier helps in reducing interference due to image signals and a higher IF frequency results in a very high image rejection ratio.

By choosing an IF greater than half the entire band to be covered it is possible to eliminate image interference. For the lower VHF band (41 to 68 MHz) the IF frequency comes to 13 MHz. In the upper VHF band (174 to 230 MHz) desired IF frequency is 28 MHz. In the UHF band (470 to 528 MHz), where the image problem is most serious, half of the difference of entire band results in the choice of an IF frequency of 56 MHz. But this is higher than the lowest frequency used in the lower VHF band and because of direct pick-up problems in that band, it cannot be used. Therefore, the IF frequency must be less than 41 MHz.

(ii) Pick-up Due to Local Oscillator Radiation from TV Receivers

If the output from the local oscillator of a TV receiver gets coupled to the antenna, it will get radiated and may cause interference in another receiver. This is shown in Fig. 8.6 (c). Here again advantage lies with higher IF frequency, because with higher IF there is a greater separation between the resonant circuits of local oscillator and RF amplifier circuits. Thus lesser signal is coupled from the local oscillator through the RF amplifier to the antenna circuit and interference due to local oscillator radiation is reduced.

(iii) Ease of Separation of Modulating Signal from IF Carrier at the Demodulator

For ease of filtering out the IF carrier freuency, it is desirable to have a much higher IF frequency as compared to the highest modulating frequency. In radio receivers the IF frequency is 455 KHz and the highest audio frequency is only 5 KHz. In TV receivers, with the highest modulating frequency of 5 MHz, an IF frequency of atleast 40 MHz is desirable.

(iv) Image Frequencies Should Not Lie in the FM Band

The FM band is from 88 MHz to 110 MHz. With IF frequency chosen close to 40 MHz, the image frequencies of the lower VHF band fall between 121 to 148 MHz and thus cannot cause any interference in the FM band. Higher TV channels are much above the FM band.

(v) Interference or Direct Pick-Up from Bands Assigned for other Services

Amateur and industrial applications frequency band lies between 21 to 27 MHz. If the IF frequency is chosen above 40 MHz, even the second harmonics of this band will not cause any serious direct pick-up problems.

(vi) Gain

It is easier to build amplifiers with large gain at relatively low frequencies. The TV sets manufactured some 30 years back used IF frequency as low as 12 MHz. This was mostly due to limitations of active devices available, and the poor quality of components marketed at that time. With the rapid strides made by electronics industry during the past three decades, active devices which can perform very well at high frequencies are now easily available. The quality of components and other techniques have also considerably improved. Thus the gain criteria is no longer a constraint in choosing higher IF frequencies. The merits of having high IF frequency are thus obvious and this has lead to the choice of IF frequencies close of 40 MHz. In the 625-B system adopted by India and several other countries the recommended IF frequencies are : Picture IF = 38.9 MHz, Sound IF = 33.4 MHz. It will be of interest to note that sets manufactured in USA have picture IF = 45.75 MHz and sound IF = 41.25 MHz. In the British 625 line system, because of channel bandwidth difference, the picture IF = 39.5 MHz and sound IF = 33.5 MHz.

Video Detector. Modulated IF signals after due amplification in the IF section are fed to the video detector. The detector is designed to recover composite video signal and to transform the sound signal to another lower carrier frequency. This is done by rectifying the input signal and filtering out unwanted frequency components. A diode is used, which is suitably polarized to rectify either positive or negative peaks of the input signal. Figure 8.7 shows a simplified circuit arrangement of a video detector. Note the use of an L-C filter instead of the usual RC configuration employed in ratio receiver detectors. This is to avoid undue attenuation of the video signal while filtering out carrier components. The video signal shown in Fig. 8.7 is of correct polarity for feeding to the cathode of picture tube after one stage of video amplification.

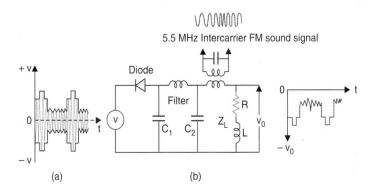

Fig. 8.7. (a) *Last IF amplifier output (modulated IF signal)*
(b) Video detector and sound signal separation circuit.

Video Amplifier. The picture tube needs video signal with peak-to-peak amplitude of 80 to 100 volts for producing picture with good contrast. With an input of about 2 volts from the detector, the video amplifier is designed to have a gain that varies between 40 to 60. A contrast control which essentially is gain control of the video amplifier is provided on the front panel of the receiver to adjust contrast between black and white parts of the picture. A large constrast makes the picture hard, whereas too low a value leaves it weak or soft.

The video amplifier is dc coupled from the video detector to the picture tube, in order to preserve the dc component for correct brightness. However, in some video amplifier designs, on account of complexities of a direct coupled amplifier, ac coupling is instead used. The dc component of the video signal is restored by a diode clamper before feeding it to cathode or grid of the picture tube. In video amplifiers that employ tubes, one stage is enough to provide the desired gain. In transistor amplifier designs, a suitable configuration of two transistors and a driver often becomes necessary to obtain the same gain.

Chapter 8

Besides gain, response of the amplifier should ideally be flat from dc (zero) to 5 MHz to include all essential video components. This needs rigorous design considerations because the band of frequencies to be covered extends from dc through audio range to radio frequencies. A loss in gain of high frequency components in the video signal would reduce sharpness of the picture whereas a poor low frequency response will result in loss of boundary details of letters etc. It is also essential that phase distortion in the amplifer is kept to a minimum. Excessive phase distortion at low frequencies results in smear effect over picture details. Thus the video amplifier of a television receiver needs careful design to achieve desired characteristics. Various wide-banding techniques are employed to extend bandwidth of the amplifier.

8.5 PICTURE TUBE CIRCUITRY AND CONTROLS

The output from the video amplifier may be fed either at the cathode or control grid of the picture tube. In either case a particular polarity of the video signal is essential for correct reproduction of picture details. In most cases cathode drive is preferred. The grid is thus left free to receive retrace blanking pulses to ensure that no retrace lines are seen on the screen for any setting of the brightness control. Figure 8.8 shows the passage of video signal from video detector to the picture tube.

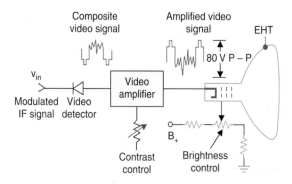

Fig. 8.8. *Passage of video signal from detector to picture tube.*

8.6 SOUND SIGNAL SEPARATION

The picture and sound signals on their respective carriers are amplified together in the IF section. On application of the two signals to the video detector, the picture IF (38.9 MHz) acts as the carrier and beats with the sound carrier (33.4 MHz) and its associated FM side-band frequencies, to produce difference *i.e.,* 5.5 MHz ± 50 KHz components.This is called inter-carrier beat signal and is in effect the second conversion of sound carrier frequency. The resultant product, however, retains its original FM modulation.

If amplitude variations in the FM modulated difference signal of 5.5 MHz is to be avoided to suppress audio signal distortion, the amplitude of sound IF carrier (33.4 MHz) together with its side bands must be attenuated by about 20 db below the picture IF carrier level. This is achieved by shaping the lower skirt of the IF section response in such a way that the sound IF lies at — 26 db (5 per cent of the maximum) on the voltage gain axis of the IF response. This is clearly shown in Fig. 8.5, where a small pedestal has been created at 33.4 MHz to achieve the desired objective. Note that any drift in the local oscillator frequency has no effect on the inter-carrier sound beat frequencies. This is so, because any shift in the local oscillator frequency, shifts both the picture and sound IFs by the same amount. The video detector circuit is modified (see Fig. 8.7) by providing a resonant trap circuit to isolate the sound signal. In some receivers the inter-carrier sound signal is separated after one stage of amplification in the video amplifier.

8.7   SOUND SECTION

As shown in the receiver block diagram (Fig. 8.2), the relatively weak FM sound signal, now on a carrier frequency of 5.5 MHz is given at least one stage of amplification before feeding it to the FM detector. This stage is a tuned amplifier, with enough bandwidth to pass the FM sound signal. This tuned amplifier is known as sound IF. The FM detector is normally a ratio detector or a discriminator preceded by a limiter. The characteristics of a typical FM detector are shown in Fig. 8.9. As shown, the audio output is proportional to deviations from the carrier frequency. The frequency of audio signal depends on the rate of frequency deviation. At the output of FM detector, a de-emphasis circuit is provided that has the same time constant (50 μs) as that of the pre-emphasis circuit employed at the sound transmitter. This restores the amplitude of higher audio frequencies to their correct level. The audio signal receives atleast one stage of amplification before it is passed on to the audio output (power) amplifier. The volume and tone controls form part of the audio amplifiers. These are brought out at the front panel of the receiver. The power amplifier is either a single ended or push-pull configuration employing tubes or transistors. Special ICs have been developed which contain FM demodulator and most parts of the audio amplifier. These are fast replacing discrete circuits hitherto used in the sound section of the receiver. The audio amplifier feeds into one or two loudspeakers provided at a convenient location along front panel of the receiver.

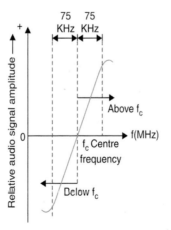

Fig. 8.9. Response curve of an FM detector.

Automatic Gain Control (AGC)

AGC circuit controls gain of RF and IF stages to deliver almost constant signal voltage to the video detector, despite changes in the signal picked up by the antenna. The change in gain is achieved by shifting the operating point of the amplifying devices (tubes or transistors) used in the amplifiers. The operating point is changed by a bias voltage that is developed in the AGC circuit. Any shift in the operating point changes gm (mutual conductance) of the tube or power gain of the transistor circuit which in turn results in change of stage gain.

Sync level in the composite video signal is fixed irrespective of the picture signal details. Hence, sync pulse tops represent truly the signal strength. A peak rectifier is used to develop a control voltage which is proportional to the sync level. The composite video signal to the peak rectifier in the AGC circuit is either obtained from the output of video detector or after one stage of video amplification. The output is filtered and the dc voltage thus obtained is fed to the input (bias) circuits of the RF and IF amplifiers to control their gain. Decoupling circuits are used to avoid interaction between different amplifier stages. AGC is normally not applied to the last IF stage because at that level the signal strength is quite large and any shift in the initially chosen operating point can cause distortion because of partial operation on the non-linear portion of the device characteristics.

Since AGC voltage is proportional to the signal strength, even weak RF signals will also develop some control voltage. This when applied to the RF amplifier will tend to reduce its gain, though the stage should provide maximum possible gain for weak signals to maintain high signal to noise ratio. Therefore, the RF amplifier is not fed any AGC voltage till the signal strength attains a certain predetermined level. This is achieved by providing a voltage delay circuit in the AGC line. Such a provision is known as delayed AGC. The AGC control, as explained above is illustrated in Fig. 8.10 by a block schematic circuit arrangement.

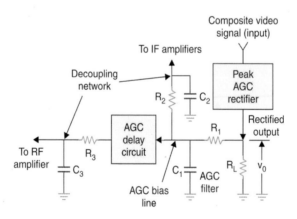

Fig. 8.10. *Block diagram of AGC system.*

Sync Separation

The horizontal and vertical sync pulses that form part of the composite video signal are separated in the sync separator. The composite video signal is either taken from the video detector output or after one stage of video amplification. A sync separator is a clipper that is suitably biased to produce output, only during sync pulse amplitude of the video signal. In some receivers, a noise gate preceds the sync separator. This suppresses strong noise pulses if present in the video signal. A sync pulse train as obtained from a sync separator is shown in Fig. 3.5.

8.8 SYNC PROCESSING AND AFC CIRCUIT

The pulse train as obtained from the sync separator is fed simultaneously to a differentiating and an integrating circuit. The differentiated output (see Fig. 3.5) provides sharp pulses for triggering the horizontal oscillator, while output from the integrator controls the frequency of the vertical oscillator. As explained in Chapter 3, pre and post equalizing pulses ensure identical vertical sync pulses both after the first and second fields.

An integrating circuit is a low-pass filter and hence sharp noise pulses do not appear at its output. However, the differentitator, being a high-pass filter, develops output in response to noise pulses in addition to the spiked horizontal sync pulses. This results in occasional wrong triggering of the horizontal oscillator which results in diagonal tearing of the reproduced picture. To overcome this difficulty, a special circuit known as automatic frequency control (AFC) circuit (Fig. 8.11) is employed. The AFC circuit employs a discriminator arrangement which compares the incoming horizontal sync pulses and the voltage that develops across the output of the horizontal deflection amplifier. The AFC output is a dc control voltage that is free of noise pulses. This control voltage is used to synchronize the horizontal oscillator with the received horizontal sync pulses.

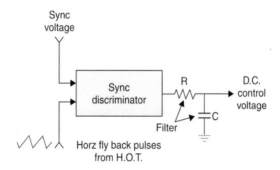

Fig. 8.11. Block diagram of AFC circuit.

8.9 VERTICAL DEFLECTION CIRCUIT

Blocking oscillators or cathode coupled multivibrators are normally employed as vertical deflection oscillators. The controlling time constants are suitably chosen to develop output corresponding to trace and retrace periods. The necessary sawtooth voltage is developed by charging and discharging a capacitor with different time constants. This capacitor forms part of the waveshaping circuit which is connected across the oscillator.

The frequency of the oscillator is controlled by varying the resistance of the RC coupling network and is locked in synchronizm by the vertical sync pulses. A part of the coupling network resistance is a potentiometer that is located on the front panel of the reciver. This is known as 'Vertical Hold Control' (see Fig. 8.2) and enables resetting of the vertical oscillator frequency when it drifts far away from 50 Hz.

The output from the oscillator-cum-waveshaping circuit if fed to a power amplifier, the output of which is coupled to the vertical deflection coils to produce vertical deflection of the beam on picture tube screen.

Chapter 8

8.10 HORIZONTAL DEFLECTION CIRCUIT

The horizontal oscillator (see Fig. 8.2) is similar to the vertical oscillator and is set to develop sweep drive voltage at 15625 Hz. However, the frequency of this oscillator is controlled by dc control voltage developed by the AFC circuit. Since the noise pulses in the control voltage are completely suppressed, most receivers do not provide any horizontal frequency (hold) control, as is normally done for the vertical oscillator. The oscillator output is waveshaped to produce linear rise of current in the horizontal deflection coils. Since the deflection coils need about one amp of current to sweep the entire raster, the output of the oscillator is given one stage of power amplification (as for vertical deflection) and then fed to the horizontal deflection coils.

Low Voltage Power Supply

The usual B + or low voltage power supply is obtained by rectifying and filtering the ac mains supply. If necssary the mains voltage is stepped up or down before rectification. Silicon diodes are normally used for rectification. In some power supply designs, which do not employ a mains transformer, voltage doubler circuits are used to raise the dc voltage. For circuits that employ transistors and integrated circuits, regulated low voltage power supplies are normally provided. While branching the dc supply to various sections of the receiver, decoupling networks are used to avoid any undue coupling between different sections of the receiver. The filament power is supplied by either connecting all the heaters in series across the ac mains or by a low voltage winding on the mains transformer.

High Voltage (EHT) Supply

As already stated in the chapter on picture tubes, an anode voltage of the order of 15 kV is needed for sufficient brightness in black and white picture tubes. This is known as HV or EHT (extra high tension) supply.

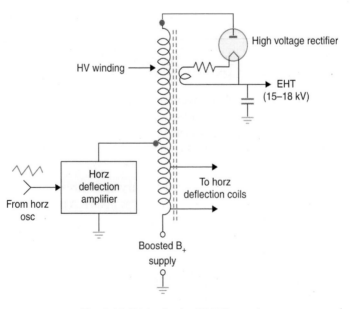

Fig. 8.12. Basic circuit of E.H.T. supply.

Chapter 8

To obtain such a high voltage by stepping up the mains voltage with a transformer is almost impossible and prohibitive in cost. A novel method used for obtaining EHT source is illustrated in Fig. 8.12. During retrace intervals of horizontal scanning, high voltage pulses of amplitude between 6 to 9 kV are developed across the primary winding of the horizontal ouptut transformer. As shown in the figure these are stepped up by an autotransformer winding to about 10 to 15 kV and then fed to a high voltage rectifier. The output of the rectifier is filtered to provide required dc voltage.

Such an arrangement does not load very much the horizontal output stage because the current demand from this high voltage source is less than 1 mA.

The horizontal output circuit is so designed, that in addition to providing EHT source, the energy stored in the horizontal deflection coils during retrace is tapped through a diode called damper diode to charge a capacitor. The voltage thus developed across the capacitor, actually adds 200 to 300 volts to normal B + voltage to give a boosted B + supply of 400 to 700 volts. This voltage is also suitable for first and second anodes of the picture tube. This arrangement makes the horizontal deflection circuit very efficient.

Review Questions

1. Draw block diagram of an RF Tuner (Front End) and explain how incoming signals from different stations are translated to common picture IF and sound IF frequencies. Illustrate your answer by choosing carrier frequencies of any channel in the VHF band.

2. What do you understand by image rejection ratio ? Explain how by providing an RF amplifier, image signal reception is greatly minimized. What are the other merits of using an RF amplifier before the frequency converter ?

3. Describe briefly the factors that influenced the choice of picture IF = 38.9 MHz and sound IF = 33.4 MHz in the 625-B monochrome television system.

4. What are the essential functions which are assigned to IF section of the receiver ? Show by sketching output voltage verses frequency response of the IF section, how vestigial sideband correction is carried out. Why is the sound signal amplitude attenuated to about 5 per cent of the maximum output voltage ?

5. Explain how composite video signal is detected ? How is the polarity of the video output signal decided ? Why is it dependent on the number of video amplifier stages ?

6. What do you understand by intercarrier sound system ? Explain why any shift in the local oscillator freuency does not effect the frequency of the intercarrier beat signal. Where and how is the intercarrier sound signal separated from the video signal ?

7. Draw a block diagram of the sound channel in a TV receiver. Explain briefly how the inter-carrier sound signal as obtained at the video detector, is processed to produce sound output. Why is a de omphasis circuit provided after the FM detector ?

8. Explain briefly how sync pulses are separated from the composite video signal and processed to synchronize the vertical and horizontal oscillators.

9. Describe briefly how EHT and boosted B + voltages are developed from the horizontal output circuit of the sweep amplifier.

10. Draw block diagram of a monochrome TV receiver and label its various sections Indicate by waveshapes the nature of signal at the input and output of each block of the receiver.

Chapter 8

9

Television Signal Propagation and Antennas

9.1   RADIO WAVE PROPAGATION

Radio waves are electromagnetic waves, which when radiated from transmitting antennas, travel through space to distant places, where they are picked up by receiving antennas. Although space is the medium through which electromagnetic waves are propagated, but depending on their wavelengths, there are three distinctive methods by which propagation takes place. These are: (*a*) ground wave or surface wave propagation, (*b*) sky wave propagation, and (*c*) space wave propagation.

(a) Ground Wave Propagation

Vertically polarized electromagnetic waves radiated at zero or small angles with ground, are guided by the conducting surface of the ground, along which they are propagated. Such waves are called ground or surface waves. The attenuation of ground waves, as they travel along the surface of the earth is proportional to frequency, and is reasonably low below 1500 kHz. Therefore, all medium wave broadcasts and longwave telegraph and telephone communication is carried out by ground wave propagation.

(b) Sky Wave Propagation

Ground wave propagation, above about 1600 kHz does not serve any useful purpose as the signal gets very much attenuated within a short distance of its transmission. Therefore, most radio communication in short wave bands up to 30 MHz (11 metres) is carried out by sky waves. When such waves are transmitted high up in the sky, they travel in a straight line until the ionosphere is reached. This region which begins about '120 km above the surface of the earth, contains large concentrations of charged gaseous ions, free electrons and neutral molecules. The ions and free electrons tend to bend all passing electromagnetic waves. The angle by which the wave deviates from its straight path depends on (*i*) frequency of the radio wave (*ii*) angle of incidence at which the wave enters the ionosphere (*iii*) density of the charged particles in the ionosphere at the particular moment and (*iv*) thickness of the ionosphere at the point. Figure 9.1 illustrates the path of several waves entering the ionosphere at different incident angles. With increase in frequency, the allowable incident angle at the ionosphere becomes smaller until finally a frequency is reached, when it becomes impossible to deflect the beam back to earth. For ordinary ionospheric conditions this frequency occurs at about 35 to 40 MHz. Above this frequency, the sky waves cannot be used for radio communication between distant points on the earth. This is why no frequencies beyond about 30 MHz (11 metres) are allotted for radio communication.

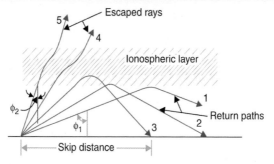

Fig. 9.1. *Ray paths for different angles of incidence (f) at the ionosphere.*

(c) Space Wave Propagation

As explained above, propagation of radio waves above about 40 MHz (which is the beginning of television transmission band) is not possible through either surface or sky wave propagation. Thus, the only alternative for transmission in the VHF and UHF bands, despite large attenuation, is by radio waves which travel in a straight line from transmitter to receiver. This is known as space wave propagation. Its maximum range, because of the nature of propagation, is limited to the line of sight distance between the transmitter and receiver.

For not too large distances, the surface of the earth can be assumed to be flat and different rays of wave propagation can reach the receiver from transmitter as shown in Fig. 9.2(a). As seen there, h_t and h_r, are the heights of transmitting and receiving antennas respectively, and d is the distance that separates them from each other. Both the direct wave AB and reflected wave ACB contribute to the field strength at the receiving antenna. Assuming the earth's surface to be perfectly reflecting, the total field strength E, due to both direct and reflected waves, for reasonably large value of d can be expressed as:

$$E^* = \frac{4\pi f h_t h_r}{d^2} E_0$$

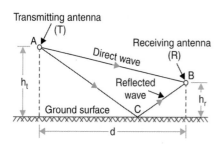

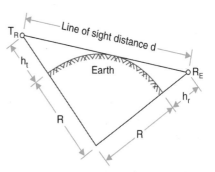

Fig. 9.2(a). *Space wave propagation. For the sake of clarity, the antenna heights have been greatly exaggerated in comparison with the distance between them.*

Fig 9.2(b). *Computation of line-of-sight distance. The height of antennas has been greatly exaggerated in comparison with R, the radius of earth.*

$$*E = \frac{2E_0}{d} \sin \frac{2\pi f h_t h_r}{d}$$ but when d is large, as is often the case, the sine of the angle can be replaced by

the angle and thus the above expression becomes

$$E = \frac{4\pi f h_t h_r}{d} E_0$$

where E_0 is the field strength at unit distance from the transmitter, and f is the frequency of the transmitted signal. The field strength varies inversely as the square of the distance between the two antennas but is directly proportional to their heights.

Various Aspects of Space Wave Propagation

(i) *Effect of Earth's Curvature.* Earth's curvature limits the maximum distance between the transmitting and receiving antennas. This is depicted in Fig. 9.2(b). The maximum line of sight distance d between the two antennas can be easily found out. Neglecting $(h_r)^2$ and $(h_t)^2$, being very small as compared to R, the radius of the earth, the line-of-sight distance $d \approx 4.22(\sqrt{h_t} + \sqrt{h_r})$ km.

where h_t and h_r are expressed in metres. In reality etectromagnetic waves are bent slightly as they glide along the surface of the earth and this increases the line-of sight distance by a small amount. It is evident that the ground coverage will increase with increase in height of both transmitting and receiving antennas. It is due to this reason that television transmitting and receiving antennas are placed as high as possible, for example atop tall buildings and on high plateaus. Another advantage is that the local noise disturbance pick-up is reduced by placing the antennas at high altitudes.

(ii) *Effect of Atmospherics.* The presence of gas molecules and water vapour affects the dielectric constant and hence the refractive index of the atmosphere. As a result, the space waves are differently refracted or reflected under varying conditions of atmosphere. This under certain conditions enables the propagation to reach points very much beyond the line of sight. Similarly under adverse weather conditions the signal attenuation increases, thereby reducing effective distance of transmission.

Occasionally the concentration of charged particles in the ionosphere increases sharply and it becomes possible for waves up to 60 MHz to return to earth. Though this enhances the range of sky wave propagation, but the exact time and place of occurrence of such phenomena cannot be predicted. Thus this phenomenon has little value for commercial operation, but does explain to some extent the distant reception of high frequency and TV signals, which occurs at times under unusual conditions.

(iii) *Effect of Obstacles.* Tall and massive objects like hills and buildings, will obstruct surface waves, which travel close to ground. Consequently, shadow zones and diffraction will result. For this reason in some areas antennas higher than those indicated by theoretical expressions are needed. On the other hand, some areas receive such signals by reflection only. Again, in some areas strong reflected signals are received besides direct signals. This can cause a form of interference known as 'ghosting' on the screen of a television receiver because of a phase difference between the two signals.

9.2 TELEVISION SIGNAL TRANSMISSION*

The RF carrier power output of commonly used VHF television transmitters varies from 10 to 50 kW. A satisfactory level of signal strength is said to exist when the image produced on the screen

---

*Television broadcast channels are given in Appendix C.

of the receiver overrides noise by an acceptable margin. Signal strength is a function of power radiated, transmitting and receiving antenna heights, and the terrain above which the propagation occurs. The acceptable signal to noise ratio at the picture tube screen is measured in terms of peak-to-peak video signal voltage (half tone), injected at the grid or cathode of the picture tube versus the r.m.s. random noise voltage at that point. A peak signal to r.m.s. noise ratio of 45 db is generally considered adequate to produce a good quality picture.

Field strength is indicated by the amount of signal received by a receiving antenna at a height of 10 metres from ground level, and is measured in microvolts per metre of antenna dipole length. The field strength for very good reception in thickly populated and built-up areas is 2500 µV/ metre for channels 2 to 4 (47 to 68 MHz), and 3550 µV/metre for channels 5 to 11 (174 to 223 MHz). For channels in the UHF band, a field strength of about 5000 µV/metre becomes necessary. This is so because of the lower sensitivity of the receiver for higher channels.

Range of Transmission

A sample calculation shows that for a transmitting antenna height of 225 metres above ground level the radio horizon is 60 km. If the receiving antenna height is 16 metres above ground level the total distance is increased to 76 km. Greater distance between antennas may be obtained by locating them on top of very tall buildings or hillocks. However, links longer than 120 km are hardly ever used for TV transmission because of limitations of radiated power, high channel frequencies and antenna heights. Thus, depending on the transmitter power and other factors the service area may extend up to 120 km for the channels in the VHF band but drops to about 60 km for UHF channels.

Booster Stations

Some areas are either shadowed by mountains or are too far away from the transmitter for satisfactory television reception. In such cases booster stations can be used. A booster station must be located at such a place, where it can receive and rebroadcast the programme to receivers in adjoining areas. Mussoorie (U.P.) is one such booster station. Its receiving and transmitting antennas are located on top of a hill. The station receives Delhi TV station (channel 4) programmes and relays them in channel 10 to the surrounding areas and regions on the other side of the hills.

9.3 INTERFERENCE SUFFERED BY CARRIER SIGNALS

In addition to thermal and man-made noise, the carrier signal must compete with various other forms of interfering signals originating from other television stations, radio transmitters, industrial radiating devices and TV receivers. When the interfering signal has a frequency that lies within the channel to which a TV receiver is tuned, the extent of interference depends only on the relative field strengths of the desired signal and the interfering signal. If the interfering signal frequency spectrum lies outside the desired channel, selectivity of the receiver aids in rejecting the interference.

(a) Co-channel Interference

Two stations operating at the same carrier frequency, if located close by, will interfere with each other. This phenomenon which is common in fringe areas is called co-channel interference. As the two signal strengths in any area almost equidistant from the two co-channel stations become

Chapter 9

equal, a phenomenon known as 'venetian-blind' interference occurs. This takes the form of horizontal black and white bars, superimposed on the picture produced by the tuned channel. These bars tend to move up or down on the screen. As the strength of the interfering signal increases, the bars become more prominent, until at a signal-to-interference ratio of 45 db or so, the interference becomes intolerable. The horizontal bars are a visible indication of the beat frequency between the two interfering carriers. Figure 9.3 shows the bar pattern that appears on the screen. The frequency of the beat note, which is equal to frequency separation between the two carriers, is usually of the order of a few hertz. This is so because most transmitters operate almost at the correct assigned frequencies. Motion of the bars, upwards or downwards occurs whenever the beat frequency is not an exact multiple of the field frequency. Co-channel interference was a serious problem in early days of TV transmission, when the channel allocation was confined to VHF band only. This necessitated the repetition of channels at distances not too far from each other. Now, when a large number of channels in the UHF band are available such a problem does not exist. The sharing of channel numbers is carefully planned so that within the 'service area' of any station, signals from the distant stations under normal conditions of reception are so weak as to be imperceptible. However, during a period of abnormal reception conditions (often during spring) when the signals from distant VHF stations are received much more strongly, co-channel interference can occur in fringe areas. The use of highly directional antennas is very helpful in eliminating co-channel interference.

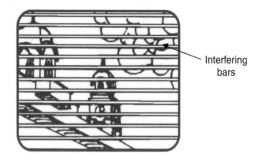

Fig. 9.3. Venetian-blind interference caused by beat frequency between picture carriers of co-channels.

(b) Adjacent Channel Interference

Stations located close by and occupying adjacent channels, present a different interference problem. Adjacent channel interference (see Fig. 8.5) may occur as a result of beats between any two of these frequencies or between a carrier and any sidebands. A coarse dot structure is produced on the screen if picture carrier of the desired channel beats with sound carrier of the lower adjacent channel. The beat pattern is more pronounced if the lower adjacent sound carrier is relatively strong and is not sufficiently attenuated in the receiver.

The next most prominent source of interference is the one produced by picture sideband components of the upper adjacent channel. The beat frequency between adjacent picture carrier is 7 MHz. Since this is far beyond the video frequency range, the resultant beat pattern is not discernible. However, the picture sidebands of the upper adjacent channel may beat with the desired channel carrier and produce an interfering image. To prevent adjacent channel interference, several sharply tuned band eliminator filters (trap circuits) are provided in the IF section of the receiver. This was explained in Chapter 8 while discussing desired IF response characteristics of the receiver. In addition to this, the guard band between two adjacent channels also minimizes the intensity of any adjacent channel interference. A space of about 150 km between adjacent channel stations is enough to eliminate such an interference and is normally allowed.

(c) Ghost Interference

Ghost interference arises as a result of discrete reflections of the signal from the surface of buildings, bridges, hills, towers etc. Figure 9.4 (a) shows paths of direct and reflected electro-magnetic waves from the transmitter to the receiver. Since reflected path is longer than the direct path, the reflected signal takes a longer time to arrive at the receiver.

The direct signal is usually stronger and assumes control of the synchronizing circuitry and so the picture, due to the reflected signal that arrives late, appears displaced to the right. Such displaced pictures are known as 'trailing ghost' pictures. On rare occasions, direct signal may be the weaker of the two and the receiver synchronization is now controlled by the reflected signal. Then the ghost picture, now caused by direct signal, appears displaced to the left and is known 'as leading ghost' picture. Figure 9.4 (b) shows formation of trailing and leading ghost pictures on the receiver screen.

Fig. 9.4 (a). Geometry of multiple path transmission.

Fig. 9.4 (b). Ghost interference.

The general term for the propagation condition which causes ghost pictures is 'multipath transmission'. Ghost pictures are particularly annoying when the relative strengths of the two signals, vary such, that first one and then the other assume control of the receiver synchronism. In such cases the ghost image switches over from a leading condition to a trailing one or vice-versa at a very fast rate. The effect of such reflected signals (ghost images) can be minimized by using directional antennas and by locating them at suitable places on top of the buildings.

9.4 PREFERENCE OF AM FOR PICTURE SIGNAL TRANSMISSION

At the VHF and UHF carrier frequencies there is a displacement in time between the direct and reflected signals. The distortion which arises due to interference between multiple signals is more objectionable in FM than AM because the frequency of the FM signal continuously changes. If FM were used for picture transmission, the changing best frequency between the multiple paths, delayed with respect to each other, would produce a bar interference pattern in the image with a shimmering effect, since the bars continuously change as the beat frequency changes. Hence, hardly any steady picture is produced. Alternatively if AM were used, the multiple signal paths can atmost produce a ghost image which is steady. In addition to this, circuit complexity and bandwidth requirements are much less in AM than FM. Hence AM is preferred to FM for broadcasting the picture signal.

9.5 ANTENNAS

In the preceding sections of this chapter, while explaining various methods of propagation, it was taken for granted that transmitters can somehow transmit and receivers have some means of receiving what is transmitted. Actually a structure must be provided, both for effective radiation of energy at the transmitter and efficient pick up at the receiver. An antenna is such a structure. It is generally a metallic object, often a rod or wire, that is used to convert high-frequency current into electromagnetic waves, and vice versa. Though their functions are different, transmitting and receiving antennas behave identically.

Radiation Mechanism

An antenna may be thought of as a short length of a transmission line. When high frequency alternating source is applied at its one end, the resulting forward and reverse travelling waves combine to form a standing wave pattern on the line. However, all the forward energy does not get reflected at the open end, and a small portion escapes from the system and is thus radiated.

The electromagnetic radiation from the transmitting antenna has two components—a magnetic field associated with current in the antenna and an electric field associated with the potential. The two fields are perpendicular to each other in space and both are perpendicular to the direction of propagation of the wave. This is illustrated in Fig. 9.5. An electromagnetic wave is horizontally polarized if its electric field is in the horizontal direction. Thus an antenna fixed horizontally produces horizontally polarized waves. Similarly a vertical antenna produces vertically polarized waves.

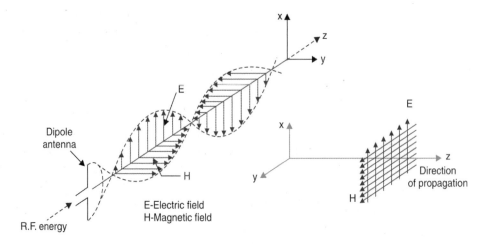

Fig. 9.5. Transverse electromagnetic wave in free space.

The amount of energy that is radiated in space by a transmission line antenna is however, extremely small, unless the wires of the line are suitably oriented and their lengths made comparable to the wavelength. If the two wires of the transmission line are opened up, there is less likelihood of cancellation of radiation from the two-wire tips and this improves the efficiency of radiation. This type of radiator is called a dipole. When total length of the two wires is half-wavelength, the antenna is called a half-wave dipole. Figure 9.6 shows the evolution of such an antenna from the basic transmission line. As shown there, the antenna is effectively a piece of quarter-wavelength transmission line bent out and open circuited at the far ends. Such a length has low impedance at the end connected to the main feeder transmission line. This in turn means that a large current will flow at the input of the half-wave dipole and efficient radiation will take place.

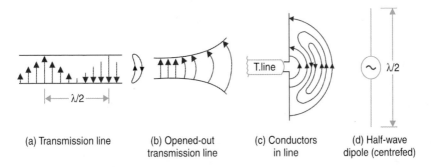

(a) Transmission line   (b) Opened-out   (c) Conductors   (d) Half-wave
                                transmission line   in line   dipole (centrefed)

Fig. 9.6. Evolution of the dipole.

With the help of Maxwell's equations it is possible to deduce expressions for the energy radiated by an antenna, the direction or directions in which it propagates and the field strength at any distance from it.* The results show that the field strength depends on the power transmitted and is inversely proportional to the distance from the radiating source. The coefficient of current (I^2_{rms}) in the expression for the radiated power has the dimensions of resistance and is called radiation resistance. In effect, radiation resistance is the equivalent resistance which dissipates the same amount of power as that radiated from the antenna when same current flows through them. For a quarter-wave antenna the radiation resistance is 36.5 ohms, and for a half-wave antenna it is 73 ohms.

Radiation Patterns of Resonant Antennas

A resonant antenna is a transmission line whose length is an exact multiple of wavelengths and is open at both ends. The current distribution and radiation patterns of such resonant wires of different wavelengths which are remote from the ground are shown in Figs. 9.7 and 9.8. As seen there, for a λ/2 dipole the radiation is maximum at right angles to it, and eventually falls to zero in line with the antenna. This may be explained by considering that at right angles to the short length of the antenna the distance from a remote point to any part of the antenna wire is practically the same. Thus reinforcement of radiation will take place in this direction.

---

*Such analysis is beyond the scope of this book.

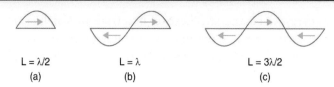

$L = \lambda/2$
(a)

$L = \lambda$
(b)

$L = 3\lambda/2$
(c)

Fig. 9.7. *Current distribution on resonant dipoles.*

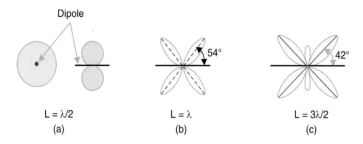

Dipole

$L = \lambda/2$
(a)

$L = \lambda$
(b)

$L = 3\lambda/2$
(c)

Fig. 9.8. *Radiation patterns (Polar diagrams) of various resonant dipoles located remote from ground.*

When the distant point lies in a direction other than normal, there will be some cancellation and finally full cancellation will take place at points that are in line with the antenna. Thus the radiation pattern cross section, as presented in Fig. 8.8 (*a*) is a figure of eight with its axis at right angles to the dipole. Moreover, exactly the same radiation pattern will exist in any other plane, and so the three-dimensional pattern is the surface of revolution obtained by rotating the cross section about an axis coinciding with the dipole. For an antenna of length equal to a whole wavelength the polarity of current, (as shown in Fig. 9.7 (*b*)) on one half of the antenna is opposite to that on the other half. As a result, the radiation at right angles from this antenna will be zero because the field due to one half fully cancels the field due to the other half of the antenna. The direction of maximum radiation exists at 54° to the antenna. The pattern acquires lobes and there are four such for this antenna. As the length of the dipole is increased to three half-wavelengths, the current distribution is changed to that of Fig. 9.7(*c*) and the radiation pattern takes the shape shown in Fig. 9.8 (*c*). As the length of the aerial wire is further increased, the number of lobes in the radiation pattern increases, but the angle of largest lobe with the direction of antenna decreases.

Nonresonant Antennas

A nonresonant antenna (Fig. 9.9(*a*)) is one which is correctly terminated and as such only a forward travelling wave exists and there are no standing waves. As shown in Fig. 9.9(*b*) the

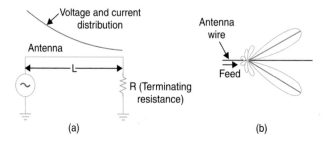

Voltage and current distribution

Antenna

L

R (Terminating resistance)

Antenna wire

Feed

(a)

(b)

Fig. 9.9. *Nonresonant antenna (a) layout and current distribution (b) typical directional radiation pattern for L = 4λ.*

radiation pattern, though similar to the corresponding resonant antenna, is unidirectional. In fact there are half the number of lobes compared to the resonant antenna. This is due to absence of the reflected wave, which otherwise combines vectorially with the forward wave to create the radiation pattern.

Ungrounded Antennas

When the antenna is very close to the ground, its radiation pattern gets modified on account of reflections from the ground. If the ground is assumed to be a perfect conductor, a true mirror image of the actual antenna is considered to exist below the ground. The overall radiation pattern is then the sum of patterns caused by an array of two nearby antennas. Some typical radiation patterns for various heights above ground are shown in Fig. 9.10.

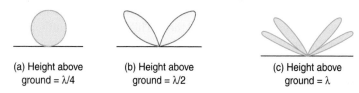

(a) Height above   (b) Height above   (c) Height above
ground = $\lambda/4$ ground = $\lambda/2$ ground = λ

Fig. 9.10. Vertical radiation patterns of an ungrounded half-wave horizontal dipole with varying heights above the ground surface.

Grounded Antennas

When one end of the antenna is actually grounded, the image of the antenna behaves as if it has been joined to the physical antenna and the combination acts as an antenna of double the size. The current distribution and radiation patterns of different earthed vertical antennas are shown in Fig. 9.11. As shown there, the voltage and current distribution on such a grounded $\lambda/4$ antenna, commonly known as the basic Marconi antenna, is the same as those of the ungrounded half-wave Hertz antenna. As is obvious Marconi antenna needs half the length as compared to Hertz antenna to produce the same radiation pattern.

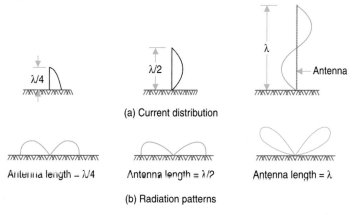

(a) Current distribution

Antenna length – $\lambda/4$ Antenna length = $\lambda/2$ Antenna length = λ

(b) Radiation patterns

Fig. 9.11. Current distribution and vertical directional patterns of grounded antennas.

Antenna Gain

As explained above, all practical antennas concentrate their radiation in some direction, to a greater or lesser extent. Thus the field (power) density in the direction is greater than what it would have been if the antenna were omnidirectional. This may be interpreted that the antenna has a gain in a particular direction. The directive gain is thus defined as the ratio of the power density in the direction of maximum radiation to the power density that would be radiated by an isotropic antenna. The gain being a ratio of powers is expressed in decibels.

Antenna Arrays

It is clear from previous discussion that radiation from different types of antennas is not uniform in all directions. Though an antenna can be suitably oriented to get maximum response in any desired direction, additional directive gain in preferred directions can be obtained by using more than one radiator arranged in a specific manner in space. Such arrangements of radiators are known as antenna arrays. The simplest type of array consists of two antennas A_1 and A_2 separated by a distance d. A special case of directivity is obtained when $d = \lambda/4$ and the currents in the two antennas have a phase difference of 90° between them. This results in a cardioid shaped directional pattern as shown in Fig. 9.12.

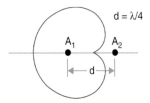

Fig. 9.12. Cardioid shaped directional pattern formed by parallel half-wave antennas.
A_1 and A_2 are the locations of two antennas with currents 90° out of phase.

A broadside array consists of a number of identical radiators equally spaced along a line and carrying same amount of current in phase from the same source. As indicated in Fig. 9.13(a), this arrangement is strongly directional at right angles to the plane of the array.

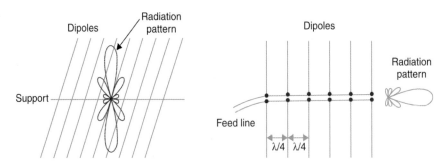

Fig. 9.13 (a). Broadside array and pattern.   *Fig. 9.13 (b). End-fire array and pattern.*

Another arrangement known as end-fire array consists of a number of conductors equally spaced in a line (Fig. 9.13(b)), carrying same magnitude of current but with a progressive phase difference between them. The directional pattern of such an antenna has directivity along the array axis in the direction, in which antenna currents become more lagging.

It is possible to combine several different arrays to obtain highly directional radiation patterns. Such combinations are often used in HF transmission/reception for point to point communication. Gains, well in excess of 50, are not uncommon with such arrangements.

Folded Dipole

As shown in Fig. 9.14 (*a*), the folded dipole is made of two half-wave antennas joined at the ends with one open at the centre where the transmission line is connected. The spacing between the two conductors is small compared with a half wave length. This antenna has the same directional characteristics and signal pick up as that of a straight dipole but its impedance is approximately 300 ohms. This is nearly four times that of a dipole, because for the same power applied, the antenna now draws half the current than it would have, in the case of a dipole. Hence the impedance $(Z_0) = 4 \times 72 = 288$ ohms for a half-wave dipole with equal diameter arms. This is generally considered as 300 ohms.

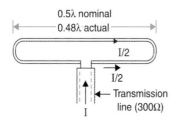

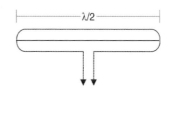

Fig. 9.14 (a). Folded dipole antenna.   *Fig. 9.14 (b). High impedance folded dipole antenna.*

If elements of unequal diameter are used, or an additional closed conductor of the same diameter is added in between the two (Fig. 9.14(*b*)), an impedance as large as 650 ohms can be obtained.

Parasitic Elements

It is not necessary for all the elements of an array to be connected to the output of the transmitter. An element connected direct to the transmitter is called a driver, whereas a radiator not directly connected is called a parasitic element. Such a parasitic radiator receives energy through the induction field of a driven element, rather than by a direct connection to the transmission line. In general, a parasitic element longer than the driver and close to it reduces signal strength in its own direction, and increases it in the opposite direction. This in effect amounts to reflection of energy towards the driver and thus, this element is called a reflector. Again, a parasitic element shorter than the driver from which it receives energy, tends to increase radiation in its own direction, and is therefore called a director. The number of directors and their lengths can be varied to obtain increased directivity and broad band response.

9.6 TELEVISION TRANSMISSION ANTENNAS

As already explained, television signals are transmitted by space wave propagation and so the height of antenna must be as high as possible in order to increase the line-of-sight distance. Horizontal polarization is standard for television broadcasting, as signal to noise ratio is favourable for horizontally polarized waves when antennas are placed quite high above the surface of the earth.

Turnstile Array

To obtain an omnidirectional radiation pattern in the horizontal plane, for equal television signal radiation in all directions, an arrangement known as 'turnstile array' is often used. In this type of antenna two crossed dipoles are used in a turnstile arrangement as shown in Fig. 9.15(a). These are fed in quadrature from the same source by means of an extra $\lambda/4$ line. Each dipole has a figure of eight pattern in the horizontal plane, but crossed with each other. The resultant field in all directions is equal to the square root of the sum of the squares of fields radiated by each conductor in that direction. Thus the resultant pattern as shown in Fig. 9.15(b) is very nearly circular in the plane of the turnstile antenna. Fig. 9.15(c) shows several turnstiles stacked one above the other for vertical directivity.

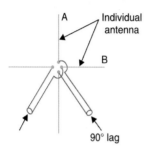

Fig. 9.15 (a). *Turnstile array.*

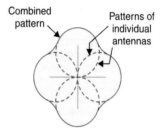

Fig. 9.15 (b). *Directional pattern in the plane of turnstile.*

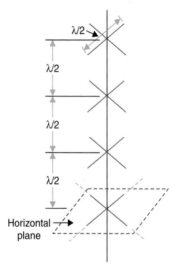

Fig. 9.15 (c). *Stacked turnstile array.*

Dipole Panel Antenna System

Another antenna system that is often used for band I and band III transmitters consists of dipole panels mounted on the four sides at the top of the antenna tower as shown in Fig. 9.16. Each panel consists of an array of full-wave dipoles mounted in front of reflectors. For obtaining unidirectional pattern the four panels mounted on the four sides of the tower are so fed that the current in each lags behind the previous by 90°. This is achieved by varying the field cable length by $\lambda/4$ to the two alternate panels and by reversal of polarity of the current.

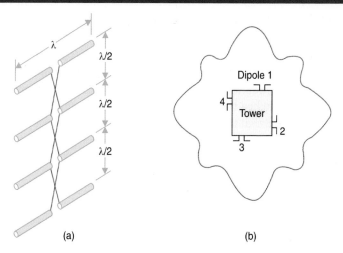

Fig. 9.16. *Dipole panel antenna system (a) panel of dipoles (b) radiation pattern of four tower mounted dipole antenna panels.*

Combining Network

The AM picture signal and FM sound signal from the corresponding transmitters are fed to the same antenna through a balancing unit called diplexer. As illustrated in Fig. 9.17, the antenna combining system is a bridge configuration in which first two arms are formed by the two radiators of the turnstile antenna and the other two arms consist of two capacitive reactances. Under balanced conditions, video and sound signals though radiated by the same antenna, do not interfere with the functioning of the transmitter other than their own.

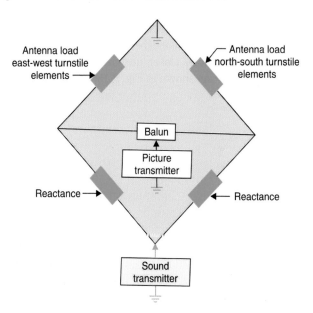

Fig. 9.17. *Equivalent bridge circuit of a diplexer for feeding picture and sound transmitters to a common turnstile array.*

Chapter 9

Chapter 9

9.7   TELEVISION RECEIVER ANTENNAS

For both VHF and UHF television channels, one-half-wave length is a practical size and therefore an ungrounded resonant dipole is the basic antenna often employed for reception of television signals. The dipole intercepts radiated electromagnetic waves to provide induced signal current in the antenna conductors. A matched transmission line connects the antenna to the input terminals of the receiver. It may be noted that the signal picked up by the antenna contains both picture and sound signal components. This is possible, despite the 5.5 MHz separation between the two carriers, because of the large bandwidth of the antenna. In fact a single antenna can be designed to receive signals from several channels that be close to each other.

Antennas for VHF Channels

Although most receivers can produce a picture with sufficient contrast even with a weak signal, but for a picture with no snow and ghosts, the required antenna signal strength lies between 100 and 2000 μV. Thus, while a half-wave dipole will deliver satisfactory signal for receivers located close to the transmitter, elaborate arrays become necessary for locations far away from the transmitter.

Yagi-Uda Antenna

The antenna widely used with television receivers for locations within 40 to 60 km from the transmitter is the folded dipole with one reflector and one director. This is commonly known as Yagi-Uda or simply Yagi antenna. The elements of its array as shown in Fig. 9.18(*a*) are arranged collinearly and close together. This antenna provides a gain close to 7 db and is relatively unidirectional as seen from its radiation pattern drawn in Fig. 9.18(*b*). These characteristics are most suited for reception from television transmitters of moderate capacity. To avoid pick-up from any other side, the back lobe of the radiation pattern can be reduced by bringing the radiators closer to each other. The resultant improvement in the front to back ratio of the signal pick-up makes the antenna highly directional and thus can be oriented for efficient pick-up from a particular station. However, bringing the radiators closer has the adverse effect of lowering the input impedance of the array. The separation shown in Fig. 9.18(*a*) is an optimum value.

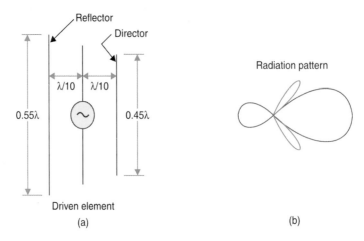

Fig. 9.18. Yagi-Uda antenna (a) antenna (b) radiation pattern.

Antenna Length

As mentioned earlier, it is not necessary to erect a separate antenna for each channel because the resonant circuit formed by the antenna is of low 'Q' (quality factor) and as such has a broad band response. For the lower VHF channels (Band I—channels 2 to 4) the length of the antenna may be computed for a middle value. While this antenna will not give optimum results at other frequencies, the reception will still be quite satisfactory in most cases if the stations are not located far away.

Though the antenna length used should be as computed by the usual expression: Wavelength $(\lambda) = \dfrac{3 \times 10^8}{f \text{ (Hz)}}$ metres, but in practice it is kept about 6 per cent less than the calculated value.

This is necessary because the self-capacitance of the antenna alters the current distribution at its ends. The small distance between the two quarter wave rods of the driver, where the lead-in line is connected can be taken as too small and hence neglected. Note that this gap does not affect the current distribution significantly.

Antenna Mounting

The receiving antenna is mounted horizontally for maximum pick-up from the transmitting antenna. As stated earlier, horizontal polarization results in more signal strength, less reflection and reduced ghost images. The antenna elements are normally made out of 1/4″ (0.625 cm) to 1/2″ (1.25 cm) dia aluminium pipes of suitable strength. The thickness of the pipe should be so chosen that the antenna structure does not get bent or twisted during strong winds or occasional sitting and flying off of birds. A hollow conductor is preferred because on account of skin effect, most of the current flows over the outer surface of the conductor.

The antenna is mounted on a suitable structure at a height around 10 metres above the ground level. This not only insulates it from the ground but results in induction of large signal strength which is free from any interference.

The centre of the closed section of the half-wave folded dipole is a point of minimum voltage, allowing direct mounting at this point to the grounded metal mast without shorting the signal voltage. A necessary precaution while mounting the antenna is that it should be at least two metres away from other antennas and large metal objects. In crowded city areas close to the transmitter, the resultant signal strength from the antenna can sometimes be very low on account of out of phase reflections from surrounding buildings. In such situations, changing the antenna placement only about a metre horizontally or vertically can make a big difference in the strength of the received signal, because of standing waves set up in such areas that have large conductors nearby. Similarly rotating the antenna can help minimize reception of reflected signals, thereby eliminating the appearance of ghost images.

In areas where several stations are located nearby, antenna rotators are used to turn its direction. These are operated by a motor drive to set the broad side of the antenna for optimum reception from the desired station. However, in installations where a rotating mechanism is not provided, it is a good practice to connect the antenna to the receiver before it is fixed in place permanently and proceed as detailed below:

   (i) Try changing the height of the antenna to obtain maximum signal strength.

   (ii) Rotate the antenna to check against ghost images and reception of signals from far-off stations.

Chapter 9

Chapter 9

(*iii*) When more than one station is to be received, the final placement must be a compromise for optimum reception from all the stations in the area. In extreme cases, it may be desirable to erect more than one antenna.

Indoor Antennas

In strong signal areas it is sometimes feasible to use indoor antennas provided the receiver is sufficiently sensitive. These antennas come in a variety of shapes. Most types have selector switches which are used for modifying the response pattern by changing the resonant frequency of the antenna so as to minimize interference and ghost signals. Generally the switch is rotated with the receiver on, until the most satisfactory picture is obtained on the screen. Almost all types of indoor antennas have telescopic dipole rods both for adjusting the length and also for folding down when not in use.

Fringe Area Antenna

In fringe areas where the signal level is very low, high-gain antenna arrays are needed. The gain of the antenna increases with the number of elements employed. A Yagi antenna with a large number of directors is commonly used with success in fringe areas for stations in the VHF band. As already mentioned, a parasitic element resonant at a lower frequency than the driven element will act as a mild reflector, and a shorter parasitic element will act as a mild 'concentrator' of radiation. As a parasitic element is brought closer to the driven element, then regardless of its precise length, it will load the driven element more and therefore reduce its input impedance. This is perhaps the main reason for invariable use of a folded dipole as the driven element of such an array. A gain of more than 10 db with a forward to back ratio of about 15 is easily obtained with such an antenna. Such high gain combinations are sharply directional and so must be carefully aimed while mounting, otherwise the captured signal will be much lower than it should be. A typical Yagi antenna for use in fringe areas is shown in Fig. 9.19 (*a*). In such antennas the reflectors are usually 5 per cent longer than the dipole and may be spaced from it at 0.15 to 0.25 wavelength depending on design requirements. Similarly the directors may be 4 per cent shorter than the antenna element, but where broadband characteristics are needed successive directors are usually made shorter (see Fig. 9.19 (*a*)) to be resonant for the higher frequency signals of the spectrum.

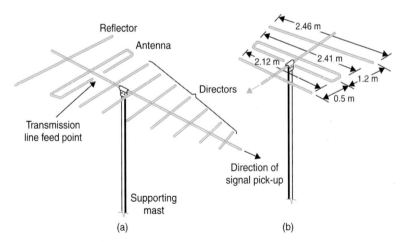

Fig. 9.19 (a) A typical Yagi antenna, (b) Channel four antenna.

In some fringe area installations, transistorised booster amplifiers are also used along with the Yagi antenna to improve reception. These are either connected just close to the antenna or after the transmission line, before the signal is delivered to the receiver.

Yagi Antenna Design

The following expressions can be used as a starting point while designing any Yagi antenna array.

Length of dipole (in metres) $\approx \dfrac{143}{f\,(\text{MHz})}$ (f is the centre frequency of the channel)

Length of reflector (in metres) $\approx 152/f\,(\text{MHz})$

Length of first director (in metres) $\approx 137/f\,(\text{MHz})$

Length of subsequent directors reduces progressively by 2.5 per cent.

Spacing between reflector and dipole $= 0.25\lambda \approx 75/f\,(\text{MHz})$

Spacing between director and dipole $= 0.13\lambda \approx 40/f\,(\text{MHz})$

Spacing between director and director $= 0.13\lambda \approx 39/f\,(\text{MHz})$

The above lengths and spacings are based on elements of 1 to 1.2 cm in diameter. It may be noted that length of the folded dipole is measured from centre of the fold at one end to the centre of the fold at the other end.

It must be remembered that the performance of Yagi arrays can only be assessed if all the characteristics like impedance, gain, directivity and bandwidth are taken into account together. Since there are so many related variables, the dimensions of commercial antennas may differ from those computed with the expressions given above. However, for single channel antennas the variation is not likely to be much. Figure 9.19 (b) shows a dimensioned sketch of channel four (61 to 68 MHz) antenna designed for locations not too far from the transmitter. It has an impedance $= 40 + j20\ \Omega$, a front to back pick up ratio $= 30$ db, and an overall length $= 0.37$ wavelength.

Multiband Antennas

It is not possible to receive all the channels of lower and higher VHF band with one antenna. The main problem in using one dipole for both the VHF bands is the difficulty of maintaining a broadside response. This is because the directional pattern of a low-band dipole splits into side lobes at the third and fourth harmonic frequencies in the 174 to 223 MHz band. On the other hand a high-band dipole cut for a half wavelength in the 174 to 233 MHz band is not suitable for the 47 to 68 MHz band because of insufficient signal pick-up at the lower frequencies. As a result, antennas for both the VHF bands generally use either separate dipoles for each band or a dipole for the lower VHF band modified to provide broadside unidirectional response in the upper VHF band also.

Diplexing of VNF Antennas

When it is required to combine the outputs from the lower and upper VHF band antennas to a common lead-in wire (feeder) it is desirable to employ a filter network that not only matches the signal sources to the common feeder but also isolates the signals in the antennas from each other. Such a filter arrangement is called a 'diplexer'. Its circuit with approximate component values for bands I (47 to 68 MHz) and III (174 to 263 MHz) is given in Fig. 9.20 (*a*). The manner in which it is connected to the two antennas is shown in Fig. 9.20 (*b*). Similarly a triplexer filter can be employed when three different antennas are to feed their outputs to a common socket in the receiver. The combining arrangement can be further modified to connect the output from a UHF antenna to the same feeder line.

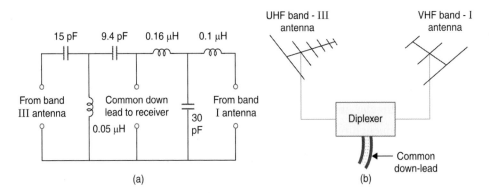

Fig. 9.20. *Diplexing antenna outputs (a) diplexer network, H.P.-L.P.-filter combination (b) diplexer connections.*

Conical Dipole Antenna

The VHF dual-band antenna pictured in Fig. 9.21 (*a*) is generally called a conical or fan dipole. As shown in Fig. 9.21 (*b*), this antenna consists of two half-wave dipoles inclined at about 30° from the horizontal plane, similar to a section of a cone. In some designs a horizontal dipole is provided in between the two half-wave dipoles. The dipoles are tilted by about 30° inward towards the wavefront of the arriving signal. This as shown in the figure results in a total included angle of 120° between the two conical sections in the broadside direction. A straight reflector is provided behind the conical dipoles.

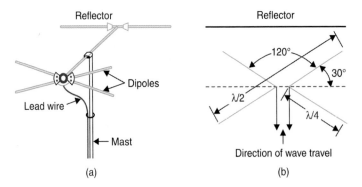

Fig. 9.21. *VHF fan (conical) dipole with reflector (a) pictorial view, (b) spacing of elements.*

With the dipole lengths chosen for channel 2, this antenna is extensively used as a receiving antenna to cover both the VHF bands. The antenna resistance is about 150 ohms.

The response pattern of the antenna contains only one major lobe on all the channels. This is so because for the 174-223 MHz band, the tilting of the dipole rods shifts the direction of the split lobes produced at the third and fourth harmonic frequencies so that they combine to produce a main forward lobe in the broadside direction. This is an improvement over the conventional dipole where an element cut for the low frequencies will have a multilobed pattern on the higher channels, and an element cut for the high frequencies will have a poor response on the lower channels. Though, one conical antenna array may be adequate for all VHF channels, sometimes three or four such arrays are stacked high for better and more uniform reception.

In-line Antenna

Another combination antenna which is known as in-line antenna is shown in Fig. 9.22. It consists of a half-wave folded dipole with reflector for the lower VHF band, that is in line with the shorter half-wave folded dipole meant for the upper VHF band. The distance between the two folded dipoles is approximately one-quarter wavelength at the high-band dipole frequency. This is the length of the line connecting the short dipole to the long dipole, where the transmission line to the receiver is connected. The directivity pattern of the in-line antenna is relatively uniform on all VHF channels, with a unidirectional broadside response. Its input resistance is about 150 ohms.

UHF Antennas

The basic principle of antennas for picking up signals from stations that operate in the UHF band is more or less the same as that in the VHF band. However, on account of higher attenuation suffered by the UHF signals, it becomes necessary to have very high gain and directive antennas. Besides this, higher gain is also necessary because receivers are less sensitive and tend to be more noisy at these frequencies than at lower frequencies. Therefore at microwave frequencies, some special type of antennas are used, in which the optical properties of reflection, refraction and diffraction are utilized to concentrate the radiated waves for higher directivity and more gain. Though a large number of microwave antennas for specific applications have been developed, the two types that find wide application for television reception are briefly described below.

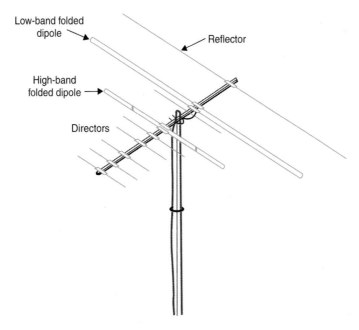

Fig. 9.22. In-line YAGI antenna array for lower and upper VHF bands.

Bow-Tie or Di-Fan Antenna

This di-fan half-wave dipole is the simplest type of UHF antenna as the basic Yagi is for the VHF band. As shown in Fig. 9.23, the dipoles are triangular in shape made out of metal sheet, instead of rods. This unit has a broad band response with radiation pattern resembling the figure of eight. When a screen reflector is placed at its back the response becomes unidirectional. For greater gain two or four sets of dipoles can be put together to form an array. Note the use of a mesh screen reflector instead of a rod as used in VHF antennas. Screen reflectors are more efficient than rods but their big size and bulk makes it impossible to use them in VHF antennas.

Parabolic Reflector Antenna

In this type of antenna (see Fig. 9.24) the dipole is placed at the focal point of a parabolic reflector. The principle is the same as that of parabolic reflectors of the headlights of a vehicle though in an inverse way. The incoming electromagnetic waves are concentrated by the reflector towards the dipole. This provides both high gain and directivity. Note that instead of using an entire parabolic structure only a section is used. The use of such a reflector provides a gain of 8 db over that of a resonant half-wave dipole.

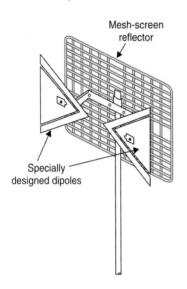

Fig. 9.23. Fan dipole UHF antenna.

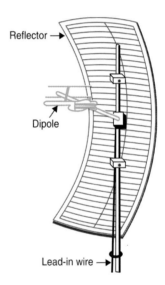

Fig. 9.24. Parabolic reflector antenna.

In areas where both VHF and UHF stations are in operation, combination antennas serve to simplify reception problems from all the channels. Various combinations of different VHF and UHF antennas are in use. One such combination consists of a low-band conical antenna for VHF signals and a broad-band fan dipole for the microwave frequency region. A single lead-in line delivers signals to the receiver through the use of a special coupling device which is mounted directly on the antenna itself.

9.8 COLOUR TELEVISION ANTENNAS

The requirements to be met by colour television antennas are somewhat different than those for monochrome receivers. In monochrome receiver antennas, the emphasis is on higher gain while it may vary from channel to channel because of the wide frequency range. This, in itself, is no problem. In fact manufacturers generally design antennas to deliver more gain on higher frequency channels than on lower channels in order to compensate for higher transmission losses and lower receiver sensitivity at very high frequencies. However, in such antennas the gain not only varies from channel to channel but also from one end of the channel to the other. As an illustration Fig. 9.25 shows the response curve of a typical wide range monochrome antenna for channel 4, *i.e.*, from 61 to 68 MHz. As shown there, the gain changes by about 4 db from beginning to end of the channel. If this antenna is used for colour TV reception from the same channel, the colour signal spectrum, which lies around 66.68 MHz will receive almost 2 db less gain than the video carrier and most of its sidebands. While it is true that the channel sound signal spectrum receives even lesser gain than the colour components but this does not affect the receiver reproduction. This is so because the picture contrast and sound volume controls provided in the receiver can be varied independently to get desired picture and sound outputs. However, the reduction in gain of colour signal frequencies cannot be separately compensated for and this results in poor colour picture quality. In colour television, the relative phase angle of the combined colour difference signal phasor determines the colour in the picture. Any change in gain is accompanied by a phase shift of the incoming signal. Thus a change in gain in the region of colour signal spectrum tends to change the hues in the picture. For example a slight shift of the colour signal phasor can turn red colour to orange and yellow to green. In fact, if a large phase shift occurs, no colours may get reproduced. Therefore the most important requirement of a colour TV antenna is a flat response over the entire channel. As labelled along the response curve in Fig. 9.25, the antenna output should not vary by more than one db over the frequency range of any one channel for satisfactory reproduction of colour details.

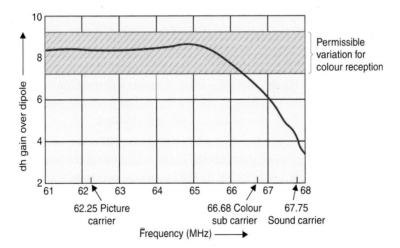

Fig. 9.25. *Typical response curve of a wide range antenna for channel 4.*

Log Periodic Antennas

The stringent requirement of almost flat response besides high gain over any single channel has led to the development of a relatively new class of broadband antennas. The most popular of this type is the log periodic antenna. The name log periodic stems from the fact that the impedance of the antenna has a periodic variation when plotted against logarithm of frequency. Figure 9.26 (*a*) illustrates this periodic nature of the antenna impedance. Such a behaviour results from the geometric relationship between the relative lengths of the antenna elements and the distances which separate them from each other. This naturally results in the antenna getting larger and larger as the distance from the smallest element increases.

The basic construction of a log periodic antenna consisting of a six element array is illustrated in Fig. 9.26 (*b*). As shown there, the largest dipole is at the back and each adjacent element is shorter by a fixed ratio typically 0.9. Also the distance between the dipoles becomes shorter and shorter by a constant factor which is typically 35 per cent of quarter wave spacing. As a result, the resonant frequencies for the dipoles overlap to cover the desired frequency range. All the dipoles are active elements without parasitic reflectors or directors. The active dipoles, as shown in the figure, are interconnected by a crossed wire net which transposes the signal by 180°.

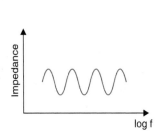

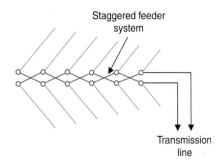

Fig. 9.26 (a). Periodic nature of the impedance of a log periodic antenna when plotted on a logarithmic scale.

Fig. 9.26 (b). Constructional details of a log periodic antenna.

When this antenna is pointed in the direction of the desired station, only one or two of the dipole elements in the antenna react to that frequency and develop the necessary signal. All the other elements remain inactive, *i.e.*, do not develop any signal at that particular frequency. However, for any other incoming channel some other elements will resonate to develop the signal. Thus only one or two elements combine to deliver the signal from any one channel. Such an arrangement results in a uniform gain response over each channel.

When the largest dipole is cut for channel 2, the array will cover all the low-band VHF channels as antenna resonance moves towards the shorter elements at the front. However, for the high-band VHF channels from 174 to 223 MHz the elements operate as 3 λ/2 dipoles. They are angled in as a 'V' to line up with the split lobes in the directional response for third harmonic resonance. Figure 9.27 shows such a log periodic antenna.

When the largest dipole is cut for the lowest channel in the UHF band of 470 to 890 MHz, the array can cover all the UHF channels. The 'V' angle is not necessary in the UHF array because this frequency range is less than 2 : 1. The UHF antenna array can be mounted along with the VHF array where a U/V splitter, *i.e.*, a diplexer network connects the two antennas to a common transmission line for the downlead. It may be noted that the antenna described above is only one type of log periodic antenna out of a wide variety, quite different in appearance.

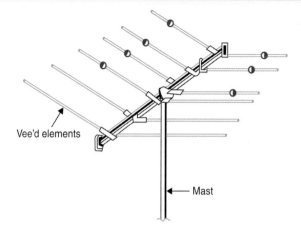

Fig. 9.27. A colour log periodic antenna. The elements are vee'd
to eliminate dual phase problems.

When colour transmission is to be received from only one channel, there is no need for a specially designed antenna. For example, the antenna designed for monochrome reception on channel four only can also be used with good results for receiving colour transmission from the same channel. However, the elements of the antenna must be cut and spaced accurately to ensure almost uniform gain over the entire channel. The antenna shown in Fig. 9.19 (b) can be used with colour receivers for receiving colour transmission from channel four.

9.9 TRANSMISSION LINES

A transmission line is used for delivering the antenna signal to the receiver. The desirable requirements of a transmission line are:

(i) the losses along the line should be minimum.

(ii) there should be no reflection of the signal on the line.

(iii) the line itself should not pick up any stray signals. To prevent this the line should be balanced or shielded or both.

The two main types of transmission lines are the two wire parallel conductor type and the concentric (co-axial) type. Flat twin-lead, tubular twinlead, open wire line and co-axial type transmission lines are shown in Fig. 9.28. Flat-twin lead and tubular-twin lead transmission lines are constructed in the form of a plastic ribbon and are generally called twin-lead either flat or tubular. These together with the open wire line though balanced are not shielded lines. A line is balanced when each of the two conductors has the same capacitance (or voltage) to ground. The balance corresponds to the dipole antenna itself which has balanced signals of opposite phase in the arms. The connections of a balanced line between the antenna and receiver are shown in Fig. 9.29 (a). The balanced line is connected to the two ends of a centre-tapped antenna input transformer. Then any in-phase stray field cutting across both the wires of a balanced line will induce an equal voltage in each line. The consequent currents tend to induce voltages of opposite polarity in the centre tapped input transformer which cancel each other. However, the antenna signal from the dipole has opposite phases in the two sides of the line and the voltages that are induced in the input transformer reinforce each other and consequently a large signal voltage is delivered to the receiver from the secondary side of the input transformer. A shielded line, *i.e.*, the coaxial cable is completely enclosed with a metal sheath or braid that is grounded to serve as a shield for the inner conductor. The shield prevents stray signals from inducing current in the centre conductor. Usually the shield is grounded to the receiver chassis. With only one conductor the line is unbalanced. The connections of a co-axial transmission line between the antenna and receiver are shown in Fig. 9.29 (*b*). Though the line is unbalanced, a balanced transformer (balun) can be used at the input of the receiver for converting the input signal from unbalanced to balanced form, if necessary. It may be noted that if the two inner conductors (insulated) are used within the shield then the line is both balanced and shielded. Shielded lines generally have more capacitance and higher losses. The attenuation is caused by I^2R losses in the a.c. resistance of the line. This reduces the amplitude of the antenna signal delivered by the line to the receiver. The longer the line and higher the frequency, the greater is the attenuation. The characteristic impedance of the line which results from uniform spacing between the two conductors is the same regardless of length of the line.

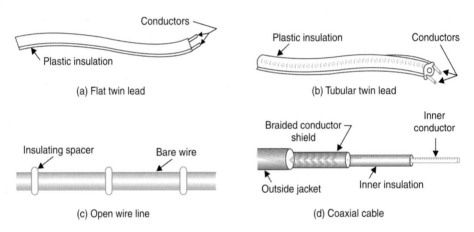

Fig. 9.28. Transmission lines.

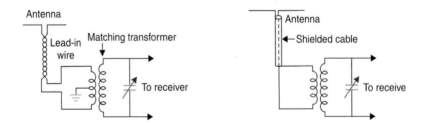

Fig. 9.29 (a). Connections from antenna to
receiver input terminals (balanced).

Fig. 9.29 (b). Unbalanced match.

Flat Twin-Lead

The flat parallel wire is one of the most popular transmission lines in use for the VHF range. The wires are encased in a plastic ribbon of polyethylene which is strong, flexible and unaffected by sunlight, water or cold. The characteristic impedance ranges from 75 ohms to 300 ohms. As stated earlier this line is balanced. It is, however, unshielded and therefore not recommended for noisy locations. It should not be run close to power lines to avoid pick-up of 50 Hz hum. It should also be kept away from large metal structures which can alter the balance of the line. Since most receivers have a balanced input impedance of 300 ohms, the 300 ohms twin-lead (spacing about 1 cm, wire gauge 20 to 22) is convenient for impedance matching. The losses in a flat twin lead are much greater when the line is wet.

Tubular Twin-Lead

In this type the two parallel conductors are embedded in a polyethylene plastic tubing with air as dielectric for most of the inside area. Though expensive it has low losses and is especially suited for the UHF band of frequencies. The twin line is enclosed in a strong plastic jacket for protection against adverse weather conditions.

Open-wire Line

As shown in Fig. 9.28 (c), this line is constructed with low loss insulating spacers between the parallel bare-wire conductors. The open wire line causes least attenuation because air is the dielectric between conductors. However, the characteristic impedance is relatively high. With a spacing of about 1.5 cm the impedance of this line is about 450 ohms.

Co-axial Cable

This line consists of a central conductor in a dielectric that is completely enclosed by a metallic shield which may be a tubing or a flexible braid of copper or aluminium. A plastic jacket moulded over the line provides protection. Because of the grounded shield the coaxial cable is immune to any stray pick-ups. With an outside diameter of about 1 cm the characteristic impedance is 75 ohms. Because of higher attenuation and higher costs, a shielded line is used only when the surrounding noise is very severe or where multiple lines must be run close to each other. In cable distribution systems, coaxial cable is a necessity despite its high losses. The losses in this system are compensated by the use of distribution amplifiers. Special connectors are available for terminating coaxial lines. Foam coaxial cable is also available. The use of foam as dielectric reduces the attenuation by about 20 per cent at 100 MHz.

Characteristic Impedance

When a transmission line has a length comparable with a wavelength of the signal, the characteristic impedance of the line depends on the small inductance of the conductors and the distributed capacitance between the conductors. It can be shown that the characteristic impedance $Z_0 = \sqrt{L/C}$ ohms, where L is the inductance per unit length and C the capacitance per unit length. The closer the conductor spacing, the greater is the capacitance and smaller the Z_0 of the line.

Resonant and Non-resonant Lines

When a line is terminated by a resistive load equal to Z_0 of the line, there is no reflection and maximum power transfer takes place from source to load. Such a line is non-resonant because there are no reflections. A line terminated by Z_0 then becomes effectively an infinitely long line because there is no discontinuity at the load. The length of the line is not critical. When such a line having $Z_0 = 300$ ohms is connected to the 300 ohms antenna input terminals of the receiver, there is no reflection and maximum energy is delivered to the receiver from the antenna. Because of correct termination the line can be cut to any length without any loss in match of impedances. However, more length will produce more I^2R losses.

When the line is not terminated by Z_0, there will be reflections and standing waves will be set up in the line. This effect makes the line resonant. The signal strength at any point on the transmission line will now be a function of the length of the transmission line unlike the case with non-resonant lines. The ratio of the voltage maximum to that of voltage minimum along the line is defined as the voltage standing wave ratio, abbreviated VSWR. Note that for a line terminated in Z_0, the VSWR is one. In a resonant line, the greater the mismatch, the higher the VSWR which is greater than one. The extreme cases of high VSWR correspond to a short or an open terminated line. If one touches a resonant line, the added hand-capacitance can mean much more or much less signal delivery depending on where one touches the line.

The use of transmission lines as resonant circuits is illustrated in Fig. 9.30. The maximum impedance is at the point of highest voltage on the line, say at the open end of an equivalent quarter-wave section of the line and minimum at the point of highest current, say at the short

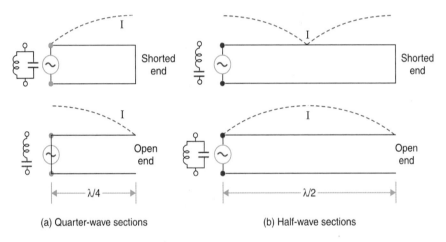

(a) Quarter-wave sections   (b) Half-wave sections

Fig. 9.30. Transmission-line sections as resonant circuits.

circuited end of an equivalent quarter-wave line. Note that the impedance at any point equals the ratio of voltage to current. As shown in the figure, a quarter-wave section shorted at the end is equivalent to a parallel-tuned circuit at the generator side because there is a high impedance across the terminals at the resonant frequency. For a line-length shorter than a quarter-wave, the line is equivalent to an inductance. The open quarterwave section provides a very low impedance at the generator side of the line. A line less than a quarter-wave makes the line appear as a capacitance. The half-wave sections however repeat the impedances at the end of the line to furnish the same impedance at the generator side. The main features of quarter-wave ($\lambda/4$) and half-wave ($\lambda/2$) sections are given in the table below.

Length	Termination	Input impedance	Phase shift
Quarter-wave	shorted	Open circuit	90°
Quarter-wave	open	Short circuit	90°
Half-wave	shorted	Short circuit	180°
Half-wave	open	Open circuit	180°

Such transmission lines are often called STUBS. A stub can be used (*i*) for impedance matching, (*ii*) as an equivalent series resonant circuit to short an interfering *r.f.* signal, and (*iii*) for phasing signals correctly in antennas. A quarter-wave line produces a phase change of 90° whereas a half-wave section shifts the phase by 180°. To reduce interference, an open λ/4 stub at the interfering signal frequency can be used. One side is connected across the antenna input terminals of the receiver, while the open end produces a short at the receiver input one quarter-wave back.

A 300 ohms twin-lead is designed to have almost a constant impedance for all the channels in a band. This is used to connect the antenna output to the input of the receiver. Matching the impedance of a multiband antenna to the characteristic impedance of the line is not critical because an impedance mismatch of 2.5 to 1 results in a one db loss of the signal.

Quarter-wave Matching Section

A quarter-wave section can be used for matching two unequal impedances Z_1 and Z_2. The characteristic impedance Z_0 of the quarter-wave section should then be, $Z_0 = \sqrt{Z_1 Z_2}$. This is iliustrated in Fig. 9.31, where an antenna with an impedance equal to 75 ohms is matched to a 300 ohms transmission line by a quarter-wave matching section. This section should then have an impedance equal to $\sqrt{Z_1 Z_2} = \sqrt{75 \times 300} = 150$ ohms. The required length of the quarter-wave

section can be estimated using the expression λ/4 (metres) = $v\,\dfrac{80}{f(\text{MHz})}$, where '*v*' is the velocity

factor which varies between 0.6 to 0.8 depending on the type of lead-in wire used.

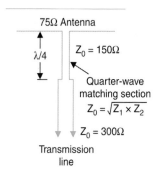

Fig. 9.31. *Use of a quarter-wave section for matching antenna to transmission line.*

Balun (Balancing Unit)

Two quarter-wave sections of the type discussed above can be combined to make a balancing and impedance transforming unit. This is illustrated in Fig. 9.32 (*a*). This is called a Balun and is used for matching balanced and unbalanced impedances. Two quarter-wave lines each having an impedance equal to 150 ohms are connected in parallel at one end, resulting in 75 ohms impedance across points A and B. Either A or B can be grounded to provide an unbalanced impedance at the ungrounded point with respect to ground. At the other end the two 150 ohms quarter-wave sections are connected in series to provide 300 ohms impedance between points C and D. The quarter wavelength of the line isolates the ground point from C or D, allowing a balanced impedance with respect to ground. Either side of the Balun can be used for input with the other side as output. It is usefully employed for matching a 72 ohms coaxial line to the 300 ohms receiver input or in the reverse direction, *i.e.*, a 300 ohms twin-lead to a 72 ohms unbalanced input. As illustrated in Fig. 9.32 (*b*) bifilar windings are used to simulate the 150 ohms transmission line sections. The windings, as illustrated in Fig. 9.32 (*c*), are on a ferrite core to increase the inductance, thus making the line electrically longer.

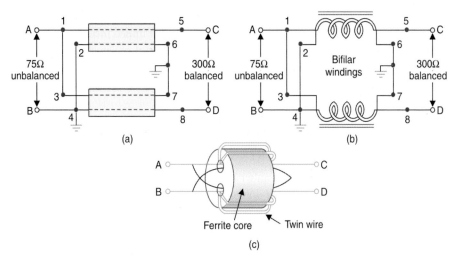

Fig. 9.32. *Balun to match between 75W unbalanced and 300W balanced impedances (a) with l/4 matching sections, (b) equivalent transformer, (c) constructional details.*

9.10 ATTENUATION PADS

In cases where excessive antenna signal causes overloading, the signal can be attenuated without introducing any mismatch by using resistive networks called pads. Two different pad configurations are illustrated in Fig. 9.33, where component values for different attenuations are given in separate charts. The resistors used in such pads are low wattage carbon resistances. Wire-wound resistors are not used because of their inductance. The use of a resistance matching pad has the advantage of providing both attenuation and impedance match that is independent of frequency.

Attenuation	R_1	R_2
6 db	47Ω	390Ω
10 db	82Ω	220Ω
20 db	120Ω	68Ω

Attenuation	R_3	R_4
6 db	22Ω	100Ω
10 db	33Ω	51Ω
20 db	56Ω	15Ω

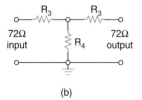

Fig. 9.33. *Resistance pads (a) balanced 'H' pad for 300W twin lead (b) unbalanced 'T' pad for 72W coaxial line.*

REVIEW QUESTIONS

1. Describe briefly the different methods by which radio waves of different frequencies are propagated. Why space wave propagation is the only effective mode of radiation above about 40 MHz ?

2. What do you understand by line-of-sight distance in space wave propagation ? What are the effects of atmospherics and obstacles on space waves ? Why is it necessary to keep both the transmitting and receiving antennas as high as possible for television ?

3. What do you understand by wave polarization ? Why is horizontal polarization preferred for television and FM broadcasts ? Why is TV transmission limited to about 100 km ? What are booster stations and under what conditions are they employed ?

4. Describe briefly co-channel and adjacent channel interference effects. Discuss the techniques employed to eliminate such interference in fringe areas.

5. What is a ghost image and what causes it to appear on the receiver screen along with the reproduced picture ? Differentiate between leading and trailing ghost pictures. Why is AM preferred over FM for picture signal transmission ?

6. Describe briefly radiation mechanism from an antenna. Explain the evolution of a dipole for effective radiation. Sketch approximate radiation patterns for ungrounded resonant antennas of lengths $\lambda/4$, $\lambda/2$ and $3\lambda/2$ and justify them.

7. Define directional gain and front to back ratio as applied to receiving antennas. What is an antenna array ? Sketch radiation patterns for (i) a broadside array and (ii) end-fire array. How are these patterns modified when the antennas are very close to the ground ?

8. What is the function of a reflector and a director in a Yagi antenna. Explain why such antenna configurations are highly directional. What is the effect of increasing the number of director elements ?

9. What are the special requirements of a fringe area television antenna and how are these achieved ? Give constructional details of a typical fringe area antenna and explain the precautions that must be taken while mounting it.

10. Give constructional details of a turnstile antenna and explain by drawing radiation pattern its suitability for television transmission. Draw the circuit of a diplexer arrangement employed for feeding the output from both picture and sound signal transmitters to the same antenna.

11. Why is it not possible to use the same antenna for reception for both lower and upper VHF channels ? Describe any one type of multiband array commonly employed to cover all the channels in the VHF band.

12. Describe briefly the basic principle of bow-tie (di-fan) and parabolic reflector type of antennas commonly employed for reception from UHF television channels.

13. Explain why an antenna used for colour TV reception must deliver almost constant output voltage over any one channel.

14. Sketch a typical log periodic antenna and explain its special characteristics. Why are its elements bent in a 'V' shape ?

15. Describe with suitable sketches various types of lead-in wires used for connecting the antenna to the TV receiver. What is the essential difference between balanced and unbalanced lines and how are they connected to the receiver ? Why is a coaxial cable preferred for connecting a UHF antenna ?

16. What is a stub ? Explain how quarter-wave line sections can be used for providing an impedance match between a low impedance antenna and 300 ohms lead-in line.

17. What is a 'Balun' ? Give its constructional details and explain how it can be used as an impedance matching network between two different impedances at high frequencies.

18. Under what conditions does it become necessary to use an attenuator pad between the transmission line and receiver ? Draw 'H' and 'T' pad configurations and explain how besides providing a match between the line and receiver the desired attenuation is also achieved.

10

Television Applications

Television, by its use in broadcasting has opened broad new avenues in the fields of entertainment and dissemination of information. The not-so-well-known applications are in the area of science, industry and education, where the television camera has contributed immeasurably to man's knowledge of his environment and of himself. The television camera is probably best described as an extension of the human eye because of its ability to relay information instantaneously. Its capability to view events occurring in extremely hazardous locations has led to its use in areas of atomic radiation, underwater environments and outer space. Some of its important applications which are of direct interest to our society are described briefly in this chapter.

10.1 TELEVISION BROADCASTING

Broadcasting means transmission in all directions by electromagnetic waves from the transmitting station. Broadcasting, that deals mostly with entertainment and advertising, is probably the most familiar use of television. Millions of television sets in use around the world attest to its extreme popularity. Most programmes produced live in the studio are recorded on video tape at a convenient time to be shown later. Initially television transmission was confined to the VHF band only but later a large number of channel allocations were made in the UHF band also. The distance of transmission, as explained earlier, is confined to line of the sight between the transmitting and receiving antennas. The useful service range is up to 120 km for VHF stations and about 60 km for UHF stations. Television broadcasting initially started with monochrome picture but around 1952 colour transmission was introduced. Despite its complexity and higher cost, colour television has become such a commercial success that it is fast superseding the monochrome system.

10.2 CABLE TELEVISION

In recent years master antenna (MATV) and community antenna (CATV) television systems have gained widespread popularity. The purpose of a MATV system is to deliver a strong signal (over 1 mV) from one or more antennas to every television receiver connected to the system. Typical applications of a MATV system are hotels, motels, schools, apartment buildings and so on.

The CATV system is a cable system which distributes good quality television signal to a very large number of receivers throughout an entire community. In general, this system feeds increased TV programmes to subscribers who pay a fee for this service. A CATV system may have many more active (VHF and UHF) channels than a receiver tuner can directly select. This requires use of a special active converter in the head-end.

(a) MATV

The block diagram of a basic MATV system is shown in Fig. 10.1 (a). One or more antennas are usually located on roof top, the number depending on a available telecasts and their direction. Each antenna is properly oriented so that all stations are received simultaneously. In order to allow a convenient match between the coaxial transmission line and components that make up the system, MATV systems are designed to have a 75 Ω impedance. Since most antennas have a 300 Ω impedance, a balun is used to convert the impedance to 75 ohms. As shown in the figure, antenna outputs feed into a 4-way hybrid. A hybrid is basically a signal combining linear mixer which provides suitable impedance matches to prevent development of standing waves. The standing waves, if present, result in ghosts appearing in an otherwise good TV picture.

The output from the hybrid feeds into a distribution amplifier via a preamplifier. The function of these amplifiers is to raise the signal amplitude to a level which is sufficient to overcome the losses of the distribution system while providing an acceptable signal to every receiver in the system. The output from the distribution amplifier is fed to splitters through coaxial trunk lines. A splitter is a resistive-inductive device which provides trunk line isolation and impedance match.

Coaxial distribution lines carry television signals from the output of splitters to points of delivery called subscriber tap-offs. The subscriber taps, as shown in Fig. 10.1 (b), can be either transformer coupled, capacitive coupled or in the form of resistive pads. They provide isolation between receivers on the same line thus preventing mutual interference. The taps look like ac outlets and are normally mounted in the wall. Wall taps may be obtained with 300 Ω output 75 Ω output and a dual output. The preferred method is to use a 75 Ω type with a matching transformer. The matching transformer is usually mounted at the antenna terminals of the receiver and will have a VHF output and a UHF output. Since improperly terminated lines will develop standing waves, the end of each 75 Ω distribution cable is terminated with a 75 Ω resistor called a terminator.

(b) CATV

Formerly CATV system were employed only in far-fringe areas or in valleys surrounded by mountains where reception was difficult or impossible because of low level signal conditions. However, CATV systems are now being used in big cities where signal-level is high but all buildings render signals weak and cause ghosts due to multipath reflections. In either case, such a system often serves an entire town or city. A single antenna site, which may be on top of a hill, mountain or sky-scraper is chosen for fixing antennas. Several high gain and properly oriented antennas are employed to pick up signals from different stations. In areas where several signals are coming from one direction, a single broad based antenna (log-periodic) may be used to cover those channels. Most cable television installations provide additional services like household, business and educational besides commercial TV and FM broadcast programmes. These include news, local sports and community programmes, burgler and fire alarms, weather reports, commercial data retrieval, meter reading, document reproduction etc. Educational services include computer aided instructions, centralized library services and so on. Many of the above options require extra subscription fee from the subscriber.

Chapter 10

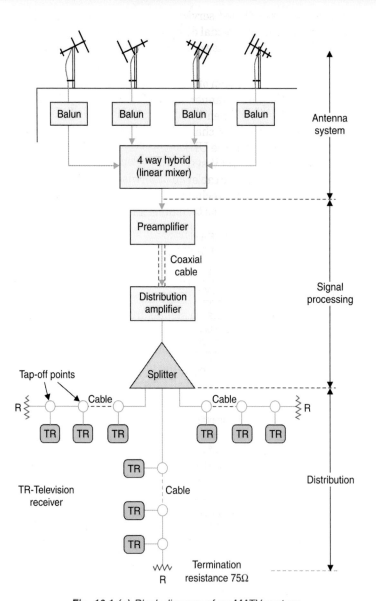

Fig. 10.1 (a) Block diagram of an MATV system.

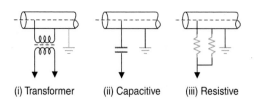

Fig. 10.1 (b) Subscriber taps of different types.

Since several of the above mentioned service need two-way communication between the subscriber and a central processor, the coaxial distribution network has a large number of cable pairs, usually 12 or 24. This enables the viewer to choose any channel or programme out of the many that are available at a given time.

CATV Plan. Figure 10.2 shows the plan of a typical CATV system. The signals from various TV channels are processed in the same manner as in a MATV system. In fact, a CATV system can be combined with a MATV set-up. When UHF reception is provided in addition to VHF, as often is the case, the signal from each UHF channel is processed by a translator. A translator is a frequency converter which hterodynes the UHF channel frequencies down to a VHF channel. Translation is advantageous since a CATV system necessarily operates with lengthy coaxial cables and the transmission loss through the cable is much greater at UHF than at VHF frequencies. As in the case of MATV, various inputs including those from translators are combined in a suitable mixer. The set-up from the antennas to this combiner is called a head-end.

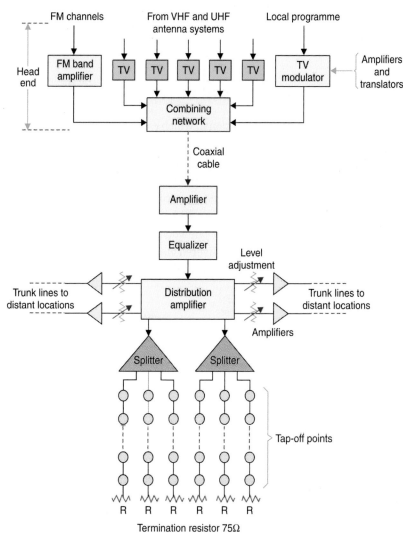

Fig. 10.2. A simplified block diagram of a CATV system.

Chapter 10

Further, as shown in the figure the CATV outputs from the combiner network are fed to a number of trunk cables through a broadband distribution amplifier. The trunk cables carry signals from the antenna site to the utilization site (s) which may be several kilometres away. Feeder amplifiers are provided at several points along the line to overcome progressive signal attenuation which occurs due to cable losses. Since cable losses are greater at higher frequencies it is evident that high-band attenuation will be greater than low-band attenuation. Therefore, to equalize this the amplifiers and signal splitters are often supplemented by equalizers. An equalizer or tilt control consists of a bandpass filter arrangement with an adjustable frequency response. It operates by introducing a relative low-frequency loss so that outputs from the amplifiers or splitters have uniform relative amplitude response across the entire VHF band.

The signal distribution from splitters to tap-off points is done through multicore coaxial cables in the same way as in a MATV system. In any case the signal level provided to a television receiver is of the order of 1.5 mV. This level provides good quality reception without causing acompanying radiation problems from the CATV system, which could cause interference to other installations and services.

10.3 CLOSED CIRCUIT TELEVISION (CCTV)

Closed circuit television is a special application in which camera signals are made available only to a limited number of monitors or receivers. The particular type of link used depends on distance between the two locations, the number and dispersion of receivers and mobility of either camera or receiver. Figure 10.3 illustrates various link arrangements which are often used. The simplest link is a cable where video signal from the camera is connected directly through a cable to the receiver. A television monitor, which is a receiver, without RF and IF circuits, is only required for reception in such a link arrangement. About one volt peak-to-peak signal is required by the monitor. Since the video signal is normally delivered via cables and even when transmitted, it is over a limited region and for restricted use, CCTV neede not follow television broadcast standards.

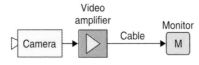

(a) Direct camera link to one monitor

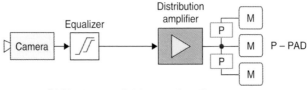

(b) Direct camera link to several monitors

(Figure 10.3 Contd. .....)

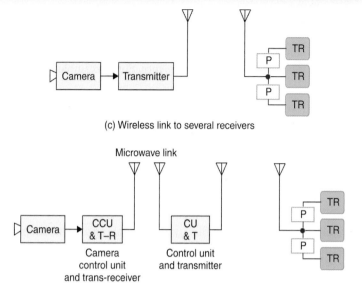

(c) Wireless link to several receivers

(d) Output of a remotely controlled camera feeding several
TV receivers located at a distance

Fig. 10.3. *Commonly used closed circuit television (CCTV) systems.*

CCTV Applications

There are numerous applications of CCTV and a few are briefly described here.

(*i*) *Education.* One instructor may lecture to a large number of students sitting at different locations. Similarly close-ups of demonstration experiments and other aids can be shown on monitors during these lectures.

(*ii*) *Medicine.* Several monitors and camera units can be installed to observe seriously ill patients in intensive care units. In medical institutions, operations when performed can be shown to medical students without their actually gathering around the operation table.

(*iii*) *Business.* Television cameras can be installed at different locations in big departmental stores to keep an eye over customers and sales personnel.

(*iv*) *Surveillance.* In banks, railway yards ports, traffic points and several other similar locations, closed circuit TV can be effectively used for surveillance.

(*v*) *Industry.* In industry CCTV has applications in remote inspection of materials. Observance of nuclear reactions and other such phenomena would have been impossible without television. Similarly television has played a great role in the scanning of earth's surface and probing of other planets.

(*vi*) *Home.* In homes a CCTV monitor finds its application in seeing the caller before opening the door.

(*vii*) *Aerospace and Oceanography.* Here a wireless link is used between the transmitter and receiver. In some applications camera is remotely controlled over a microwave radio link. As shown in Fig. 10.3 (*c*), for aerospace and oceanography a carrier is used for transmitting the signal and a complete receiver is then necessary for reception.

Chapter 10

10.4 THEATRE TELEVISION

Television programmes can be shown to a large audience in theatres. Similarly cinematographic films can be telecast for viewing on television receivers. Some examples of such applications are as follows.

(i) TV Programmes in Theatres

Special programmes can be shown on a large screen by optical projection in a theatre where spherical mirrors and reflectors are used to enlarge the image. With about 80 KV on the final anode of the picture tube, there is enough light to show pictures on a standard theatre screen. The same idea can be used for projecting TV programmes at home on a small screen.

(ii) Film Recorders

Film recorders produce a cinematographic film by photographing a television picture displayed on the screen of a picture tube. For doing so the film has to be pulled down frame by frame during successive blanking intervals. A video tap recording can only be rebroadcast in countries where the same TV standards are in use, whereas film recordings offer a ready means of exporting programmes to other countries using different standards.

(iii) Telecine Machines and Slide Projectors

Many television programmes originate from 35 mm and 16 mm photographic cinema films. Slides are also often used in TV programmes. Therefore, telecine machines and slide projectors form part of the television studio equipment for transmitting motion pictures and advertisement slides. Telecine machines are cinema projectors equipped with mirror or prism reflector arrangement for focusing pictures, as produced by them, on the face of a TV camera. Slide scanners also have a similar optical arrangement for transmitting still from different slides. For high utilization of the projector camera chain, an optical multiplexer is often used. This switches or directs one of the several optical image sources to the lens of a single camera, thus enabling the use of one TV camera for receiving programmes from three or four film and slide projectors. For the accompanying sound signal pick-up, the usual optical or magnetic track playback facility is incorporated in the multiplexer setup.

An additional problem of using telecine projectors is the difference in frame rate of motion pictures and television scanning. Motion pictures are taken at the rate of 24 frames/sec but while screening, each frame is projected twice to reduce flicker. This amounts to an effective frame rate of 48/sec. However, in TV transmission, while the frame rate is 25, the field rate is 50 on account of interlaced scanning. Thus, there is a difference of one picture frame/sec between the two and if not corrected, would cause a rolling bar on the raster besides loss of some signal output. In order to overcome this discrepancy, the film in the telecine projector is pulled down by the shutter mechanism at the rate of 25 frames/sec. It is achieved by a suitable speed correction in the drive mechanism. This naturally results in a little faster movement of scenes and objects on the television screen but the distortion caused is so small that it is hardly noticeable. The corresponding small increase in the pitch of reproduced sound also goes undetected.

Similarly in the 525 line system where the frame rate is 30/sec, it becomes necessary to suitably modify the film rate of 24 to achieve compatibility between the two in order to prevent rolling and loss of any picture information. The necessary correction is carried out by projecting one frame three times and the following frame two times. This sequence is repeated alternately by means of an intermittent shutter mechanism of 3 : 2 pull-down cycle for the film. The pull-down is carried out by a shutter which operates at the rate of 60 pulls per second. Thus, out of a total of 24 frames, 12 are projected three times while the rest only two times. This then makes the film rate of 24 frames equal to the scanning rate of $(12 \times 3 + 12 \times 2 = 60)$ 60 fields per second.

The speed alteration and sequencing explained above makes direct scanning of motion pictures possible through any TV network. While slides are used for stills, small advertisement films are recorded before hand on video tapes for telecasting when required.

(iv) Pay Television

Special programmes like first-run films, sports events and cultural programmes that are normally not broadcast on the usual TV channels because of their higher cost are made available to television subscribers either through the cable television network or on special channels. At the receiver a special decoder is used to receive the picture. The decoder also has the provision to register an extra charge for such special screenings. This service is optional but a fixed charge is made for initial installation.

10.5 PICTURE PHONE AND FACSIMILE

This is another fascinating application of television where two people can see each other while talking over the telephone line. A picture phone installation includes a unit that contains a small picture tube and a miniature TV camera. The highest modulating frequency in picture-phone services is normally limited to about 1.5 MHz.

Facsimile is another application of electronic transmission of visual information, usually a still picture, over telephone lines. Since there is no motion, a slow scanning rate is employed. Facsimile is employed for sending copies of documents over telephone lines.

10.6 VIDEO TAPE RECORDING (VTR)

Video tape recording was introduced in 1956 and it proved to be a vast improvement over the earlier method of recording motion pictures taken from the screen of the television receiver. Video tape retains the 'live' quality of broadcasting and has the capability of being edited and duplicated without any delay. The other advantages are (i) immediate playback capability, (ii) convenience of repeating the recorded material as many times as the viewer wishes, and (iii) ease of duplication for distribution to a large number of users.

The video signal can be recorded on a magnetic tape for picture reproduction in a similar way as the audio tape is used for reproduction of sound. In an audio recorder, the plastic tape (mylar) that has a very fine coating of ferric oxide is made to move in physical contact with the tape head. Any electrical signal applied to the tape head magnetizes the magnetic particles on the tape, as it passes across the head. For each cycle of the signal, two tiny bar magnets are produced on the tape and the length of bar magnets is inversely proportional to the frequency of applied signal. Thus, on a recorded tape these bar magnets form a chain with like poles adjacent to each other. When a recorded tape moves across the playback head, there is a change in flux linkages with the

Chapter 10

head and hence a voltage is developed across its coil terminals. This technique of recording and reproducing audio signals is illustrated in Fig. 10.4.

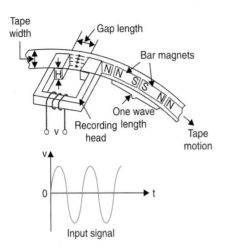

Fig. 10.4(a) *Electrical signal recorded in the form of bar magnets on the magnetic tape.*

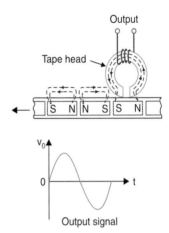

Fig. 10.4(b) *Development of signal during playback.*

Audio Range

Each head has a certain gap length. As the recording frequency increases, the length of the bar magnets decreases. A limiting frequency is reached when at a given tape speed, total length of the two adjacent bar magnets becomes equal to the gap length. At this frequency, output of the playback head will be zero since each bar magnet will produce equal and opposite voltage in the coil, with the result that no net flux passes through core of the head. For a tape speed of 19 cm (7.5″) per second and with a gap length of 6.3 microns (0.00025″) the usable frequency comes to about 15 KHz and is enough for audio recording.

Audio Signal Dynamic Range

Since output voltage from a playback head is directly proportional to the rate of change of flux, for every doubling of frequency the output voltage will become twice. In other words, every time the frequency gets halved (one octave lower), the output falls by 6 db. Assuming that the entire audio range occupies 9 octaves, output at frequencies that lie in the lowest octave will be 54 db below the output in the highest octave. This discrepancy is got over by providing equalizing circuits in the playback amplifier. The equalizing network is designed to have characteristics where the output voltage falls by 6 db per octave to allow for the rising response at the playback head.

AC Bias

If the signal to be recorded is applied directly to the record head, the output will be highly distorted on account of non-linearity of B-H curve of the core material around its zero axis. This difficulty is solved by superimposing the recording signal on a high frequency ac voltage. The amplitude of the high frequency bias is so chosen that its positive and negative peaks lie within the linear portions of the B-H curve. As shown in Fig. 10.5, the two outputs (marked X and Y) add up to give linear output with improved signal to noise ratio. The ac bias frequency is kept fairly high so that the beat signal between the highest signal frequency and bias frequency does not fall in the audio range. An ac bias frequency close to 60 KHz is considered adequate and is normally employed in audio tape recorders.

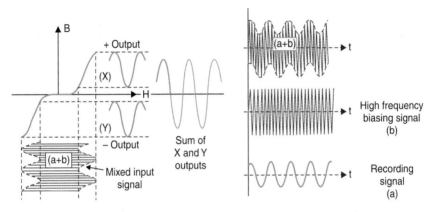

Fig. 10.5. *Effect of a.c. (ultrasonic) biasing in audio tape recording.*

Video Recording

The above introduction to audio recording will enable a better appreciation of the special problems of recording video signals on a magnetic tape.

Video Frequencies

For recording up to the highest video frequency of 5 MHz if the head gap is kept at 0.00025″ (6.3 microns) a tape speed of the order of 39 metres (1300 inches) per second would be necessary. With the head fixed as in audio recording and moving the tape at such a high speed would result in excessive wear and tear besides mechanical instability. Decreasing the gap below about 6 microns to lower the tape speed is not possible because of technological problems. However, it is not essential that only tape should move and head remains stationary. It is the relative speed of tape and head which is responsible for the output voltage. Hence the relative speed is increased by moving the tape head in opposite direction to the tape movement. In practice tape heads are mounted on the periphery of a drum which rotates around an axis relative to the direction of tape movement. This enables reduction in tape speed to as low as 7.5 inches ($\approx$ 20 cm) per second.

Video Signal Dynamic Range

As described earlier, output voltage from a playback head is directly proportional to the rate of change of flux. In case of video frequencies consisting of dc to 5 MHz, theoretically there would be infinite number of octaves. Even if frequencies down to 25 Hz are required to be retained (dc can be recovered by clamping), the band between 25 Hz to about 5 MHz would occupy nearly 17 octaves. This means that the average output on account of frequencies in the lowest octave will be about 100 db below the output in the highest octave. It is very difficult to handle such a large dynamic range where low frequencies run into noise levels. The octave problem could be solved by translating video frequencies to higher frequencies by amplitude modulation. For example, a carrier frequency of 10 MHz would provide sidebands in the range of 5 to 15 MHz. This covers only two octaves. However, amplitude modulation is not used because of the following reasons:

(*i*) At the speeds used in video recording, it is difficult to keep good contact with the head and this results in amplitude variations. Such amplitude variations are reproduced as noise.

(*ii*) In order to avoid distortion it is necessary to use an ac bias having a frequency atleast four times the carrier frequency. However, such a high frequency (4 × 10 MHz = 40 MHz) is not desirable on account of high frequency heating and other such losses.

Keeping in view the above drawbacks FM is used for video recording. A carrier frequency of about 6 MHz is often employed. Its amplitude is kept quite large thus making use of any ac bias unnecessary.

FM is insensitive to small amplitude variations. However, if present due to improper contact between the tape and head, such variations can be removed by the use of amplitude limiters. The dynamic range with FM recording is also quite small despite the fact that it has a wider bandwidth. This is so because FM sideband frequencies which have significant amplitudes occupy only a few octaves.

An additional advantage of FM is that it becomes possible to record and transmit even the dc component of the video signal. Audio signals which accompany the scene are recorded on the same tape by a separate head and played back by normal audio tape recording and playback techniques.

Scanning Methods

The two methods commonly used for video recording are called transverse scanning and helical scanning. In transverse scanning four quadruplex heads are used for recording and reproduction while in helical scanning either one or two heads are employed.

Transverse (Quadruplex-head) Scanning. Transverse scanning, though expensive is superior, and is therefore preferred in professional video tape recorders. Figure 10.6 (*a*) illustrates

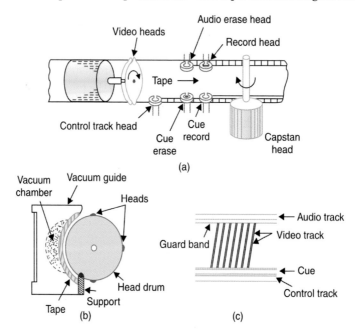

Fig. 10.6. A transverse recording system (a) basic principle of transverse recording (b) vacuum guide and head wheel (c) tape format.

the unidirectional path of the tape past a rotating head wheel which has 4 record/reproduce heads affixed to it at accurately spaced 90 degree intervals. Thus the heads rotate transversely across the tape while it is pulled slowly from one spool to another. Each head comes in contact with the tape as the previous one leaves it. The net result is a large relative motion between the head gaps and tape surface.

In order to achieve good contact between the tape and head (see Fig. l0.6 (b)) the tape is made to move in a curvature. The curvature is of the shape of head travel and this is given to the tape by a vacuum pump arrangement to ensure that each head maintains a proper fixed pressure contact with the oxide surface.

Because of tape movement and relative arc of the recording heads, the recorded track is somewhat slanted (see Fig. 10.6(c)) in the direction of tape travel. At the top and bottom edges of the tape, sync and audio signals are recorded respectively. A switching arrangement transfers signal to the active head (the head which is in contact with the tape) at the appropriate moment. A small guard band appears between any two slanted recording tracks on the tape. A full track erase head is used before the tape goes to the drum that carries the recording heads.

Synchronization. While recording, the completion of one video track must correspond to one field of picture scanning. Similarly during replay the video head must track it accurately otherwise reproduced picture will get severely distorted. Therefore speed regulation of the tape mechanism is very critical. The control signals recorded at the edges of the tape should be synchronized and thus locked with head rotation. This is done by servo control methods. The television signal to be recorded has its own pulse train but this cannot be used directly on the tape since it needs drastic changes before use for synchronizing and control of drive speeds.

Servo Control System. A signal in the form of pulses is generated for each rotation of the video head. This as shown in Fig. 10.7 is done either by a bulb and a photocell or by a small magnet and coil combination. The drum has a small slit through which light from a bulb falls

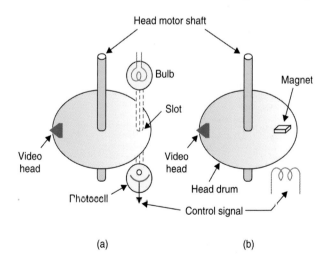

Fig. 10.7. Generation of control signal for video head speed regulation (a) photocell method, (b) magnetic method.

on a photocell once during each revolution of the drum. In the second method, a small magnet fixed on the drum, induces a small voltage pulse as it passes over a fixed coil once during each revolution. The drum pulses thus generated are compared with the incoming field sync pulses obtained during recording. The comparator produces an error signal corresponding to the difference in phase and frequency between the two signals. This is amplified and applied to eddy current brakes provided on the head motor. The brakes adjust the motor speed to provide necessary synchronization. On replay the drum pulses are compared with 50 Hz pulses derived from supply mains. The error signal thus derived from the comparator is used for controlling speed of the head motor. In addition to this the drum pulses are also used to ensure that the head runs accurately at the centre of the recorded track. Maximum video output is the indication of correct video head tracking.

The use of four recording heads together with switching arrangements, a vacuum system for good tape contact with recording heads and elaborate synchronizing facilities contribute to the high cost of this system. However, it is justified because professional VTRS which employ quadruplex system of recording with a full bandwidth of 5 MHz provide such good quality pictures that they look like a live programme.

Helical Scan Recording. Smaller and low-priced video tape recorders employ 0.5″ or 1″ tape and provide a bandwidth up to about 2.5 MHz. Such recorders normally use one head and a relatively simple drive mechanism. In this system of recording the tape is wrapped around a drum inside which the head rotates. The video head protrudes through a horizontal slit in the drum to come in contact with the tape. Figure 10.8(*a*) shows the mechanical layout of this method of recording. The video head rotates at 50 revolutions per second such that one field of picture information is recorded in one revolution. As shown in the figure, the tape comes in contact at the upper edge of the drum and leaves it at its bottom edge. Thus the recorded track is in the form of a helix and this gives it the name of helical scanning. While a single video head is used for both recording and playback, the tape passes before a full track erase head before it goes around the drum. A typical one inch tape format is shown in Fig. 10.8(*b*).

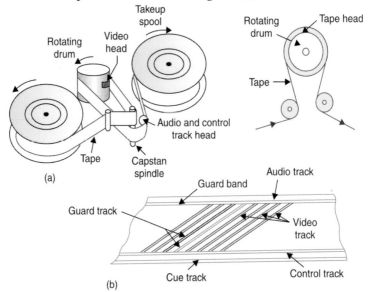

Fig. 10.8. Helical scanning system (a) single head wrap scanning (b) 1″ tape format.

It consists of an audio track and guard band at the top and audio cue, guard and control tracks at the bottom. The video tracks across the tape are at an angle of 5° to the tape length axis separated by 4 mil (0.004″) guard bands.

Quadruplex Head Recording and Playback Circuits. Figure 10.9 shows the basic block schematic arrangements of recording and playback in a quadruplex head VTR. In record mode the composite video signal of about 1 V P-P amplitude, as obtained from a camera set up, feeds into a two stage wide-band video amplifier in the video tape recorder. The output from this amplifier is fed to the FM modulator via a pre-emphasis network and driver.

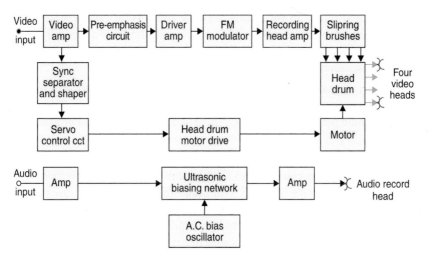

Fig. 10.9(a) *Simplified block diagram of a quadruplex head video tape recording system.*

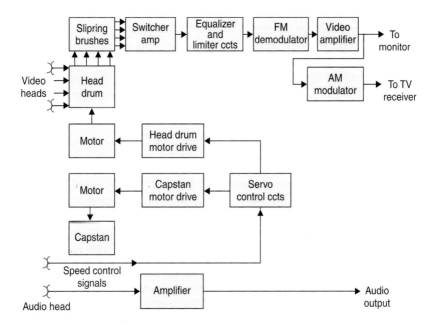

Fig. 10.9(b) *Simplified block diagram of a quadruplex head video tape system in the playback mode.*

Chapter 10

The modulator output is amplified by the recording head amplifier and fed to the record heads through slip-ring brushes. The accompanying audio signal is amplified, given an ac bias, and then fed to the audio head. As shown in Fig. 10.9 (a) a sync separator circuit is also connected across the output of the wide-band video amplifier. The sync pulses, after separation, are suitably shaped and used to synchronize and control the head drum motor speed by servo control techniques.

During playback (see Fig. 10.9 (b)) the video head outputs are collected through slip-ring brushes. These are then fed to an electronic switcher, which selects and amplifies the signal from the head that is in contact with the tape. The selection takes place during horizontal retrace blanking intervals so that the switching transients are not visible. The selected output is fed to the FM demodulator through an equalizing network and several limiter stages. The detected output is amplified by a four stage video amplifier before feeding it to a monitor or an amplitude modulator. After modulation with the carrier frequency of any one of the band I channels, the incoming programme can be viewed on a TV receiver.

The speed control signals are recovered from the corresponding tape tracks and processed for driving the capstan motor and synchronizing the head drum motor drive. The audio signal, as obtained from the audio playback head, is amplified and fed to the monitor or a modulator for use in a TV receiver.

Video Disc Recorder. Optical video disc recording is a recent development. It uses a laser beam in its pick up system. The video disc is 12″ (301.6 mm) in diameter and is made of transparent plastic. Since its reverse surface is coated with aluminium to reflect the laser beam, it has an appearance of a metallic disc. Sound and image signals are stored in tiny pits located in a substrate 1.1 mm from the surface. There are close to 14 billion pits on one surface of a disc. The reflection of the laser beam from the disc is intermittently interrupted according to the distribution and width of these pits and the reflected laser beam is converted into electrical signals. Since there is no friction from a stylus as in the case of conventional audio pickups, the sound and image quality of the laser beam type video disc is almost permanent. The rotational speed of the disc is so controlled that the relative beam velocity is constant from the outermost edge to the innermost end. One revolution of each track forms one frame of picture and one side of the disc can record up to 54000 frames. Both, one hour and half an hour duration dises are now available. In the later type of video disc, sound is FM modulated and recorded in two separate channels. Therefore, the two channels may be used for recording and reproduction of bilingual programmes or high fidelity stereo music.

10.7 TELEVISION VIA SATELLITE

The conventional methods for extended coverage of TV by microwave space communication and coaxial cable links are relatively expensive. Geostationary communication satellites launched into synchronous orbits around the earth in recent years have enabled not only national but also international television programmes to be relayed between a number of ground stations around the world. Three artificial satellites placed in equatorial orbits at 120° from each other cover practically the whole populated land area of the world.

High power, highly directive land based transmitters transmit wideband microwave signals to the geostationary satellite above the transmitter. Each microwave channel has a bandwidth of several tens of megahertz and can accommodate many TV signals and thousands of telephone channels or suitable combinations of these. The satellites usually powered by solar batteries receive the transmission, demodulate and amplify it and remodulate it on a different carrier before transmitting again. The transmitting antenna on the satellite, by the use of a suitable reflector, can direct the radiated beam to a narrow region on the earth and economize on power to provide a satisfactory service in the desired area. Higher power satellites can provide large power flux densities so that smaller size antennas can be used for reception. For national distribution the transmission is downwards from a wide angle antenna so that the whole national area is 'illuminated' by the transmission if possible. For international distribution the transmission is also towards the other one or two satellites (which are in line of sight direction) from highly directive antennas. The demodulation-amplification-remodulation transmission process is repeated in the second satellite. The final 'down channel' transmission is received (in the same or a different country) by a large cross-section antenna and processed in low noise receivers and finally reradiated from the regular TV transmitters.

There are a number of 'INTELSAT'* satellites over the Atlantic, Pacific and Indian Oceans operating as relay stations to some 40 ground stations around the world. The international system of satellite communication caters to the continental 625/50 and the American 525/60 systems. As television standards differ from country to country, the transmitting station adopts the standards of the originating country. The ground station converts the received signal with the help of digital international conversion equipment to the local standards before relaying it.

Frequency modulation is used for both 'up channel' and 'down channel' transmission. FM, though it needs a larger bandwidth, offers good immunity from interference and requires less power in the satellite transmitter.

Frequency Allocation

The frequency bands recommended for satellite broadcasting are 620 to 790 MHz, 2.5 to 2.69 GHz, and 11.7 to 12.2 GHz on a shared basis with other fixed and mobile services. The satellite antenna size and the RF power naturally depend upon the frequency of operation. Space erectable antennas are used for the 620-790 MHz band, with the size limited to about 15 metres, while rigid antennas are used both for 2.5 and 20 GHz bands, the size being limited to about 3 metres.

For the ground terminal, the maximum diameter of the antenna is restricted by the allowable beamwidth and frequency. The cost and complexity of the receiver increases with increase in frequency.

Extended Coverage of Television

Besides the use of satellites for international TV relaying, satellites can be used for distributing national programmes over extended regions in large countries because of their ability to cover large areas. For this, satellites can be used in three following ways.

(*i*) *Rebroadcast System.* In this system emission from a low power satellite is received with the help of a high sensitivity medium size (about 9 m) antenna on a high sensitivity low noise earth station. The received satellite programme is rerelayed over the high power terrestial transmitter for reception on conventional TV receivers. This method is suitable for metropolitan areas where a large number of TV receivers are in operation. This method also enables national hook-up of television programmes on all distantly located television transmitters. For a large country like India, to do so by microwave or coaxial cable links would be expensive.

---

*INTELSAT stands for INternational TELecommunication SATellite-consortium.

(*ii*) *Limited Rebroadcast System.* In rural areas where clusters of villages and towns exist, and the receiver density is moderate, low power transmitters can be used to cover the limited area. This reduces the ground segment cost by eliminating the need for special front-end equipment, dish antennas and convertors for each receiver.

The SITE (Satellite Instructional Television Experiment) programme conducted by India in cooperation with NASA of USA in 1975-76 was a limited rebroadcast system. A high power satellite ATS-6 (Application Technology Satellite) positioned at a height of 36,000 km, in a geostationary synchronous orbit with sub-satellite longitude of 33° East was used for beaming TV programmes over most parts of the country. TV programmes from the earth station at Ahmedabad were transmitted to ATS-6 at 6 GHz FM carrier with the help of a 14 m parabolic dish antenna. The FM carrier had a bandwidth of 40 MHz. The 'down transmission' from the satellite was done from a 80 W FM transmitter at 860 MHz. The transmitted signal consisted of a video band of 5 MHz and two audio signals frequency modulated on two audio subcarriers of 5.5 MHz and 6 MHz. This enabled transmission in two different languages and reception of any one of these. A block diagram of this system is shown in Fig. 10.10.

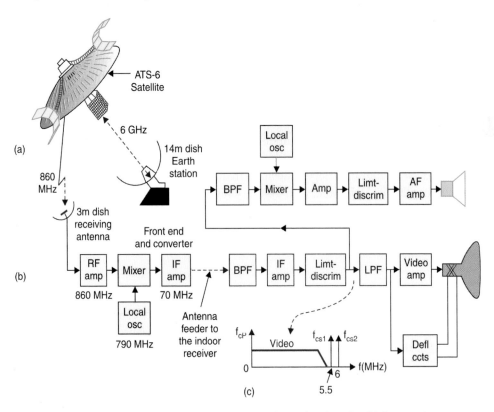

Fig. 10.10. *Direct reception system from a satellite. (a) ATS-6 Satellite (b) front end converter and receiver block diagram (c) base band frequency spectrum.*

(*iii*) *Direct Transmission.* Direct reception of broadcast programme is the only possibility in areas remote from terrestial broadcast stations. In this system the cost of reception is very high even with high power satellite transmissions. This is so because of the need for a special antenna to receive the signals and front-end convertor unit to modify the signals into conventional broadcast standards. Advances in technology reducing the cost of low noise front-end for receivers may make direct individual reception feasible in the near future. Japan was the first country to launch a medium scale satellite ('YURI') in April 1978 for experimental purposes towards direct reception. It radiates two colour channels in the 12 GHz band. The additional receiver equipment consists of 1 to 1.6 metres parabolic dish antenna and a front-end convertor to feed UHF-AM TV signals to the conventional receiver.

Cost of Satellite Communication

The cost of satellite communication would be very much lower than it is but for the limited life of the satellite. The life is limited because a geosynchronous satellite using high gain antenna requires close control of both its position and altitude in 'Orbit'. The position and altitude control rockets require fuel that has to be put in once for all before launch. Thus for a given payload, the longer the life the heavier is the satellite and correspondingly expensive. All communication satellites are therefore designed for a maximum operating life limited by its positioning fuel capacity. This of course has an advantage too. Successive generations of communication satellites can incorporate the latest developments in electronics and communication technology, packing much more capacity into satellites of comparable size. It is noteworthy that the cost per channel of one hop satellite communication has decreased over the last decade by a factor of more than ten. We can therefore hope that with advances in technology direct reception at reasonable costs will become a reality in a not too distant future.

10.8 TV GAMES

Television games is a relatively new application of digital electronics and IC technology to TV products. The first of the solid-state games used Transistor-Transistor Logic (TTL). The earlier set-ups provided logic for playing question answer games on the television screen. Later paddle type games were developed which included generation of sounds to give a touch of reality to the game being played. Then colour was added to the display and the challenge of contest amongst players increased by programming the game electronics to adapt to the player's skills. Now with the development of microprocessors (μP) further sophistication has become possible, where, for example in card games, players can compete against the computer and against each other.

Though logic can be developed to play almost any game but most common and commercially available games include tennis, soccer, squash and rifle shooting. The receiver screen shows the game in progress and also displays its current score. As the game proceeds the score display is updated properly. As an illustration, if the game being played is tennis, the screen (see Fig. 10.11) will show the court lines, the net, the rackets and the ball. The movement of the ball is fully shown as it is hit from one side to the other. In addition, individual scores are displayed on both sides of the court. Suitable sounds are generated when any contact occurs and these are reproduced on a loudspeaker in synchronism with the action. In some games like football, players are also shown along with movements of the ball.

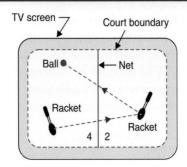

Fig. 10.11. Tennis game display on a receiver screen.

Functional Organization of TV Games

Any television game consists of two parts-the game unit and TV receiver. The game unit contains complete electronic circuitry necessary for generating signals, which when fed to the television receiver display the current status of the game on the screen. Functionally a TV game unit consists of three sub-units or blocks. These are: (*i*) player or user's control unit, (*ii*) game cum control circuit logic unit and (*iii*) RF oscillator and modulator section. Figure 10.12 shows these blocks or units and also the manner in which the modulated game video signal is fed to the receiver.

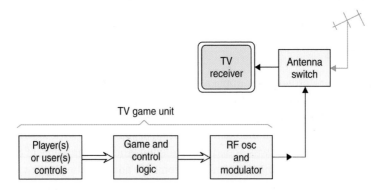

Fig. 10.12. Functional blocks of a TV game system.

The player control block contains various controls available to players for playing the game. In addition, it has the associated circuitry for generating corresponding command signals to initiate various actions.

The game and control logic section is the heart of any TV game. It produces video signals necessary for displaying game characters and game field on the screen. The characters may be simple paddles, bats, rackets or complex figures representing men, women and other objects necessary for the game. The interface circuitry for both player and game-action control forms part of the character and field video generation circuits. This section also provides logic circuits for game-playing rules, score display and totalling during the game.

The control and logic section also inserts sound signals at appropriate points in the game by pulse detection gating. In addition to all this it has a sync generation section which develops both horizontal and vertical sync pulses needed to time the composite video signals correctly.

The composite video signal which contains full game and sync information feeds into the VHF oscillator-modulator block. The oscillator is set for channels 3 or 4 and its output after modulation provides input signal for the TV receiver. It is normally fed to the receiver input terminals through a special antenna isolation switch. The output signal level from the oscillator-modulator unit is kept low to avoid any interference to other television sets operating in the vicinity.

Development of TV Game Circuits

Earlier TV game units, built with TTL consisted of different general purpose chips interconnected on a printed circuit board. These games being simple in nature needed a limited number of printed circuit boards. However, with time, the number of games offered by the same game unit increased and this made the units more complex. The corresponding complexity of electronic circuits required to implement these games made the TTL hardware too bulky and unmanageable. Accordingly manufacturers of TV games have now switched over to the use of dedicated (special purpose) ICs and microprocessors for designing all types of complex games and their logic circuits.

Dedicated ICs for TV Games

Single customer built LSI chips are now available which contain almost all the circuitry which goes into the making of a game unit. Besides a modulator, a clock-generator and a regulated power supply, such ICs need only a few discrete components like resistors and capacitors to produce multi game units. TV game industry has made big strides during the past decade and a large number of dedicated n-channel MOS chips, both for 625/50 scan (PAL-Colour) and 525/60 scan (NTSC-Colour) systems are now available. These include a choice of ball and paddle games with true game rules, realistic courts, and individual player identification. The battle games offer all the thrils and excitements of real battle scenes.

The organization of a TV game unit employing a specially built IC is illustrated in Fig. 10.13. As shown there, all outputs (command signals) from the user's panel and clock generator chip feed into this IC. The various outputs from the IC are combined in a video summer unit to form a composite video signal for feeding onto the modulator unit.

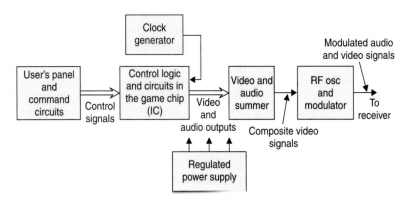

Fig. 10.13. Block diagram of a TV game system employing a dedicated IC chip.

In order to illustrate various functions performed by such ICs, a 28 lead dual-in-line dedicated IC package is shown in Fig. 10.14. It contains logic and controls for six selectable games which can be played by one or two persons with vertical paddle motion. The games

Chapter 10

Pin No.	Designation	Function
1	NC	No connection
2	V_{SS}	Negative supply input, nominally 0 V (ground)
3	Audio output	Three different audio tones for hit, boundary, reflection and score can be selected. The sound output lasts for about 35 ms
4	V_{CC}	Positive dc supply, + 7 V to + 9 V
5	Ball angles	(i) Input set to logic '1' (open circuit) selects ± 20° rebound angles
		(ii) Input set to logic '0' (V_{ss} i.e. ground) selects ± 20° and ± 40° rebound angles
6	Ball output	The ball video signal is output at this pin.
7	Ball speed	(i) Input set to logic '1' selects a low speed of ball motion
		(ii) Input set to logic '0' selects a higher speed of ball motion
8.	Manual service	With input logic at '1' the game stops after each service. However, when logic is switched to '0', the circuitry changes to automatic service mode
9, 10	Player outputs	Right Player (RP) and Left Player (LP) video outputs are available at these pins
11, 12	Right and Left Paddle/ ball position (location)	As shown in the figure an R-C network connected to each of these pins enables vertical position control of the paddle/ball through a 10 K potentiometer.
13	Bat size	(i) Logic 1—large paddle/bat size
		(ii) Logic 0—small paddle/bat size
14, 15	NC	No connection
16	Sync output	Standard horizontal and vertical sync and blanking pulses aie available at this pin.
17	Clock input	The output of the master clock (chip) is fed at this pin (Frequency = 2 MHz)
18 to 23	Game Selection	To select a particular game the corresponding switch (see figure) is set for logie '0' i.e. connected to V_{SS}. Other game selection switches remain at logic '1' i.e. open circuit.
24	Score and field output	The score and field output video signals are available at this pin.
25	Game Reset	The input switch is momentarily connected to V_{ss} (logic 0) to reset the score counters and to start a new game. Normally this pin connection stays at logic '1'.
26	Shot input	This input is driven by a positive pulse indirectly obtained from the user's panel to indicate a 'shot'
27	Hit input	This pin is also driven by a positive pulse triggered by the shot input if the target is hit.
28	NC	No connection

Chapter 10

which can be played are tennis, soccer, squash, practice (one man squash) and two rifle shooting games. It features automatic on-screen score display from 0 to 15, sound generation for hit, boundary and service, selectable paddle size, ball speed, two different rebound angles, automatic or manual ball service and visually defined areas for all ball games. The video signal output is suitable for black and white display on a standard domestic TV receiver. The functions of various pin connections on this IC (see Fig. 10.14) are as follows.

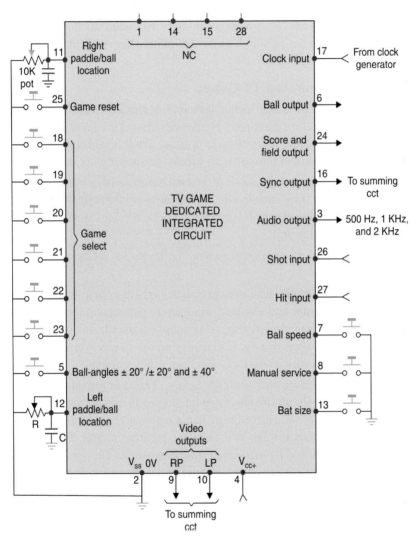

Fig. 10.14. Block diagram of a TV game dedicated IC. (Note: The pin numbers and their locations are arbitrary).

TV Games with Colour Display

Initially, designers of TV games were hesitant to provide colour display because of the complications of colour signal generation and modulation. However, now with the availability of separate ICs for such purposes, addition of colour needs little more than adding one or two integrated circuits to the schematic. National Semiconductor's LMI 889 is one such IC which accepts luminance (brightness), syne, chrominance (colour) and audio inputs and produces an RF modulated composite video signal. This IC includes two RF oscillators which are tuned to VHF low-band channels 3 and 4. Either output can be selected by applying a voltage to the external R-L-C tank circuit. The sound oscillator is isolated from the rest of the IC, and can be externally frequency modulated with a varactor diode or by switching a capacitor across the tank circuit. The crystal controlled colour subcarrier oscillator feeds two chroma modulators with quadrature signals for generating (B-Y) and (R-Y) colour-difference signals. Two RF modulators then add video, chroma and sound to the selected carrier frequencies.

Microprocessor (µP) Controlled TV Games

Some of the available games have a µP as the basic control element and use plug in ROM (Read Only Memory) cartridges to store game sets. Many side benefits are accrued by incorporating a µP into a particular video game. Most obvious of these are the versatility of the design, the multiplicity of games available and the ease with which a particular game design can be modified.

A µP controlled game may include, as its functional components, a microprocessor, Random Access Memory (RAM), Read Only Memory (ROM), Cassettes or other secondary magnetic storage medium besides a key board or other player control blocks, a video interface, a modulator and course a receiver as the display unit.

A block schematic of such a system is shown in Fig. 10.15. A brief description of the various blocks follows:

(i) *Display Unit.* The television receiver is used as the display unit. The receiver handles the signal fed to it in the usual and displays appropriate patterns at desired positions on its screen. The rate of display is made fast enough to maintain the illusion of continuity as is the normal practice in television broadcasts. The display on the screen is organised by the games system in a standard format which is usually 150 rows × 250 columns.

(ii) *Player Controls/Keyboard and Interface.* This unit provides a link between the system and players. The knobs on the keyboard are moved to initiate various actions. For instance in a tennis gams the rate and direction of displacement of the control knob will decide how quickly and in which direction the bat will move for hitting the ball. The outputs from the keyboard are analog in nature. The associated interface accepts the serial analog inputs and converts them into parallel digital form for processing by the µP (microprocessor). Similarly other command knobs generate appropriate analog signals which are necessary for a particular game.

(iii) *Memory—RAM (Random Access Memory) and ROM (Read Only Memory).* As shown in the figure the memory consists of two blocks—RAMs and ROMs. The RAMs store information temporarily which continuously updated or changed on receipt of commands from the players. The ROMs store fixed instructions for the µP for its processing the received data and sending out command signals to associated units. The ROMs also contain instructions for generating video signals for fixed patterns. For example in a football game, ROM stores the dot pattern of a football. When the game is in progress, it feeds this data continuously to the µP along with the information it receives about speed and direction of motion of the ball.

(*iv*) *Row and Column Counters.* The receiver screen is divided into 150 rows and 250 columns. This is done to determine the position of the object on the screen. The row and column counters are used to determine the row and column position of any object. To display a particular object at a certain place on the screen, the counters are suitably set by the μP. The counters in turn feed digital signals to the PVI (Programmable Video Interface) for generating corresponding video signals. For example, to give the impression of a horizontal movement of any object, the μP on receipt of such information from the memory, continuously changes the column counter to indicate the next horizontal position. Accordingly the column counter continuously feeds digital command signals to the PVI for generation of varying video signals to flash the object at its correct location on the screen.

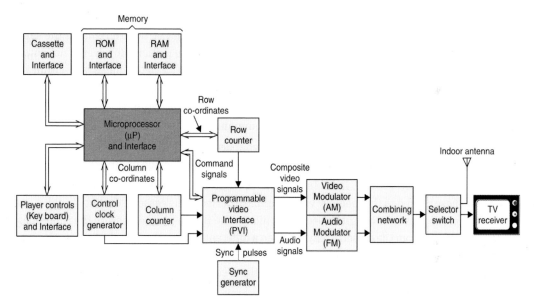

Fig. 10.15. *Simplified block schematic of a microprocessor controlled TV games setup.*

(*v*) *Programmable Video Interface (PVI).* This unit generates video and audio signals on receipt of digital commands from the μP both directly and indirectly. The μP continuously feeds the PVI with dot pattern output of various objects and indicates their motion and location through Row and Column counters. The PVI also receives input from the clock generator to produce audio signals to indicate various sound outputs at appropriate moment. The sync pulses are also added here to form a composite video signal.

(*vi*) *Sound Generator.* In order to make playing of TV games more realistic, suitable sounds are generated for various events like service, rebounds, shots, etc. The audio signals are processed within the PVI. The μP is programmed to control a square wave generator (clock generator) whose fundamental frequency, f_0, is usually about 7800 Hz. For each different occurrence a special bit (= 1) is set on. An 8-bit number (n) in the PVI determines the type of sound to be generated like a rattle, collision, a shrill whistle etc. This as stated above is determined by the μP and depends on the nature of game and the occurrence of different sounds while the game is in progress. The PVI on receipt of a particular 8-bit number sends an audio signal whose frequency is given by

$$f = \frac{f_0}{(n+1)} = \frac{7800}{(n+1)} \text{ Hz}$$

The sound output can have a frequency as low as 30 Hz. The audio signal is either fed directly to a separate loudspeaker or is reproduced on the receiver loudspeaker in the usual way.

(vii) Microprocessor (μP). The heart of the TV games system is the microprocessor. It processes the data fed to it and generates digital output which is used by the PVI to generate proper video and audio signals continuously. The memory unit (RAMs and ROMs) inform the μP how to process the data and send appropriate command signals. For example in a tennis game, as the player moves the control knob in an effort to hit the ball it generates a particular control signal. On receipt of this information via the memory unit, the μP decides whether row and column locations of the racket and ball overlap or not (for a contact) at the same instance. If contact is made the direction of the ball is reversed. The ball's dot structure remains the same while its address location changes continuously to show the ball in motion. However, if no contact is made the μP decides whether the point where the ball landed (X, Y co-ordinates) are within the acceptable areas, that is, inside the court or outside it and accordingly gives credit.

The output signal is also used to update the score on the screen. Similarly the μP decides the nature of the sound to be produced and sends a command signal for its generation.

The above explanation is a simplified view of what actually goes on in the system. The actual process involved is more complex and requires detailed and complicated programming.

(viii) Cassette Recorder and Interface. A large number of games can be stored in cassettes. These are fed to the memory via a suitable interface. The cassette output is analog and the interface converts it into digital form. This allows the RAMs to readily store the contents of the required 'game' from the cassette player. When a different game is desired, the tape is advanced or rewound till the desired game appears on the tape head. Information about one or two games is permanently stored in the memory unit and thus these can be played without any input from an external cassette receiver.

(ix) VHF Modulator. The video and audio outputs from the PVI together with sync pulses can be fed directly to the receiver through a cable. However, in modern systems these signals modulate VHF carriers of a particular channel (usually 2 or 3) as is normally done in a TV transmitter. The modulated output is fed to the receiver input terminals via a coaxial cable. In some designs the modulated signal is radiated through a small antenna. The receiver antenna intercepts the radiated signal and processes it in the usual way to reproduce visual display on the screen and sound in the loudspeaker.

REVIEW QUESTIONS

1. Draw the block diagram of a MATV system and explain how television signals are picked up from several stations and distributed to various locations in an apartment building or hotel. How is the impedance match maintained at different subscriber tap points ?

2. Describe the main merits and applications of a CATV system. Draw a typical layout of this system of signal distribution and label all the blocks. Why are amplifiers and equalizers required along trunk distribution lines ?

3. How is a CCTV system different from regular TV broadcasts ? Enumerate various applications of this system of television. Describe with suitable block diagrams various methods employed to feed/transmit video signal to different monitors/receivers.

4. Discuss special problems of video tape recording and explain how these are overcome for recording video signals on a magnetic tape. What is a basic difference between transverse and helical scan recording ? Explain with suitable diagrams the basic difference between these two methods of video recording.

5. Draw block diagrams illustrating record and playback modes of a quadruplex head VTR system. Label all the blocks and explain sequence of operations both for recording and playback. How is scanning and speed synchronization achieved in such a recording and reproduction system ?

6. Describe briefly various systems which can be employed for distributing national television programmes by a satellite over extended regions in large countries like India. What is the function of a special front-end used along with TV receivers for direct reception from a satellite ?

7. Describe with a block diagram the functional organization of a TV game set-up and explain the use of dedicated ICs for processing and control of analog signals generated at the user's panels.

8. Draw a schematic block diagram of a TV games set-up which employs a microprocessor for processing input data and generating digital outputs for the PVI.

Video Detector

From antenna to the input of video detector of a television receiver, it is all radio frequency circuitry, and is similar to a superhetrodyne AM radio receiver. It is at the video detector that picture signal is extracted from the modulated intermediate carrier (IF) frequency. This, after detection is amplified and fed to the picture tube cathode or grid circuit for reproduction of the picture.

The sound signal which is frequency modulated with the sound IF carrier frequency is translated in the detector to another carrier frequency, which is the difference of the picture IF and sound IF, *i.e.*, 38.9 – 33.4 = 5.5 MHz. The intercarrier FM sound signal thus obtained is separated at this stage or after the first video amplifier. It is then amplified and detected before feeding to the audio section of the receiver.

11.1 VIDEO (PICTURE) SIGNAL DETECTION

The video detector is essentially a rectifier cum high frequency filter circuit to recover video signal from the modulated carrier. Semiconductor diodes are used exclusively for detection and need about 2 volts or more of IF signal for linear detection without distortion. The signal to the detector is fed from the output of last IF amplifier stage. Either polarity of this signal can be rectified by suitably connecting the diode, since both sides of the modulated envelope have the same amplitude variations. This choice depends on the number of video amplifier stages used and the manner in which the video signal is injected in the picture tube circuit. It should be noted, however, that polarity is not important in an audio system because the phase of ac audio signal for the loudspeaker does not matter in reproduction of sound, but a polarity inversion of video signal driving the picture tube would produce a negative picture.

The detector may use either series circuit or shunt circuit, the basic forms of which are shown in Fig. 11.1. The series circuit arrangement is preferred because it is more suited for impedance match between the last IF amplifier output and input of the video amplifier.

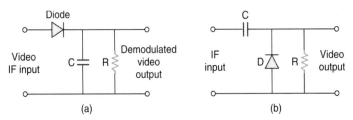

Fig. 11.1. Basic detector circuits (a) series (b) shunt.

In most television receivers a positive going signal is obtained at the output of the detector because this is the correct polarity for cathode injection in the picture tube after one stage of video amplification. The choice of polarity is also influenced by the type of AGC used. In some transistor receivers a negative going signal is developed, where a two stage video amplification is employed for feeding the picture signal to the cathode of picture tube. A schematic representation of the two types is illustrated in Fig. 11.2.

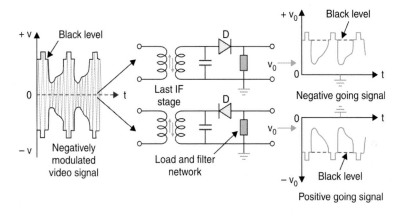

Fig. 11.2. *Production of negative and positive going video signals from a negatively modulated video signal.*

11.2 BASIC VIDEO DETECTOR

The basic circuit of a video detector employing a diode is shown in Fig. 11.3 where a parallel combination of C, a small capacitor and R_L, a large resistance constitutes the load across which

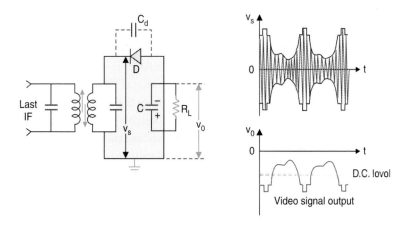

Fig. 11.3. *A simple diode detector and filter circuit.*

rectified output voltage v_0 is developed. Note that the load is connected to anode of the diode to develop a negative output voltage with respect to ground. The diode conducts during negative half cycle of the input to charge 'C' up to a potential almost equal to the peak signal voltage v_S. The difference is due to diode trop, since the forward resistance of the diode though small is not zero. On the downward and positive half of each carrier cycle the diode becomes non-conducting, and during this interval some of the charge on capacitor 'C' decays through R_L, to be replenished at the next negative peak. The time constant of $R_L C$ network is kept large compared with the time period of the applied IF signal. Then the circuit settles down to a condition, in which short current pulses flow through the diode only during the tip of each IF cycle to replenish the charge lost. The time constant of the network must, however, allow the capacitor voltage to follow the comparatively slow variations of the envelope of the modulated carrier. This condition can be shown to occur approximately when

$$\frac{1}{R_L C} = \omega_m \times \frac{m}{\sqrt{1 - m^2}}$$

where ω_m is the highest modulating frequency and m is the modulation index. With the highest modulating frequency of 5 MHz and assuming an average modulation index of about 0.4, the time constant $(R_L C)$ comes to nearly 0.08 µs. The period t_C of the carrier (IF = 38.9 MHz) frequency = 0.025 µs and that of the highest modulating frequency (5 MHz) = 0.2 µs. In practice a time constant close to half the period of the highest modulating frequency is chosen for effective and distortion free detection. This, in our case, is 0.1 µs and is nearly equal to the calculated value of 0.08 µs. This time constant (0.1 µs) is seen to be much higher than the IF time period and thus an output voltage that is very nearly equal to that of the envelope of the AM wave is ensured. If the $R_L C$ time constant is made too small, the output waveform will have a large IF ripple content which is not desirable. However, if the $R_L C$ product is kept too large it will not affect the negative going half-cycle of the envelope waveform, but will cause distortion of the positive going movements of the modulated envelope. This is known as positive peak clipping.

Choice of R_L and C

When the diode is conducting, it is obvious that some part of the output voltage is dropped across R_d, the series diode forward resistance. The detector efficiency then depends on the ratio of R_L / R_d, where R_L is the load resistance. Thus, for higher efficiency, a diode with a small forward resistance must be chosen and R_L should be kept as high as possible. As R_L is increased, C has to be reduced to maintain the correct time constant. The smaller the value of C the more significant becomes C_d, the anode to cathode capacitance of the diode. A reference to Fig. 11.3 will show that C and C_d form a potential divider across the input circuit. During positive half cycles of the applied carrier voltage, when the diode does not conduct, the entire signal voltage splits across these two capacitors. This results in some unwanted positive half of the applied voltage being developed across the load capacitor C. This reduces the net negative-going output and hence the efficiency. Therefore in an effort to increase R_L, C cannot be reduced too much because then most of the unwanted positive going voltage would develop across R_L and reduce the net output voltage greatly. Since the voltage divides itself in inverse proportion to the capacitances, and the larger voltage appears across the smaller of the two capacitors in series, the ratio C / C_d should be as large as possible. If C is to be decreased to use a high value of R_L, then C_d must be very small.

Choice of the Diode

For the time constant fixed at 0.1 μs, R_L varying between 2.7 K and 5 K in parallel with C, between 15 pF and 30 pF are commonly employed as load network for the detector in television receivers. Since R_L is only a few kilo ohm, the forward resistance of the diode must be as small as possible. The diode capacitance C_d must also be very small because the value of C chosen for the network is only around 20 pF. Some of the semiconductor diodes that meet the above requirements are 0A79, IN69 and IN64, and are often used in video detector circuits.

It would be instructive to compare the values of R_L and C chosen for the TV receiver detector with that of a broadcast radio receiver. The corresponding values for a radio receiver detector are : R_L = 500 K and C = 100 pF (see Fig. 11.4) and are based on the same considerations as explained for the television detector.

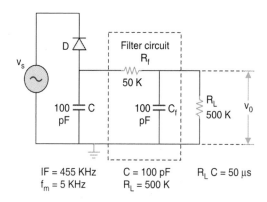

Fig. 11.4. *Detector and filter circuit of a radio receiver.*

11.3 IF FILTER

The detector output voltage, v_0 consists of three components—(*i*) the required video signal, (*ii*) an IF ripple voltage superposed on the video waveform and (*iii*) a dc component of amplitude almost equal to the average amplitude of the AM wave. The IF carrier component is removed by passing the signal through an IF filter. The dc component, if not required, is blocked by inserting a coupling capacitor in series with the signal path. However, the dc component forms a useful source of AGC voltage and represents the average brightness of the scene.

A brief survey of the filter circuit used along with a radio receiver detector will be helpful before looking into the special problems involved in detection of video signals. Such a detector circuit with provision to filter out IF ripple frequency is shown in Fig. 11.4 R_f is chosen to be very much greater than X_{cf} (reactance of C_f) at the intermediate frequency of 455 KHz. Most of the IF voltage gets dropped across R_f and v_0 is practically ripple free. C_f being small acts as an open circuit for the audio signal. R_f and R_L then form a potential divider for the desired audio signal, and a part of this signal is also lost across R_f. However, since $R_f << R_L$ (nearly 1/10th) the loss of audio signal is very small. The filter configuration shown is standard in radio receivers.

The RC filter circuit used in radio receivers is not practicable for video detectors because of very low values of load resistance. A suitable value of R_f to attenuate IF would seriously attenuate the video signal as well, because of small value of R_L across which the video output voltage develops. A series inductor is therefore used in place of R_f and the RC filter is thus replaced by an LC filter (see Fig. 11.5). At the IF frequency (38.9 MHz), the inductor L_f has much higher reactance than that of the shunt capacitor C_f but at the video frequencies the

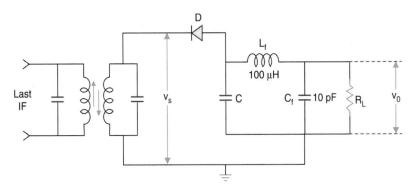

Fig. 11.5. Basic video detector and fitter circuit.

reactance of L_f is much lower as compared to that of C_f. As labelled in Fig. 11.5, $L_f = 100\ \mu H$ and $C_f = 10\ pF$ have been chosen to illustrate the filter action. At IF frequency the ratio of the reactance of series inductor to that of shunt capacitor (X_{Lf}/X_{Cf}) is nearly equal to 60. This means an attenuation of about 35 db for the ripple voltage and is considered adequate. However, the attenuation of the video signal will be different for different frequency components of the composite signal. The higher video frequencies suffer more attenuation than the lower ones. In order to overcome this discrepancy the load is modified to include an inductor in series with R_L. The reactance of this compensating coil (L_C) increases with frequency and thus the magnitude of the complex load increases to counteract the additional drop across L_f, the series inductor. A typical value of such a compensating coil is of the order of 200 µH. Such an arrangement also takes care of the input capacitance of the following video amplifier which would otherwise tend to attenuate high frequency components of the video signal. The filter circuit thus modified, ensures low, and almost constant attenuation over the entire frequency range. This is known as video bandwith compensation. The bandwidth must extend up to 6 MHz to include sound intercarrier frequency of 5.5 MHz and its FM sidebands besides the video signal.

Figure 11.6, shows a practical video detector circuit. It employs a load resistance = 3.9 K and a compensating coil = 250 µH. Its output is dc coupled to the video driver. The resistors R_1 and R_2 form a voltage divider across the 12 V dc supply to fix necessary forward bias at the base of the emitter follower (driver). The capacitor C_1 provides effective ac bypass across R_2.

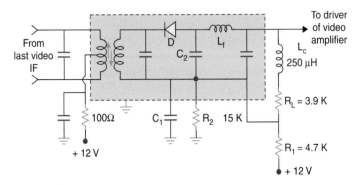

Fig. 11.6. A practical video detector circuit.

Filter Circuit Modifications

A somewhat non-linear behaviour of the diode results in production of a series of harmonics and beat frequencies at output of the detector. It is possible for these unwanted products to get coupled to the tuner and go through a zero beat as the tuner is tuned (through the corresponding RF range) to produce what are known as 'tweets'. Such interference is usually confined to channels that lie between 80 to about 180 MHz. To eliminate these interferences, the diode and part of the filter circuitry are enclosed in the 'can' of last picture IF stage coupling transformer for effective screening (see Fig. 11.6).

In some detector designs self-resonant chokes that are tuned to suppress specific troubling frequencies are inserted in series with the signal path. Figure 11.7 shows such a circuit configuration. Note that a major part of filter capacitors is provided by the stray and wiring capacitances and the 'wired in' (physical) capacitors are much less in value and range from 5 pF to 10 pF. Both the series inductors L_{fa} and L_{fb} are made to resonate at the desired frequencies by their self-capacitance and no physical capacitors are actually needed. In some cases the filter capacitor C_f across the load is provided by the input capacitance of the video amplifier stage so that no 'wired in' C_f appears in the circuit.

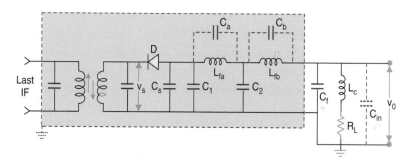

Fig. 11.7. *A modified video detector circuit.*

11.4 DC COMPONENT OF THE VIDEO SIGNAL

The video detector output includes a dc component which must be preserved for a true representation of the transmitted picture. Therefore with dc coupling employed between the detector and the video amplifier and video amplifier to the picture tube, all shades from white to grey get correctly reproduced. However, in some receivers ac coupling is used between the detector and video amplifier. The insertion of a coupling capacitor in series with the signal path completely removes the dc component. The video signal waveform then settles down with equal areas on either side of the zero voltage line. This is illustrated in the waveforms drawn in Fig. 11.8. Note that this results in lesser contrast between two lines having different brightness levels. Similarly an increase in the average brightness of the transmitted scene results at the receiver in a depression of the black level, so that the reproduced range in brightness is less than that of the original scene. Notwithstanding this disadvantage ac coupling is sometimes used because of certain other merits. However, the dc level of the video signal can be reinserted by what is known as 'dc reinsertion' technique before injecting it at the grid or cathode of the picture tube. This is fully explained in Chapter 14.

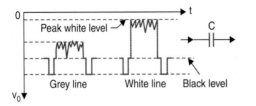

Fig. 11.8 (a). Video detector output for two different lines, one grey and the other white. Note the black level is same for both the lines.

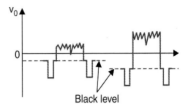

Fig. 11.8 (b). Effect of a.c. coupling. The black level is now different and thus the d.c. component is lost.

11.5 INTERCARRIER SOUND

In addition to recovering the composite video signal, rectifying action of the diode in the video detector also results in frequency translation of the sound IF signal. The strong picture IF carrier at 38.9 MHz acts as a local oscillator and heterodynes with the attenuated sound IF carrier at 33.4 MHz to produce a difference frequency of 5.5 MHz. The resulting new IF together with its FM sidebands is known as intercarrier sound signal. The video detector circuit is modified (see Fig. 11.9) to extract the sound signal. As shown in the figure, a parallel tuned circuit, commonly known as sound IF trap is inserted in the signal path. It is tuned somewhat broadly with a centre frequency of 5.5 MHz. This offers a high impedance to the sound component of the detected signal to remove it effectively from the video signal path. A tuned secondary circuit delivers the intercarrier IF to sound section of the receiver. In some TV receivers the intercarrier sound signal is allowed one stage of amplification in the video amplifier and then separated through a trap circuit. It may be noted that in colour receivers two separate diode circuits are used, one as a 5.5 MHz sound convertor and the other for video signal detection. This done to reduce interference in colour pictures due to the beat note produced at the difference frequency of sound carrier and colour subcarrier frequencies.

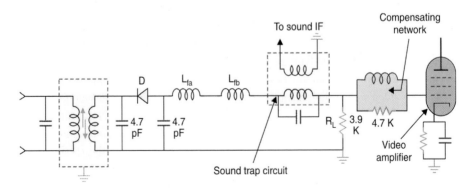

Fig. 11.9. Video detector circuit with intercarrier sound trap.

11.6 VIDEO DETECTOR REQUIREMENTS

It is now obvious that in practice the video detector diode load is more complex than a simple shunt combination of resistance and capacitance because of the many functions it is required to perform. The requirements are as follows:

(a) The detector load must provide a suitable impedance as seen through the diode, at input of the detector to tune and damp secondary of the last IF coupling circuit correctly.

(b) The detector load must remove from the output, the IF content in the signal as much as possible. For this purpose the load usually includes one or two low-pass filter sections.

(c) The detector load should have a trap circuit (a series rejector circuit) for separating the intercarrier sound signal.

(d) The detector load must also include a provision to boost the higher video frequencies to compensate for the loss due to input capacitance of the video amplifier.

It is obvious from these requirements that rigorous theoretical design of such a diode detector is very complex and therefore, in practice the design is usully reached by empirical methods based on filter circuit theory.

11.7 FUNCTIONS OF THE COMPOSITE VIDEO SIGNAL

Figure 11.10 illustrates various paths for the composite video signal as obtained from the video detector. We can consider that the signal is coupled to several parallel branches for different functions. Therefore, each circuit can be operated independently of the others. For instance clipping the sync pulses in the sync-separator stage does not interfere with the video amplifier supplying signal to the picture tube. Similarly, the AGC circuit rectifies the video signal for developing AGC bias. With the same video signal the video amplifier provides complete video signal to the picture tube for reconstruction of the televised scene. In some receiver designs a cathode or emitter follower is used to isolate the video detector from these circuits. In many receivers the video signal for the sync-separation circuit is tapped after one stage of video amplification.

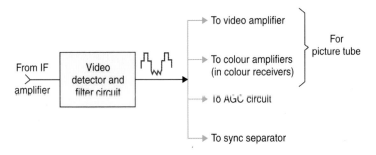

Fig. 11.10. Functions of the composite video signal.

REVIEW QUESTIONS

1. Draw the basic circuit of a video detector and explain design criteria for the choice of time constant of the load circuit. What determines the polarity of the diode in the detector circuit ?

2. Describe the factors that influence the choice of R_L and C in the load circuit. Suggest suitable values of R_L and C for the 525 line system where $IF = 45.75$ MHz and $f_m = 4$ MHz.

3. Explain why is it necessary to employ and L-C instead of R-C filter to remove IF ripple from the detected output. Give typical values of L and C and justify them.

4. Why is the filter circuit generally modified to include self-resonant inductors in the signal path ? What precautions are taken to prevent undesired harmonics from reaching the tuner ?

5. Explain how compensation is provided in the detector load circuit to extend its bandwidth. Why is it necessary to have a bandwidth of nearly 6 MHz ? Sketch the complete circuit and label all components.

6. How is the FM sound signal at 33.4 MHz translated to a new carrier frequency of 5.5 MHz and separated from the composite video signal ?

7. Give a practical video detector circuit imcorporating the following features: (*i*) effective IF filtering, (*ii*) suppression of harmonics, (*iii*) separation of the intercarrier sound signal, and (*iv*) frequency compensation. Give typical values of all the circuit components.

Chapter 11

12

Video Section Fundamentals

The amplitude of composite video signal at the output of video detector is not large enough to drive the picture tube directly. Hence, further amplification is necessary, and this is provided by the video amplifier. The manner in which video signal is applied to the picture tube (cathode or control grid) decides the type of video section circuitry. The video signal on application to the picture tube varies the intensity of its beam as it is swept across the screen by deflection circuits. The gain control of the video amplifier constitutes the contrast control, whereas the brightness control forms part of the picture tube circuit.

12.1 PICTURE REPRODUCTION

Figure 12.1 shows how the input video signal voltage for one line, impressed between grid and cathode of the picture tube results in reproduction of picture elements for that line. It should be noted that the results are the same when reversed video signal is applied between cathode and control grid. In fact the video signal should so align itself that its black level drives the grid voltage to cut-off. Any grid voltage more negative than that is called blacker than black and this part of the video signal corresponds to sync-voltage amplitude. At the grid of picture tube sync pulses really have no function, but these are used in the synchronizing section of the receiver to time deflection circuits for vertical and horizontal scanning.

12.2 VIDEO AMPLIFIER REQUIREMENTS

In order to produce a suitable image on the screen of picture tube, the video amplifier must meet the following requirements:

(i) Gain

The video signal must be strong enough to vary the intensity of the picture tube scanning beam to produce a full range of bright and dark values on the screen. This is illustrated in Fig. 12.1, where the signal amplitude is large enough to provide the desired contrast between white and dark parts of the scene being televised. However, with a video signal having smaller peak-to-peak variations, the brightness extends from dark, at cutoff bias, to some shade of grey, with the result that there is less contrast between dark and light areas. Figure 12.2 illustrates the light variations produced when the signal amplitude is reduced to about half as compared to the signal amplitude required for full contrast. Any further reduction in the video signal amplitude will result in a washed-out picture, and there will be little difference between dark and light areas of the picture. A video signal amplitude of about 75 volts peak-to-peak is needed to obtain a picture with full

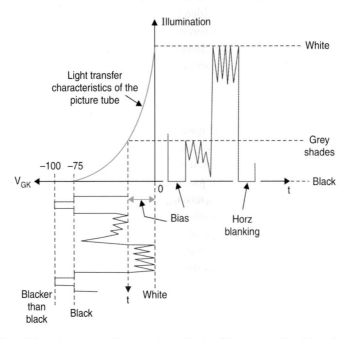

Fig. 12.1. *Light output with correct amplitude of the composite video signal.*

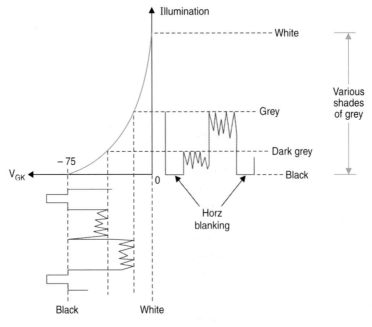

Fig. 12.2. *Effect of insufficient video signal amplitude on brightness variations in the picture.*

contrast. Some colour picture tubes need higher signal amplitudes of the order of 150 V peak-to-peak, for proper reproduction of the picture. With a detector output betwen 2 to 4 volts in all tube receivers, a gain between 25 to 50, at the video amplifier is considered adequate. A single pentode valve can develop this gain and most TV receivers using tubes have one stage of video amplification between video detector and picture tube. However, in transistor receivers where detector output seldom exceeds 2 volts, a two to four stage video amplifier becomes necessary to fully modulate the beam of the picture tube.

(ii) Bandwidth

As explained in an earlier chapter higher frequencies are needed to reproduce horizontal information of the picture. The lowest frequency for picture information in the horizontal direction can be considered as 10 KHz when the camera beam scans all white and all black lines alternately. Note that the active line period has been taken as 50 μs instead of the actual period of 52 μs. Similarly when the beam scans half white and half black lines in succession, the video frequency generated is 20 KMz. This is illustrated in Fig. 12.3 where different alternate black and white widths have been closen to demonstrate the generation of high frequency signal. Thus, for reproducing very minute details, a very high video frequency would be necessary, but keeping in view the limitations of channel bandwidth, the upper limit has been fixed at 5 MHz.

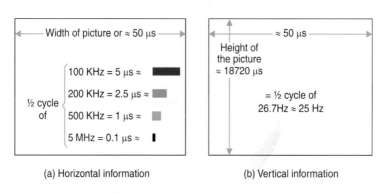

(a) Horizontal information   (b) Vertical information

Fig. 12.3. *Relationship between picture size and video frequencies.*

The signal frequencies corresponding to picture information scanned in the vertical direction are much lower compared with those for reproduction within a line. If the video voltage it taken from top to bottom through all the horizontal lines in a field, the variation will correspond to a half-cycle of a signal with a frequency of approximately 25 Hz. When the brightness of the picture varies from frame to frame the resultant signal frequency is lower than 25 Hz. However, this is considered as a change in dc level corresponding to a change in the brightness of the scene; and this can become almost zero Hz (i.e., dc) when the average brightness does not change over a long period of time. Ideally then, the video amplifier response should be linear from dc to the highest modulation frequency of 5 MHz. This is possible only when the video amplifier is direct coupled.

(iii) Frequency Distortion

The gain at high frequencies falls-off because of shunting effect of the device's output capacitance, stray capacitances and input capacitance of the picture tube. When ac coupling is employed the gain decreases at low frequencies on account of increasing reactance of the coupling capacitor.

Chapter 12

This inequality in gain at different frequency components of the signal is called frequency distortion. Excessive frequency distortion cannot be tolerated because it changes picture information. If high frequency content of the video signal is lost due to poor high frequency response the rapid changes between black and white for small adjacent picture elements in the horizontal line cannot be reproduced. This results in loss of horizontal detail. For example in the test pattern shown in Fig. 12.4, the individual black lines close to the centre will loose their identity and instead appear as an illdefined black patch. Similarly small details such as individual hairs of a person's eyebrows do not appear clearly. However, in a close up of the same face, where each area of the picture gets enlarged, the sharpness of each detail improves because a relatively low frequency range is required for proper reproduction of details. Frequencies from about 100 KHz down to 25 Hz represent the main parts of the picture information, out of which frequencies from 100 KHz to about 10 KHz correspond to black-and-white information of most details in the horizontal direction and frequencies from 10 KMz down to 25 Hz represent changes of shading in the vertical direction. If the low frequency response is poor, the picture as a whole is weak with poor contrast. The lettering, if any, is not solid and the average brightness appears to be changing gradually from top to bottom of the raster, instead of complete change of brightness in the actual scene.

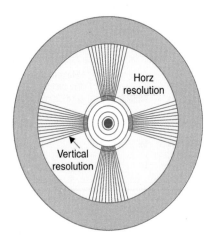

Fig. 12.4. Test chart for determining vertical and horizontal resolution.

(iv) Phase Distortion

Phase distortion is not important in audio amplifiers, because the ear does not detect changes in relative phases of the various frequency components present in a given sound signal. However, it is important in video amplifiers, since phase shift implies time shift, which in turn means position shift in the reproduced visual image. The resultant shift in relative positions of the various picture elements is detected by the eye as distortion. Therefore relative phases of all the frequency components present in the video signal must be preserved. The time delay due to phase shift is not harmfull if all frequency components have the same amount of delay. The only effect of such uniform delay would be to shift the entire signal to a later time. No distortion results because all components would be in their proper place in the video signal waveshape and so also in the reproduced picture. Therefore, the phase angle delay should be directly proportional to the frequency or all frequency components must have the same time delay. This is illustrated in Fig. 12.5 (*a*) and (b). It should be noted that the signal inversion of exactly 180° in any one stage of the video amplifier does not mean phase distortion. There is no time delay, but only a polarity reversal.

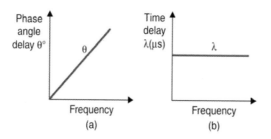

Fig. 12.5. *Phase response of the amplifier:*
(a) Phase angles proportional to frequency.
(b) Corresponding time delay which is constant.

Figure 12.6 shows frequency response of an RC coupled amplifier where f_L and f_H are the lower and upper 3 db down frequencies. The corresponding phase shift angles at these frequencies are +45° and −45° relative to midband frequencies. Phase distortion is very important at low video

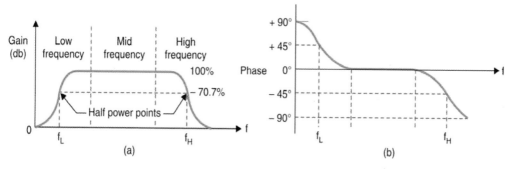

Fig. 12.6. *(a) Frequency and (b) phase response of a practical amplifier.*

frequencies because here even a small phase delay is equivalent to a relatively large time delay. As an illustration consider an amplifier designed to have f_L = 2.5 Hz. The phase shift at 2.5 Hz = 45° and that at $10 f_L$ (25 Hz) it is nearly 6°. The relative time delay between this frequency and

midband frequencies would be $\dfrac{6}{360} \times \dfrac{10^6}{25} \approx 660$ µs, which in turn would mean that the picture

information due to the two frequency components (2.5 Hz and 25 Hz) would get displaced with

respect to each other by nearly $\dfrac{660}{64} \approx 10$ lines. To correct this discrepancy even if phase shift at

25 Hz is made = 1°, the corresponding time delay would be 120 µs and the picture information will get displaced by about 1.5 lines on the raster. The eyes are very sensitive to time delay errors and see this as 'smear' on the picture. At very high video frequencies the effects of phase distortion are not as evident on the screen because the time delay at these frequencies is relatively small. For example, if f_H, the upper corner frequency of the amplifier is set at 5 MHz, the corresponding time

delay with respect to midband frequencies is only $\dfrac{45}{360} \times \dfrac{1}{5 \times 10^6} \approx \dfrac{1}{40}$ th of a µs. The consequent

Chapter 12

displacement of the picture elements is too small to be detected. Thus, a video amplifier with flat frequency response up to the highest useful frequency, has negligible time delay distortion for very high video frequencies.

(v) Amplitude Distortion of Nonlinear Distortion

If the operating point on the transfer characteristics of a device for a given load and signal amplitude is not carefully chosen, amplitude distortion occurs where different amplitudes of the signal receive different amplification. This can result in limiting and clipping of the signal or in weak signal output. If sync pulse voltage gets compressed, synchronization may be lost, because the video amplifier usually provides composite video signal for the sync separator. Very often some gain has to be sacrificed to avoid amplitude distortion.

(vi) Manual Contrast Control

It should be possible to vary amplitude of the video signal for optimum setting of contrast between white and black parts of the picture. Any control that varies the amount of ac video signal will operate as a contrast adjustment. Therefore contrast varies when gain is varied in either picture IF section or video amplifier. However, such a control is not possible in the IF section because all receivers employ automatic gain control circuits to maintain almost a constant voltage at the output of the video detector. Also, with intercarrier sound, any change of gain in the IF section would affect the sound volume. Therefore, contrast control is provided in the video amplifier, and this in effect, is the gain control of video amplifier.

12.3 VIDEO AMPLIFIERS

It is obvious from the preceding discussion that video amplifiers must meet several exacting demands and this calls for careful and rigorous design considerations. The wide-band requirement starting from almost dc to several MHz with minimum phase distortion is perhaps the most stringent requirement. Both direct-coupled and RC coupled configurations are used and each type has its own merits and demerits. Both types need high frequency compensation, and this is met by shunt-peaking and series-peaking techniques. Though a dc amplifier does not need any low frequency compensation, the RC coupled amplifier employs special low frequency boost techniques to extend the bandwidth at the lower end of its response.

A basic RC coupled amplifier configuration, which applies to both tubes and transistors, is shown in Fig. 12.7. It is designed to work under class 'A' operation. The gain of this amplifier

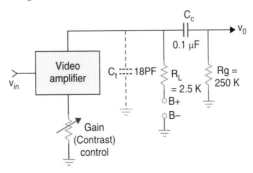

Fig. 12.7. Basic R.C. coupled amplifier

Note: *The component values shown are of a typical video amplifier employing a pentode.*

falls off rapidly at high frequencies because of shunting effect of inter-electrode, input, and stray capacitances in parallel with the load resistance R_L. Video stages use low value of R_L compared with audio amplifiers because of large bandwidth requirements. Typical values are 2 KΩ to 8 KΩ in tube versions. Although gain is reduced, lowering R_L extends the high frequency response. The effect of C_t does not become important until its reactance is low enough to become comparable with the resistance R_L. Then, the shunt reactance X_{ct} lowers the impedance Z_L, and this causes a fall in the gain at higher frequencies. The lower the R_L, and smaller the shunt capacitance C_t, better is the high frequency response. However, practical video amplifiers generally use a relatively higher value of R_L with peaking coils to boost the gain at higher frequencies. A typical arrangement known as 'shunt peaking' is shown in Fig. 12.8. Here a small inductor, L_0 is connected in series with R_L. This peaking coil resonates with C_t to boost the gain at high frequencies (see Fig. 12.8 (b)), where the response of the uncompensated RC coupled amplifier would normally drop off. The resistance of the coil is very small and so it does not effect dc voltages and response at the middle frequencies. A peaking coil is effective for frequencies above about 400 KHz.

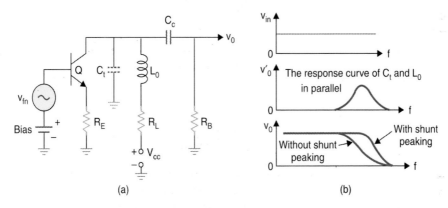

Fig. 12.8 (a). Video amplifier employing shunt coil (peaking) frequency compensation.

Fig. 12.8 (b). Frequency response of the amplifier.

Another arrangement to extend the high frequency response is known as 'series peaking' compensation (Fig. 12.9). In this circuit, L_C is in series with the two main components of C_t. At

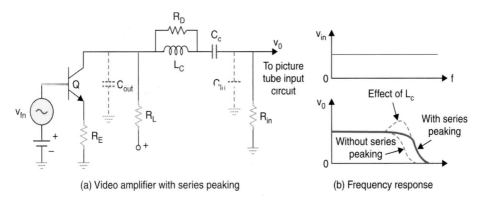

(a) Video amplifier with series peaking

(b) Frequency response

Fig. 12.9. Video amplifier employing series peaking coil compensation.

Chapter 12

one side of L_C is C_{out} of the video amplifier and on the other side is C_{in} of the next stage or input capacitance of the picture tube. This arrangement reduces shunting capacitance across R_L which results in more gain, while L_C resonates with C_{in} to provide a rise in voltage across C_{in} at high frequencies. A series peaking coil usually has a shunt damping resistance such as R_D, the function of which is to prevent oscillations or ringing in the coil with abrupt changes in signal.

The circuit of Fig. 12.10 combines shunt and series peaking which results in more gain, extended high frequency response and improved transient behaviour.

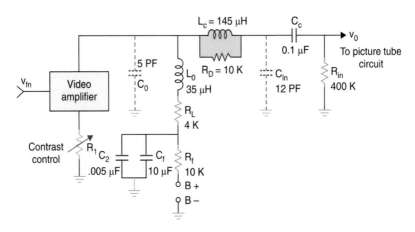

Fig. 12.10. *Video amplifier employing both shunt and series peaking for high frequency compensation and a special decoupling circuit to boost low frequency response. Note that the component values shown in the circuit are for a tube amplifier.*

The low frequency response is affected by increased reactance of the coupling and bypass capacitors. This can be improved by using largest possible values of these capacitors. In addition, the decoupling filter $R_f C_f$ in the B_+ supply line (Fig. 12.10) can be used to boost gain and reduce phase shift distortion at very low frequencies. The capacitor C_f offers large reactance at low frequencies and then R_f in series with R_L becomes the effective load. The increased load results in higher gain for low frequencies. This rise in gain compensates for the reduction in gain caused by reactance of the coupling capacitor C_C. Furthermore, phase shift caused by shunt capacitor C_f is opposite to phase shift caused by series capacitor C_C. As a result C_f tends to correct phase distortion introduced by the coupling capacitor. A small paper or ceramic capacitor C_2 is normally provided across C_f to bypass very high video frequencies because C_f, being a large value electrolytic capacitor, has a non-negligible inductance and fails to provide a bypass at these frequencies.

Gain Control

The change in gain of the video amplifier to provide contrast control can be effected in different ways. A common method is to vary the cathode/emitter resistance R_1. This resistance is left unbypassed and its variation alters the negative feedback which in turn changes the gain of the amplifier. The change in the value of R_1 also varies the bias and the consequent shift in the operating point can introduce amplitude distortion. To overcome this problem, in many video amplifier designs, a potentiometer is provided at the output of the video amplifier to alters the magnitude of the video signal applied to the picture tube. This method is the same as volume control in an audio amplifier.

5.5 MHz Sound Trap

The video amplifiers usually have a trap circuit, tuned to the intercarrier sound frequency of 5.5 MHz to keep the sound signal out of the picture signal. If sound signal is not separated at the video detector, the trap circuit can be modified to deliver the intercarrier sound signal to the sound IF amplifier. This is illustrated in Fig. 12.11 (a) where L_1 and C_1 constitute the trap circuit and winding L_2 coupled to L_1 serves as take-off point for the 5.5 MHz sound signal.

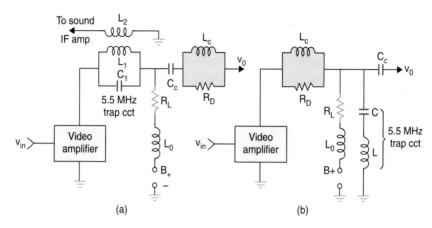

Fig. 12.11. *Video amplifier circuits (a) with parallel resonant trap in series with the output. The trap circuit also serves as the sound take-off circuit (b) with series resonant 5.5 MHz trap in shunt with the load.*

As shown in the figure, the trap circuit is in series with the output and when tuned to resonance at 5.5 MHz, offers maximum impedance. Therefore, the sound signal is removed because maximum voltage across the trap is developed at this frequency.

In Fig. 12.11 (b) another trap circuit arrangement is shown. Here L and C form a series resonant circuit tuned to 5.5 MHz. This trap circuit is in shunt with the load, and at resonance, provides practically a short circuit to the intercarrier sound signal. This prevents it from appearing across the picture tube input.

If the 5.5 MHz sound signal together with its sidebands is not fully suppressed it causes beat interference which results in diagonal lines on the picture having small wiggles. The weave in the lines is the result of frequency variations in the FM sound signal. This effect is also called 'wormy' picture. The interference disappears when there is no voice or music, leaving just the straight lines corresponding to the 5.5 MHz carrier without modulation. The trap circuits are tuned for minimum interference in the picture.

12.4 BASIC VIDEO AMPLIFIER OPERATION

Before attempting to discuss complete video section circuits it is desirable to recapitulate the operation of a basic RC coupled amplifier, which after adding compensating elements serves as the video amplifier. The circuit of such an amplifier employing a pentode is drawn in Fig. 12.12 (a). The voltage drop across R_K provides grid bias and ensures class 'A' operation. The load resistance R_L provides output voltage between the plate and ground. While R_S is the screen voltage dropping resistance. The bypass capacitor C_S, connected between screen grid and ground provides an effective short at all frequencies of interest. C_K connected across R_K is a bypass for the ac component of the current to prevent degeneration of the input signal.

The input signal waveform and corresponding plate current and plate voltage waveforms are shown in Figs. 12.12 (*b*) to (*d*). When the input signal swings to the extreme positive, maximum current flows and the plate voltage swings to its minimum value. Again, when the input signal attains its maximum negative value, the plate current becomes minimum and plate voltage attains its maximum positive value. Thus, the output voltage is 180° out of phase with respect to the input voltage. The dc component of the plate voltage is blocked by coupling capacitor (C_C) and the amplified ac component then appears across the output terminals of the amplifier.

An RC coupled amplifier employing a transistor in the common emitter configuration performs in the same way as its tube counterpart, but with a difference, that instead of voltage it needs current drive for its operation. Normally potentiometer biasing is provided for setting the operating point. The voltage and current polarities for an *n-p-n* transistor are the same as with a vacuum tube. However, with a *p-n-p* transistor all signal polarities are negative with respect to ground.

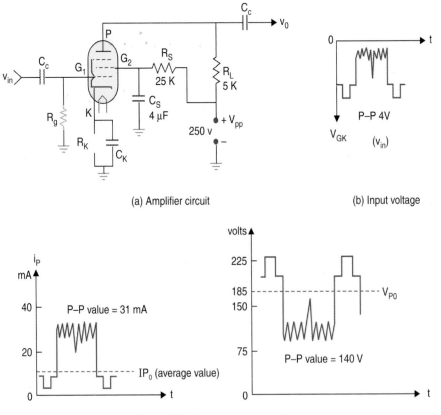

(a) Amplifier circuit

(b) Input voltage

Fig. 12.12. R.C. coupled amplifier.

12.5 COMPARISON OF VIDEO SIGNAL POLARITIES IN TUBE AND TRANSISTOR CIRCUITS

As explained earlier a negative going signal is a must when the video signal is injected at the cathode. Similarly positive going signal is required for feeding at the grid of the picture tube. If opposite polarity is used the result will be a negative picture, in the same sense as a photographic negative. Besides this, the polarity of video signal with respect to ground also affects the biasing of the device employed in the video amplifier. These aspects are examined in greater detail both for tube and transistor amplifiers using grid and cathode modulation of the picture tube.

Grid Modulation of the Picture Tube

(i) *Transistor Circuits.* The circuit arrangement shown in Fig. 12.13 is of a direct coupled video amplifier employing a *n-p-n* transistor to grid modulate the picture tube. The resistors R_1 and R_2 together with R_E, that is bypassed by C_E, provide forward biasing at the base-emitter junction. This biasing arrangement provides good stability of the operating point against device replacement, temperature variations, dc supply changes and ageing of circuit components.

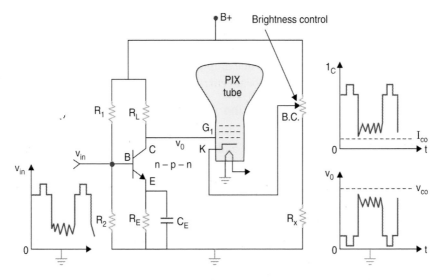

Fig. 12.13. Direct coupled video amplifier circuit to grid modulate the picture tube.

Since a positive going signal is needed at the collector, a negative going video signal is required to the base. The value of emitter resistance R_E is chosen to set the operating point 'Q' for minimum collector current in the absence of input signal. In fact the base-emitter bias is so set that when video signal is applied, the sync tips make the base most positive. Thus the collector current becomes maximum for sync tip levels and the collector acquires minimum positive potential. This is the correct polarity for obtaining minimum beam current of the picture tube. Since the operating point of the transistor is set close to cut-off bias, the collector voltage, in the absence of any video signal is highest, which makes the grid of the picture tube least negative with respect to cathode. Therefore, with no video voltage drive, though the transistor draws a minimum current, the picture tube beam current is maximum.

Notice that the cathode of picture tube is returned to the brightness control which sets the cathode at a voltage which is more positive with respect to ground than the grid. Brightness control is achieved by varying net negative voltage between grid and cathode. The resistance R_X fixes the minimum limit of bias and prevents the possibility of a net positive voltage on the grid.

If a *p-n-p* transistor is employed the base-emitter junction must be enough forward biased, with no input signal, so that the positive going sync tips drive the bias backwards towards the I_b = 0 point. Note that the input signal is all positive and the least positive level is reached on peak-white and the transistor then passes maximum current. The maximum collector current results in least negative collector voltage which in turn makes the picture tube grid less negative with respect to its cathode and the beam current increases to reproduce the peak-white values of the picture. It may also be noted that in the absence of any video signal the transistor collector is at a minimum negative potential and the corresponding beam current is large. Similarly a *p-n-p* transistor needs a dc source of opposite polarity than a *n-p-n* configuration. This necessitates interchanging of the locations of orightness control and resistor R_X to make sure that the picture tube grid never attains a positive potential.

(ii) Vacuum Tube Circuit. In the case of a vacuum tube a positive supply voltage is needed and a positive drive to the grid is necessary to increase the plate current. Therefore, the working conditions of a vacuum tube video amplifier are exactly similar to those of a *n-p-n* transistor. However, the potentials needed are much higher and the biasing technique is somewhat different. Thus in a vacuum tube amplifier, with no input signal the tube draws a minimum current, and the picture tube beam current is maximum.

Cathode Modulation of the Picture Tube

When the video signal is injected at the cathode of picture tube a negative going signal is needed at the anode/collector of the video amplifier and this necessitates a positive going signal at the grid/base of the amplifier.

(i) Tube Circuit. The necessary circuit details of the video section and the signal waveforms at the input and output of the video amplifier are illustrated in Fig. 12.14. The tube is biased close to V_{GK} = 0 point so that maximum current corresponds to peak-white and minimum plate current occurs on sync pulse tips. The anode voltage waveform is then negative going as required for correct cathode modulation. The picture tube beam current is again high with no signal at input of the amplifier.

(ii) Transistor Circuits. Fig. 12.15 shows a *p-n-p* configuration and associated waveforms. The transistor input circuit is back biased towards I_b = 0 μA so that on peak-whites the base current and hence the collector current is very small. Minimum collector current results in maximum negative voltage at the collector, so that once again the necessary negative going collector signal is derived for the picture tube cathode. In the absence of any input video signal the collector voltage is more negative and hence the signal beam current is maximum.

If a *n-p-n* transistor is employed in the video amplifier designed for cathode modulation, the signal polarities both at the input and output terminals are exactly the same as shown in Fig. 12.14 for a vacuum tube configuration. Therefore the picture tube beam current will be high with no input signal to the amplifier.

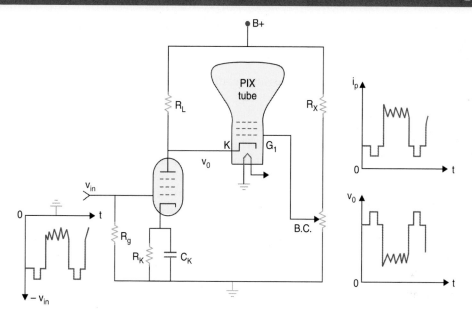

Fig. 12.14. Direct coupled video amplifier circuit to cathode modulate the picture tube.

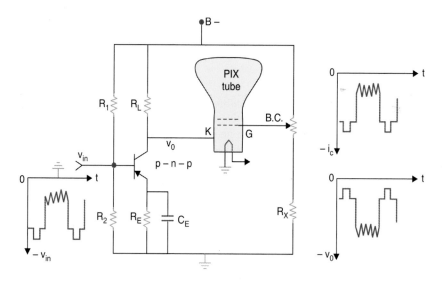

Fig. 12.15. Direct coupled transistor (p-n-p) video amplifier circuit
to cathode modulate the picture tube.

These results may be summarized as follows:

(i) The sense of the video signal relative to black level seen either at the input or output terminals is the same in tube and transistor circuits.

(ii) When the tube or n-p-n transistor is called upon to deliver maximum plate/collector current the p-n-p transistor has to pass minimum collector current and vice versa.

(iii) The picture tube beam current is maximum with no input video signal both for grid modulation and cathode modulation when dc coupling is used.

Chapter 12

12.6 RELATIVE MERITS OF GRID AND CATHODE MODULATION OF THE PICTURE TUBE

The main points to be considered in assessing relative merits of the two systems are as follows:

(i) Picture Tube Input Characteristics

This is a measure of the beam current change for a given change in the video signal drive voltage. The advantage lies with cathode modulation. The beam current is determined by the grid-cathode voltage and by the positive voltage on the first anode with respect to cathode. With grid modulation, the only factor which determines the change in beam current results from a given change in video signal voltage between the grid and the cathode. No other electrode voltage is effected with the applied voltage. However, when cathode modulation is applied a second factor influences the beam current. This is the voltage between the first anode and cathode. Thus as the video signal input voltage moves, from black-level towards peak-white the cathode moves more negative, not only to the control grid but also to the first anode. As stated above the voltage on the first anode and cathode itself have a marked influence upon the beam current of the picture tube. Thus the change in beam current brought about by effective reduction in grid-to-cathode. negative bias, is further augmented due to increase in positive voltage between the first anode and cathode. However, with grid modulation the voltage between the first anode and cathode remains constant. If follows then, that for a given change in the input signal voltage, the change in beam current is less with grid modulation than with cathode modulation. For the same input, the beam current is about 20% greater on peak-white with cathode modulation, or the video amplifier gain can be smaller for the same beam current.

(ii) Feed to the Sync Separator

Most commonly used sync separator circuits employ either a tube or a transistor, which is cut-off during the picture information part of the video signal but is driven into conduction by the sync pulses. A negative going video signal is then needed at the grid of the tube or base of the *n-p-n* transistor being used as a sync separator. With cathode modulation the right polarity is available at the anode/collector of the tube/transistor. It is not so when grid modulation is employed and thus cathode modulation has the added advantage over grid modulation of automatically providing the right polarity of the video signal needed to drive the sync separator. However, if a *p-n-p* transistor is used for sync separation, the grid modulation will provide signal with the correct polarity. With transistor receivers employing a two stage video amplifier, the question of feeding the sync separator is no longer important since both positive and negative polarities of the video signal are always available.

(iii) Safety of Picture Tube in the Event of Video Amplifier Failure

With tube receivers or the ones employing an *n-p-n* transistor for video signal amplification the advantage lies with cathode modulation if direct coupling is used. Should the emission of the video amplifier tube fail or the *n-p-n* transistor stop conducting, the plate/collector current drops to zero and the cathode of the picture tube attains a positive potential equal to B_+ supply. This immediately cuts-off the beam current and thus no damage is caused to the picture tube. With grid modulation such a fault will make the grid highly positive causing excessive beam current. The shorts between plate and cathode (collector and emitter) are rare and need not be considered. When a *p-n-p* transistor is employed in the video amplifier the effects would be opposite.

(iv) Cathode to Heater Voltage Stress

In tube receivers using series heater arrangement the picture tube is placed at or near the ground end of the chain to give it. maximum protection against possibility of damage due to short circuits across the heater line. Thus the heater is at negative end of the dc supply. With grid modulation the cathode voltage of the picture tube is at a higher potential than when cathode modulation is used. The stress is therefore more in the case of grid modulation. However, in modern picture tubes the likelihood of cathode heater breakdown has been minimized. Also with capacitive coupling the voltages get reduced, and therefore this point does not very much influence the choice of modulation method.

In conclusion, it may be said, that the video amplifier is one of the most important section of the receiver. It not only amplifies the video signal which extends from almost dc to a very high video frequency of 5 MHz, it also acts as the source for feeding the video signal to the sync separator and automatic gain control circuits. In most cases the sound signal is also separated at this stage after amplification. It would then be desirable to take up a detailed consideration of the video section design and circuitry. The next two chapters are devoted to these aspects of video amplifiers.

REVIEW QUESTIONS

1. What are the essential requirements that a video amplifier must meet for faithful reproduction of picture details ?

2. How does phase distortion in the video signal affect the quality of the picture ? What causes 'smear' in the picture and how can this be minimized ?

3. Draw the circuit diagram of an RC coupled amplifier employing an *n-p-n* transistor in common emitter configuration and explain its operation as a voltage amplifier. Sketch its frequency response and explain why the gain falls-off both at very high and low frequencies.

4. Describe briefly with circuit diagrams the techniques employed to extend the bandwidth of an RC coupled amplifier to accommodate full range of the video signal.

5. Why are trap circuits provided in video amplifiers to attenuate frequency spectrum occupied by the FM sound signal ? What is the undesired effect of sound signal on the reproduced picture ?

6. Draw simple circuit diagram of a dc coupled video amplifier that feeds the grid of the picture tube. Sketch suitable input and output voltage waveforms and justify that the chosen polarity of the video signal will result in correct reproduction of the picture on the screen. Identify the location of the brightness control in the circuit drawn by you.

7. Discuss relative merits of cathode and grid modulation of the picture tube. Explain why cathode modulation is considered superior to grid modulation.

Chapter 12

13

Video Amplifiers— Design Principles

The choice of basic amplifier that can be modified to meet the requirements of a video amplifier is restricted to direct coupled and RC coupled configurations. The techniques employed to achieve broad-band characteristics are same for tube and transistor amplifiers. However, their mode of operation and impedance levels are quite different from each other. Therefore, while explaining design fundamentals, video amplifiers employing tubes and transistors are considered separately.

13.1 VACUUM TUBE AMPLIFIER

The basic circuit of a video amplifier employing RC coupling, together with its ac equivalent circuits valid for medium, high and low frequency regions are shown in Fig. 13.1. The capacitor C_0 represents output capacitance of the tube, C_s stray shunt and wiring capacitance and C_i input capacitance of the picture tube circuitry. All the three capacitances are effectively in parallel, and when added constitute $C_t = C_0 + C_s + C_i$. The total shunt cpacitance seldom exceeds 20 pF and thus acts as an open circuit to frequencies in the low and midband ranges. The coupling capacitor C_C is chosen to be quite large to provide nearly a complete ac short, even at very low frequencies.

Gain Expressions

It is an easy matter to deduce gain expressions for the three frequency regions from the corresponding equivalent circuits of the basic amplifier configuration (Fig. 13.1). The results are summarized below as a starting point for explaining wide-banding techniques.

(i) *Gain at midband* (A_{mid}) (see Fig. 13.1 (b))

$$A_{(mid)} = -g_m R \parallel \approx -g_m R_L \qquad \qquad ...(13.1)$$

where $R \parallel = R_L \parallel R_g \parallel r_p \approx R_L$, because load resistance in video amplifiers seldom exceeds 10 K-ohms. The minus sign signifies phase reversal of 180°.

(ii) *Gain at high frequencies* (see Fig. 13.1 (c))

$$A_{(HF)} = -g_m Z_t$$

where $Z_t = R_L \parallel X_{C_t}$.

Substituting for $X_{C_t} = \dfrac{1}{\omega C_t}$ and simplifying yields

$$|A_{(HF)}| = \frac{A_{(mid)}}{\sqrt{1+(\omega C_t R_L)^2}}\ \angle -\theta_H$$

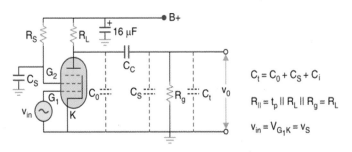

$$C_t = C_0 + C_S + C_i$$

$$R_{ii} = t_p \parallel R_L \parallel R_g = R_L$$

$$v_{in} = v_{G_1 K} = v_S$$

Fig. 13.1 (a). Basic R.C. coupled amplifier.

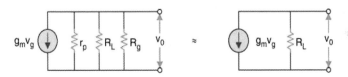

Fig. 13.1. (b). A.c. equivalent circuit valid at medium frequencies.

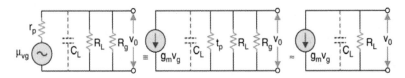

Fig. 13.1 (c). A.c. equivalent circuit valid at high frequencies.

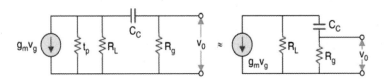

Fig. 13.1 (d). A.c. equivalent circuit valid at low frequencies.

Defining the upper 3 db frequency as

$$f_H = \frac{1}{2\pi C_t R_L}\ \text{or}\ \omega_H = \frac{1}{C_t R_L}$$

$$|A_{(HF)}| = \frac{A_{(mid)}}{\sqrt{1+\left(\dfrac{\omega}{\omega_H}\right)^2}}\ \angle -\tan^{-1}\left(\dfrac{\omega}{\omega_H}\right)\qquad\qquad ...(13.2)$$

At $\omega = \omega_H$, $R_L = X_{C_t}$ and $\angle\theta_H = -45°$ (relative to phase-shift at midband).

Note that at $f = f_H = \dfrac{1}{2\pi C_t R_L}$, $Z_t = 0.707$ of its midband value and thus the gain at this frequency falls to become -3 db with respect to the midband gain.

(iii) Gain at Low Frequencies $(A_{(LF)})$ (see Fig. 13.1 (d)). Proceeding in the same way as for the high frequency gain expression, and after a little manipulation, the gain in the low frequency region can be expressed as

$$| A_{(LF)} | = \frac{A_{(mid)}}{\sqrt{1 + \left(\dfrac{1}{\omega C_C R_g}\right)^2}} \angle + \theta_L \qquad \text{where} \quad \frac{r_p R_L}{r_p + R_L} + R_g \approx R_g.$$

Defining the lower 3 db down frequency as

$$f_L = \frac{1}{2\pi C_C R_g} \quad \text{or} \quad \omega_L = \frac{1}{C_C R_g}$$

the gain expression can be written as

$$| A_{(LF)} | = \frac{A_{(mid)}}{\sqrt{1 + \left(\dfrac{\omega_L}{\omega}\right)^2}} \angle + \tan^{-1}\left(\frac{\omega_L}{\omega}\right) \qquad \qquad \text{...(13.3)}$$

At $\omega = \omega_L$, $R_g = X_{CC}$ and $\theta_L = +45°$ (relative to phase shift at midband).

Note that at $f = f_L = \dfrac{1}{2\pi C_C R_g}$, $Z_t = 0.707$ of its midband value, and the gain at this frequency again falls to -3 db with respect to midband gain. The frequencies f_H and f_L are known as corner frequencies and the gain at these frequencies is 70.7% of the midband value.

Bandwidth

The bandwidth of an amplifier is defined as

$$BW = (f_H - f_L) \approx f_H = \frac{1}{2\pi C_t R_L} \qquad \qquad \text{...(13.4)}$$

In an RC coupled amplifier, even when R_L is made as low as 0.5 K-ohms, f_H seldom exceeds 3 MHz. However, it is not possible to make the load resistance (R_L), too small, because the gain requirement (gain = $g_m R_L$) of the video amplifier is not fully met. Therefore, some other means have to be devised to extend the frequency range upto 5 MHz without unduly reducing R_L.

13.2   HIGH FREQUENCY COMPENSATION

The bandwidth is normally extended by making the plate load complex in such a way that its magnitude increases with increase in frequency. Thus, the compensation technique is aimed at pushing up the upper -3 db frequency f_H, which normally would occur at a relatively low frequency due to the presence of shunt capacitance C_t. Negative feedback is also applied to increase the bandwidth but this results in some loss of gain. The various HF compensation techniques are as follows:

(a) Shunt Inductance Peaking

A small inductor of the order of 50 to 250 μH is added in series with the load resistor R_L. Though connected in series with R_L, the coil is in fact a part of the shunt plate circuit. This is llustrated in Fig. 13.2, where the compensated amplifier configuration together with its high frequency equivalent circuit is drawn. As shown there, the effective circuit, in shunt with the signal path, consists of C_t in parallel with series combination of R_L and inductor L_{px}. The inductor increases the net plate circuit impedance at high frequency end of the frequency range being handled, and thus partly compensates for the decreasing reactance of C_t. This results in shifting the higher corner frequency f_H to a still higher value to enhance bandwidth of the amplifier.

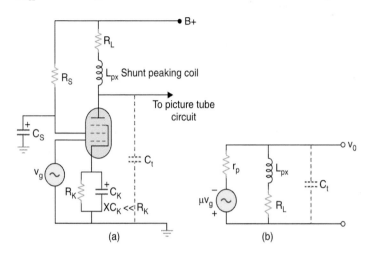

Fig. 13.2. Video amplifier employing shunt peaking (a) Circuit (b) Equivalent circuit.

The combined impedance of R_L, L_{px} and C_t can be expressed as

$$Z_t = \left[\frac{(X_{L_{px}})^2 (X_{Ct})^2 + R_L^2 (X_{Ct})^2}{(R_L)^2 + (X_{L_{px}} - X_{Ct})^2} \right]^{1/2} \qquad ...(13.5)$$

As stated earlier the gain at f_H where $X_{Ct} = R_L$, falls to 70.7% of the midband value. Therefore, to extend the midband range, Z_t should increase at this frequency to yield a gain equal to the midband gain. This can be readily achieved by setting $X_{L_{px}} = \dfrac{R_L}{2}$ and $X_{Ct} = R_L$ (at f_H) in equation (13.5). This, on substitution, yields

$$Z_t = \left[\frac{\left(\dfrac{R_L}{2}\right)^2 \times (R_L)^2 + (R_L^2)^2}{(R_L)^2 + \left(\dfrac{R_L}{2} - R_L\right)^2} \right]^{1/2} = R_L \qquad ...(13.6)$$

Thus, the midband gain extends to f_H, where, without compensation it was down by 3 db.

Procedure for fixing R_L and L_{px}. It is necessary to first determine the total shunting capacitance (C_t) and the highest frequency (f) up to which flat response is desired before fixing the values of R_L and L_{px}. The highest frequency of interest in the 625 line system is 5 MHz. The value of C_t can be estimated from the data of the tube chosen for the amplifier, and by measuring stray capacitances if necessary.

With both f and C_t known, the values of R_L and L_{px} can be found as under.

Chapter 13

$$R_L = \frac{1}{2\pi f C_t} \left(\text{from } f_H = \frac{1}{2\pi C_t R_L} \right) \qquad \text{...(13.7)}$$

Since $X_{L_{px}}$ is to have a value equal to $R_L/2$,

$$\therefore \qquad 2\pi f L_{px} = 0.5 R_L = \frac{0.5}{2\pi f C_t} \text{ (from equation 13.7)}$$

or

$$L_{px} = \frac{0.5}{4\pi^2 f^2 C_t} \qquad \text{...(13.8)}$$

On substituting $R_L^2 = \frac{1}{4\pi^2 f^2 C_t^2}$ in equation (13.8)

we get

$$L_{px} = 0.5 \, C_t R_L^2 \qquad \text{...(13.9)}$$

This equation can be written in a general form as

$$L_{px} = n C_t R_L^2 \qquad \text{...(13.10)}$$

where 'n' can be made to have any value of $\lessgtr 1$, in order to vary frequency response in the region close to the new value of f_H. It may be noted that the circuit will resonate if 'n' exceets unity.

Effect of varying L_{px}. If gain versus frequency plot of such an amplifier is drawn for different values of 'n', it is revealing to note, that for values of 'n' greater than 0.5, the response has a peak which becomes more pronounced as 'n' increases. Furthermore, increasing the value of inductance (*i.e.*, n) increases the amplitude of the hump and also steepens the rate at which the gain falls off above f_H. It is characteristic of compensating coils that while they lift the response curve in the desired region, the subsequent fall-off of gain is more rapid than in the uncompensated circuits. Too steep an edge in the response curve is not a desirable characteristic, since it can give rise to a tendency to produce overshoot or transients. In fact no single value of 'n' can give (*i*) constant gain throughout the pass-band, (*ii*) linear phase response, and (*iii*) fast transient response without overshoot, all at the same time. It can be shown, that for optimum frequency response, a value of $n = 0.414$, for least phase distortion; $n = 0.322$ and for critical damping $n = 0.25$ is necessary.

In the practical development of a particular circuit, it is usual to start-off with an inductor of value $L_{px} = 0.5 \, R_L^2 C_t$ and then experiment with larger or smaller inductors until the desired response is obtained. It may be noted that time delay due to phase shift at high frequencies is very small and if linear phase characteristics are not obtainable while satisfying other requirements, it will not cause any problems.

(b) Series Inductance Peaking

In this arrangement the compensating coil is inserted in series with C_C, which means that the inductor is in series with the signal path, rather than in shunt with it. Figure 13.3 shows this circuit arrangement with its equivalent circuit. In practice, the coil is fixed very close to the plate pin of the tube, and in this position it effectively separates the total shunt capacitance C_t, into two parts, with C_0 on the tube side and $(C_s + C_i)$ on the other side of the coil. As seen in the equivalent circuit, this arrangement takes the form of a low-pass filter. The value of L_{py} is so chosen, that the filter passes all frequencies within the required video band, but offers a rising attenuation above the upper limit of this frequency hand.

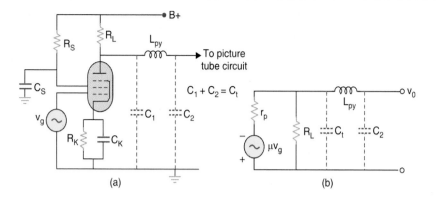

Fig. 13.3. *Video amplifier with series compensation (a) Circuit, (b) Equivalent circuit.*

Since the total shunt capacitance gets divided into two parts, it is possible to choose a higher value of R_L, because C_0 is only across R_L and not C_t as was the case in shunt compensation. A 50% increase in R_L becomes possible, *i.e.*, Eqn. (13.7) can be modified to become

$$R_L = \frac{1.5}{2\pi f_H C_t} \qquad \qquad ...(13.10)$$

This results in higher gain of the amplifier.

Choice of L_{py}. It is obvious that the behaviour of the filter will be affected by the disposition of the total shunt capacitance across the input (shown as C_1) and across the output (shown as C_2) of the resultant filter configuration. A typical value of C_2/C_1 is 0.75. A useful basic design formula for the inductor is given by

$$L_{py} = nR_L{}^2C_t \qquad \qquad ...(13.11)$$

where n varies between 0.5 and 1. With a ratio of $C_2/C_1 = p = 0.75$, a value of $n = 0.67$ is commonly used, since it gives optimum frequency response.

Though with series compensation more gain is possible and a better rise time performance results, but the fall-off in gain just beyond and upper edge of the required band is much steeper. This can cause excessive overshoot and even oscillations. This tendency towards ringing can be reduced by connecting a resistance in parallel with L_{py}. A typical practical value of such a damping resistance is $5R_L$ and varies between 15 and 20 K-ohms.

(c) Combined Shunt and Series Peaking Coils

Shunt and series inductance compensation can be combined to get a peformance slightly superior to that of the series peaking circuit. The corresponding amplifier configuration with its equivalent circuit is shown in Fig. 13.4. The following formulae may be used as a guide to establish approximate values of L_{px} and L_{py}

$$L_{px} = n_x R_L{}^2 C_t \text{ and } L_{py} = n_y R_L{}^2 C_t \qquad \qquad ...(13.12)$$

where n_x and n_y are the corresponding values of n and are dictated by the value of p, *i.e.*, (C_2/C_1). As a starting point, approximate values can be determined with the help of the following chart:

	p	n_x	n_y
(i) Linear frequency response	0.6	0.14	0.58
(ii) Optimum phase response	0.72	0.1	0.46
(iii) Critical damping	0.8	0.063	0.39

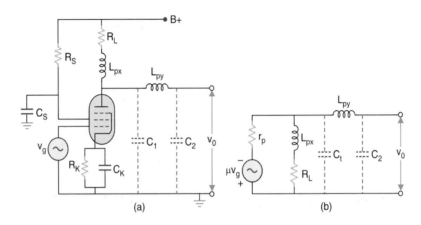

Fig. 13.4. *Video amplifier employing both shunt and series compensation*
(a) Circuit (b) Equivalent circuit.

It may be noted that in any case it would be necessary to test experimentally with various inductors to achieve the desired response. This is best done by using a visual display system*. In addition, the transient response may be checked by feeding a square-wave signal to the amplifier, and measuring the rise-time of the output waveform with a cathode-ray oscilloscope.

Cathode Compensation. The basic principle of this method is to apply negative feedback over the low and middle frequency regions, but arrange to remove it progressively in the HF region. Because of negative feedback the overall gain gets reduced but it results in a considerable increase in bandwidth. The simplest of the various possible circuit arrangements is shown in Fig. 13.5 (a), where the value of C_k has been so chosen, that it completely bypasses R_k at high video frequencies, but at medium and low frequencies, its reactance becomes comparable with R_k. This results in negative feedback which increases as the frequency decreases. In fact at medium and low frequencies, the reactance of C_t is large compared with R_L, and that of C_k is large compared with R_k, thus the amplifier effectively performs as one with a plate load of R_L and an unbypassed cathode resistance R_k. However, at higher frequencies when the shunting effect of C_t on R_L becomes appreciable, the reactance of C_k becomes comparable with R_k, and this reduces feedback to improve gain and maintain the frequency response.

Condition for Maximum Flatness of Frequency Response. Gain of the above amplifier that employs cathode degeneration can be expressed in the form

$$\frac{A_{(\text{mid})}}{A_{(\text{HF})}} = 1 + j\omega C_t R_L + g_m R_k \frac{1 + j\omega C_t R_L}{1 + j\omega C_k R_k} \qquad \qquad ...(13.13)$$

---

*Necessary details of visual display system are given in Chapter 28.

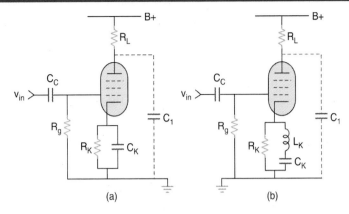

Fig. 13.5. *Widebanding by cathode compensation (a) Reactance of C_K comparable to R_K at medium and low frequencies, (b) L_K and C_K resonate at the upper edge of the video frequency band.*

Differentiating Eqn. (13.13) with respect to ω and equating this equal to zero yields

$$R_L C_t = R_k C_k \qquad \qquad ...(13.14)$$

This is the condition that must be met for maximum flatness of the frequency response. As already stated the improvement in bandwidth occurs at the expense of overall gain. However, this sacrifice is worth it, because with negative feedback the amplifier gain becomes more stable and in addition there is a considerable reduction of distortion in the output of the amplifier.

Another cathode compensation method is shown in Fig. 13.5 (*b*), where an inductor L_k in series with C_k shunts the cathode resistor. The values of L_k and C_k are so chosen, that the combination exhibits series resonance at the upper edge of the video frequency band. At resonance the very low impedance of the series tuned circuit effectively short circuits the feedback resistor and negative feedback is virtually reduced to zero. Typically, the ratio of cathode to plate circuit time constants falls in the range of 0.5 to 2.0.

13.3 LOW FREQUENCY COMPENSATION

In video amplifiers that employ ac coupling a large coupling capacitor is normally used to obtain a fairly low value of f_L. No special low frequency compensation is thought necessary because of the annoying effects of too good a low frequency response. This aspect is fully explained in the next chapter.

Direct Coupled Video Amplifier

When direct coupling is used the question of low frequency compensation does not arise but the problems of high frequency response are the same as with RC coupled amplifiers. Though such amplifiers can amplify changes in dc level, they have other inherent problems of drift, need for a highly regulated power supply and the high voltage dc source for adjustments of voltages at the grid and cathode in the absense of ac coupling. All this adds to cost and therefore in many cases partial dc coupling is preferred.

Selection of Tubes for Video Amplifiers

The ability to provide high gain and to handle signals up to 5 MHz are the primary considerations in selection of tubes for use as video amplifiers. To achieve a high gain and large signal swings without excessive distortion, pentodes and beam power tetrodes having high current and large power dissipation ratings are preferred.

Figure of Merit

The figure of merit of a high frequency tube is defined as the product of gain and bandwidth.

This can be expressed as:

$$\text{Gain} \times \text{bandwidth} = A_{(mid)} \times (f_H - f_L)$$

Substituting for $A_{(mid)} = g_m R_L$ and setting $(f_H - f_L) \approx f_H = \dfrac{1}{2\pi C_t R_L}$ we get,

$$\text{Figure of Merit} = g_m \times R_L \times \frac{1}{2\pi C_t R_L} = \frac{g_m}{2\pi C_t} \qquad \text{...(13.15)}$$

Obviously, larger the value of this expression, better is the tube for use as a video-amplifier. However, because of large power needs, and the consequent large electrode structure, the output capacitance (C_0) of such tubes cannot be made very small. This reduces the figure of merit of such tubes (f_H reduces) and it becomes necessary to provide HF compensation to achieve the desired bandwidth. PCL84 is one such tube which has been specially designed for use in TV receivers. The pentode section of this tube is used as a video amplifier whereas the triode section is connected as a cathode follower for feeding video signal to AGC and sync circuits.

Video Amplifier Circuit

Based on the design criteria developed in the earlier sections of this chapter, video amplifier design employing tube PCL84 has been more or less standardized. A typical circuit is drawn in Fig. 13.6 with component values labelled on it. The tube employs a load resistance to the order of 4 K-ohms and with a steady plate current close to 18 mA, the amplifier delivers enough peak-to-peak video signal to produce a full contrast picture.

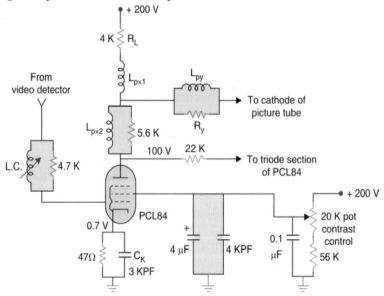

Fig. 13.6. Complete video amplifier circuit with gain control in the screen grid circuit.

The main design features of this video amplifier are summarized below :

(*i*) The plate circuit contains both shunt and series compensating coils, that are mutually coupled to provide adequate frequency broadbanding.

(*ii*) Cathode compensation is also provided by using a small cathode bypass capacitor.

(*iii*) The amplifier is designed for a full gain of about 30. The screen grid voltage is varied to contol the gain and this serves as the contrast control.

(*iv*) The amplifier is dc coupled and has excellent low frequency response. In many designs partial dc coupling is used for optimum results.

13.4 TRANSISTOR VIDEO AMPLIFIER

Gain requirement from both tube and transistor video amplifiers is usually the same, and varies between 25 to 60, depending on the video signal amplitude available at the output of video detector and the transfer characteristics of the picture tube. Transistor video amplifiers are almost always direct coupled. This not only solves the gain and phase shift problems at low frequencies, but also makes the use of large coupling capacitors unnecessary. Direct coupling in transistor circuits does not present any serious problems so far as dc supply is concerned, because the magnitudes of voltage needed are much less than in tube circuits. However, the output transistor must have a V_{CC} supply of the order of about 150 volts for delivering a video signal of nearly 75 volts peak-to-peak to modulate the picture tube.

Transistors for Video Amplifiers

The output capacitance of transistors is comparable with that of tubes, but because transistors operate at lower impedances, the frequency and phase response in the collector circuit of an RC coupled transistor amplifier remains unaffected up to a higher frequency than in the tube plate circuits.

The input capacitance of bipolar transistors is in general, much greater than that of tubes, but its effect on the previous collector circuit, can be nullified by using interstage emitter followers.

In a transistor amplifier the upper frequency limit is determined not by stray and shunt capacitances, but by the reduction in current gain, as the cut-off frequency of the transistor is approached. Low power transistors amplify up to very high frequencies and the problem is one of limiting the bandwidth rather than extending it. However, this remark is not applicable to power transistors. The high signal voltage, that the last video amplifier stage is expected to deliver with restricted load resistance, needs large collector current operation. This in turn needs a transistor of about 2 watt rating with a high breakdown voltage. When the above two conditions are met, the desired gain at high frequencies cannot be easily achieved, because it is difficult to make transistors with lesser collector to base capacitance and high junction breakdown ratings. Therefore, it becomes necessary to use peaking coils in the collector circuits of transistor video amplifiers to extend the high frequency range.

Amplifier Configuration

One high gain, high frequency transistor could by itself provide the required gain and bandwidth, but input and output impedance requirements make a single stage transistor amplifier difficult to design. Therefore, all video amplifier configurations are preceded by a driver stage, connected as an emitter follower. The driver connects the video detector to the output stage and meets the following requirements:

(*i*) It presents a high input impedance to allow the use of a high detector load (about 5 KΩ). This ensures higher detector efficiency and more output voltage.

(*ii*) It has low output impedance which facilitates matching to the input of the video amplifier transistor.

(*iii*) Since the gain of the driver is less than one, it has a large bandwidth to ensure full transmission of video and intercarrier sound signals.

(*iv*) The driver output is in phase with its input, and thus provides the correct polarity of video signal for cathode modulation of the picture tube after one stage of amplification.

13.5 TRANSISTOR CIRCUIT ANALYSIS

Common emitter circuit arrangement is the best suited configuration as a video amplifier, because of its moderate input and output impedances, high voltage and current gains besides a large power output. Figure 13.7 shows hybrid-pi model of a transistor in the common emitter mode. The equivalent circuit (model) has been simplified by reflecting, collector to base junction capacitance (C_{cb} or C_μ), to the input loop and neglecting the high collector to base resistance (r_b' c, or r_μ).

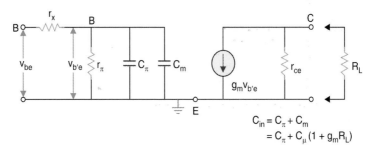

Fig. 13.7. Hybrid-pi(p) model of a transistor in common emitter configuration.

In the equivalent circuit (See Fig. 13.7):

(*i*) r_x is the base spreading resistance expressed as a lumped parameter. Its value varies between 50 to 150 ohms.

(*ii*) r_π is base to emitter junction resistance. Note that be common emitter input resistance in the '*h*' parameter model, *i.e.*, hie = $r_x + r_\pi$.

(*iii*) g_m (mA/V) is a constant of proportionality between collector current and base-emitter voltage. It varies with collector current and is governed by the relation

$$g_m = \frac{I_C}{V_T}, \text{ where } V_T = \frac{KT}{V} \text{ (}V_T = 0.026 \text{ V at room temperature.)}$$

(*iv*) C_π is the sum of diffusion capacitance and emitter to base junction capacitance.

(*v*) β is the short circuit current gain of the transistor at low frequencies. It varies with collector current and falls-off rapidly at collector currents beyond 10 mA.

$$\beta = g_m \times r_\pi \qquad \qquad ...(13.16)$$

when β decreases, both g_m and r_π are effected, and get reduced.

    (vi) r_{ce} is collector to emitter resistance.

    (vii) C_m (reflected Miller capacitance) $= C_\mu(1 + A_v)$

where A_v is the voltage gain of the stage.

    (viii) C_{in} (total input capacitance) $= C_\pi + C_\mu(1 + A_v)$ ...(13.17)

    (ix) R_L is external load resistance.

    (x) v'_{be} is the effective voltage between base and emitter.

f$_\beta$ (Half Power Frequency)

The circuit of Fig. 13.7 takes the form shown in Fig. 13.8 (a) when R_L is set equal to zero. Note that r_{ce} disappears from the circuit once R_L is made zero.

    The input side of the circuit has a single time constant, consisting of r_π in parallel with C_{in}. From this, 3 dB down frequency

$$f_H = f_\beta = \frac{1}{2\pi r_\pi C_{in}}$$ ...(13.18)

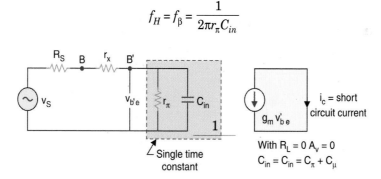

Fig. 13.8 (a). Equivalent circuit for the calculation of short-circuit CE current gain.

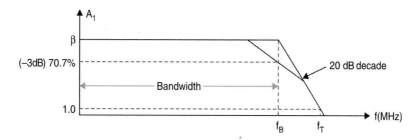

Fig. 13.8 (b). Short-circuit CE current gain vs frequency.

The short circuit current gain, A_{is}, can be expressed as

$$A_{is} = \frac{\beta}{1 + \dfrac{f}{f_b}}$$

At $f \ll f_\beta$, $A_{is} = \beta$ and for $f \gg f_\beta$, $A_{is} \approx \dfrac{\beta f_\beta}{f}$ ...(13.19)

When f is set equal to f_β, the short circuit current gain drops by 3 db (see Fig. 13.8 (b)). This frequency f_β, at which the short circuit current gain becomes 70.7% of its maximum value is the half-power or—3 db frequency. The frequency range up to f_β is then referred to as the bandwidth of the circuit.

f_T (Unity Current Gain Frequency)

f_T is defined as the frequency at which A_{is} attains a magnitude equal to unity, that is:

$$1 = \frac{\beta f_\beta}{f_T} \text{ (from Eqn. 13.19)}$$

or

$$f_T = \beta f_\beta \qquad \qquad ...(13.20)$$

Substituting values of β and f_β from (13.16) and (13.18) in (13.20), we get

$$f_T = \frac{g_m}{2\pi C_{in}} \qquad \qquad ...(13.21)$$

Since at short circuit $A_v = 0$, Eqn. (13.17) reduces to

$$C_{in} = C_\pi + C_\mu$$

$\therefore$

$$f_T = \frac{g_m}{2\pi(C_\pi + C_\mu)} \qquad \qquad ...(13.22)$$

or

$$\omega_T = \frac{g_m}{(C_\pi + C_\mu)} \approx \frac{g_m}{C_\pi} \mid \text{ since } C_\pi \gg C_\mu. \qquad \qquad ...(13.23)$$

The above expression shows that f_T is a function of the transistor parameters only. Since f_T controls the gain at high frequencies it is also known as 'Figure of Merit' of the transistor.

Most of the above parameters are listed in transistor manuals. However, some parameters, if not given, can either be measured or calculated from the various relations given above.

Voltage Gain of the Basic Amplifier

The circuit of the basic amplifier employing a BJT (transistor) in common emitter configuration is shown in Fig. 13.9 (a). In its equivalent circuit (Fig. 13.9 (b)), biasing resistance R_B has not been included, because its shunting effect on input impedance of the transistor is negligible. Similarly r_{ce} being very large, in comparison with R_L has been neglected.

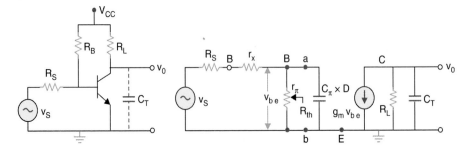

Fig. 13.9. *Basic transistor amplifier in CE configuration (a) Circuit (b) Equivalent circuit.*

The Factor 'D'

Before proceeding to find voltage gain, it will be useful to define a factor D, to which, the corner frequency of the amplifier is related.

From Eqn. (13.17), $C_{in} = C_\pi + C_\mu (1 + A_v)$

A_v at midband $= g_m R_L$

$\therefore \qquad C_{in} = C_\pi + C_\mu(1 + g_m R_L) \approx C_\pi + C_\mu g_m R_L$

Since $\qquad \omega_T = \dfrac{g_m}{C_\pi}$ (from Eqn. 13.23)

$\therefore \qquad C_{in} = C_\pi(1 + C_\mu R_L \omega_T)$ ...(13.24)

$\qquad\qquad = C_\pi \times D$

where $\qquad D = (1 + C_\mu R_L \omega_T)$ ...(13.25)

Voltage Gain

From the equivalent circuit of the amplifier (Fig. 13.9 (b)), the expression for voltage gain at any frequency (ω) can be expressed as

$$A_v = \frac{-\beta R_L}{(r_\pi + r_x + R_S)[1 + j\omega D C_\pi (r_\pi \,||\, \overline{R_S + r_x})]}$$

If follows from this expression that

$$|A_{(mid)}| = \frac{\beta R_L}{(r_\pi + r_x + R_S)}$$ ....(13.26)

Gain Bandwidth Product

The corner frequency occurs when the reactance of C_{in} ($= C_\pi \times D$) equals R_{th}, where R_{th} is the Thevenin's equivalent of the circuit (Fig. 13.9 (b)) to the left of points a and b.

$$R_{th} = \frac{r_\pi (R_S + r_x)}{r_\pi + R_S + r_x}$$ ...(13.27)

where R_S is the source resistance.

On equating $X_{C_{in}} = R_{th}$ and some manipulation we get :

$$f_H = \frac{\beta f_\beta}{B} \quad \text{or} \quad \omega_H = \frac{\beta \omega_\beta}{B}$$

where $\omega_\beta = \dfrac{1}{r_\pi C_\pi}$ and the other factor

$$B = \frac{r_x + r_\pi + R_S}{R_S + r_x}$$ ...(13.28)

f_H (= bandwidth) can also be expressed as

$$f_H = \frac{B}{D} \times \frac{f_T}{\beta} \quad (\text{since } f_T = f_\beta \times \beta)$$

or $\qquad\qquad \omega_H = \dfrac{B}{D} \times \dfrac{\omega_T}{\beta}$ ...(13.29)

Chapter 13

Finally, Gain × Bandwidth = $A_{(mid)} \times \omega_H$

$$= \frac{\beta R_L}{r_\pi + r_x + R_S} \times \frac{B}{D} \times \frac{\omega_T}{\beta} \quad \text{(from Eqs. (13.26) and (13.29))}$$

Substituting for B from Eqn. (13.28)

$$G \times B \cdot W = \frac{\omega_T}{D} \times \frac{R_L}{R_S + r_x} \qquad\qquad ...(13.30)$$

13.6 GUIDELINES FOR BROAD-BANDING

Equation (13.30) serves as a guideline for explaining the means to extend high frequency response of the amplifier with or without sacrificing midband gain. This is explained by considering separately all the constituents of the Gain-Bandwidth expression.

(a) The gain-bandwidth product increases with decrease of source resistance R_S. Thus a reasonable first step while designing a video amplifier is to choose the lowest possible value of R_S. This requirement is readily met, since the driver stage is an emitter follower, and its output resistance can be made very low without any appreciable loss in gain.

(b) ω_T ($f_T = \beta f_\beta$) does not stay constant and drops-off both at very low and high values of emitter current. However, there is a range of I_E (emitter current) over which it stays high and substantially constant. Therefore, it is advantageous to fix the transistor operation, in this region, as far as possible.

(c) Once the transistor and its operating point (I_E or I_C) have been chosen, r_x gets fixed and cannot be varied.

(d) If C_μ, that forms part of factor D is decreased, the bandwidth will increase with no corresponding loss in gain. However, the value of C_μ is dictated by the V_{CC} supply chosen or the maximum collector voltage rating of the transistor. Therefore, C_μ is more or less fixed and cannot be changed for extending high frequency region of the amplifier.

(e) The last variable is R_L, that can be changed to control bandwidth. But, this too cannot be varied much because of large peak-to-peak output voltage required to modulate the picture tube, and the maximum permissible dissipation of the transistor.

Output Circuit Corner Frequency

Input capacitance of the picture tube, together with transistor output and wiring capacitances easily add up to about 15 pF and very much limit the value of R_L. In fact the net output capacitance (C_T) of the video amplifier in parallel with R_L provides another corner frequency which turns out to be lower than the input circuit corner frequency. This, then controls the bandwidth of the amplifier. As stated above, decreasing R_L will push up the output corner frequency, but the load resistance cannot be made too small because it will increase the power dissipation of the device.

13.7 FREQUENCY COMPENSATION

It is clear from the above discussion that a high f_T and high power dissipation transistor is necessary for video amplifiers. These two requirements, though necessary, are mutually contradictory to a large extent.

In the past these requirements were met by cascoding, where the power dissipation was shared equally by the two transistors employed in such a configuration.

With advances in technology, transistors with high power dissipation and reasonably high f_T have now become available. However, the configuration still requires some high frequency compensation and is normally provided by a shunt or peaking coil in the collector circuit.

This and other relevant details are explained by a design example.

Video Amplifier Design Data

Output voltage	75 V (p-p)
Bandwidth	5 MHz
Detector output voltage	2 V (p-p)
Voltage gain	40
Configuration	Common Emitter
Coupling	Direct
V_{CC} supply	150 V
Transistor	BF 178

Transistor parameters at I_C = 15 mA are : r_x = 50 ohm, r_π = 200 ohms, C_π = 100 pF, β = 20, f_T = 120 MHz, C_μ at 150 V = 1.25 pF, max collector dissipation = 1.7 watts, minimum collector emitter breakdown voltage = 145 V.

Choice of R_L and Operating Point

To avoid non-linear distortion due to saturation and cut-off, the best course for such a large output is to draw several load lines on the characteristics of the chosen transistor and calculate distortion for each load by the usual three or five point analysis. This would help to decide the optimum value of R_L. Note that too small and too large a value of load resistance is not acceptable for reasons already explained.

Such an exercise on the characteristics of BF178 led to the following results:

$$I_C \approx I_E = 15 \text{ mA}$$
$$R_L = 4.9 \text{ K}$$
$$V_{CE} \text{ (min) inclusive of drop across } R_E = 30 \text{ V}$$
$$(R_E \text{ is the emitter resistance})$$

This leaves 120 V to accommodate the output signal with enough margin for the blanking excursion.

For class 'A' operation the following relations are valid:

$$P_{max} = \frac{V_{CC}^2}{4(R_L + R_E)} \qquad \qquad ...(13.31)$$

$$I_E = \frac{V_{CC}}{2(R_L + R_E)} \qquad \qquad ...(13.32)$$

Choosing $R_E = 100$ ohms and allowing a 10% limit in V_{CC} variations, *i.e.*, $V_{CC\,max} = 165$ V, P_{max} from eqn. (13.31) = 1.36 watts and $I_C \approx I_E$ from eqn. (13.32) = 15 mA.

The calculated value of I_C checks with that found graphically. Similarly the calculated value of P_{max} is within the max. permissible dissipation. However, the transistor would need a suitable heat sink and this is always provided.

Amplifier Circuit

The circuit of the amplifier is drawn in Fig. 13.10 (*a*) and its equivalent circuit valid at high frequencies is shown in Fig. 13.10 (*b*). Besides other circuit elements the two corner frequencies (break points) are labelled as f, 3 db (in) and f, 3 db (out) in the equivalent circuit.

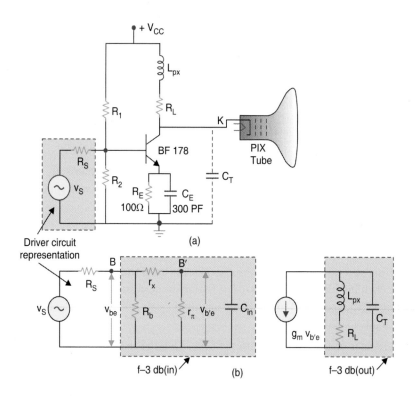

Fig. 13.10. Transistor video amplifier (a) Circuit (b) Equivalent circuit valid at high frequencies.

Voltage Gain

Equation (13.26) can be modified to include the effects of biasing network and the inadequately bypassed R_E. Putting $R_b = R_1 \mid\mid R_2$ (biasing network resistors) and adding $(1 + \beta)R_E \approx \beta R_E$ (reflected emitter resistance) to r_π, the new voltage gain can be calculated. With the given design values even if R_b is taken as 7 K it can be neglected in comparison with the other shunting resistances. With this assumption the new voltage gain

$$| A_v(\text{mid}) | = \frac{\beta R_L}{R_s + r_x + r_\pi + \beta R_E} \qquad \qquad ...(13.33)$$

Assuming $R_s = 150\ \Omega,\ | A_v(\text{mid}) | \approx 40$

This nearly checks with the midband gain computed from the approximate relation valid with feedback:

$$| A_v(\text{mid}) | \approx \frac{R_L}{R_E} = \frac{4.9\,K}{100} = 49$$

Input Circuit Corner Frequency $f_{3\,\text{db (in)}}$

$$f_{3\,\text{db (in)}} = \frac{1}{2\pi C_{in} R_{th}} \qquad \qquad ...(13.34)$$

where
$$C_{in} = C_\pi + C_\mu(1 + A_v) = 100 + 1.25(1 + 40) = 150 \text{ pF}$$

Equation (13.27) can be modified to include βR_E. This takes the form

$$R_{th} = r\pi \parallel (R_s + r_x + \beta R_E) = \frac{r\pi(R_s + r_x + \beta R_E)}{r\pi + R_s + r_x + \beta R_E} \qquad \qquad ...(13.35)$$

On substituting numerical values, $R_{th} = \dfrac{200(150 + 50 + 2000)}{200 + 150 + 50 + 2000} \approx 180\ \Omega$

$$\therefore \qquad f_{3\,\text{db (in)}} = \frac{1}{2\pi C_{in} R_{th}} = \frac{10^{12}}{2\pi \times 150 \times 180} \approx 5.8 \text{ MHz}$$

This result shows that the input circuit does not need any compensation and would safely transmit up to the highest modulating frequency of 5 MHz.

Output Circuit Corner Frequency ($f_{3\,\text{db (out)}}$)

The total collector network capacitance in a well laid out receiver would by typically as follows:

Picture tube cathode and leads	= 7 pF
Heat dissipator	= 3 pF
BF 178 output capacitance	= 3 pF
Total capacitance C_T	= 13 pF

$$\therefore \quad f_{3\,\text{db (out)}}(\text{uncompensated}) = \frac{1}{2\pi R_L C_T} = \frac{1}{2\pi \times 4.9\,K \times 13 \text{ pF}} \approx 2.6 \text{ MHz}.$$

Frequency Compensation

The above result shows that the output circuit has a lower corner frequency and hence would determine the extent of compensation needed to push this corner frequency to about 5 MHz.

The design criteria for calculating the values of peaking coils are the same as used in tube circuits. For shunt compensation:

From equation 13.9 we have
$$L_{px} = 0.5\,C_T R_L{}^2$$
On substituting $R_L = 4.9$ kΩ and $C_T = 13$ pF
$$L_{px} = 0.5 \times 13 \times 10^{-12} \times (4.9 \times 10^3)^2 \approx 158 \times 10^{-6} \text{ H} = 158\ \mu\text{H}$$
An inductor of this value will form part of the collector load to provide necessary compensation.

Broad-Banding by Negative Feedback

In addition to shunt compensation the emitter resistance R_E is bypassed by a very small (300 pF) capacitor C_E (see Fig. 13.10 (a)) to provide negative feedback at medium and low video frequencies. This not only improves the bandwidth but also ensures stable operation of the amplifier. The value of C_E is determined by the consideration that emitter and collector network time constants should be approximately equal.

13.8 VIDEO DRIVER

The output resistance of the driver together with the input resistance and capacitance of the video transistor form a network whose time constant should be compatible with the required bandwidth. Assuming 3 db point at 5 MHz and with C_{in} = 180 pF

$$R_{(out)} \leq \frac{10^{12}}{2\pi \times 5 \times 10^6 \times 180} \approx 150 \text{ ohms}$$

Therefore, ideally, ignoring R_{in}, the output resistance of the driver stage should not exceed this value. This checks with the value of R_S used while determining midband gain and input corner frequency.

However, in most practical designs. a value of 500 ohms can be used, because of increase in bandwith available by partially decoupling the emitter resistance of the output transistor.

The emitter follower will have a gain nearly equal to 0.9. This will feed about 1.8 V video signal to the output transistor which in turn will deliver ≈ 75 (p-p) (gain ≈ 40) for the picture tube.

Video Detector Loading

A high frequency transistor like BF 184 having β = 75 and f_T = 300 MHz, if employed as an emitter follower with R_E = 470 ohms will have R_{in} = 75 × 470 = 35 KΩ, and C_{in} = 5 pF. This is acceptable to a video detector circuit having a 3.9 KΩ load resistance.

Video Driver Biasing

The bias in the driver stage must be carefully set to permit maximum collector voltage swing. With a high input signal, the output will clip if the transistor cuts off or if V_{CE} reaches zero. The result is loss of detail in dark grey or white parts of the picture and a buzzing tone in the sound output. For this reason the upper biasing resistance is often a pre-set variable resistor. The video detector diode is invariably direct coupled to the driver and thus held at the same steady bias voltage as the base of the emitter follower. The biasing network is often designed to provide a small forward bias on the diode to reduce distortion on small input signals.

13.9 CONTRAST CONTROL METHODS

Contrast control is a manual control for setting level of the video signal fed to control grid or cathode of the picture tube. Its setting determines the ratio of light to dark in the picture.

Contrast Control in Vacuum Tube Video Amplifiers

(*a*) *Cathode Network Control.* In this method negative feedback is applied in one form or the other, taking into account the biasing requirements of the tube. Figure 13.11 (*a*) shows one such method, where the adjustment of the contrast control does not change the operating point of the amplifier. This maintains a constant black level of the video signal.

(*b*) *Plate Network Control.* In this arrangement (Fig. 13.11 (*b*)) the contrast control potentiometer regulates the magnitude of the video signal to the picture tube. The stray capacitance of the potentiometer and its connecting leads can reduce high frequency response of the amplifier. To minimize shunt capacitance the control is usually mounted close to the video amplifier with the shaft mechanically coupled to the front panel of the receiver. In addition, the capacitors shown along with the potentiometer, provide frequency compensation to maintain the same frequency response, at different settings of the contrast control.

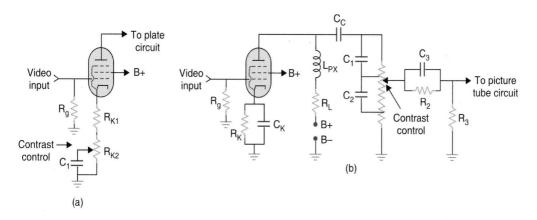

Fig. 13.11. *Contrast control circuits in vacuum tube video amplifiers*
(a) Control network in cathode circuit, (b) Control network in plate circuit.

Contrast Control in Transistor Circuits

(*a*) *Base Network control.* A contrast control technique that maintains a constant black level is shown in Fig. 13.12 (*a*). If the values of R_3 and R_4 are chosen to give a voltage that is equal to the black level of the video signal at the emitter of Q_1, the black level of the signal fed to Q_2 will remain constant over the contrast control range. Because of bandwidth requirements, R_2 should not be higher than 1 K-ohm. The parallel value of R_3, R_4 should be about one quarter of the value of R_2 thus giving a contrast range of about 5 : 1.

(*b*) *Emitter Network Control.* Fig. 13.12 (*b*) shows one type of emitter network contrast control. It is a degeneration control. When the arm of potentiometer R_4 is at ground, its resistance is unbypassed causing maximum feedback. This results in small video output. Any variation of the arm towards the emitter reduces feedback to deliver more output. This provides the desired contrast control. C_1R_1, and C_2R_2 are video peaking networks which cause higher gain at low contrast settings for high frequencies, making the picture sharper.

(*c*) *Collector Network Control.* As shown in Fig. 13.12 (*c*) the 25-K frequency-compensated potentiometer operates like a volume control. Setting of contrast control determines the peak-to-peak amplitude of the video signal taken from the collector of the video amplifier and coupled to the cathode of picture tube. Because of the high impedance level of the collector network, stray capacitances place severe restrictions on the circuit layout. The compensation should be so provided that for any setting of the contrast control almost equal time constants are obtained in the two arms of the connecting network.

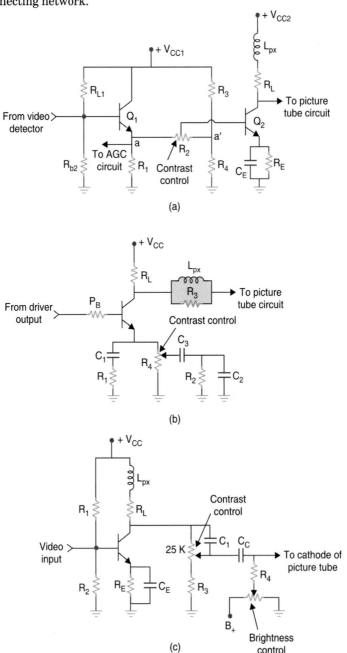

Fig. 13.12. *Contrast control circuits in transistor video amplifiers*
(a) Base network control (b) Emitter network control (c) Collector network control.

13.10 SCREEN SIZE AND VIDEO AMPLIFIER BANDWIDTH

It may not be immediately apparent but the size of the screen on which the picture is reproduced will also govern how much finer details the picture should possess. On a small screen the number of active lines that carry video information are very close to each other. This disables the eye to distinguish fine details, unless the viewer comes very close to the screen. The reason stems from the fact, that unless the two adjacent objects subtend an angle of one minute or more at the observer's eye, they cannot be seen as distinct units. With small screens the distance necessary for the eye to resolve details is so short that the viewer normally never comes that close to the screen.

Manufacturers of small screen TV receivers take advantage of this fact and design video amplifiers with a bandwidth that is much less than 5MHz. This reduction, in the highest modulating frequency to be reproduced, also makes full bandwidth in the RF and IF tuned amplifiers unnecessary. All this results in considerable reduction in the overall cost of the receiver which is a big factor in a highly competitive consumer goods industry.

REVIEW QUESTIONS

1. Draw small signal equivalent circuit of an R.C. coupled amplifier employing a vacuum tube and derive expressions to show that its bandwidth $\approx \dfrac{1}{2\pi R_L C_t}$, where R_L is the load resistance and C_t the total shunting capacitance.

2. Describe briefly the methods normally employed to extend the bandwidth of an R.C. coupled amplifier to meet video signal requirements. Show that in order to extend midband to the higher corner (– 3 db) frequency, the required value of shunt peaking coil inductance, $L_{px} = nC_t R_L^2$, where $n < 1$.

3. What are the relative merits of shunt and series compensation techniques employed to extend the bandwidth of an amplifier. Illustrate your answer by drawing equivalent circuits of amplifiers employing the two types of compensation.

4. Describe different methods of low frequency compensation. What are the special problems of direct coupled amplifiers ?

5. Draw equivalent circuit of a grounded emitter amplifier and derive expressions for voltage gain and gain-bandwidth product.

6. Design a video amplifier to meet the following requirements :

   Output voltage = 60 V (p-p), bandwidth = 4 MHz, detector output voltage 1.2 V (p-p), voltage gain = 50, configuration—common emitter, coupling—direct, V_{CC} supply = 100 V, Transistor-BF 177. Its approximate parameters are :

   r_x = 50 ohms, r_π = 200 Ω, c_π = 100 pF, β = 25, f_T = 120 MHz, C_π at 100 V = 1 pF, maximum collector dissipation = 0.8 watt. Take total collector network cpacitance = 16 pF. Assume any other data if necessary.

7. Describe with suitable circuit diagrams different methods of contrast control used in both tube and transistor video amplifiers. Mention relative merits of each type.

14

Video Amplifier Circuits

The essential requirements of video section circuitry and general wideband techniques were explained in the previous two chapters. The merits of cathode over grid modulation of the picture tube were brought out in an earlier chapter. This is now the most preferred method of feeding video signal to the picture tube unless there are strong reasons in favour of grid modulation. However, there are several methods of coupling the video amplifier to the picture tube. This, together with other relevant circuit details, is discussed in this chapter.

Various Coupling Methods. Though it cannot be denied that for near perfect reproduction of the transmitted picture, dc link has to be maintained between the video detector and picture tube, but dc coupling has its own problems which when fully taken care of add to the cost of the receiver. Therefore in many video amplifier designs full dc coupling is dispensed with yet maintaining optimum reception which is subjectively acceptable. The various possible coupling arrangements between the video amplifier and picture tube may be classified as:

(a) DC coupling

(b) Partial dc coupling

(c) AC coupling with dc restoration

(d) AC coupling

Though the circuit details differ from chassis to chassis, typical circuits of each type are examined to identify merits and demerits of the various types of coupling.

14.1 DIRECT COUPLED VIDEO AMPLIFIER

A commonly used video amplifier employing PCL 84 (pentode-triode) is shown in Fig. 14.1. The video signal is directly coupled from video detector to cathode of the picture tube. The main features of this circuit are as follows:

(i) Frequency Compensation

The plate circuit contains both shunt and series peaking coils to provide enhanced high frequency respouse. Additional broadbanding is achieved by using a small (0.003 µF) cathode bypass (C_k) capacitor. The network L_1R_1 in the grid circuit of the tube provides frequency compensation to offset its input capacitance.

(ii) Contrast Control

Gain of the amplifier is controlled by varying dc voltage (potentiometer R_6) at screen grid of the pentode. This becomes the contrast control. The need for two decoupling capacitors at the screen grid arises from the fact that electrolytic capacitors have a small self-inductance which is sufficient to introduce considerable reactance at high frequencies. Therefore, to provide adequate decoupling at high frequencies the 4 µF electrolytic capacitor (C_1) is shunted by a small 0.005 µF capacitor (C_2) as a high frequency bypass.

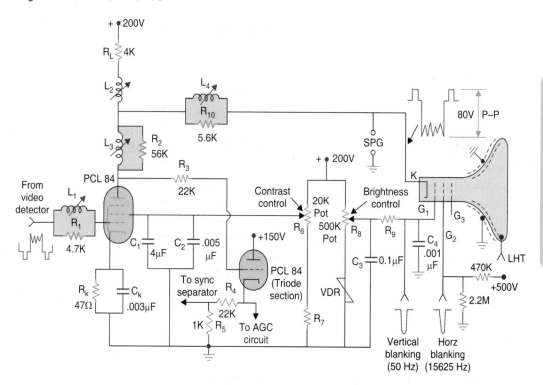

Fig. 14.1. Direct coupled video amplifier.

(iii) Sync and AGC Take-off Points

The triode section of PCL 84 is connected as a cathode follower. The resistors R_4 and R_5 form a potential divider at the cathode of the triode to feed necessary video voltage to the sync separator circuit. AGC circuit is fed directly from output of the cathode follower. The use of cathode follower avoids any loading effects from sync separator and AGC circuits and thus fully isolates the video amplifier from these circuits. In the absence of such a provision some additional capacitance appears across the output of the amplifier and tends to lower its high frequency response.

(iv) Flyback Suppression Pulses

Field and line flyback suppression pulses are injected at the control grid and first anode of the picture tube through isolating networks. These negative going pulses are of sufficient amplitude to cut-off the beam current during flyback intervals.

Chapter 14

(v) Brightness Control and Switch-off Spot

Brightness control is achieved by varying positive voltage at the grid (G_1) of the picture tube with potentiometer R_8 that is connected in series with a VDR (voltage dependent resistance) across B_+ supply. The VDR has a special function to perform. When the receiver is switched-off the time-base circuits stop immediately and the picture tube spot assumes central position on the screen. The cathode and grid potentials rapidly becomes equal to chassis potential since the B_+ voltage disappears. However, the picture tube's cathode remains hot for some time and keeps emitting electrons. At the same time the EHT capacitance formed by the aquadag coating of the tube does not immediately discharge since the resistance of the EHT circuit is very high. For a few moments, therefore, beam current continues to flow with zero grid-cathode bias and no deflection fields. A bright spot known as 'switch-off spot' appears on the screen centre, which can in due course burn a small portion of the phosphor coating on the screen. Suppression of the switch-off spot is brought about in this circuit by the VDR which forms the lower arm of the brightness control potential divider. When the receiver is switched off and the B_+ voltage disappears, the resistance of the VDR immediately becomes very high. This allows the charge on the associated capacitor C_3 to remain for a short time so that the grid is momentarily positive with respect to cathode. The result of this is that a high beam current passes for a brief instant and this discharges the EHT smoothing capacitor. This happens as the normal B_+ voltage is decreasing and before the raster finally collapses. Thus the heavy beam current is spread over the screen face and not concentrated on a central spot.

14.2 PROBLEMS OF DC COUPLING

Direct coupling, though very desirable has the following stringent requirements, which, when provided for add very much to the cost of the receiver.

(a) Regulated EHT Supply

Regulation of normal type of EHT systems used in most television receivers is not good. Full contrast range to be handled by the picture tube with dc coupling puts a heavy demand on the high voltage supply. On signal levels that correspond to peak whites, excessive beam current flows and this results in a drop of voltage. In turn this tends to an instantaneous increase in deflection sensitivity so that the picture expands (blooms) as the voltage falls. The increase in deflection sensitivity is due to the fact that as EHT voltage falls, forward velocity of the electrons, that constitute beam current, decreases. The electrons then spend a longer time under the deflection field, and are deflected more by a given field than they would normally be. Therefore, either a well regulated EHT system should be provided, or the natural range of brightness levels, which occur in the original transmitted picture, should be artificially restricted at the receiver. The latter, that is, partial loss of dc component of the video signal is lesser of the two evils and involves a commercial compromise. This is explained in another section of this chapter. However, as stated earlier, in terms of absolute picture fidelity, both the complete retention of dc component and a well regulated EHT system are necessary.

(b) Beam Current Limiting

In a dc coupled video stage, for cathode injection, the picture tube can be driven to a high brightness level if the input signal is removed. This increase in beam current is expensive both in high voltage supply source and the life of the picture tube.

A simple circuit which limits the mean beam current to a pre-determined value (without limiting the contrast range) is shown in Fig. 14.2 (a). With this circuit arrangement, dc coupling is maintained only on low key scenes, where it is most important. In this circuit,

$$V_3 = (I_d + I_b)R_1 = I_d R_1 + I_b R_1,$$

The diode will remain conducting and clamp the picture tube cathode to collector of the amplifying transistor, so long as $I_b R_1 < V_2$.

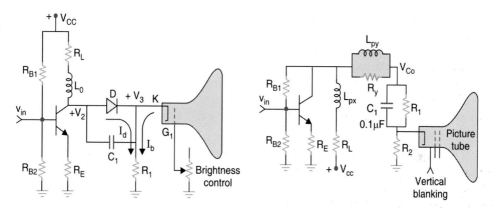

Fig. 14.2 (a). Beam current limiting.   Fig. 14.2 (b). Partial dc coupling.

At the threshold of limiting,

$$I_b \times R_1 = V_2 = V_3$$

Therefore

$$I_d = 0$$

Beyond this threshold $V_2 < V_3$ and the capacitor C_1 charges to $V_3 - V_2$, that is, the picture tube drive is now ac coupled. The picture tube, therefore, receives an additional back-bias proportional to the excess mean drive.

It may be noted that it is not possible to establish precise limiting threshold because the mean value of V_2 varies with the picture content. However, by a suitable choice of the value of R_1, excess of picture tube EHT current and the consequent blooming (or breathing) of the picture are prevented. In video circuits, that employ partial dc coupling, beam current is automatically reduced, making use of a diode unnecessary.

(c) Other Direct Coupling Problems

Besides the need for a regulated EHT supply and beam current limiting, the complexity in direct coupled amplifiers arises on account of the following :

(i) It is necessary to have a stable (regulated) B_+ source to avoid any drift in the output of the amplifier.

(ii) The reflections from any passing aeroplanes result in a steep rise and fall of input signal at the antenna of the receiver. This, despite an efficient fast acting AGC, causes a momentary flutter of the reproduced picture. It occurs because of very good low frequency response of the amplifier that extends down to zero Hertz.

Chapter 14

(*iii*) There is a possibility of heater to cathode insulation breakdown during picture high-lights because of high dc voltage, equal to the plate voltage that appears on the cathode of the picture tube.

(*iv*) For cathode injection, if the detector output is directly coupled to the video amplifier, the latter must be biased to conduct heavily when no signal is present. This is expensive both in B_+ current and life of the device.

14.3 PARTIAL DC COUPLING

As mentioned earlier the solution to direct coupling problems, as a compromise, lies in attenuating the dc voltage before applying it to picture tube and providing a low frequency bypass to reduce some of its annoying effects. This arrangement is a common feature in most television receivers and is known as 'partial dc coupling'. The relevant portion of a transistor video amplifier is shown in Fig. 14.2 (*b*). In this circuit

(*i*) C_1, R_1 and R_2 constitute dc attenuator and low frequency filter circuit. The long time constant circuit R_1, C_1 in series with video signal path to the picture tube cathode makes the aeroplane flutter effect less annoying. The capacitor C_1 (0.1 μF) fails to bypass very low frequencies, with the result, that R_1 provides series attenuation in the signal path to offending low frequency pulsations. As obvious, low frequency components of the video signal get attenuated by a factor $R_2/(R_1 + R_2)$, and this is what gives the circuit the name 'partial dc coupling'.

(*ii*) The values of R_1 and R_2 are chosen so as to considerably reduce the dc voltage that reaches the cathode of picture tube. This not only attenuates the low frequency content of the

video signal but the consequent reduction in the working dc voltge $\left(V_{CO} \times \dfrac{R_2}{R_1 + R_2} \right)$ at the cathode

reduces the magnitude of dc voltages required at the accelerating and focusing anode of the picture tube. This in turn results in a saving in the cost of power supply circuit.

(*iii*) The reduction in the cathode voltage reduces any possibility of heater cathode break-down of the picture tube.

(*iv*) Since the dc component is partly removed by potential divider action of R_1 and R_2, the difference in the mean level brightness from scene to scene is reduced. This restricts the overall range to be handled by the tube and hence limits the maximum demand that is made on the EHT system.

Video Amplifier Circuit

A transistorised video amplifier circuit with emitter follower drive and partial dc coupling is shown in Fig. 14.3. The salient features of this circuit are :

(*i*) Signal from the video detector is dc coupled to the base of Q_1. This transistor combines the functions of an emitter follower and CE amplifier. The high input impedance of emitter follower minimizes loading of the video detector. The sync circuit is fed from the collector of this transistor, where as signal for the sound section and AGC circuit is taken from the output of the emitter follower.

(ii) The output from the emitter follower is dc coupled to the base of Q_2. This is a 5 W power transistor, with a heat-sink mounted on the case. The collector supply is 140 V, to provide enough voltage swing for the 80 V P-P video signal output.

(iii) In the output circuit of Q_2, contrast control forms part of the collector load. The video output signal is coupled by the 0.22 μF (C_2) capacitor to the cathode of picture tube. The partial dc coupling is provided by the 1 M (R_2) resistor connected at the collector of Q_2.

(iv) The parallel combination of L_1 and C_1 is tuned to resonate at 5.5 MHz to provide maximum negative feedback to the sound signal. This prevents appearance of sound signal at the output of video amplifier.

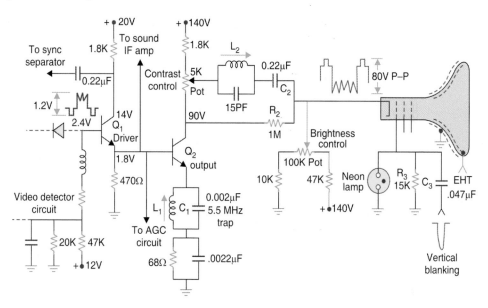

Fig. 14.3. *Video amplifier employing partial dc coupling.*

(v) The neon bulb in the grid circuit provides protection of a spark-gap since the neon bulb ionizes and shorts to ground with excessive voltage. The 'spark gaps' are employed to protect external receiver circuitry from 'flash overs' within the tube. The accumulation of charge at the various electrodes of the picture tube results in the appearance of high voltages at the electrodes, which if not discharged to ground, will do so through sections of the receiver circuitry and cause damage.

(vi) Note that dc voltages at the base and emitter of the two transistors have been suitably set to give desired forward bias.

(vii) Vertical retrace blanking pulses are fed at the grid of the picture tube through C_3, and the grid-return to ground is provided by R_3.

(viii) Brightness control. The adjustement of average brightness of the reproduced scene is carried out by varying the bias potential between cathode and control grid of the picture tube. In the circuit being considered a 100 KΩ potentiometer is provided to adjust dc voltage at the cathode. This bias sets correct operating point for the tube and in conjunction with the video blanking pulses cuts-off the electron beam at appropriate moments.

The setting of grid bias depends upon the strength of signal being received. A signal of small amplitude, say from a distant station, requires more fixed negative bias on the grid than a strong signal. The dependency of picture tube grid bias on the strength of the arriving signal is illustrated in Fig. 14.4. For a weak signal, the bias must be advanced to the point where combination of the relatively negative blanking voltage plus the tube bias drives the tube into cut-off. However, with a strong signal the negative grid bias must be reduced, otherwise some of the picture details are lost. Since the bias of the picture tube may required an adjustment for different stations, or under certain conditions from the same station, the brightness control is provided at the front panel of the receiver.

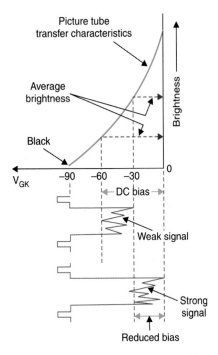

Fig. 14.4. Optimum setting of contrast control for different amplitudes of the video signal.

The effects of brightness and contrast controls described earlier overlap to some extent. If setting of the contrast control is increased so that the video signal becomes stronger, then the brightness control must be adjusted to meet the new condition, so that no retrace lines are visible and the picture does not look milky or washed out. Too small a value of the negative grid bias allows average illumination of the scene to increase thus making part of the retrace visible. In addition, the picture assumes a washed out appearance. Too low a setting of the brightness control, which results in a high negative bias on the picture tube grid, will cause some of the darker portions of the image to be eliminated. Besides this overall illumination of the scenes will also decrease. To correct this latter condition, either the brightness control can be adjusted or the contrast control setting can be advanced until correct illumination is obtained. If the brightness control is varied over a wide range the focus of the picture tube may be affected. However, in the normal range of brightness setting made by the viewer, changes in focus do not present any problem.

It is now apparent that despite the fact that video signal, as received from any television station, contains all the information about the background shadings of the scene being televised, an optimum setting of both contrast control and brightness control by the viewer is a must to achieve desired results. Many viewers do not get the best out of their receivers because of incorrect settings of these controls. However, to ensure that retrace lines are not seen on the screen due to incorrect setting of either contrast or brightness control, all television receivers provide blanking pulses on the grid electrode of the picture tube.

14.4 CONSEQUENCES OF AC COUPLING

As already explained, dc coupling though desirable, adds to the cost of receiver. Partial dc coupling does not reduce fully the circuit complexity and other side effects of dc coupling. This suggests the use of ac coupling from video detector to picture tube. However, before doing so, it would be desirable to review the consequences of ac coupling. Figure 14.5 (a) shows video signals for two lines taken at different moments from a television broadcast. One signal represents a line from a predominently white picture while the other belongs to a black background. As they come out of the video detector, their sync pulses are aligned to the same level. When amplified by a dc coupled amplifier, they get inverted but retain their common blanking level (Fig. 14.5 (b)). At the picture tube, with a suitable fixed bias, black levels of the two signals automatically align themselves along the beam current cut-off point. This happens because of different dc contents in the two signals. Thus dc components of video signals enable scenes with different background shadings to be correctly reproduced on the raster without having to change the setting of brightness control.

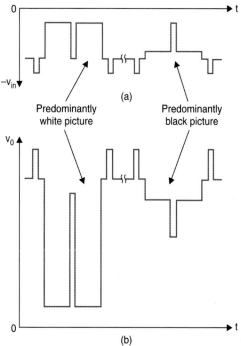

Fig. 14.5. Video amplifier signal waveforms for two different pictures
(a) Input voltage (b) Output voltage.

Now consider that these signals are passed through a capacitor as would be the case if ac coupling were employed. This is illustrated in Fig. 14.6 by an equivalent circuit and associated waveshapes. In the equivalent circuit (see Fig. 14.6 (a)) dc component of each signal has been represented by a battery and the ac component by a generator. The combined signal feeds into an RC circuit. On application of any composite signal the coupling capacitor will charge to a value equal to the battery voltage. However, the ac video content will cause the capacitor to charge and discharge as the applied voltage exceeds and falls around the dc voltage across the capacitor. Thus, while the dc component is blocked by the capacitor, the current which continuously flows through the load resistance develops an ac voltage drop across it. The resulting output waveshapes and their locations along with corresponding input waveshapes are shown in Figs. 14.6 (b) and (c). Note that while the output waveshapes are almost identical to their input counterparts, their sync and blanking levels no longer align with each other. Each signal has set itself around the zero axis as a consequence of ac coupling. This leads to the following undesired effects.

(i) Visible Retrace Lines

The retrace lines become visible because the blanking pulses do not remain at a constant level and may not have enough amplitude to cause retrace blanking. Most modern receivers, irrespective of the coupling employed, incorporate special vertical and horizontal retrace blanking circuits as a means of preventing retrace lines from becoming visible.

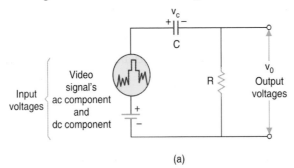

(a)

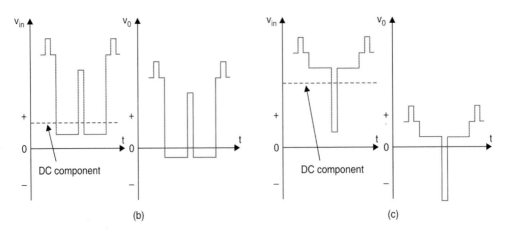

(b)   (c)

Fig. 14.6. Effect of RC coupling on the dc component of video signals
*(a) equivalent circuit of video signal and RC network (b) a predominantly
white scene, and (c) a predominantly dark scene.*

(ii) Possible Loss of Sync

This occurs due to loss of dc component. The sync pulse amplitudes now vary with picture content and this can lead to inadequate amplitude of sync pulses at the output of sync separator. Therefore, if synchronization is to remain stable, the dc level must be restored (sync pulses must line up) before the video signal is fed to the sync separator circuitry.

(iii) Loss of Average Brightness of the Scene

This means that bright and dark scenes may not be easily distinguishable. With loss of dc component the average brightness information is lost. Thus signals from different brightness backgrounds will lose this identity and reproduce pictures with a background of some grey shade.

(iv) Poor Colour Reproduction

In colour television a change in the luminance (brightness) signal will cause a change in the brightness of a colour. Therefore loss of dc component of the video signal will result in poor colour reproduction.

14.5 DC REINSERTION

As explained earlier, relative relationship of ac signal to the blanking and sync pulses remains same with or without the dc component. Furthermore, brighter the line, greater is the separation between the picture information variations and the associated pulses. As the scene becomes darker, the two components move closer to each other. It is from these relationships that a variable bias can be developed to return the pulses to the same level which existed before the signal was applied to the RC network.

DC Restoration with a Diode

A simple transistor-diode clamp circuit for lining up sync pulses is shown in Fig. 14.7 (a). The V_{CC} supply is set for a quiescent voltage of 15 V. In the absence of any input signal the coupling capacitor 'C' charges to 15 V and so the voltage across the parallel combination of resistor R and diode D will be zero. Assume that on application of a video signal, the collector voltage swings by 8 V peak to peak. The corresponding variations in collector to emitter voltage are illustrated in Fig. 14.7 (b). The positive increase in collector voltage is coupled through C to the anode of diode D, turning it on. Since a forward biased diode may be considered to be a short, it effectively ties (clamps) the output circuit to ground (zero level). In effect, each positive sync pulse tip will be clamped to zero level, thereby lining them up and restoring the dc level of the video signal.

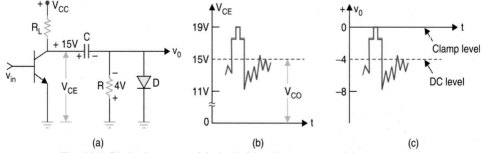

Fig. 14.7. Diode dc restorer (a) circuit (b) collector voltage (c) output voltage.

Chapter 14

In the case under consideration the diode will cause the coupling capacitor to charge to a peak value of 19 V. However, during negative excursion of the collector voltage the capacitor fails to discharge appreciably, because the diode is now reverse biased and the value of R has been chosen to be quite large. The average reverse bias across the diode is -4 V, being the difference between the quiescent collector voltage and peak value across the capacitor. Note that as the input video signal varies in amplitude a corresponding video signal voltage appears across the resistor R and it varies in amplitude from 0 to -8 V (peak to peak). This, as shown in Fig. 14.7 (c), is the composite video signal clamped to zero.

Similarly as and when the average brightness of the scene varies the capacitor C charges to another peak value thereby keeping the sync tips clamped to zero level.

Reversing the diode in the restorer circuit will result in negative peak of the input signal being clamped to zero. This would mean that the dc output voltage of the circuit will be positive. The video signal can also be clamped to any other off-set voltage by placing a dc voltage of suitable polarity in series with the clamping diode.

Limitations of Diode Clamping

It was assumed while explaining the mechanism of dc restoration that charge across the coupling capacitor C does not change during negative swings of the collector voltage. However, it is not so because of the finite value of RC. The voltage across C does change somewhat when the condenser tends to discharge through the resistor R. Another aspect that merits attention is the fact that whenever average brightness of the picture changes suddenly the dc restorer is not capable of instant response because of inherent delay in the charge and discharge of the capacitor. Some receivers employ special dc restoration techniques but cost prohibits their use in average priced sets.

14.6 AC COUPLING WITH DC REINSERTION

Figure 14.8 is the circuit of a video amplifier employing ac coupling with dc restoration. The coupling capacitor C_1 offers negligible reactance to the high frequency content of the video signal and it gets coupled to the picture tube grid directly. With no input signal, C_2 charges to the steady dc potential existing at 'X' through R_4, R_6 and R_7. With arrival of video signal the potential at 'X' falls during negative swing of the collector voltage. This causes C_2 to discharge through R_4, R_3, V_{CC} source, R_7 and diode D_1. However during positive voltage swing at the collector, C_2 fails to regain its charge because during this interval the diode is reverse biased and R_6 has been chosen to be too large. Thus the reduced voltage across C_2 is maintained at this level during intervals between sync tips because of the relatively large value of C_2 and associated resistors. The resultant difference of potential between 'X' and V_{C2}, that effectively appears between 'Y' and ground (see Fig. 14.8) gets applied to the grid via isolating resistance R_5. This amounts to restoring dc component of the video signal, which otherwise is blocked by the coupling capacitor. When the average brightness of the scene increases the video sync tips move further away from the picture signal content and the point 'X' then attains a new less positive potential during sync tip intervals. This further lowers the potential across the capacitor C_2. The enhanced difference in potential between the point 'X' and the new value of V_{C2} gets applied to the grid of the tube. This, being positive, reduces the net negative voltage between the grid and cathode and the scene then moves to a brighter area on the picture tube characteristics. Any decrease in average brightness of the scene being televised will have the opposite effect and net grid bias will become more negative to reduce background illumination of the picture on the raster. Thus the diode with the associated components serves to restore the dc content of the picture signal and the difference in potential between 'X' and 'Y' serves as a variable dc bias to change the average brightness of the scene.

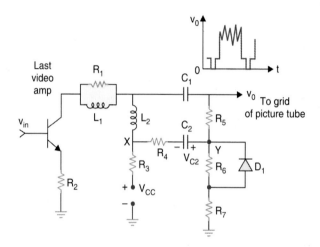

Fig. 14.8. Practical dc restorer circuit.

14.7 THE AC COUPLING

Cost is a strong determining factor in the design of commercial television receivers. If it is possible to reduce the cost of a set without impairing the picture quality too much, then a sacrifice in quality for cost is justifiable and is made in some receiver designs. Therefore many receivers use only ac coupling. In other words the dc component is removed from the signal and never reinserted.

The AC Coupled Video Amplifier

Figure 14.9 shows an ac coupled video amplifier. The coupling capacitor (0.22 μF) and resistance of the brightness control network constitute the ac coupling network. The contrast control is located in the emitter circuit of the first video amplifier. It is also ac coupled to the output video amplifier. The amplifier employs the usual broadbanding techniques. It has a sound trap (resonant) circuit in the emitter lead of the output transistor. An interesting feature of this circuit is the provision of a spot-killer switch. This switch opens when the receiver is switched off. Its operation removes dc voltage at the cathode of picture tube reducing grid-cathode voltage to zero. The residual beam current increases and quickly discharges the EHT smoothing capacitor thereby reducing intensity of the switch-off spot.

Chapter 14

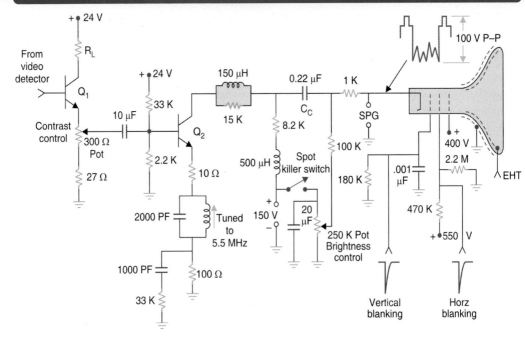

Fig. 14.9. *AC coupled video amplifier circuit.*

14.8 VIDEO PREAMPLIFIER IN AN IC CHIP

The use of integrated circuits between video detector and video output amplifier is very common in all solid state TV receivers. TBA 890 is one such dedicated IC which performs the following functions. Figure 14.10 shows circuit connection at various pins of this IC.

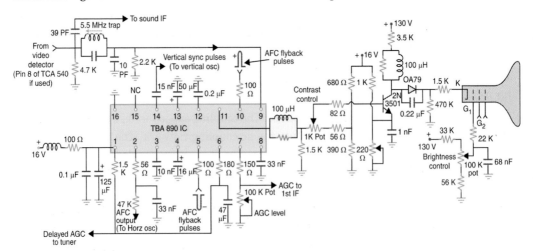

Fig. 14.10. *Video output amplifier driven by the IC TBAS90.*

(i) First Video Amplifier

The output from video detector feeds at pin 9 and this is also the sound take-off point. The video preamplifier employs atleast two stages of differential amplifiers and is preceded by a driver stage to provide impedance matching. The voltage gain from this stage is about 70 db. Input signal from the detector is clamped at 3 V and the video output is obtained at pin 11 of the IC. The output drives video output transistor 2N 3501 through a contrast control network as shown in the figure. The V_{CC} supply to the IC is a stabilized + 16 V derived/from the low voltage (LV) rectifier and filter network.

(ii) Sync Separator

The sync separator receives input from the video preamplifier and is suitably biased to deliver clean sync pulses. The circuit also employs a noise suppression circuit. Integrated vertical sync pulses are fed to the vertical oscillator through a capacitor from pin number 14 on the IC. The amplitude of the vertical sync output is around 11 V.

(iii) AFC Circuit

The horizontal AFC circuit employs a single ended discriminator. It derives sync pulse input from the sync separator and flyback pulses of opposite polarity from the horizontal output transformer at pins 4 and 10. The vertical blanking pulses are also added at pin 10 from the vertical output transformer. The AFC control voltage is available at pin 2 and is fed to the input of horizontal oscillator through an anti-hunt filter circuit. AFC output voltage ranges from 2 to 10 volts.

(iv) AGC Circuit

The IC includes a keyed AGC circuit and receives flyback pulses through pin 10 along with the AFC circuit. The AGC output voltage varies from 1 to 12 V and is fed to the IF section from pin 7 as shown in the figure. Delayed AGC voltage for the tuner is available at pin 6 and its amplitude varies from 0.3 to 12 V.

Chapter 14

REVIEW QUESTIONS

1. Enumerate the various coupling methods employed between video detector and picture tube. Why does dc coupling add to the cost of the receiver ?

2. Describe the main features of the dc coupled video amplifier shown in Fig. 14.1.

   What is a 'switch-off' spot ? Explain how its undesirable effect on the screen is minimized by using a VDR in the brightness control circuit.

3. Explain briefly the essential requirements, which must be met, while providing dc coupling in the video section of the receiver. Describe with a suitable circuit diagram how a diode can be used to limit beam current of the picture tube to a safe upper limit.

4. What do you understand by partial dc coupling ? Explain with a circuit diagram how some of the annoying features of dc coupling are almost eliminated by partial dc coupling. Justify that it is a reasonable compromise between cost and quality.

5. Describe the consequences of ac coupling. Show with a suitable circuit diagram and illustrations how dc component of the video signal is restored back by diode clamping in an otherwise ac coupled video amplifier.

6. Explain how the removal of dc component from the video signal affects the overall contrast range and necessitates repeated adjustments of the brightness control.

7. State the arguments that are advanced in favour of employing ac coupling in the video section of a TV receiver. Draw the circuit diagram of a typical coupled video amplifier and explain its main features.

15

Automatic Gain Control and Noise Cancelling Circuits

Automatic gain control (AGC) circuit varies the gain of a receiver according to the strength of signal picked up by the antenna. The idea is the same as automatic volume control (AVC) in radio receivers. Useful signal strength at the receiver input terminals may vary from 50 μV to 0.1 V or more, depending on the channel being received and distance between the receiver and transmitter. The AGC bias is a dc voltage proportional to the input signal strength. It is obtained by rectifying the video signal as available after the video detector. The AGC bias is used to control the gain of RF and IF stages in the receiver to keep the output at the video detector almost constant despite changes in the input signal to the tuner.

15.1 ADVANTAGES OF AGC

The advantages of AGC are:

(a) Intensity and contrast of the picture, once set with manual controls, remain almost constant despite changes in the input signal strength, since the AGC circuit reduces gain of the receiver with increase in input signal strength.

(b) Contrast in the reproduced picture does not change much when the receiver is switched from one station to another.

(c) Amplitude and cross modulation distortion on strong signals is avoided due to reduction in gain.

(d) AGC also permits increase in gain for weak signals. This is achieved by delaying the application of AGC to the RF amplifier until the signal strength exceeds 150 μV or so. Therefore the signal to noise ratio remains large even for distant stations. This reduces snow effect in the reproduced picture.

(e) Flutter in the picture due to passing aeroplanes and other fading effects is reduced.

(f) Sound signal, being a part of the composite video signal, is also controlled by AGC and thus stays constant at the set level.

(g) Separation of sync pulses becomes easy since a constant amplitude video signal becomes available for the sync separator.

AGC does not change the gain in a strictly linear fashion with change in signal strength, but overall control is quite good. For example, with an antenna signal of 200 μV the combined RF and IF section gain will be 10,000 to deliver 2 V of video signal at the detector output, whereas with an input of 2000 µV, the gain instead of falling to 1000 to deliver the same output, might attain a value of 1500 to deliver 3 V at the video detector.

Basic AGC Circuit

The circuit of Fig. 15.1 illustrates how AGC bias is developed and fed to RF and IF amplifiers. The video signal on rectification develops a unidirectional voltage across R_L. This voltage must be filtered since a steady dc voltage is needed for bias. R_1 and C_1, with a time constant of about 0.2 seconds, constitute the AGC filter. A smaller time constant, will fail to remove low frequency variations in the rectified signal, whereas, too large a time constant will not allow the AGC bias to change fast enough when the receiver is tuned to stations having different signal strengths. In addition, a large time constant will fail to suppress flutter in the picture which occurs on account of unequal signal picked up by the antenna after reflection from the wings of an aeroplane flying nearby. With tubes, a typical AGC filter has 0.1 µF for C_1 and 2 M for R_1.

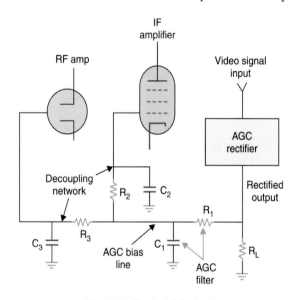

Fig. 15.1. Basic AGC circuit.

For transistors, typical values are 20 kΩ for R_1 and 10 µF for C_1. The filtered output voltage across C_1 is the source of AGC bias to be distributed by the AGC line. Each stage controlled by AGC has a return path to the AGC line for bias, and thus the voltage on the AGC line varies the bias of the controlled stages.

15.2 GAIN CONTROL OF VT AND FET AMPLIFIERS

The gain of a vacuum tube or FET amplifier can be determined from the equation $A_V = g_m Z_L$, where g_m is the transconductance of the device and Z_L the impedance of the load. The impedance is determined by the components used in the tuned circuit and does not lend itself to simple manipulation. However, g_m of both tubes and field-effect transistors can be controlled by varying their bias. Hence all AGC systems vary the bias of RF and IF stages of the receiver to control their gain. The variation of g_m with control grid voltage for a vacuum tube is illustrated in Fig. 15.2(a).

The transconductance is smaller near cut-off but increases as the bias decreases towards zero. It also decreases if operation of the tube is brought close to saturation. The region near cut-off bias is used for AGC operation. The self-bias is chosen to fix the operating point for high gain. The AGC voltage which is negative gets added to it and shifts the operating point to change the gain. The resulting change in gain compensates for variations in the input signal thereby maintaining almost constant signal amplitude at the output of video detector. Figure 15.2 (b) shows how the control grid is returned to the AGC line for negative bias. In tube circuits a negative bias varying between – 2V and – 20V is developed by the AGC circuit depending on the strength of incoming signal. In order to obtain high gain and minimum cross modulation effects, pentodes with remote cut-off characteristics are preferred in video IF amplifier circuits.

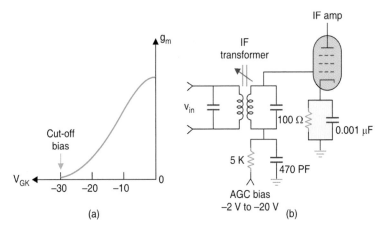

Fig. 15.2. AGC action in tube circuits (a) plot of g_m v_S grid bias, (b) method of returning AGC bias to the control grid.

15.3   GAIN CONTROL OF TRANSISTOR AMPLIFIERS

The transistor is a current controlled device. Therefore, in transistor amplifiers it is desirable to consider power gain instead of voltage gain for exploring means of their gain control by AGC techniques. The power gain (G) of a transistor amplifier may be determined by the equation,

$$G \approx \frac{\beta^2 R_L}{R_{in}}$$

where β and R_{in} are the short circuit current gain and input resistance of the transistor respectively. Here again it is convenient to vary one of the parameters of the transistor in order to control overall amplification of the receiver. The magnitude of β depends on the operating point of a transistor which is established by the base to emitter (V_{BE}) forward bias. Shifting the operating point, both towards collector current cut-off and collector current saturation causes a decrease in β which in turn reduces the power gain. Figure 15.3 shows the effect of change in V_{BE} on collector current and power gain of an amplifier employing a silicon transistor. A number of conclusions may be drawn from the curves shown in the figure. First, the amount of change in V_{BE} which is necessary to shift the operating point of the transistor from cut-off to saturation is small, only 0.4 to 0.5 V. This is much smaller as compared to about 30 V in a vacuum tube. Second, at some optimum value of forward bias (0.7 V in this case) power gain of the amplifier is maximum and does not change much for small variations in the bias voltage. However, the gain decreases as the bias is either increased (shifting the operating point towards saturation) or decreased (moving it towards cut-off).

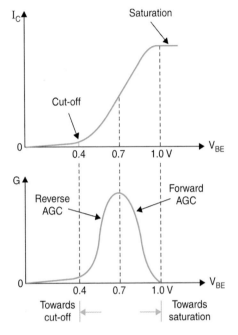

Fig. 15.3. Variation of collector current (I_C) and power gain (G) of a transistor amplifier as its base to emitter voltage (V_{BE}) is varied.

15.4 TYPES OF AGC

If the operating point is shifted towards saturation for controlling the amplifier gain, it is called forward AGC. An AGC system which operates by shifting the operating point towards cut-off is referred to as reverse AGC.

In many TV receiver designs, either forward or reverse AGC is exclusively employed for affecting gain control. However, in some receivers both forward and reverse AGC are simultaneously employed in different parts of the RF and IF amplifier chain. It may be noted that receivers which use either reverse or forward AGC do not operate the amplifiers at peak gain but fix the no-signal operating point at such a value that the stage gain may be increased or decreased without having to move to the other side of the power gain peak.

Reverse AGC

The power gain curve in Fig. 15.3 is not symmetrical, that is, the reverse AGC region of the curve falls off more rapidly than does the forward AGC region. This means that reverse AGC will require a smaller change in voltage for full gain control than will forward AGC. However, operation in this region, which is close to cut-off makes the receiver more susceptible to overload and cross modulation distortion on strong signals.

Chapter 15

The circuit of Fig. 15.4 (a) is of a single stage transistor (n-p-n) IF amplifier employing reverse AGC. The voltage divider formed by R_1 and R_2 provides a suitable fixed forward bias from the V_{CC} supply. The resistor R_3 and capacitor C_1 constitute the AGC decoupling network.

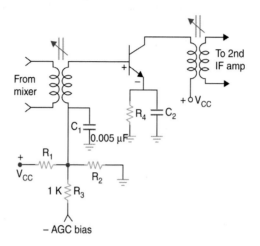

Fig. 15.4 (a). Tuned amplifier with reverse AGC.

Forward AGC

Forward AGC is often preferred for controlling gain of video IF amplifiers because it is more linear in its control action. Besides a change in β, the input resistance (R_{in}) of the transistor also decreases with increase in forward bias. This results in a power mismatch between the tuned IF transformer and the transistor, thereby providing an additional control on power gain.

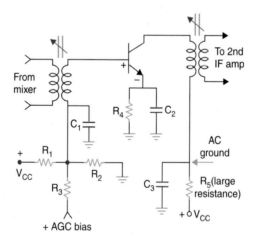

Fig. 15.4 (b). Tuned amplifier with forward AGC.

A single stage tuned IF amplifier employing forward AGC is shown in Fig. 15.4 (b). In this case, for any increase in signal strength the base to emitter forward bias must increase to shift the operating point of the transistor towards saturation. Similarly a decrease in signal strength would require a decrease in forward bias. To achieve this, the AGC system must deliver a positive going voltage to the base of amplifier. If a *p-n-p* transistor is used the forward AGC system would develop a negative going voltage proportional to the signal strength. As shown in the figure, amplifiers employing forward AGC often use a large resistor (R_5) in series with the collector circuit. When AGC voltage varies to increase collector current, effective V_{CE} decreases, thereby allowing the transistor to approach saturation quickly for faster AGC action.

AGC is applied to the tuner and 1st and 2nd IF stages but not to the third or last IF stage because amplitude of the input signal to the third IF amplifier is quite large and any shift in the chosen optimum operating point by the application of AGC would result in amplitude distortion. Another reason for not applying AGC bias to the last IF stage is the fact that the AGC control is proportional to stage gain and this is more suited to RF and first two IF stages because the RF signal amplitude here is quite small and these stages can be designed for more gain without any appreciable distortion.

15.5 VARIOUS AGC SYSTEMS

An 'average' or simple AGC system, as used in radio receivers, is not suited for control of gain in TV receivers. The average value of any video signal depends on brightness of the scene besides signal strength and so is not a true representation of the RF signal picked up at the antenna. For example, a dark scene would develop more AGC bias as compared to a white one, the signal strength remaining the same. This, if used to control the gain of the receiver, would tend to make dark scenes more dark and white ones more bright.

With the present system of transmission, the carrier is always brought to the same level when synchronizing pulses are inserted irrespective of the average level of the video signal. The amplitude of the sync level would change only if the signal strength changes. The sync amplitude level, then can serve as the true reference level of the strength of the picked up signal. The system based on sampling the sync tip levels is known as 'Peak' AGC system. Either the modulated picture IF carrier signal or the detected video signal can be rectified by the AGC stage to supply AGC bias voltage. In most receivers, however, the AGC circuit uses signal from the video amplifier because the higher signal level allows better control by AGC bias. It is necessary to provide dc coupling between the video amplifier and AGC system in order to keep the pedestals of the video signal aligned. The peak rectifier output then will be a true measure of the signal picked up by the antenna.

Peak AGC System

A typical peak detector circuit is shown in Fig. 15.5, where a separate diode is used to rectify the signal which is fed to it through capacitor C_1 from the output of the last IF amplifier. During positive half cycles of the modulated video signal, diode D_1 conducts and the capacitor C_1 charges to peak value of the input signal with the polarity marked across the capacitor. During periods other than sync pulse intervals the diode is reverse biased and no current flows through it. However, the capacitor tends to discharge through secondary winding of the IF transformer and R_1. Time constant of the discharge path is 270 µs and this being much greater than the line period of 64 µs, the capacitor discharges only partially and regains charge corresponding to the sync tip (peak) amplitude on each successive sync pulse. Thus the current that flows through R_1

is proportional to the peak value of the modulated video signal and the voltage drop across it becomes the source of AGC bias.

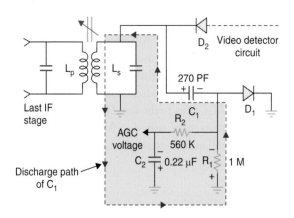

Fig. 15.5. *Peak AGC system.*

Negative voltage drop across R_1 is filtered by R_2 and C_2 to remove 15625 Hz ripple of the horizontal sync pulses. The output AGC voltage thus obtained is fed to IF and RF stages as an AGC control voltage.

Drawbacks of non-keyed AGC. The peak AGC system which is also called the non-keyed AGC system suffers from the following drawbacks, though it measures the same signal strength.

(*a*) The AGC voltage developed across the peak rectifier load tends to increase during vertical sync pulse periods because the video signal amplitude remains almost at the peak value every time the vertical sync pulses occur. This results in a 50 Hz ripple over the negative AGC voltage and reduces gain of the receiver during these intervals. The reduced gain results in weak vertical sync pulse which in turn can put the vertical deflection oscillator out of synchronism causing rolling of the picture. To overcome this drawback a large time constant filter would be desirable to filter out the 50 Hz ripple from the AGC bias. But with too large a time constant the AGC voltage fails to respond to fast changes like aeroplane flutter and quick change of stations.

(*b*) In fringe areas noise pulses develop an additional AGC voltage which tends to reduce the overall gain. This effect is more pronounced for dark scenes. The net effect is that S/N ratio further deteriorates and this results in a lot of snow on the picture.

(*c*) Even when the input signal strength is quite low, a small AGC voltage gets developed and this reduces the gain of the receiver, when actually, maximum possible gain is desired for a satisfactory picture and sound output.

To overcome these drawbacks special AGC circuits known as 'keyed' or 'gated' AGC circuits have been developed and are used in almost all present day television receivers. The problem of reduction of gain with weak input signals is resolved by using 'delayed' AGC action.

Keyed AGC System

In this system, the AGC rectifier is allowed to conduct only during horizontal sync pulse periods, with the help of flyback pulses derived from the output of the horizontal deflection circuit of the receiver. Video signal is also coupled to AGC rectifier to produce AGC voltage proportional to signal strength. However, AGC tube or transistor is generally biased to cut-off so that it conducts only for the short time the keying pulse is applied. This ensures that the rectifier conducts only when the blanking and sync pulses are on. As shown in Fig. 15.6 (a) the flyback pulses are generated during the retrace period of horizontal sweep circuit. Thus the time of flyback pulses corresponds to the time of sync and blanking, assuming that the picture is in full synchronization. The gating or AND function means that both inputs must be 'on' at the same time to produce AGC output.

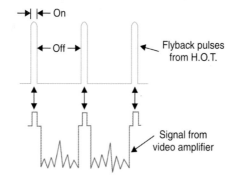

Fig. 15.6 (a). Keying pulses at horizontal sync rate for AGC circuit.

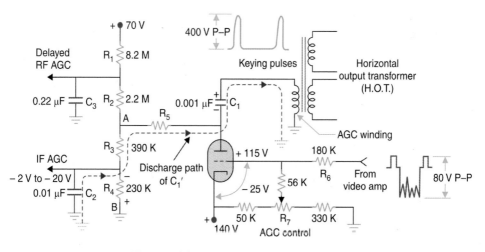

Fig. 15.6 (b). Typical triode keyed AGC circuit.

The basic circuit of a keyed AGC system employing a triode is shown in Fig. 15.6 (b). Video signal is directly coupled from video amplifier to the grid of AGC tube. Because of dc coupling the grid is at + 115 volts and so the cathode is maintained at + 140 volts to develop a grid bias voltage equal to – 25 V (115 – 140). Without any video signal on the grid the tube stays beyond cut-off. No dc potential is applied to the plate of the triode, but instead, positive going flyback pulses derived from the horizontal deflection output transformer are fed through C_1, which drive the plate positive during the sync pulse periods. The video signal of large amplitude drives the grid positive and clamps it at almost zero potential when the signal rises towards blanking level, so that the tube can conduct as the plate is pulsed positive at that time. When plate current flows, it charges C_1 with the plate side negative. The path for the charging current includes cathode to plate circuit in the tube, C_1 and AGC winding on the horizontal output transformer (H.O.T.). Between pulses when the tube does not conduct, C_1 partially discharges through the path marked on the diagram to charge C_2, the AGC filter capacitor. The time constant of the filter circuit is much larger than 64 μs and therefore a relatively steady dc voltage is developed to serve as the AGC source. Since the tube conducts only during horizontal sync pulse periods, even when the vertical sync pulse train arrives, AGC voltage developed across the points A and B represents true signal strength and has no tendency to vary during vertical sync periods. The potentiometer R_7 is adjusted for optimum grid bias. The function of R_6 is to isolate the AGC tube form the video amplifier which supplies video signal. The chain of resistors (R_1 through R_4) is for delayed AGC action which is explained in a subsequent section of this chapter.

Transistor Keyed AGC

A basic keyed AGC circuit designed to develop a positive AGC voltage is shown in Fig. 15.7. The p-n-p transistor Q_1 is biased to cut-off under no signal conditions. Negative going retrace pulses are applied to the collector circuit. The base-emitter junction gets forward biased when the video signal at the base approaches its minimum value. This corresponds to sync pulse periods and it is then that the collector is pulsed 'on' by the flyback pulses. The keying pulses are fed to the collector of the keyer transistor via diode D_1. When the transistor conducts, current flow through R_1, R_2, Q_1, D_1, winding on H.O.T. and C_1 thereby charging it with the polarity marked on it. The voltage developed across C_1 is the AGC output voltage. Diode D_1 prevents discharge of C_1 during the time between keying pulses. In the absence of D the charge on C_1 will forward bias the collector to base junction of Q_1 and allow the capacitor to discharge resulting in loss of AGC voltage. However, the diode has the correct polarity to couple negative flyback pulses from AGC winding to the collector of the transistor. The capacitor C_1 and resistors R_3 and R_4 form a filter to develop a steady dc voltage for controlling overall gain of the receiver.

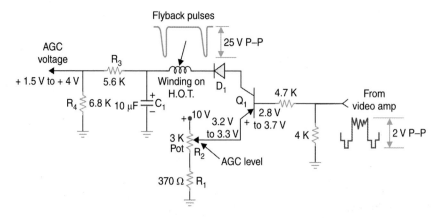

Fig. 15.7. *Typical transistor keyed AGC circuit.*

It may be noted that in comparison to vacuum tube circuits, the peak-to-peak amplitude of the keying pulses and video signal is much smaller. Typical transistor circuit values are 25 V peak-to-peak for flyback pulses and 2, V p-p for the video signal. For an *n-p-n* transistor the polarity of both the inputs is opposite to that needed for a *p-n-p* transistor.

15.6   MERITS OF KEYED AGC SYSTEM

(*a*) A long time constant to filter out 50 Hz ripple is no longer necessary because conduction takes place only during the horizontal retrace periods and no undue build up of voltage occurs during vertical sync intervals. The relatively short time constant filter, used to remove 15625 Hz ripple, enables the AGC bias to respond to flutter and fast change of stations, thereby ensuring a steady picture and sound output.

(*b*) AGC voltage developed is a true representation of the peak of fixed sync level and thus corresponds to the actual incoming signal strength.

(*c*) Noise effects are minimized because conduction is restricted to a small fraction of the total line period.

15.7   DELAYED AGC

The picture produced on the raster should be as noise free as possible. This is achieved by low-noise circuits. However, despite careful circuit design, each stage in the receiver contributes some noise. The cumulative effect of this, if not checked, would be a noisy picture. Noise effect can be overcome by high amplification of the incoming RF signal.

The amplified signal will then swamp out effects of stage noise as it is processed by the receiver. It is, therefore, necessary to operate the RF amplifier at maximum gain, particularly for weak RF signals. In the circuits discussed so far, the AGC voltage is fed not only to the first and second IF amplifiers, but also to the RF amplifier. Thus the negative AGC bias would reduce the gain of RF amplifier even for low-level RF signals. This undesirable effect is overcome by delaying the AGC voltage to the RF amplifier for weak RF signals. The technique used which is a delay in voltage, not in time, is called delayed AGC.

Delayed AGC Circuit

In Fig. 15.6 (*b*) delay action is achieved through the voltage divider R_1, R_2, R_3 and R_4 tied between $B+$ supply and ground. This places the RF AGC take-off point (grid of the RF amplifier) at approximately + 70 V to ground. Since the cathode of the RF amplifier tube returns to ground, the grid to cathode circuit acts like a diode and is forced into conduction. Since a conducting diode may be considered to be a short, the AGC take-off point is clamped to approximately zero volt because the cathode of the amplifier returns to ground. The bias clamp will continue to conduct until the input signal becomes sufficiently strong to provide a negative AGC voltage which is large enough to overcome the forward bias applied to the bias clamp diode and turn it off. This will restore normal AGC action to the RF amplifier for input signal amplitudes higher than a predetermined level. It can be set by varying the potentiometer R_7.

Another method of obtaining delayed AGC is to use a separate diode for clamping action. This method is explained later while discussing a typical tube AGC circuit.

15.8 NOISE CANCELLATION

The need for minimizing noise set-up in AGC and sync separator circuits has given rise to several methods of noise cancellation. Three commonly used methods are described. In Fig. 15.8 (a) diode D_2 is used as a switch which opens in the presence of noise preventing it from reaching the video amplifier. The necessary forward bias enabling it to pass noise free video signals is set by the potentiometer R_3. When a strong noise signal arrives the diode gets reverse biased thereby stopping noise pulses from reaching the video amplifier. Since the video signal to both AGC keyer and sync separator is obtained from the output of video amplifier, noise pulses are prevented from reaching these circuits.

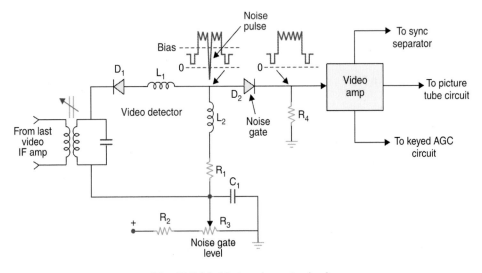

Fig. 15.8 (a). *Diode noise gate circuit.*

In the noise cancellation circuit of Fig. 15.8(b) the signal inputs to the control grid (G_1) suppressor grid (G_3) and plate (P) of a pentode are applied in such a way that the tube conducts only if all the three inputs are present at the same time. This arrangement is often referred to as a coincidence gate or an AND gate. Noise cancellation is achieved by setting the control grid bias in such a way that it goes to cut-off when a strong noise pulse arrives.

Special tubes with sharp cut-off characteristics have been developed for use in TV receivers and 6HA7 is one such tube—a twin pentode, where one section is for AGC and the other for sync separation. As shown in Fig. 15.8 (b) the cathode, control grid and screen grid are common to both the sections. The control grid serves as the noise gate for both the circuits. However, there are two suppressor grids and two plates to separate sync output from one pentode and AGC voltage from the other.

Another noise cancellation circuit is shown in Fig. 15.8 (c), where noise is eliminated by using a separate noise gate. It separates noise from the composite video signal, amplifies it and then adds it to the inverted composite video signal. The noise gate is a grounded base amplifier, normally set to cut-off. Any incoming noise pulse of sufficient amplitude counteracts the fixed reverse bias and sets the amplifier into conduction. Thus the noise pulse is amplified without inversion of its polarity. The gain of the noise gate amplifier is set equal to that of the video amplifier. Since the two noise pulses are equal in amplitude but opposite in polarity, they cancel on addition. Thus the AGC and sync separator circuits remain immune to incoming noise pulses. This system is frequently used in solidstate receivers.

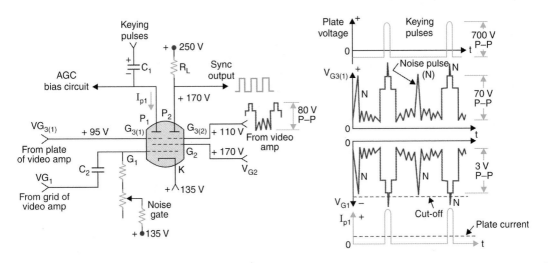

Fig. 15.8 (b). Keyed AGC and sync separator circuits with a common noise gate.

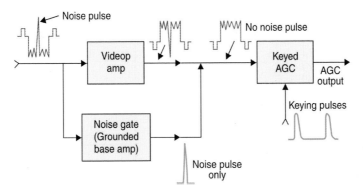

Fig. 15.8 (c). Noise cancellation by a separate noise gate amplifier.

15.9 TYPICAL AGC CIRCUITS

As already explained there is a considerable difference between the methods of controlling gain of vacuum tube and transistor amplifiers. Transistor keyed AGC circuits are complex than their tube counterparts. In some circuits three or more transistors are used to develop the required AGC voltage. There is no standard approach to such circuits and they may vary from chassis to chassis. Some typical AGC circuits employing tubes and transistors are discussed.

Keyed AGC with a Twin-Triode

The keyed AGC circuit shown in Fig. 15.9 employs a twin-triode for generating AGC voltage. One section (V_1) of the tube is used for developing a dc voltage proportional to the peak value of the input video signal. The second triode (V_2) is connected for keyed AGC action. The video signal is dc coupled to the grid of V_1 through a frequency compensating network. It is connected as a cathode follower with R_3 as its load resistance. The time constant of R_3 in parallel with C_3 is chosen to be quite large as compared to the line period of 64 µs. Therefore, the voltage which develops across this network is a dc voltage proportional to the peak value of sync pulses. This is direct coupled to the grid of V_2 which conducts during flyback pulse intervals to charge the capacitor C_2. In between sync pulses C_2 discharges to develop a negative voltage at point A with respect to ground. The potentiometer R_5 is varied to adjust optimum bias voltage for V_2 to permit sufficient conduction for developing suitable AGC voltage with a known input signal strength. Note that any strong noise pulse will make G_2 positive with respect to K_2 causing clamping action. Thus noise pulses are prevented from developing any AGC voltage. The resistors R_7 and R_8 form a potential divider and the voltage which develops across the filter circuit R_8–C_5 is fed to the IF section of the receiver.

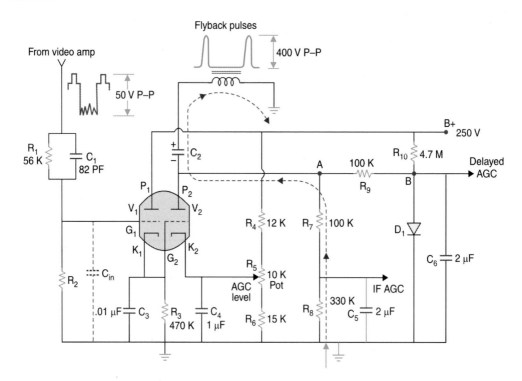

Fig. 15.9. Keyed AGC with delay circuit for RF amplifier.

The values of resistors R_7 through R_{10} are so chosen that at low RF signal levels D_1 remains forward biased from $B+$ supply. Thus point B of the AGC circuit is clamped to ground and no AGC voltage gets applied to the RF amplifier. Any change in the negative potential at A affects the potential at point B through the isolating resistor R_9. When the input signal strength increases, say when another strong channel is selected, point A becomes more negative with the result that at a predetermined voltage level the potential at B changes to become negative. This reverse biases diode D_1 removing its clamping action. Thus full AGC voltage becomes available on the RF amplifier AGC bias line. This then amounts to a delay in applying AGC to the tuner section of the receiver.

AGC Circuit for Solid-State Receivers

A typical circuit is shown in Fig. 15.10 where transistor Q_1 is keyed to develop AGC bias and Q_2 serves as the AGC amplifier. The video signal is dc coupled at the base of Q_1 and its emitter is fed with a suitable dc voltage to reverse bias the emitter-base junction in the absence of any video signal. The flyback pulses are fed at the collector of this transistor through the diode. The pulsed current completes its path back to the winding on the H.O.T. through C_1 which then charges with positive polarity towards ground. R_5-C_1 acts as the AGC filter and this voltage is dc coupled to the base of Q_2. The forward-bias on the emitter-base junction of this transistor varies with the voltage across C_1 and thus controls the collector current through load resistance R_{10}. A strong input signal at the antenna will develop more negative voltage across C_1, thereby, reducing the forward bias on Q_2. This in turn will decrease its collector current making the collector more positive. The receiver employs forward AGC control on the RF and IF amplifiers and the voltage developed at the collector is then of the right polarity to decrease gain of these stages. R_7-C_3 constitute an additional filter circuit for the RF bias line. R_8 is the isolating resistance and R_9-C_4 acts as filter for IF bias line. Another transistor can be added to this circuit to achieve delay action for the RF amplifier.

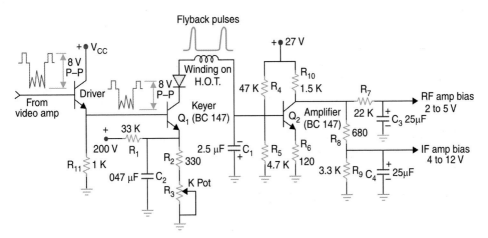

Fig. 15.10. Keyed automatic gain control circuit with AGC amplifier.

Improved AGC Circuit

An improved AGC circuit is shown in Fig. 15.11. The bias for the two IF amplifiers is developed by Q_2, the AGC 'keyer'. For the RF amplifier, bias is supplied by Q_3, the AGC driver. It will be seen that there are different modes of AGC operation, depending on the level of incoming signal. On very low signals, no AGC bias is developed, and the controlled IF and RF stages are fixed biased.

On medium-level signals the AGC keyer turns on and biases the two IF amplifiers. As the signal level increases further the AGC driver is also turned on, biasing the RF amplifier, while the bias on the controlled IF amplifiers continues to increase. A point is reached beyond which the bias of the IF amplifiers is clamped. When the incoming signal level becomes very high, both RF bias and IF bias again increase to reduce the gain of the controlled amplifiers.

The circuit operates in the following manner. The negative going (sync) composite video signal from the video detector is fed to the video driver Q_1 (p-n-p) which is connected as an emitter follower and develops at its emitter a negative-going composite video signal, superimposed on a + 3.8 V dc level. This is dc coupled to the base of Q_2 through R_{15}, an isolating resistor. The AGC keyer Q_2 is normally at cut-off. This is so because its emitter-base junction is reverse biased and no dc voltage is applied at its collector. The transistor Q_3 is also normally gated to cut-off. Therefore, with no or on very low-level RF signals, both Q_2 and Q_3 remain in cut-off.

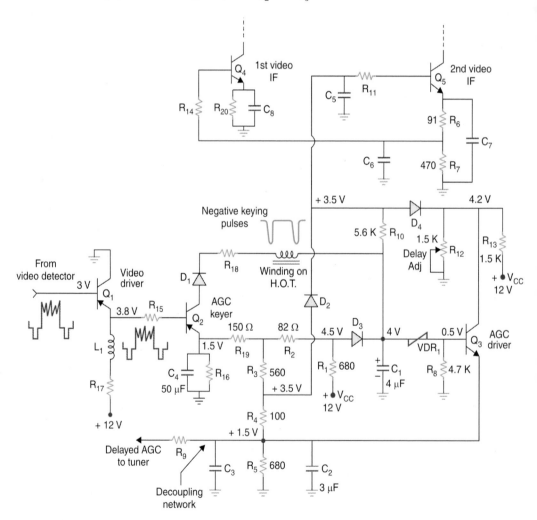

Fig. 15.11. Improved AGC circuit. Note that the voltages shown are with no signal applied.

The series dc voltage divider consisting of R_1 through R_5 develops about + 3.5 V at the junction of R_3 and R_4 from the 12 V dc supply. This potential is coupled via diode D_2 to the base of transistor Q_5 in the second IF amplifier. Bias for the base of Q_4 (first video IF amplifier) is derived from the junction of R_6 and R_7 located in the emitter lead of Q_5. The operating points of both Q_4 and Q_5 are chosen for forward AGC control.

From the same V_{CC} source another circuit is completed through D_3, VDR_1 and R_8, thereby providing about 4 V across C_1 and 0.5 V at the base of Q_3. The voltage drop across R_5, the emitter resistance of Q_3 is about 1.5 V on account of current which flows through it from the V_{CC} source. This is enough to keep Q_3 in cut-off. The voltage drop across R_5 (1.5 V) is fed as bias voltage to the RF amplifier through the decoupling network R_9-C_3. Thus with very low input signal levels the IF and RF amplifiers operate at a fixed bias chosen to provide maximum gain.

When the incoming RF signal level increases sufficiently the negative sync pulse amplitude at the emitter of Q_1 overcomes the reverse-bias on the emitter-base junction of Q_2. As a result, the negative-going horizontal keying pulses injected in the collector circuit of Q_2 from the H.O.T. cause collector current flow during sync pulse intervals. This results in a higher voltage across C_1 and thus D_2 is reverse biased. The increased potential on C_1 is now coupled through R_{10} and R_{11} to the base of Q_5. The enhanced forward bias on Q_5 while increasing its collector current takes it to a region of reduced power gain. Similarly the increased voltage drop across R_7 is enough to shift the operating point of Q_4 to reduce its gain. Thus the gain of both, 1st and 2nd video IF amplifiers is reduced. Note that the increased positive AGC voltage appearing across C_1 is prevented from reaching the emitter of Q_2 by diode D_3 and will therefore not effect the current flow in this transistor. Similarly diode D_1 which couples keying pulses to the collector of Q_2 prevents application of positive AGC voltage in its collector circuit. For the signal levels just considered the AGC driver transistor Q_3 remains cut-off and no additional AGC bias for the RF amplifier is developed.

When the incoming RF signal level increases sufficiently, the large positive AGC voltage developed across C_1 decreases the resistance of the voltage dependent resistance VDR_1, with the result that a higher dc voltage is applied to the base of Q_3. This turns on Q_3 and the resultant increased positive voltage across its emitter resistance (R_5) furnishes additional RF AGC bias. Potentiometer R_{12} sets the collector voltage level of Q_3 thus determining the condition for conduction of this transistor. R_{12} therefore acts as the tuner AGC delay control.

With still higher signal levels, IF AGC voltage will increase unitl it becomes 0.5 V more positive than the voltage across R_{12}. D_4 then conducts clamping the IF AGC voltage at that level. Any added increase in the received RF signal will now cause the RF amplifier AGC voltage to increase rapidly. Finally on very strong signals, when the tuner AGC voltage exceed 7 V, Q_3 conducts heavily causing the voltage drop across R_5 to rise rapidly. The additional voltage drop across R_5 raises the potential at point A to such a level the diode D_2 again gets forward biased. Thus the IF AGC will again increase causing further reduction of gain in the IF section.

15.10 AGC ADJUSTMENTS

Some receivers do not have any AGC adjustments. Other receivers have as many as three adjustments that affect the operation of the AGC system. In case a noise cancellation control forms part of the AGC circuit, it should be first advanced till the noise cancellation circuit becomes inoperative. This will ensure that no sync inversion occurs while other adjustments are made. After all other AGC adjustments have been made the noise threshold control should be advanced to the point of most stable sync.

Many receivers have a tuner AGC delay control. This control should be adjusted on medium-strength signals. As the tuner delay is increased the signal will become noise free. With too much delay, strong signals will cause overloading of amplifiers. Indications of overload are buzz in the sound and bending of the picture.

The third type of control is an AGC level control, sometimes called the 'threshold control'. This adjustment sets the voltage to which the sync tips must rise to give AGC action. The effect of this control is to adjust the detector level. In the absence of service notes the output from the detector must be estimated. This normally lies between 2 to 5 volts. However, manufacturers' instructions, if available, should be followed for all adjustments.

AGC Circuit in an IC Chip

Weak signal sections of the receiver which commonly use integrated circuits include video IF, video detector, AGC, AFC, video preamplifier and sound strip. The various functions performed by one such IC (TBA 890) were discussed in the previous chapter. The BEL CA3068 is another IC which has complete video IF subsystem and tuner AGC for monochrome and colour receivers. This integrated circuit consists of nearly 39 transistors, 10 diodes, 67 resistors and 18 capacitors. Figure 15.12 is a simplified block diagram of CA3068. Note that the tuned circuit filters and decoupling resistors and capacitors are connected externally at the corresponding pins of the integrated package.

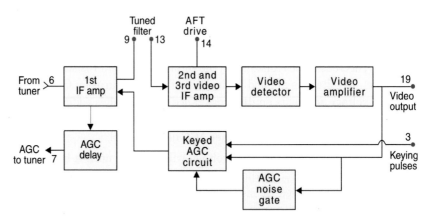

Fig. 15.12. Simplified block diagram of the AGC section in IC CA3068 (BEL).

The IF output from the tuner is fed at pin 6 to the first video IF amplifier. Horizontal flyback pulses are needed for AGC action. The dynamic range of the IF AGC is 55 db. Delayed AGC for the tuner is available at pin number 7, with a AGC voltage variation from 2.2 V to 4.5 V. Besides noise immunity circuits, the IC employs a zener diode as an RF bias clamp to prevent application of excessive AGC bias. The chip also has a provision for 'service switch' which can be used to isolate the AGC stage for fault finding.

REVIEW QUESTIONS

1. What are the advantages of using AGC in television receivers ?

2. Describe the basic principle of automatic gain control and show how it is applied to tube and transistor amplifiers.

3. What is meant by 'forward AGC' and 'reverse AGC' in transistor tuned amplifier circuits ? Why is forward bias control preferred to reverse bias method of gain control ?

4. What is the basic principle of peak AGC system ? Explain how the control voltage is developed and applied to IF and RF amplifier stages of the receiver.

5. What is delayed AGC and how is it developed ? Why is delayed AGC applied only to the RF amplifier and somtimes to the first IF amplifier of the receiver ? Why is AGC not applied to the last IF amplifier ?

6. Describe with a simple circuit the basic principle of a 'keyed AGC' system. How does it overcome the shortcomings of a 'non keyed' (peak AGC) control system ?

7. A keyed AGC system employing tubes is shown in Fig. 15.9. Explain how the AGC voltage is developed and applied to RF section of the receiver. What is the function of diode D_1 ?

8. Draw circuit diagram of a keyed AGC system employing transistors and having a noise gate. Explain how the AGC voltage is developed and amplified.

9. Explain briefly various types of noise gates used to suppress noise pulses in the video signal before it is applied to the AGC and sync separator circuits.

10. The circuit of an improved keyed AGC system is shown in Fig. 15.11. It is designed to operate in different modes depending on the level of incoming RF signal. Explain step by step how the control voltage is developed and applied to RF and IF sections of the receiver.

Chapter 15

16

Sync Separation Circuits

The synchronising pulses generally called 'sync' are part of the composite video signal as the top 25 percent of the signal amplitude. The sync pulses include horizontal, vertical and equalizing pulses. There are separated from the video signal by the sync separator. The clipped line (horizontal) and field (vertical) pulses are processed by appropriate line-pulse and field-pulse circuitry. The sync output thus obtained is fed to the horizontal and vertical deflection oscillators to time the scanning frequencies. As a result, picture information is in correct position on the raster. The sequence of operations is illustrated in Fig. 16.1 by a block schematic diagram.

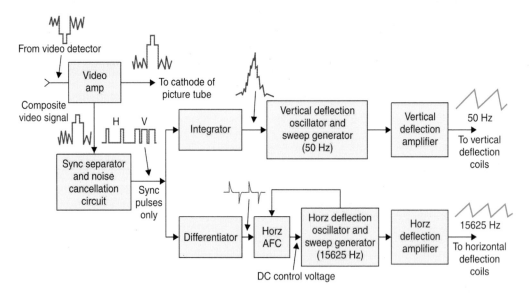

Fig. 16.1. Block diagram of the sync separator and deflection circuits in a television receiver.

16.1 SYNC SEPARATOR—BASIC PRINCIPLE

The problem of taking off the sync pulses from the video waveform is a comparatively simple one, since the action consists of merely biasing the device used in the circuit, in such a way, that only the top portions of the video signal cause current flow in the device. This is readily achieved by self-biasing the tube or transistor used in the circuit.

Two basic circuits, one employing a tube and the other a transistor are shown in Fig. 16.2 to illustrate this method of sync separation. Self-biasing or automatic bias means that the dc bias voltage is produced by the ac signal itself. The requirements are to charge the input capacitor by rectifying the input signal while it approaches its maximum value and have an RC time-constant long enough to maintain the bias on the capacitor between peaks of the ac input signal. The video signal is normally obtained from the video amplifier and coupled to the input of the sync-separator circuit. In Fig. 16.2 (a) video signal is fed to the grid of the triode with sync pulses as the most positive part of the waveform. In the quiescent state there is no

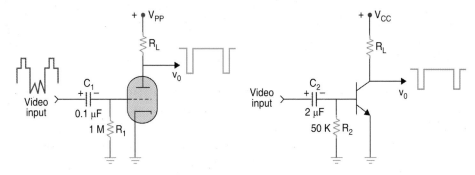

Fig. 16.2 (a). Basic sync separator circuit employing a triode.

Fig. 16.2 (b). Basic sync separator circuit employing a transistor.

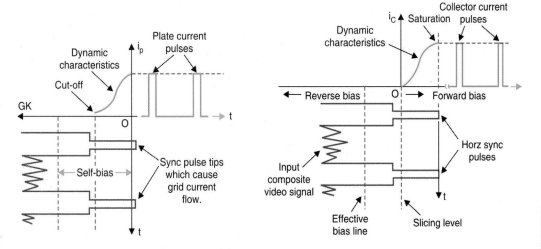

Fig. 16.2 (c). Illustration of tube circuit operation.   **Fig. 16.2 (d).** Illustration of transistor circuit operation.

bias on the tube. On the arrival of the signal, first few pulses drive a heavy grid current and the capacitor C_1 quickly charges up with the grid side negative. Between peaks of the input signal, C_1 discharges slightly through R_1. The R_1, C_1 time constant is made large enough to keep C_1 charged to about 90 percent of the peak positive input. The effect is to develop an automatic negative bias so that the operating point sweeps back form $V_{GK} = 0$ to a point well beyond cut-off. After a few pulses the tube settles down into a steady bias such that it is completely cut-off except during the positive going sync pulses. The tube then operates under 'class-C' condition i.e., biased well beyond cut-off. This is illustrated in Fig. 16.2 (c). To maintain a steady bias, the sync tip levels of the video signal make the grid slightly positive every cycle to cause a small grid current flow, which quickly replenishes the charge lost by the capacitor through the grid leak resistance R_1. The average negative bias developed across the capacitor C_1 then controls the plate current during sync pulse periods, which in turn gives rise to corresponding negative-going voltage pulses at the plate. Between pulses, the plate voltage rises to $+ V_{PP}$ since there is no voltage drop across the plate load resistor.

The corresponding transistor circuit is shown in Fig. 16.2 (b) where C_2, R_2 coupling provides self-bias between the base and emitter of the transistor. The capacitor charges because of the base current that flows during sync amplitude levels of the composite video signal. The circuit operation is illustrated in Fig. 16.2 (d) where the negative voltage developed across C_2 reverse biases the emitter-base junction in such a way that only positive sync voltage drives the transistor into conduction to produce sync output in the collector circuit. The sync amplitude varies between V_{CC} and collector voltage corresponding to the maximum collector current.

The bias voltage values in transistor circuits are less than in tube circuits because the base-emitter junction requires only a fraction of a volt as forward bias, to produce collector current output. The RC time constant is the same (about 0.1 second) as in tube circuits but with large C and small R because of the lower input resistance of a transistor.

16.2 SYNC SEPARATOR EMPLOYING A PENTODE

The triodes have relatively large interelectrode capacitances and therefore their performance in sync-separator circuits is inferior to that of pentodes. Accordingly a pentode is preferred to a triode in such circuits. A typical sync separator employing a pentode is shown in Fig. 16.3.

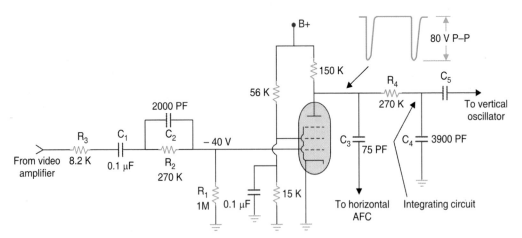

Fig. 16.3. Vacuum tube (pentode) sync separator circuit. Note that the grid bias voltage with a normal input signal is about – 40V.

The components and DC source voltage are so chosen that both plate and screen-grid voltages are low. This limits the dynamic region of the tube to a narrow range, with a cut-off bias of the order of about 5 volts. With the input voltage on, the plate voltage attains a minimum value during sync tip intervals. The actual minimum plate voltage obtained, and hence the sync peak-to-peak amplitude, depends on the value of plate load resistance and peak plate current. It should be noted (see Fig. 16.2 (c)) that the effect of driving the tube into grid current on each sync pulse tip is to clamp the sync pulse tops to zero (ground) potential, so that the sync tips align at the same level despite wide variations in the inter-pulse periods. This in effect amounts to restoration of dc component of the video signal, which is lost due to the presence of series coupling capacitor C_1.

One may ask as to why the video signal is not dc coupled from video amplifier to sync-separator. The reason is that changes in dc conditions, both because of variations of the average brightness of the scene and signal amplitude, when stations are changed, are bound to take place over a period of time and would make it very difficult to arrange for a constantly efficient sync separator action. The ac coupling, with its own automatic and flexible dc restoration function, provides the signal in the form, where it becomes easy to slice-off sync pulses of equal amplitude. The resistance R_3 provides isolation between the sync separator and video amplifier circuits. Because of saturation occurring at the sync tip level of the video signal, any noise resting on the sync tips does not get reproduced in the output.

16.3 TRANSISTOR SYNC SEPARATOR

A sync separator employing an *n-p-n* transistor is shown in Fig. 16.4. The capacitor C_1 tends to charge up to the peak input signal voltage less the base-emitter forward voltage drop. There is a marked difference in the signal voltage amplitude necessary to drive tube and transistor sync separators, since a transistor requires only about 0.6 V or so at the base to produce collector current output. An input video signal of the order of 5 to 10 V is all that is necessary to give rise to output sync pulses whose amplitude approaches about 50 volts peak-to-peak.

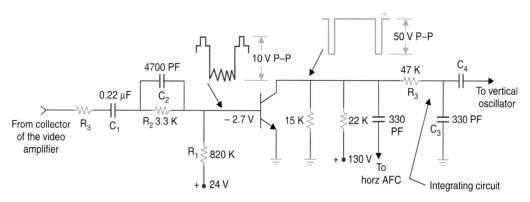

Fig. 16.4. *Transistor sync separator circuit. Note that the reverse bias voltage with a normal input signal is about – 2.7V.*

As shown in Fig. 16.4 a transistor sync separator often employs a fairly high value of collector load resistance with the result that the transistor bottoms (saturates) at a very low value of base (input) current. Full amplitude sync pulses are thus developed for a wide range of input signal strength levels. In general, a higher value of load resistance reduces the peak drive current needed to bottom the transistor but leads to broadening of the output pulses. This is due to charge storage effect, where the charge carriers stored in the base region take longer to dispel, if the collector load resistance is increased. This necessitates the use of transistors having fast switching capability which implies that the transistor must have a high upper cut-off frequency. In sync separators which employ switching transistors, a small forward bias is sometimes given to ensure that tips of the pulses drive the transistor well into the bottoming condition to produce good clean output pulses.

The factors which must be kept in view while designing a transistor sync separator can be summarized as follows:

    (i) To achieve a reasonable voltage gain the β of the transistor should be large.

    (ii) A transistor with a small output leakage current must be chosen because the leakage current lowers the collector voltage and thus reduces net amplitude of the sync output.

    (iii) To ensure a steep front edge of sync pulses a high frequency transistor is desirable.

    (iv) Since the transistor is off most of the time a low power transistor can be employed.

16.4   NOISE IN SYNC PULSES

Noise pulses are produced by ignition interference from automobiles, arcing brushes in motors, and by atmospheric noise. The noise is either radiated to the receiver or coupled through the

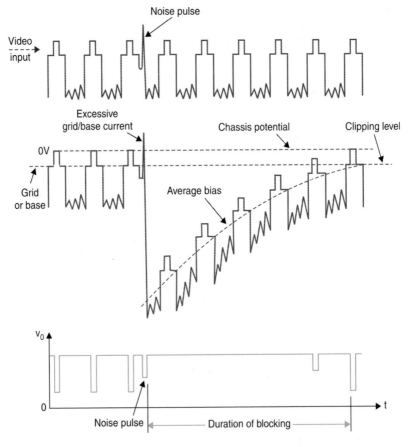

Fig. 16.5 (a). Effect of a strong noise pulse on sync output.

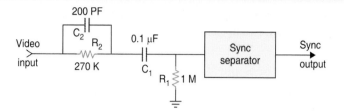

Fig. 16.5 (b). Double time-constant bias circuit at the input of a sync separator.

power line. Especially with weak signals the noise can act as false synchronizing pulses. Furthermore, when noise pulses have much higher amplitude than the sync voltage, large grid/base current flows which charges the coupling capacitor to a much higher voltage than is normal and this results in a noise set-up. Because of the long time constant of self-biasing network, the sync separator is held much beyond cut-off for a period which depends on the amplitude and width of the noise pulse and the time constant of the input circuit. As shown in Fig. 16.5 (a) the sync separator gets blocked and there is weak or no sync output till the bias returns to its normal value. During strong noise periods, the picture does not hold still until synchronization is restored again. Thus in order to reduce the effect of noise, sync circuits generally employ one or more of the following techniques :

(i) Double Time-constant for Signal Bias

The time constant of the grid/base leak-bias circuit, at the input of sync separator, must be long enough, to maintain bias from line to line and through the time of vertical sync pulses in order to maintain a constant clipping level. As stated earlier, a time constant of the order of 0.1 second is adequate for this purpose. But such large values would result in long blocking on strong noise pulses. In a similar way too long a time-constant will tend to increase the negative bias then its normal value during vertical sync intervals when the composite video signal voltage stays close to its peak value. This results in shortening of horizontal signal pulses soon after the vertical pulse train during each field. However, if the time constant is made too short to overcome the above drawbacks, this may not maintain bias between sync pulses, specially during the vertical sync pulse time. The result may be inadequate sync separation during and immediately after the vertical sync pulses. Therefore the problem of reducing the effect of high frequency noise pulses without changing the time constant of the average bias network is solved by providing a double time constant circuit at the input of the sync separator. The circuit configuration is shown in Fig. 16.5 (b). It may be noted that the two sync separator circuits described earlier (Fig. 16.3 and Fig. 16.4) also have double time constant circuits at their inputs. The network R_1, C_1 provides the normal grid/base leak-bias, with a time constant of 0.1 second for the sync signal. The small capacitance C_2 (200 pF) and resistance R_2 (270 KΩ) provide a short time-constant. The double time constant thus provided enables the negative bias to change quickly to reduce the effect of noise pulses in the input to the sync separator. The capacitor C_2 being small can quickly charge when noise pulses produce grid/base current thus increasing the bias for noise. The change in voltage across C_1 and C_2 is inversely proportional to their capacitance values. Therefore a noise pulse will charge C_2 to a voltage 500 times more than that across C_1. Since $R_2 C_2$ time constant is 50 µs, C_2 can discharge through R_2 between sync pulses. Thus the bias is maintained at the normal value for sync pulses.

Another advantage which accrues by the addition of a short time-constant circuit is that it acts like a frequency response compensator and maintains the sharp rise time of input sync pulses.

(ii) Sync Clipper after Sync Separator

The sync output limited by saturation does not result in sharp sync pulses. The clipping level is also not uniform because of shift in self-bias caused by sharp noise pulses. Many sync-separator circuits have a sync clipper stage where sync output of the separator is clipped and amplified. The purpose is to provide sharp sync-pulses with equal and high amplitude, free from noise pulses and without any camera signal. Clipping in successive stages allows the top and bottom of sync pulses to remain sharp and prevents noise pulses from having higher amplitude than the sync.

(iii) Use of Noise Cancellation Circuits

The use of a double time constant circuit as a noise suppressor is only suitable for non-recurring noise. If noise pulses are periodic, the small charge on C_1 (see Fig. 16.5 (b)) contributed by each noise pulse will be cumulative and noise set-up will still occur. Therefore, in many sync separator circuits some form of noise suppression switch is provided. Several such noise cancellation circuits which were described in the previous chapter along with AGC circuits are also used in sync separator circuits.

16.5 TYPICAL TUBE SYNC SEPARATOR CIRCUIT

The separator circuit shown in Fig. 16.6 employs a pentode as a sync-separator followed by a triode which provides gain and ensures sharp edged sync output. The video signal is coupled to

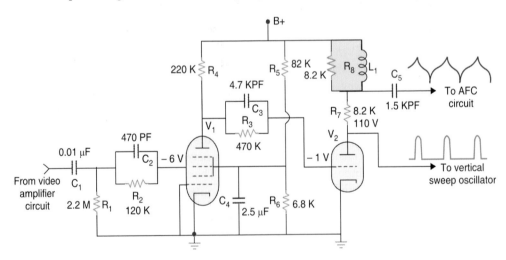

Fig. 16.6. A typical vacuum tube sync separator circuit.

the separator through a cathode follower from the video amplifier. A double time-constant circuit is provided at the input to reduce the effect of noise pulses on clipping level. The triode also operates on self-bias developed through network R_3, C_3, which couples the pentode to the triode. Since the amplitude of the sync output at the plate of the pentode is almost steady, the time-constant R_3, C_3 has been chosen to be quite small (2.5 ms). This is adequate, both for a steady auto-bias generation and suppression of occasional excessive noise pulse amplitudes.

The load of the triode consists of a small inductor L_1 and an 8.2 KΩ resistor in series. Sync pulses to the vertical oscillator are fed from plate of the triode whereas the input of the horizontal automatic frequency contol (AFC) circuit is connected to the junction of L_1 and resistance R_7. The sharp peaked pulses that develop across the inductor when triode conducts are of the desired shape and amplitude for feeding to the AFC circuit. The coil L_1 is shunted by a resistance R_8 to suppress onset of self oscillations when shock excited by sharp plate current pulses.

In some AFC circuits, discussed in the next chapter, a balanced sync pulse output is necessary. This can be easily obtained by inserting a suitable resistor in the cathode lead of the triode. The two outputs, one at the plate and the other at the cathode of the triode are then 180° out of phase with respect to each other.

16.6 TRANSISTOR NOISE GATE SYNC SEPARATOR

A transistorized sync-separator employing a noise gate is shown in Fig. 16.7. The noise gate transistor Q_2 is in series with the emitter of the sync separator Q_1. The tansistor Q_2 is so biased that it normally stays in saturation. The sync separator then operates normally and its emitter current completes its path through Q_2, which has only about 0.6 V between its collector and emitter when saturated. Any sharp noise pulse in the video signal gets coupled to the base of Q_2 through diode D_2 and this cuts-off the noise gate transistor Q_2. In the absence of any collector current through Q_2, the emitter of Q_1 rises to + 20 volts and this blocks the transistor Q_1. Hence no sync separation occurs during the noise pulse interval. It may be noted that normal amplitude of the negative video signal fed at the input of Q_2 is not sufficient to turn if off. However, if a noise spike with a negative amplitude beyond the sync tip appears, it turns the noise gate off. However, noise pulses which are not longer than sync tip will not cause the noise gate to turn off.

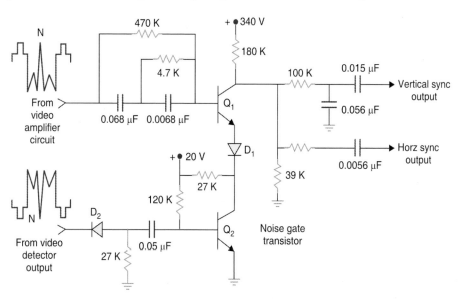

Fig. 16.7. *Transistor sync separator with a noise gate.*

16.7 IMPROVED NOISE GATE SYNC SEPARATOR

The circuit of Fig. 16.8 is another example of a typical sync clipper cum noise invertor. Q_1, an n-p-n transistor, acts as the sync separator. In the absence of any signal, the base-to-emitter voltage is zero and the transistor is off. When a positive-going signal appears at its input, the transistor is turned on. As shown along the circuit a 4 V (p-p) positive composite video signal is applied at the sync input terminals through the isolating resistor R_6. This turns Q_1 on and base current flows drawing the vertical sync pulses through C_1, D_1, emitter-base junction and D_2. The resulting current charges the capacitor C_1 with the polarity marked across it. Since the sync pulse is the highest component of the composite video signal, the voltage developed across C_1 at the end of the vertical sync pulse (which is 160 µs wide) is approximately 4 V. When the vertical sync pulse passes, the base of Q_1 is held negative with respect to the emitter and it is cut off again. Capacitor C_1 tends to discharge through R_1 during the interval of 18.84 ms between vertical sync pulses but the time-constant $C_1 R_1$ is so large (220 ms) that C_1 can discharge only about 8 percent of the voltage across it. When the next vertical sync pulse arrives, its peak is approximately 8 percent more positive than the change on C_1. Thus Q_1 is again turned on during the sync pulse interval and this part of the input signal gets amplified to develop a 20 V p-p negative going sync pulse at the collector.

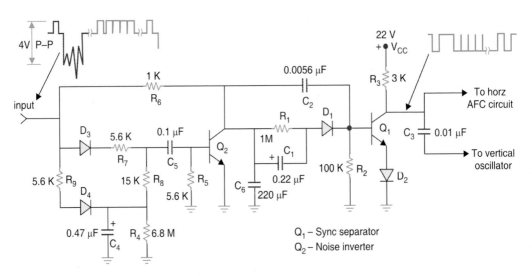

Fig. 16.8. *Improved transistor noise gate sync separator.*

After the vertical pulse interval, the horizontal sync pulses which are 4.7 µs wide, arrive. Because the charge on C_1 is still high to keep Q_1 in cut-off, the horizontal sync pulses are provided another path through C_2 to turn it on. Base current again flows during the horizontal sync pulse interval, charging capacitor C_2. In the 59 µs interval between consecutive horizontal sync pulses, C_2 discharges through R_2 ($R_2 C_2 = 560$ µs) by about 12 percent only and thus keeps Q_1 in cut-off state. When the next horizontal sync pulse arrives, it is sufficiently positive to turn Q_1 'on' again. The negative horizontal sync pulses developed at the collector of Q_1 are again of about 20 V peak-to-peak. Thus there are two separate input paths for the vertical and horizontal sync pulses, ensuring that both are amplified by Q_1, while eliminating the video information between sync pulses.

The fact that the positive horizontal sync pulse is coupled directly to the base of Q_1, back biases diode D_1 during the horizontal sync pulse interval. This prevents the base current of the transistor from discharging C_1. Diode D_2 prevents transistor conduction in the reverse direction when the negativegoing signal adds to the charge on the base input capacitors to exceed the base-emitter voltage rating.

Noise Invertor. As shown in Fig. 16.8 the noise invertor transistor Q_2 together with its associated circuitry is in parallel with the input to the sync separator. This offsets the effects of high amplitude noise pulses in the following manner. The diode D_4 rectifies the normal incoming positive going composite video signal and acts as a peak detector on account of the long time constant of R_4C_4 ($R_4C_4 = 3.2$ sec). The resulting dc voltage across C_4 is equal to peak of the sync pulses. This positive voltage gets applied to the cathode of D_3 through resistors R_8 and R_7. In turn this reverse biases D_3 under normal signal conditions and no signal appears through D_3 at the base of Q_2. It stays in cut-off and does not affect the normal operation of the sync separator. However, when a sudden noise spike appears as shown in the figure, voltage across C_4 cannot change quickly, and the diode D_3 will couple the spike through C_5 to the base of Q_2. The transistor Q_1 then conducts heavily and effectively shorts its collector circuit to ground. This results in shorting of the entire input signal to the sync separator, including video information and sync pulses, during time of the noise spike. As soon as the noise pulse disappears, however, Q_2 cuts off again and the next video signal and sync pulses will be uneffected. Capacitor C_6 prevents any dc short across Q_2.

16.8   SYNC AMPLIFIER

An amplifier is not always used before a sync separator. It will depend on the type of transistor used as a separator and the polarity of the sync pulses at the output of the video detector. If the video detector output is taken from its anode and its amplitude is sufficiently high, it can be fed directly to a *p-n-p* type sync separator. This eliminates the need for a sync amplifier before the separator. If an *n-p-n* type transistor is employed, as in Fig. 16.8, it is necessary to reverse the polarity of a negative-going video signal at the output of the detector before it can be coupled to the separator. In that case the output of a common emitter video amplifier may be used as input to the sync separator.

IC Sync Circuit

In some receivers sync-separator function is performed in an integrated circuit. This IC is a part of the sync-AGC module that combines noise inversion, sync separation and AGC.

Noise cancellation and amplification are first performed in the IC. The resultant noise free video signal emerges at a particular pin of the IC and is applied to a time-constant network that is kept outside the IC. This enables the use of different time-constant network to suit a particular receiver design. The output from the network is fed back to the IC where sync separation takes place. Over 20 volts of separated sync is available, both positive and negative going at different pins. The IC also receives gating pulses from the horizontal output circuit for AGC operation.

Chapter 16

REVIEW QUESTIONS

1. Draw the basic circuit diagram of a sync separator employing a triode with grid leak bias. Comment on choice of the time-constant of the biasing circuit. Why is a pentode preferred to a triode in tube sync-separator circuits ?

2. Draw the circuit of a sync separator employing a p-n-p transistor. Show input and output waveforms. Why are high-frequency-transistors with small output leakage current employed in sync circuits ?

3. What are the undesirable effects of high amplitude noise pulses in sync-separator circuits ? Explain how the use of a double time-constant biasing circuit overcomes the effect of noise in input signal to a sync separator. Why is a sync clipper often employed after the sync separator ?

4. Explain the operation of sync separator shown in Fig. 16.6. What is the use of inductor in plate circuit of the triode ?

5. Draw the circuit diagram of a typical transistor noise gate sync separator and explain its operation.

6. The circuit of an improved noise gate sync separator is shown in Fig. 16.8. Describe its operation and in particular explain how the noise invertor transistor Q_2 cancels the effect of noise pulses.

17

Sync Processing and AFC Circuits

The receiver has two separate scanning circuits, one to deflect the electron beam, of the picture tube, in the vertical direction and the other in the horizontal direction. Each scanning circuit consists of a waveform generator, *i.e.* oscillator and a power output stage. The synchronizing pulses obtained from the sync-pulse separator are used to control the vertical and horizontal deflection oscillators, so that picture tube is scanned in synchronism with the original picture source at the transmitter. The horizontal sync pulses hold the line structure of the picture together by locking in the frequency of the horizontal oscillator; and the vertical sync pulses hold the picture frames locked-in vertically by triggering the vertical oscillator. The equalizing pulses help the vertical synchronization to be the same in even and odd fields for good interlacing.

17.1 SYNC WAVEFORM SEPARATION

This means separating the vertical and horizontal sync pulses. It is the difference in the pulse time duration of the horizontal (line) and vertical (field) sync pulses which makes it possible to separate them. The horizontal sync pulse with a width of 4.7 µs and repeated at 15625 Hz represents a high frequency signal, whereas the vertical sync pulse with a total width of 160 µs which repeats 50 times in a second, is relatively a very low frequency signal. Therefore, the vertical and horizontal sync pulses can be separated from each other by RC filters. A low-pass filter, connected across the incoming sync pulse train, will develop appropriate trigger pulses

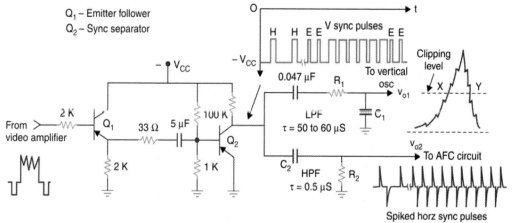

Fig. 17.1. *Separation of vertical and horizontal sync pulses.*

281

for synchronizing the vertical oscillator. Similarly, a high-pass filter will deliver sharp differentiated pulses for the horizontal oscillator from the same pulse train. This, as shown in Fig. 17.1, is done simultaneously by feeding output from the sync pulse separator to the low-pass and high-pass filter configurations, connected in parallel. The resistor R_1 and capacitor C_1 constitute the low-pass filter, also known as integrating circuit. The high-pass filter, also known as differentiating circuit, consists of C_2 and R_2 and has a very small time constant.

The integrated output across C_1 that builds up 50 times during one second is used for triggering the vertical oscillator. However, the spiked (differentiated) output that develops across R_2 is fed to the automatic frequency control (AFC) circuit, the output of which is employed for holding the horizontal oscillator at the correct frequency. The use of the AFC circuit ensures correct synchronization even in the presence of noise pulses. The serrated vertical sync pulses also develop only a spiked output across R_2, because the time constant of the circuit R_2C_2 is relatively too small to produce any appreciable voltage across R_2.

The sync separator shown in Fig. 17.1 is preceded by an emitter follower to isolate it from the video amplifier. The small amount of forward bias at the base of Q_2 ensures good bottoming and thus clean output sync pulses are fed to the filter circuits. The output waveshapes from the integrating and differentiating circuits are shown alongside corresponding filter configurations.

17.2 VERTICAL SYNC SEPARATION

The time constant of the low-pass filter circuit (see Fig. 17.1) is chosen to be much larger than the width of each serrated vertical pulse. This value is not very critical and a time constant R_1C_1 of about ten times the serrated pulse width is adequate. When the combined sync waveform, beginning with horizontal and equalizing pulses appears at the input of such a circuit, the capacitor C_1 charges along an exponential curve governed by the time constant of the filter circuit. Since this time constant is very large compared with the duration of the horizontal and equalizing pulses, the voltage output across C_1 during these intervals of the input wave is a very small fraction of the ultimate value. This is shown in the output waveform of the integrator. When the trailing edges of the horizontal or equalizing pulses appear, the small charge stored on the capacitor discharges along an exponential curve governed by the same time constant. The overall result is a very small toothed voltage across the capacitor, which lasts for a time comparable to the width of each horizontal or equalizing pulse, and thus the amplitude of the filter output voltage is negligible during the horizontal and equalizing sync periods.

However, when the vertical sync pulse (serrated) arrives, cumulative charging of the capacitor occurs, because the duration of each serrated vertical sync pulse is long compared with the gaps between serrations. Consequently, the charge accumulated from the first input serration (29.7 µs) has little opportunity to discharge during the following notch (2.3 µs). The next broad pulse adds to the charge already built-up across the capacitor which again is only partially discharged during the next gap. The five broad pulses which constitute the vertical sync pulse thus cause a gradual increase in voltage across the capacitor, with small spikes superimposed on it. At the output of the low-pass filter, therefore, the voltage rise appears to be almost smooth corresponding to the vertical pulse. This pulse amplitude is substantially greater than the small spikes caused by the horizontal and equalizing pulses. As soon as the vertical sync pulse has passed the integrated output pulse decays almost to zero during the post-equalizing pulse period, and stays at this level during the horizontal pulse train that follows.

The purpose and effectiveness of the equalizing pulses is apparent from the plot shown in the figure. The inclusion of equalizing pulses, before and after the field pulses, ensures identical integrated output despite insertion of field waveform in the middle of one line for one field, and at the end of the line on alternate fields.

In many vertical sync control circuits the integrated field pulse waveform is clipped by a reverse biased diode, along the line XY as shown in Fig. 17.1,and thus the voltage changes across C_1 due to line pulses are not seen by the field oscillator circuit at all.

Cascaded Integrator Sections

A very large time constant for the integrating circuit removes horizontal sync pulses, and also reduces the vertical sync amplitude across the integrating capacitor. This rising edge of the sync pulse is then not sharp and this can lead to incorrect triggering of the oscillator.However, when the R_1C_1 time constant is chosen to be relatively small the horizontal sync pulses cannot be filtered out and serrations in the vertical pulse produce notches in the integrated output. Thus though the vertical sync pulse rise is quite sharp, the output voltage attains the same amplitude at different charging excursions. This can also lead to wrong synchronization and so the notches must be filtered out. The resulting outputs with large and small time constant are illustrated in Fig. 17.2 (a).

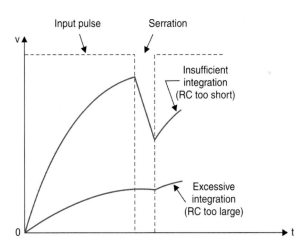

Fig. 17.2 (a). Effect of time-constant on vertical sync output.

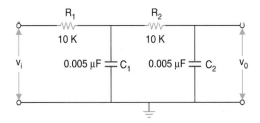

Fig. 17.2 (b). Two-section integrator for vertical sync.

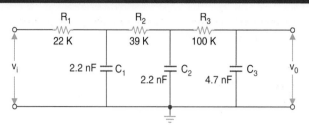

Fig. 17.2 (c). Three-section integrator for vertical sync.

To overcome the above described discrepancy, most receivers employ a two-section integrating circuit with each RC section having a time constant of about 50 μs. Such a circuit is illustrated in Fig. 17.2 (b). The operation of the circuit can be considered as though the R_1C_1 section provides integrated voltage across C_1 that is applied to the next integrating section R_2C_2. The overall time constant for both sections together is large enough to filter out horizontal sync pulses while the shorter time constant of each section allows the integrated voltage to rise more sharply because each integration is performed with a time constant of 50 μs. In some designs even a three section integrator is provided and such a configuration is illustrated in Fig. 17.2 (c).

17.3 HORIZONTAL SYNC SEPARATION

The high pass filter circuit and the differentiated output are shown in Fig. 17.1 along with the sync separator circuit. The time constant of this circuit (R_2C_2) is kept much smaller (normally 1/10th) than the width of the horizontal pulse. A time constant between 0.5 μs to 1 μs is often employed.

The physical action of the differentiator is as follows. When a leading edge of the incoming pulse train is applied to the C_2, R_2 circuit, the initial voltage across the capacitor C_2 is zero, and so full amplitude of the leading edge appears across the resistor R_2 and the output wave follows almost exactly the shape of the input leading edge. When the flat top of the input rectangular wave is reached, on further charging of the capacitor occurs, and the circuit discharges along an exponential curve governed by the time constant of the circuit. Since this time-constant is very short compared to the duration of input pulse, the discharge completes itself before the trailing edge of the pulse arrives.

When the trailing edge of the input pulse occurs, it produces another pulsed output (see Fig. 17.1) of opposite polarity to that of the first pulse. Since the trailing edge component extends in opposite direction to the leading edge output, it has no effect on the triggering of the horizontal oscillator.

The equalizing and vertical sync pulses (notched) produce two leading edge components when they occur during each line scanning interval of 64 μs. The extra leading-edge pulses occur when the horizontal oscillator is insensitive to sync pulses and hence have no effect on its frequency. As was explained in Chapter 3, the leading edge of each horizontal sync pulse, alternate equalizing pulses, and alternate serrations of the vertical sync pulse are correctly timed to initiate the horizontal retrace periods. Moreover as explained above, the differentiating circuit is insensitive to the flat top portions of the rectangular waves, that is, the amplitude of the differentiated output is independent of the duration of the input pulse. There is, therefore, no particular response to the vertical sync pulses, except at their edges and they as such have no effect on the triggering of the horizontal oscillator.

17.4 AUTOMATIC FREQUENCY CONTROL

The direct use of the incoming sync pulses to control vertical and horizontal sweep oscillators, though simple and most economical is normally unsuitable, because of their susceptibility to noise disturbance arising from electrical apparatus and equipment operating in the vicinity of the television receiver. The noise pulses extending in the same direction as the desired sync pulses cause greatest damage when they arrive during interval between the sync pulses. Noise pulses which are most troublesome, possess high amplitude, are of very short duration and tend to trigger the oscillator prior to its proper time.

When the vertical oscillator is so triggered the picture moves vertically, either upwards or downwards, until proper sync pulses in the signal again assume control. If the horizontal oscillator is incorrectly triggered a series of lines in a narrow band will be jumbled up, giving the appearance of streaking or tearing across the picture. The horizontal sweep system of the receiver is effected more by noise pulses than the vertical system.

To understand this fully, it is necessary to examine the nature of interfering voltages. The energy of the noise pulses is distributed over a wide range of frequencies. For a peak to occur, the phase relationship amongst various frequencies must be such as to permit their addition to form a high amplitude pulse. This condition, however, usually exists only for a brief interval which explains the small width of these pulses.

The high frequency noise pulses, when they reach the vertical sync separator, *i.e.* low-pass filter, get suppressed along with the line sync and equalizing pulses because of large time constant of the circuit. The presence of this low-pass filter (the integrating network) is mainly responsible for greater immunity to noise pulses enjoyed by the vertical system. This explains why no special circuit is used between the sync separator and the vertical oscillator. However, when a wide noise pulse is received, it contains enough energy to cause off-time firing of the vertical oscillator, but the annoyance caused to the viewer on account of occasional rolling of the picture is seldom great.

On the other hand, circuit leading to the horizontal oscillator, being a high-pass filter, passes noise pulses readily along with the narrow horizontal sync pulses. This results in serious interference with the normal functioning of the horizontal sweep oscillator which, in turn, results in frequent 'tearing' of the picture. In order to ensure that the horizontal oscillator operates at the correct frequency, and is basically immune to noise pulses, all horizontal deflection oscillators are controlled by some form of a circuit known as the automatic frequency control circuit (AFC circuit).

The AFC circuit receives sync pulses and output from the horizontal oscillator simultaneously and compares them regarding their phase and frequency. The discriminator in the AFC circuit develops a slowly varying voltage, the magnitude of which depends on deviation of frequency of the horizontal oscillator from its correct frequency. In case the oscillator frequency is equal to the incoming horizontal sync frequency, *i.e.* 15625 Hz, no output voltage is developed. The AFC circuit output is filtered by a low-pass filter to obtain an almost dc voltage, which then controls the frequency of the horizontal sweep oscillator. Thus the use of AFC circuit and a low-pass filter at its output, eliminates the effect of sharp noise pulses. The use of this indirect method of frequency control results in excellent horizontal oscillator stability and immunity from noise interference.

AFC Operation

The block schematic arrangement of a frequently used AFC circuit for the horizontal deflection oscillator is illustrated in Fig. 17.3. Horizontal sync voltage and a fraction of the horizontal deflection voltage, normally taken from the horizontal output circuit, and suitably processed to form horizontal flyback pulses, are coupled in the sync discriminator. The discriminator consists of two diodes and associated circuitry. It detects the difference in frequency and develops a dc output voltage proportional to the difference in frequency between the two input voltages. The dc control voltage indicates whether the oscillator is 'on' or 'off' the sync frequency. The greater the difference between the correct sync frequency and the oscillator frequency, larger is the dc control voltage. This dc control voltage is fed to a large time constant filter, the output of which is used to control the oscillator frequency. The shunt by-pass capacitor of this low-pass filter eliminates the effect of noise pulses. A large time constant filter could not be used directly, in the horizontal system ; because while suppressing noise pulses it would have prevented the desired horizontal sync pulses from reaching the horizontal sweep oscillator. This explains the need and use of the AFC circuit.

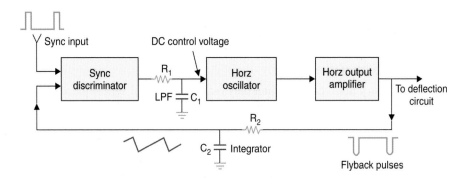

Fig. 17.3. Block diagram of the horizontal AFC system.

The automatic frequency control circuit is generally called flywheel sync, sync lock, stabilized sync or horizontal AFC, based on the technique employed to develop the control voltage. Most present day receivers use either a push-pull or a single-ended phase discriminator AFC circuit for sensing any error in the horizontal oscillator frequency.

17.5   AFC CIRCUIT EMPLOYING PUSH-PULL DISCRIMINATOR

A typical circuit arrangement employing push-pull phase discriminator is shown in Fig. 17.4. The sync pulses of equal amplitude but of opposite polarity are obtained from the phase splitter circuit and coupled to the diodes D_1 and D_2, through capacitors C_1 and C_2 respectively. R_1 and R_2 are of equal value and act as load resistors to the two diodes. The diodes are so connected that the application of sync pulses of opposite polarity forward biases them simultaneously. The horizontal deflection waveform, phase reversed and coming effectively from a voltage source is applied at the point marked 'A' in the circuit diagram. This is actually obtained from a winding (L_1) on the horizontal output transformer in the form of flyback pulses and then processed by the integrating network R_6, C_3.

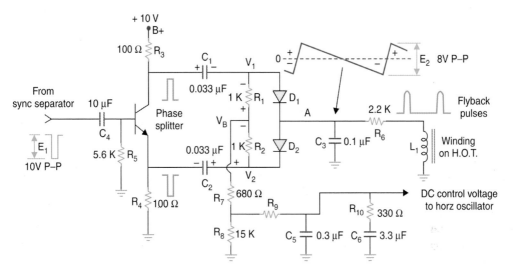

Fig. 17.4. *Push-pull AFC circuit.*

It is convenient to study the functioning of this circuit under three different conditions: (*a*) when sawtooth feedback voltage is only present, (*b*) when only sync pulses are present and (*c*) when both sawtooth and sync pulses are simultaneously present.

(*a*) When only sawtooth voltage is applied (no sync pulses) D_1 conducts during the negative and D_2 during the positive half of the sawtooth cycle, charging C_1 and C_2 with polarities shown. When voltage peaks have passed C_1 discharges via R_3, B_+, B_-, (ground) R_8, R_7 and R_1. Similarly C_2 discharges through R_2, R_7, R_8, B_- and R_4. This raises the lower end of R_2 (marked V_2) above ground potential and the upper end of R_1 (marked V_1) below ground potential. Equilibrium is reached when $| V_1 | = | V_2 | = \frac{1}{2} E_2$ (peak-to-peak) of the applied sawtooth voltage. Thus V_B equals zero and the upper end of R_7 stays at ground potential. When put in another way this means that the two equal and opposite currents flowing through R_8 develop a net zero voltage across it. In steady state, small pulses of make-up current flow at the positive and negative peaks of the sawtooth through D_2,C_2 and C_1,D_1 respectively.

(*b*) When only sync signal is present, the positive going sync pulses are coupled from the collector of the sync splitter through C_1 to the anode of D_1. At the same time negative going sync pulses are coupled via C_2 to the cathode of D_2. As a result of these pulses, current flows along the path C_1, D_1, D_2, C_2, R_4, B_-, B_+, and R_3 thereby charging the capacitors C_1 and C_2 to approximately peak value of the sync pulses with polarities shown across them. During the time between sync pulses, the capacitor C_1 discharges through R_3, B_+, B_-, R_8, R_7 and R_1. At the same time the capacitor C_2 discharges via R_2, R_7, R_8 and R_4. Two equal but opposite voltages, caused by equal capacitor discharge currents develop across R_8. These voltages cancel each other leaving a net zero voltage across R_7 and R_8. Thus V_B equals zero and this point continues to be at ground potential. Note that V_1 and V_2 are of such polarity that both the diodes are reverse biased during the time between sync pulses.

(c) Now if sync and sawtooth voltages are applied simultaneously, as they would be under normal conditions, three cases are possible—(i) pulses in synchronism, i.e. oscillator frequency is correct, (ii) oscillator frequency is more than the sync frequency, and (iii) oscillator frequency is less than the sync frequency.

(i) For this case as shown in Fig. 17.5 (a), sync pulses and sawtooth voltage arrive symmetrically, i.e. the sync pulses occur when the sawtooth is passing through its zero point during retrace. As a result the circuit behaves as though each signal were applied to the phase detector independent of each other. Thus the sync pulses deposit equal charges on C_1 and C_2, causing smaller but identical make-up currents to be drawn from the sawtooth source. The magnitudes of V_1 and V_2 remain equal ($V_B = 0$) and no control voltage is developed across R_8. Therefore, the frequency of the horizontal deflection oscillator remains unchanged.

(ii) When the oscillator frequency is high (see Fig. 17.5 (b)) the sync pulses arrive late in a relative sense, i.e. when the sawtooth is already in its positive half cycle. Thus the sawtooth forward biases D_2 and reverse biases D_1 with the result that D_2 conducts harder than D_1. Thus more charge is added to C_2 and less to C_1. During the discharge periods of these capacitors unequal currents flow through R_8 and a positive error voltage is developed across it. The discharge currents also cause V_2 to increase and V_1 to decrease from their quiescent values making V_B positive. The low-pass network R_9, C_5 filters the control voltage and feeds a dc error voltage to the horizontal oscillator. This forces the oscillator frequency to return to its normal value.

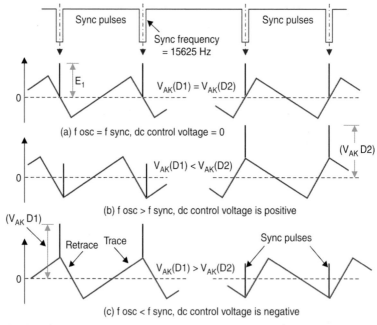

Fig. 17.5. Timing relationships between reference sync pulses and sawtooth voltage obtained from output of the horizontal deflection oscillator circuit. (a) correct oscillator frequency (b) oscillator frequency higher than sync frequency (c) oscillator frequency lower than sync frequency.

(iii) As shown in Fig. 17.5 (c) when the oscillator frequency is low, sync pulses occur during the time the sawtooth is in its negative half cycle forward biasing D_1 more than D_2. Analogous arguments establish that for this case a negative control voltage is developed across R_8. This, after filtering is applied to the horizontal oscillator causing its frequency to increase to its normal value.

Thus the sync discriminator continuously measures the frequency difference between the sawtooth and sync pulses to produce a dc correction voltage that locks the horizontal oscillator at the synchronizing frequency. In Fig. 17.4, R_{10} and C_6 constitute an anti-hunt, circuit whose function is described in a later section of the chapter.

17.6 SINGLE ENDED AFC CIRCUIT

The single ended or Gruen phase detector AFC circuit does not require push-pull sync input. Instead, as shown in Fig. 17.6 (a), negative going sync pulses of about 30V P-P are coupled through C_1 from the sync separator to the common cathode ends of the two diodes D_1 and D_2. The phase detector diodes are also fed flyback pulses obtained from a winding (L_1) on the horizontal output transformer. The pulses are passed through an integrating circuit (R_5, C_4) and converted into a sawtooth voltage of about 20V P-P before being fed at point 'X' in the circuit.

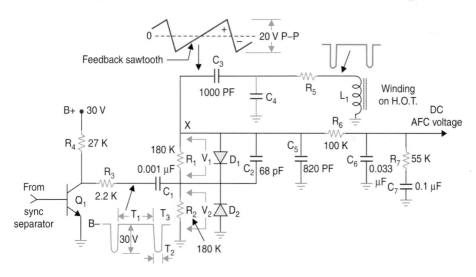

Fig. 17.6 (a). *Single ended (GRUEN) discriminator for horizontal AFC.*

Circuit Operation

Assume that only sync pulses are applied to the phase detector. During the time T_1 (see sync pulse waveform) the sync separator is cut-off and its collector voltage is equal to B_+ voltage. As a result the coupling capacitor C_1 is fully charged. When the negative sync pulse (T_2) arrives, *i.e.* Q_1 saturates, the two diodes are simultaneously forward biased. Note that the diodes are effectively in parallel for the sync input due to the large (820 pF) capacitor C_5 to ground from the anode of D_1. This enables C_1 to discharge through two different paths. One path for discharge is through R_3, Q_1, B_- and D_2 while the other is completed via R_3, Q_1, B_-, C_4, C_3 and D_1. In this process C_3 is charged with positive polarity towards C_4. Because of very short time constant during the discharge period, C_1 discharges almost completely.

During the following long period (T_3) between sync pulses both diodes are turned off and C_1 charges through two independent paths. One charging path is through B_+, R_4, R_3, C_1, and R_2. The second path is completed via B_+, R_4, R_3, C_1, R_1, C_3 and C_4. The capacitors C_3 and C_4 being practically short at the sync frequency, two equal and opposite voltage drops (V_1 and V_2) develop across R_1 and R_2 respectively producing a net zero voltage at point X (across the two diodes) with respect to ground. In addition C_3 discharges and attains its earlier status. The voltage drops V_1 and V_2 provided reverse bias across D_1 and D_2 respectively and thus permit conduction only during peak values of the applied signals.

It may be noted that a small (62 pF) capacitor C_2 is connected across D_2. This provides frequency compensation and ensures that the sawtooth voltage-drops, across D_1 and D_2 are identical in magnitude and waveshape. However, despite this precaution to correct mismatch between the components that constitute the phase detector, in practice a small voltage of the order of a few tenths of a volt does exist between point X and ground.

When only sawtooth voltage is applied at X, practically the entire voltage appears across D_1 and D_2 in series opposition and forces them into conduction alternately. When the sawtooth is positive to ground, D_1 is turned on and the current which flows through it and R_2 charges C_3. During the negative half cycle, D_1 is turned off, but D_2 conducts through R_1, C_3 and C_4 thereby discharging C_3. Once again voltage drops V_1 and V_2 are equal and opposite and zero net voltage is developed at point 'X'.

When sync pulses and sawtooth voltage are simultaneously applied, as is necessary for normal operation of any AFC circuit, three possible conditions can occur. These are illustrated in Fig. 17.6 (b) with sync waveform 'A' as the reference.

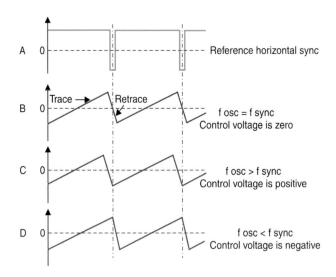

Fig. 17.6 (b). *Three possible conditions of horizontal sweep relative to sync pulse in a single ended AFC circuit.*

First (see waveform B), the sync pulses may occur when sweep (sawtooth) voltage is passing through zero. It is as if the sync pulses were applied without the sawtooth and vice versa. Consequently, no dc error voltage is produced. In a practical circuit, however, as stated earlier, a small voltage appears across the two diodes.

Second, the frequency of the oscillator may be higher than the sync pulse frequency. When this happens (see waveform C in Fig. 17.6 (b)) horizontal sync will occur during negative half cycle of the feedback sawtooth voltage. In this event D_2 conducts harder than D_1, leaving a net potential across C_3 with positive on its plate that is tied to the anode of D_1. Subsequently when C_3 discharges via D_1 and R_2, V_2 becomes greater than V_1. Thus a net positive voltage is developed across R_1 and R_2. This control voltage at X when fed after filtering to the horizontal oscillator forces its frequency to decrease.

In the third case the frequency of the oscillator may be lower than the sync pulse rate. Under this condition (waveform D) sync pulses will occur during positive half cycle of the feedback sweep voltage with the result that D_1 will conduct harder than D_2. Consequently C_3 will now attain a net negative charge on its plate tied to D_1. On discharging through D_2, R_1, the capacitor C_3 now develops a higher voltage across R_1 than R_2. Thus the net voltage ($\mid V_1 \mid -$ $\mid V_2 \mid$) across point 'X' and ground will be negative. Such a control voltage forces the oscillator to increase its frequency returning it to its normal operation.

17.7   PHASE DISCRIMINATOR (AFC) WITH PUSH-PULL SAWTOOTH

A sync discriminator in which flyback pulses of opposite polarity are coupled to the two diodes is illustrated in Fig. 17.7 (a). The flyback pulses of large amplitude are obtained from the two ends of a centre tapped winding, wound on the horizontal output transformer. These pulses are coupled at the points, marked A and B on the circuit diagram, where they are integrated by two RC networks (R_A, C_A and R_B, C_B) to develop sawtooth outputs of opposite polarity. The amplitude of the sawtooth voltage around its centre-zero axis is of the order of 20 volts.

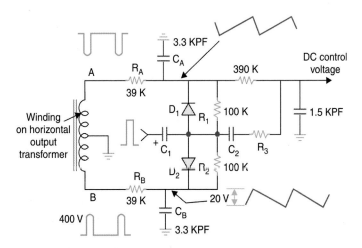

Fig. 17.7 (a). Phase discriminator circuit with push-pull sawtooth.

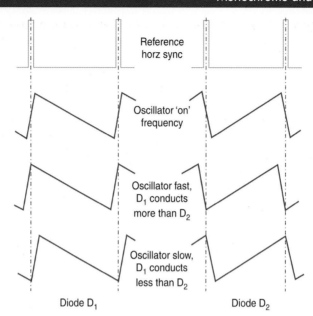

Fig. 17.7 (b). *Timing relationships between the reference sawtooth and sync pulses.*

The anodes of diodes D_1 and D_2 are tied together and positive going sync pulses from the sync separator and invertor circuit are applied at this junction. The dc control voltage that develops across the diode load resistors R_1 and R_2 depends on deviation of the horizontal oscillator frequency from that of the incoming sync pulses. The waveforms in Fig. 17.7 (b) show how the net voltage across each diode varies in accordance with variations of the oscillator frequency. As explained in the previous (single ended) circuit it is the charging and discharging action of the sync coupling capacitor C_1, through two different paths that results in the development of a net zero positive or negative control voltage. The control voltage is coupled to the grid of a reactance tube, normally the triode portion of a pentode-triode tube. Network C_2, R_3 is provided to compensate for any imbalance in the ac voltage applied to the two diodes. The pentode section together with its triode acts as a sinusoidal oscillator cum waveshaper. The reactance tube acts as a variable capacitor/ inductor across the tank circuit of the oscillator to keep the frequency of oscillations at 15625 Hz.

It may be noted that in any sync discriminator circuit it is possible by reversing the polarity of the sawtooth voltage, which is fed to the discriminator, to obtain control voltage of opposite polarity for the conditions of a fast or a slow oscillator. The dc control voltage, as obtained from the various discriminator circuits, varies between ± 2 and ± 6 volts.

17.8   DC CONTROL VOLTAGE

The time constant of the RC filter provided at the output of the discriminator, determines how fast the dc control voltage can change its amplitude to correct the oscillator frequency. A time constant much larger than 64 μs is needed for the shunt capacitor to bypass horizontal sync and sawtooth components in the control circuit while filtering out noise pulses. However, a large time constant may not permit the control voltage to change within a fraction of a second when sync is temporarily lost while changing channels. Also, if the time-constant is too large, the dc control voltage may be effected by the vertical sync pulses, causing bend at the top of the picture. A typical value of the AFC filter time-constant is about 0.005 second, *i.e.,* a period nearly equal to 75 horizontal lines.

Hunting in AFC Circuits

The filtering circuit that follows the diode section of the discriminator controls the performance of the AFC circuit. Too large a time constant makes the control sluggish, while insufficient damping, on account of too small a time constant, causes the oscillator to 'hunt' returning to the correct frequency. Excessive hunting in the AFC circuit appears as 'weaving' or 'gear-tooth' on the picture.

The manner in which the oscillator frequency deviates from the correct value on account of hunting is illustrated in Fig. 17.8 (*a*). In order to prevent this a double section filter is often used for the dc control voltage. In this network a shown in Fig. 17.8 (*b*), the R_1C_1 time constant of 0.005 sec is large enough to filter out noise, horizontal sync and flyback pulse effects. The

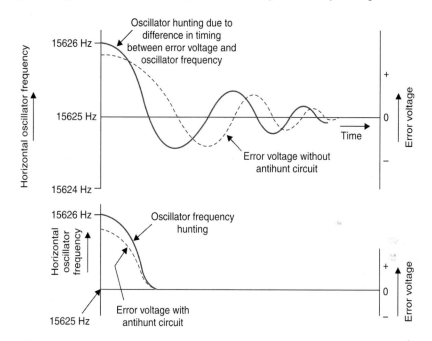

Fig. 17.8 (a). *Horizontal oscillator hunting and its correction by the antihunt circuit.*

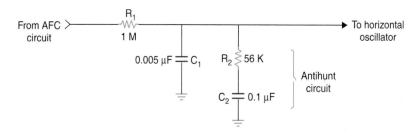

Fig. 17.8 (b). *AFC filter circuit with antihunt network.*

Chapter 17

second section, R_2 and C_2 in series, is known as the 'anti-hunt network'. The relatively low resistance of R_2 serves as a damping resistance across C_1 making the output voltage more resistive and less capacitive, thereby reducing time delay (see Fig. 17.8 (a)) in the change of control voltage.

REVIEW QUESTIONS

1. Draw basic low-pass (integrating) and high-pass (differentiating) filter configurations, which are employed to separate vertical and horizontal sync information. Comment on the choice of time constants of these circuits. Sketch accurately, output voltage waveforms of the filter circuits, when fed with a pulse train separated from the incoming composite video signal.

2. Why is a cascaded network preferred for developing vertical sync pulses ? Why is it not necessary to employ an AFC circuit for developing control voltage for the vertical deflection oscillator ?

3. Draw a basic (block schematic) AFC circuit and explain how the control voltage is developed. Explain fully how the effect of noise pulses is leiminated.

4. Draw a typical push-pull sync discriminator (AFC) circuit and explain with the help of neatly draw waveforms, how a control voltage, proportionate to the deviation of horizontal oscillator frequency is developed.

5. Why is a single-ended AFC discriminator preferred to the push-pull circuit ? Draw its circuit diagram and explain with the help of necessary waveforms, how the control voltage develops, when the oscillator frequency is (i) correct, (ii) fast and (iii) slow.

6. Why is an anti-hunt circuit used while filtering the error voltage obtained from any AFC discriminator ? Draw its circuit configuration and explain how hunting is suppressed.

18

Deflection Oscillators

In order to produce a picture on the screen of a TV receiver that is in synchronism with the one scanned at the transmitting end, it is necessary to first produce a synchronized raster. The video signal that is fed to the picture tube then automatically generates a copy of the transmitted picture on the raster.

While actual movement of the electron beam in a picture tube is controlled by magnetic fields produced by the vertical and horizontal deflection coils, proper vertical and horizontal driving voltages must first be produced by synchronized oscillators and associated waveshaping circuits. As illustrated in Fig. 18.1, for vertical deflection the frequency is 50 Hz, while for horizontal deflection it is 15625 Hz. The driving waveforms thus generated are applied to power amplifiers which provide sufficient current to the deflecting coils to produce a full raster on the screen of picture tube.

Free running relaxation type of oscillators are preferred as deflection voltage sources because these are most suited for generating the desired output waveform and can be easily locked into synchronism with the incoming sync pulses.

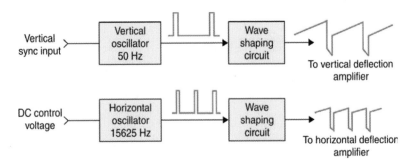

Fig. 18.1. Deflection oscillators and waveshaping.

The oscillators commonly used in both vertical and horizontal deflection sections of the receiver are:

 (*i*) Blocking oscillator,

 (*ii*) Multivibrator,

 (*iii*) Complementary pair relaxation oscillator,

 (*iv*) Overdriven sine-wave oscillator.

It may be noted that complementary pair circuits are possible only with transistors while all other types may employ tubes or transistors.

As explained earlier, both vertical and horizontal deflection oscillators must lock with corresponding incoming sync pulses directly or indirectly to produce a stable television picture.

18.1 DEFLECTION CURRENT WAVEFORMS

Figure 18.2 illustrates the required nature of current in deflection coils. As shown there it has a linear rise in amplitude which will deflect the beam at uniform speed without squeezing or spreading the picture information. At the end of ramp the current amplitude drops sharply for a fast retrace or flyback. Zero amplitude on the sawtooth waveform corresponds to the beam at centre of the screen. The peak-to-peak amplitude of the sawtooth wave determines the amount of deflection from the centre. The electron beam is at extreme left (or right) of the raster when the horizontal deflecting sawtooth wave has its positive (or negative) peak. Similarly the beam is at top and bottom for peak amplitudes of the vertical deflection sawtooth wave. The sawtooth waveforms can be positive or negative going, depending on the direction of windings on the yoke for deflecting the beam from left to right and top to bottom. In both cases (Fig. 18.2) the trace includes linear rise from start at point 1 to the end at point 2, which is the start of retrace finishing at point 3 for a complete sawtooth cycle.

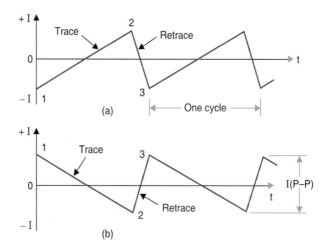

Fig. 18.2. *Deflection current waveforms (a) for positive going trace (b) for negative going trace.*

Driving Voltage Waveform

The current which flows into the horizontal and vertical deflecting coils must have a sawtooth waveform to obtain linear deflection of the beam during trace periods. However, because of inductive nature of the deflecting coils, a modified sawtooth voltage must be applied across the coils to achieve a sawtooth current through them. To understand this fully, consider the equivalent circuit of a deflecting coil (Fig. 18.3) consisting of a resistance R in series with a pure inductance L, where R includes the effect of driving source (internal) resistance. The voltage drops across R

and L for a sawtooth current, when added together would give the voltage waveform that must be applied across the coil. The voltage drop across R (Fig. 18.3 (a)) has the same sawtooth waveform as that of the current that flows through it. However, the voltage across L depends on the rate of change of current $\left(v_L = L\dfrac{di}{dt}\right)$ and the magnitude of inductance. A faster change in i_L, produces more self induced voltage v_L. Furthermore, for a constant rate of change in i_L, the value of v_L is constant. As a result, v_L in Fig. 18.3 (b) is at a

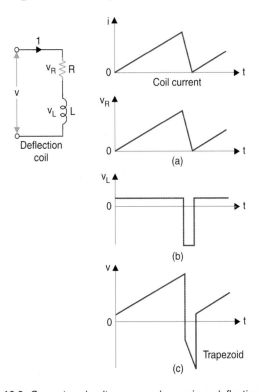

Coil current

(a)

(b)

Trapezoid

(c)

Fig. 18.3. Current and voltage waveshapes in a deflection coil
(a) voltage drop across the resistive component of coil impedance
(b) voltage drop across the inductive component of coil impedance
(c) resultant voltage 'v' ($V_R + v_L$) across input terminals of the
coil for a sawtooth current in the winding.

relatively low level during trace time, but because of fast drop in i_L during the retrace period, a sharp voltage peak or spike appears across the coil. The polarity of the flyback pulse is opposite to the trace voltage, because i_L is then decreasing instead of increasing. Therefore, a sawtooth current in L produces a rectangular voltage. This means, that to produce a sawtooth current in an inductor, a rectangular voltage should be applied across it. When the voltage drops across R and L are added together, the result (see Fig. 18.3 (c)) is a trapezoidal waveform. Thus to produce a sawtooth current in a circuit having R and L in series, which in the case under consideration represents a deflection coil, a trapezoidal voltage must be applied across it. Note that for a negative going sawtooth current, the resulting trapezoid will naturally have an inverted polarity as illustrated in Fig. 18.4.

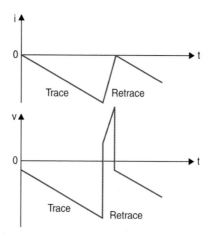

Fig. 18.4. *Inverted polarity of i and v.*

As explained above, for linear deflection, a trapezoidal voltage wave is necessary across the vertical deflecting coils. However, the resulting voltage waveform for the horizontal yoke will look closer to a rectangular waveshape, because voltage across the inductor overrides significantly the voltage across the resistance on account of higher rate of rise and fall of coil current.

Effect of Driving Source Impedance on Waveshapes

In deflection circuits employing vacuum tubes, the magnitude of R is quite large because of high plate resistance of the tube. Therefore, voltage waveshape across the vertical deflection coils and that needed to drive the vertical output stage is essentially trapezoidal. However, in a horizontal output circuit employing a tube, the waveshape will be close to rectangular because of very high scanning frequency.

When transistors are employed in vertical and horizontal deflection circuits, the driving impedance is very low and equivalent yoke circuits appear to be mainly inductive. This needs an almost rectangular voltage waveshape across the yoke. To produce such a voltage waveshape, the driving voltage necessary for horizontal and vertical scanning circuits would then be nearly rectangular. Thus the driving voltage waveforms to be generated by the deflection oscillator circuits would vary depending on deflection frequency, device employed and deflection coil impedance.

18.2   GENERATION OF DRIVING VOLTAGE WAVEFORMS

Sawtooth voltage is usually obtained as the voltage output across a capacitor that is charged slowly employing a large time constant to generate the trace period and then quickly discharged through a short time constant circuit to obtain the retrace period. The initial exponential rise of voltage across the capacitor is linear, and thus alternate charging and discharging of the capacitor at the rate of deflection frequency results in a sawtooth output voltage across it. This is illustrated in Fig. 18.5(a) where C_S is allowed to charge through a large resistance R_1 from a dc (B_+) source. The charging time is controlled by switch 'S' which is kept open during the trace period at the deflection frequency. At the end of trace, switch 'S' is closed for a time equal to the retrace period, and the capacitor discharges quickly through a small resistance R_2.

Actually, switch 'S' represents a vacuum tube or a transistor that can be switched 'on' or 'off' at the desired rate. When the tube or transistor is in cut-off state, it corresponds to 'off' position of the switch. In the 'on' state during which the active device is allowed to go into saturation, the tube or transistor conducts heavily allowing the capacitor to discharge through its very low internal resistance which corresponds to resistance R_2 shown in series with the switch. In this application the tube or transistor is called a 'discharge device' and capacitor C_S is often referred to as 'sawtooth capacitor' or 'sweep capacitor'.

As mentioned earlier the trace voltage should rise linearly. For this, only linear part of the exponential volt-time characteristic is used. To achieve this, the time constant (RC) of the circuit should at least be thrice the trace period. Same result can also be achieved by employing a higher B_+ voltage. The waveshapes shown in Fig. 18.5(b) illustrate the effect of charging time constant (RC) and source voltage B_+ on linearity and magnitude of the sawtooth voltage.

Trapezoidal Voltage Generation

As explained earlier it is often necessary to modify the sawtooth voltage to some form of a trapezoidal voltage before feeding it to the output stages for obtaining linear deflection. Figure 18.6 shows a basic circuit for generating such a voltage. It is the same circuit discussed earlier but employs a transistor as discharge switch and has a small resistance R_P (peaking resistance) in series with the sawtooth capacitor C_S. The transistor which is biased to cut-off by battery V_{BB} is driven into saturation by the incoming, large but narrow positive going pulses. It thus acts as a discharge switch to produce a fast retrace. During long intervals in-between positive pulses, C_S charges towards B_+ through the large resistance R_1 to provide trace voltage. Since the value of R_P is small as compared to R_1, voltage developed across it is quite small while C_S charges. However, on arrival of a positive pulse, Q_1 goes into full conduction thus providing a very low resistance path (small RC) for the capacitor to discharge. The high discharge current which also flows through R_P develops a large negative voltage pulse across it. This is illustrated by the waveform drawn along R_P in Fig. 18.6. As shown by another waveform, the spiked voltage across R_P adds to the sawtooth voltage across C_S to produce a trapezoidal voltage v_0 between point 'A' and ground. Note that exact charge and discharge periods must be in accordance with the synchronized vertical and horizontal scanning rates. This function is assigned to the vertical and horizontal oscillators.

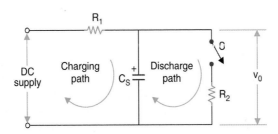

Fig. 18.5 (a). Basic circuit for generating a sawtooth voltage.

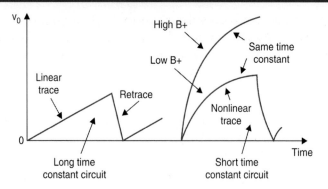

Fig. 18.5 (b). *Effect of charging time constant (RC) and source voltage (B+) on linearity and magnitude of v_0.*

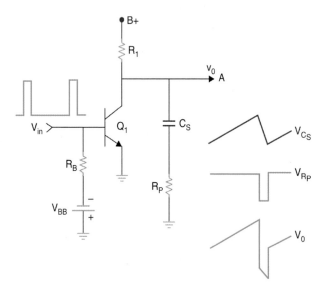

Fig. 18.6. *Generation of trapezoidal sweep voltage.*

18.3 BLOCKING OSCILLATOR AND SWEEP CIRCUITS

This oscillator may be thought of as a tuned-plate (or collector) configuration that is designed to produce an extreme case of intermittent oscillations. The feedback transformer is polarized to produce such a large amount of feedback that the cumulative action is almost instantaneous. In circuits employing tubes, the grid current that flows on account of regeneration develops such a large negative self-bias that the tube is immediately driven to much beyond cut-off. This prevents the circuit from generating continuous sinusoidal oscillations, at the natural resonant frequency of the feedback transformer, depending on its inductance and stray capacitance. Thus only a single short pulse of large amplitude is generated. The cycle is repeated when the self-bias returns to the conduction region. The number of times per second the oscillator produces pulse and then blocks itself, is the pulse repetition rate or oscillator frequency.

Similarly in a transistor blocking oscillator, feedback action switches the transistor between saturation and cut-off at a rate that is controlled by choice of time constant in the input circuit of the oscillator. This repetition rate or oscillator frequency is chosen to be 50 Hz and 15625 Hz for the vertical and horizontal deflections respectively.

Vacuum Tube Blocking Oscillator and Sweep Generator

The switch S in Fig. 18.5 (a) can be replaced by a blocking oscillator to control the charge and discharge periods of the capacitor. Such a circuit is illustrated in Fig. 18.7(a) where tube V_1 acts both as a blocking oscillator and discharge tube. The grid voltage waveform is illustrated in Fig. 18.7 (b). When the oscillator is cut-off by blocking action, C_S charges through the series resistance of R_1 and R_2 towards B_+. During oscillator pulses the tube conducts heavily and its plate resistance falls to a very low value. This provides a very low time constant path, for the capacitor C_S to discharge, with discharge current in the same direction as the normal plate current during oscillator conduction. Thus the blocking oscillator behaves like a switch where the 'on' and 'off' periods are automatically controlled by the frequency of blocking oscillator. The 'on' and 'off' periods are set equal to the retrace and trace periods respectively.

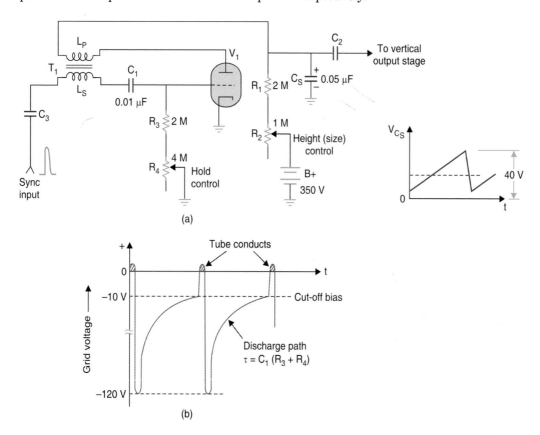

Fig. 18.7. A vacuum tube blocking oscillator (a) circuit (b) grid voltage waveform.

Frequency Control

Frequency of the oscillator can be adjusted by varying resistance R_4 which is part of resistance in the grid-leak bias circuit. A lower time constant and smaller value of R_4 will allow a faster discharge for a higher frequency. Increasing the time constant $C_1(R_3 + R_4)$ results in a lower oscillator frequency. The oscillator frequency control is adjusted to the point where sync voltage can lock the oscillator at the sync frequency to make the picture hold still. For this reason the frequency adjustment resistor R_4 is generally called the *hold control*. The range of frequency control for the vertical oscillator is usually from 40 to 60. For the horizontal oscillator, which employs an automatic frequency control, the horizontal hold adjustment is usually provided in the AFC circuit.

Height Control

The capacitor C_S is allowed to charge through R_1 and R_2 towards B_+ for a fixed interval equal to the trace period. During this period tube V_1 (Fig. 18.7(a)) remains is cut-off. Decreasing the resistance R_2 reduces time constant. Hence C_S charges at a faster rate and attains a higher voltage at the end of trace period. Thus a reduction in the value of R_2 increases amplitude of the sawtooth voltage which, after amplification, causes a larger current through the deflecting coils to increase size of the picture. Similarly, increasing resistance R_2 results in reducing size of the picture. The fixed resistance R_1 in series with potentiometer R_2 limits the range of variation for easier adjustment. This method of size control is generally used in vertical deflection circuits and is known as *height control*.

Synchronizing the Blocking Oscillator

A blocking oscillator in its free running state is not very stable and its frequency changes with variations in electrode voltages. This, however, can be easily controlled and kept constant by an external sync signal. The frequency can be synchronized either by sync pulses that trigger the oscillator into conduction at the sync frequency or by changing the grid bias with a dc control voltage. The vertical sweep oscillator is usually locked with pulses obtained by integrating the vertical sync pulses, whereas the horizontal oscillator frequency is synchronized with the dc control voltage produced by the horizontal AFC circuit.

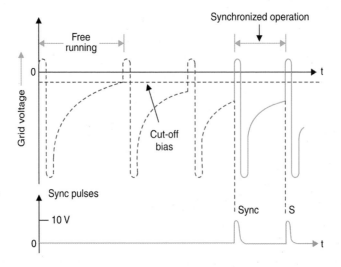

Fig. 18.8 (a). Blocking oscillator synchronization with positive trigger pulses.

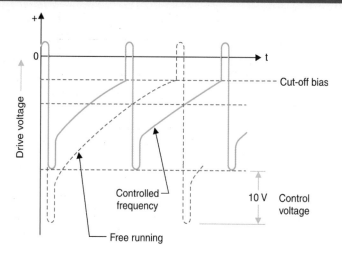

Fig. 18.8 (b). Effect of positive dc control voltage on frequency.

The waveforms of Fig. 18.8 (*a*) illustrate how a vacuum tube blocking oscillator can be synchronized by small positive pulses injected in the grid circuit of the oscillator. The sync voltage is applied in series with the grid winding of the transformer through a capacitor as shown in Fig. 18.7 (*a*). The positive sync pulses arrive at the time marked 'sync' when the declining grid voltage is close to cut-off and cancels part of the grid bias voltage produced by the oscillator. A small sync voltage is sufficient to drive the grid voltage momentarily above the cut-off voltage. This initiates plate current flow and then the oscillator goes through a complete cycle. The next positive sync pulse arrives at a similar point, in the following cycle, forcing the oscillator to begin the next cycle. As a result the sync pulses force the oscillator to operate at the sync frequency.

Free Running Frequency of the Oscillator

Operating the oscillator at the same frequency as the synchronizing pulses does not provide good triggering, because the oscillator frequency can drift, above the sync frequency, resulting in no synchronization. This is because the sync pulse will have no controlling influence when the tube or transistor has already been switched into conduction by its own bias. For best synchronization, the free-running oscillator frequency, is adjusted slightly lower than the forced or sync frequency so that the time between sync pulses is shorter than the time between pulses of the free running oscillator. Then each synchronizing pulse occurs just before an oscillator pulse and forces the tube or transistor into conduction thereby triggering every cycle, to hold the oscillator locked at the sync frequency. If the free-running frequency is too low, the sync pulses will arrive early and fail to raise the bias to a level that can cause conduction because the grid bias will still be at a large negative value.

This also explains why equalizing pulses or any noise pulses which occur at the middle of the cycle fail to trigger the oscillator. False triggering due to noise pulses that occur close to the sync pulses can be reduced by returning the grid to a positive voltage instead of the chassis ground.

Chapter 18

DC Control Voltage

A dc control voltage (see Fig. 18.8 (b)) can also be added to the existing bias voltage on the grid of the blocking oscillator tube or base of the oscillator transistor, to control its frequency. As a result of this addition, less time is needed for the bias voltage to reach its conduction level to start the next cycle. This method is often used to synchronize the horizontal oscillator frequency.

Transistor Blocking Oscillator Sweep Generator

In a blocking oscillator which employs a transistor, the method employed to turn the transistor 'on' and 'off' for generating sawtooth output is different than that employed with vacuum tubes. This frequently involves the use of sawtooth wave generated by the transistor itself, rather than by a grid-leak bias action as used in tube circuits.

Figure 18.9 is the circuit of a typical vertical blocking oscillator-sawtooth generator, where sweep capacitor C_S is in the emitter lead of the transistor. The combination of voltage divider resistors R_1 and R_2 and potentiometer R_3 that are connected across the V_{CC} supply, provide necessary forward bias to the transistor. When V_{CC} supply is switched on, the rising collector current through the primary of feedback transformer T_1 generates a voltage across its secondary, which drives the base more positive relative to emitter. This has two effects. The first is an increase in collector current, which in turn increases positive feedback at the base. This action is regenerative until saturation is quickly reached. The second effect is, that as the increasing base current drives the transistor to saturation, the capacitor C_2 discharges making the conduction period still shorter. During the brief period when Q_1 conducts heavily, the capacitor C_S charges quickly (from the emitter current) to produce retrace period of the sawtooth wave. As Q_1 saturates, the magnetic field about the transformer T_1 stops expanding. The positive voltage that was induced at the base of Q_1 then disappears. As a result, the combined effect of near-zero voltage at the base, because of potential drop across R_2 and a positive voltage on the

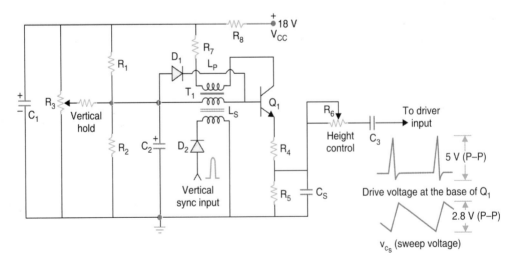

Fig. 18.9. A transistor vertical blocking oscillator.

emitter due to charge on C_S, reverse biases the base-emitter junction and the transistor immediately goes to cut-off. This sudden change in collector current and the consequent collapse of magnetic flux in the primary of T_1 induces a large reverse voltage in the secondary winding which aids in keeping the transistor in cut-off state. This voltage at the base could damage the transistor but for the protective action of diode D_1. The diode acts as a short across the secondary (base) winding of the transformer during the brief back-emf period, protecting the transistor.

After Q_1 is cut-off, base capacitor C_2 starts charging towards positive voltage at the junction of R_1 and R_2. It reaches this positive voltage level very quickly and then levels off. This is indicated by the base voltage waveform drawn along the circuit diagram. As soon as Q_1 cuts off and emitter current ceases, C_S starts discharging through R_5 producing trace portion of the sawtooth waveform. When the capacitor has discharged to the point where emitter voltage no longer keeps Q_1 cut-off, it turns on and the entire cycle is repeated.

The frequency of the oscillator is controlled primarily by the time constant R_5, C_S. Note that during discharge of capacitor C_S, Q_1 is cut-off and network R_5-C_S is isolated from rest to the circuitry. This provides excellent frequency stability.

Any change in the setting of potentiometer R_3 alters forward biasing and thus controls the instant at which Q_1 breaks into conduction. This changes frequency of the oscillator and so potentiometer R_3 acts as 'hold control'. Similarly potentiometer R_6 which controls the magnitude of sawtooth voltage that is fed to the driver is called 'size' or 'height control'. In transistor circuits the driver, which is an emitter follower, is necessary to isolate the oscillator from low input impedance of the corresponding output stage.

Synchronization

Sync pulses received as part of the composite video signal from the transmitting station to which the receiver is tuned are applied after due processing through diode D_2 to the tertiary winding of the feedback transformer. While D_2 prevents vertical oscillator waveforms from being fed back to the sync circuit, the tertiary (third winding) on T_1 provides isolation between the oscillator and sync circuit.

The positive going sync pulses drive the base more positive and turn on Q_1 a little earlier that it would have normally under free running condition. The sync pulses thus lock the vertical oscillator to the transmitter field frequency to prevent any rolling of the picture.

Another transistor blocking oscillator circuit is shown in Fig. 18.10. In this circuit sawtooth network is in the collector circuit and the sawtooth capacitor C_S charges during trace period and discharged during retrace interval. The setting of potentiometer R_2 controls initiation of conduction of the transistor and thus acts as 'hold control' over a narrow range of the oscillator frequency. The output voltage across C_S is of the order of 1 volt and is enough to operate the following driver stage.

Chapter 18

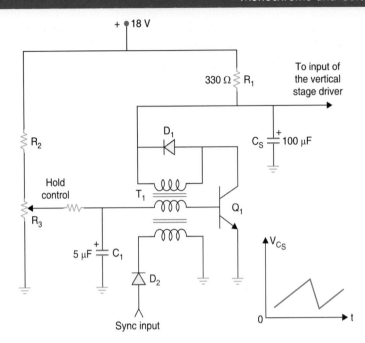

Fig. 18.10. *Vertical deflection blocking oscillator.*

18.4   MULTIVIBRATOR DEFLECTION OSCILLATORS

A multivibrator is another type of relaxation oscillator which employs two amplifier stages, where the output of one is coupled to the input of the other. This results in overall positive feedback and the circuit operates such that when one stage conducts, it forces the other to cut-off. Soon the stage that cuts off returns to conduction to force the first stage to cut-off. This sequence repeats to generate square or rectangular output with a frequency that is controlled by the coupling networks between the two amplifier stages. As in the case of a blocking oscillator the multivibrator is used as a controlled switch to charge a capacitor through a resistance to generate the required sawtooth wave output. The amplifiers may employ tubes or transistors as active devices.

Multivibrators may be classified as bistable, monostable and astable. A bistable multivibrator has two stable stages and needs two external trigger signals to complete one cycle of oscillation. The monostable type has one stable stage and completes one cycle of output with only one external pulses. However, and astable multivibrator is a free running type and does not need any external trigger pulse for its normal operation. It is this type of multivibrator that is employed as a deflection oscillator and its frequency is synchronized with the horizontal AFC voltage or vertical sync pulses. Multivibrators can also be classified on the basis of coupling between stages. The two types that are used in TV receivers are plate (or collector) coupled and cathode (or emitter) coupled.

Transistor Free Running Multivibrator

The circuit configuration shown in Fig. 18.11 (a) is of a free running collector coupled multivibrator where two common emitter amplifiers are cross coupled to provide positive feedback. Note that the base resistances are returned to V_{CC} supply to ensure precise operation of the multivibrator.

The circuit operation can be easily explained if the sequence of operations is followed from the instant when one transistor just conducts and the other goes to cut-off.

Figure 18.11 (b) illustrates the collector and base voltage waveforms for one complete cycle. At the instant marked t_0, transistor Q_2 just conducts to saturation and transistor Q_1 returns to cut-off. As this happens, the rising voltage at the collector of Q_1 charges the capacitor C_2 to V_{CC}, since, $V_{BE(sat)} \approx 0$. The charging current of C_2 flows through the base of Q_2 to complete its circuit. R_{B2} is selected to provide enough current from V_{CC} to the base of Q_2, to keep it in saturation, even when the charging current of C_2 becomes zero and C_2 charges to V_{CC}.

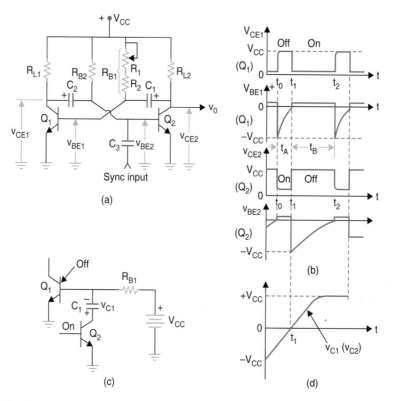

Fig. 18.11. *Free running transistor multivibrator (a) circuit (b) collector and base voltage waveforms (c) relevant circuit with Q_1 'off' and Q_2 'on' (d) charging curve of capacitor C_1 from V_{CC} towards V_{CC}.*

In a similar manner C_1 would have got charged to V_{CC} in the previous cycle when Q_1 was in saturation. Actually at $t = t_0$ the capacitor C_1 which was previously charged to V_{CC} gets earthed with its positively charged plate towards ground, the moment Q_2 goes into full conduction. As shown in Fig. 18.11(c), C_1 is then in parallel with emitter-base junction of the 'off' transistor Q_1. This puts a reverse bias on Q_1 equal to $-V_{CC}$ at $t = t_0$, which is well beyond cut-off bias of the transistor.

The capacitor C_1 now starts charging from $-V_{CC}$ towards $+V_{CC}$ as shown in Fig. 18.11 (d). At $t = t_1$ the negative voltage across C_1 reduces to zero and permits base current flow in the transistor Q_1. This action is regenerative and Q_1 instantly goes to saturation which in turn cuts-off Q_2. The time constant $R_{B1}C_1$ controls the off period $(t_1 - t_0)$ of Q_1 and can be set equal to the retrace period of the required sawtooth wave.

The moment Q_1 goes into saturation the positively charged plate of C_2 is effectively grounded through Q_1. It then begins to charge towards $+ V_{CC}$ at a rate determined by the time constant $R_{B2}C_2$. Once again when $V_{C2} = 0$, the second transition takes place to complete one cycle of operation. This cycle then repeats and permits the circuit to function as a free running multivibrator. The time constant $R_{B2}C_2$ can be made equal to the trace period. As shown in the circuit, R_{B1} consists of R_1 and R_2 where R_2 is a potentiometer to adjust the retrace period and thereby controls the frequency of the multivibrator.

The output voltage at either of the collectors is a rectangular wave with an amplitude $= V_{CC} - V_{CE(sat)}$.

Multivibrator Frequency

As shown in Fig. 18.11 (b) the total period

$$T = \frac{1}{f} = t_A \ i.e. \ (t - t_0) + t_B \ i.e. \ (t_2 - t_1).$$

A reference to Fig. 18.11 (c) will show that V_{C1} takes a time equal to t_A to return to zero from $- V_{CC}$ while charging toward $+ V_{CC}$. Then at the instant t_1, $V_{c1} = 0$ and we can write.

$$0 = V_{CC} - (V_{CC} + V_{CC}) \exp(-t_A/R_{B1}C_1)$$

This expression when solved yields $t_A = 0.69\, R_{B1}C_1$.

Similarly it can be shown that $t_B = 0.69\, R_{B2}C_2$.

$$\therefore \qquad T = t_A + t_B = 0.69\, R_{B1}C_1 + 0.69\, R_{B2}C_2.$$

Synchronization

The synchronizing pulses may be positive or negative and may be applied to the emitter, base or collector of one or both the transistors. The frequency of switching action of the astable multivibrator is kept lower than that of the synchronizing pulses, to force the transistor to switch states slightly before the free-running transition time. The sync pulses are applied at the base of the controlling transistor. The 'on-off' periods of the multivibrator are used to generate sawtoothed output across a capacitor as explained in the previous sections.

Cathode/Emitter Coupled Multivibrator

In this type of multivibrator only one RC coupling is provided between the two amplifiers. Positive feedback for regenerative action is obtained through a common resistance in the cathode/emitter leads of the two amplifiers. The oscillator action is explained by considering a cathode coupled multivibrator. As shown in Fig. 18.12 (a) the coupling from tube V_1 to V_2 is through an RC network while from V_2 to V_1 it is through the common cathode resistance R_K. When V_1 conducts the reduced plate voltage is coupled to the grid of V_2 via C_2 thereby cutting it off. Thus the cut-off period of V_2 depends on the time constant C_2 $(100\ K\Omega + R_2)$ while C_2 discharges. However, when V_2 goes into conduction, the cut-off period of V_1 depends on the time constant for charge of C_2. The charge path is through the low resistance of grid-cathode of V_2 (when grid current flows), R_K and R_{L1} while V_1 is off. As a result C_2 charges fast to provide a small cut-off period for V_1. The cathode coupled multivibrator therefore automatically produces an unsymmetrical output as V_2 must remain in cut-off for a much longer period than V_1. Thus V_2 provides the trace period and V_1 the retrace period.

Sawtooth Generation

The circuit of Fig. 18.12 (a) can be modified as shown by dotted chain lines to generate a sawtooth output for feeding into the horizontal output deflection amplifier. The sawtooth capacitor is connected in the plate circuit of V_2, which stays in cut-off for a long time and conducts for a short time. This stage then acts as a discharge tube to generate a sawtooth output as explained earlier. Figure 18.12 (b) illustrates how the sawtooth voltage output corresponds to cut-off and conduction periods of V_2. The variable resistance R_2 in the grid circuit of V_2 serves as the frequency (hold) control. The grid resistance R_1 for V_1 does not control the oscillator frequency because its grid voltage is controlled by the voltage drop across. R_k. However, this grid is best suited for frequency control because of its isolation from the oscillator voltages.

Frequency Control

The horizontal sweep oscillators are normally synchronized by the negative dc control voltage obtained from the AFC circuit. In the cathode coupled multivibrator (Fig. 18.12 (a)), negative dc control voltage is applied at the grid of V_1. The added negative grid voltage reduces plate current of V_1 when it conducts. This results in a smaller drop in its plate voltage and less negative drive at the grid of V_2. Then less time is needed for C_2 to discharge down to cut-off. This reduction in the cut-off period results in an increased multivibrator frequency. Variations in the AFC negative voltage thus apply necessary correction in the cut-off period of V_2 to keep the oscillator synchronized with the horizontal sync pulses.

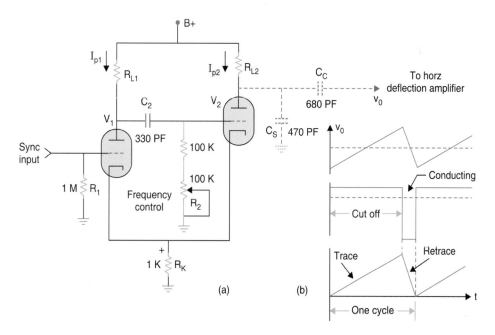

Fig. 18.12. Cathode coupled multivibrator (a) circuit (b) waveforms.

Chapter 18

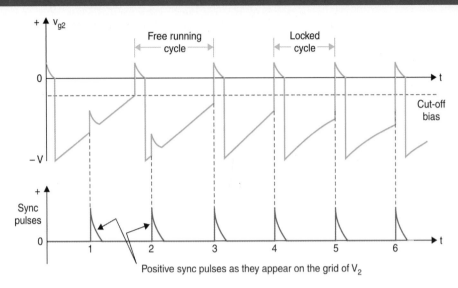

Fig. 18.13. *Grid voltage waveform of a multivibrator synchronized by sync pulses. The oscillator is pulled in to the sync frequency at time t_3.*

Multivibrator Synchronization

As stated earlier, while discussing sync processing circuits, either positive or negative sync polarity can be used with multivibrators. A Positive pulse, applied to the grid of a tube in cut-off, can cause switching action if the pulse is large enough to raise the grid voltage above cut-off. A negative sync pulse at the grid of V_1 when it is in conduction, results in a more stable operation and is generally used. In fact, the negative pulse applied to the gird of V_1 gets amplified and inverted to appear as a large positive pulse at the grid of V_2.

It is highly improbable that the first sync pulse which arrives will succeed in synchronizing the oscillator. The waveforms shown in Fig. 18.13 illustrate how the oscillator is gradually pulled into synchronism. It would be pertinent to mention here that for a blocking oscillator only a positive sync pulse can cause synchronization.

Multivibrator Stabilization

As the grid voltage approaches its cut-of value, it becomes increasingly sensitive to noise pulses which might have become part of the signal. A strong such pulse, arriving slightly before the synchronizing pulse can readily trigger the oscillator prematurely and cause rolling or tearing of the picture.

To ensure stability of operation, resonant circuits are employed in some sweep circuits. A cathode coupled multivibrator of the type shown in Fig. 18.12 (*a*) is redrawn in Fig. 18.14 (*a*), with a resonant stabilizing circuit placed in the plate circuit of tube V_1. The frequency of the resonant circuit is adjusted to 15625 Hz. The tuned circuit is shock excited by periodic switching of the tube from an 'on' to an 'off' condition. As a result the sinusoidal output of the resonant circuit modifies the voltage waveforms at the plate of V_1 and grid of V_2. This is illustrated in Fig. 18.14 (*b*), where of particular importance is the grid waveform of triode V_2. It may be noted that the grid voltage now approaches cut-off very sharply and only a very strong noise pulse will be able to trigger the second triode prematurely.

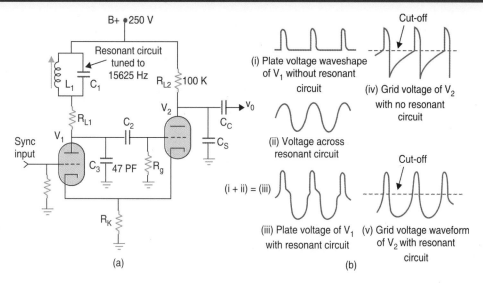

Fig. 18.14. Cathode coupled multivibrator with a stabilizing resonant circuit
(a) circuit (b) waveforms.

18.5 COMPLEMENTARY-SYMMETRY RELAXATION OSCILLATOR

A complementary-symmetry relaxation oscillator, designed to drive the vertical deflection output circuit, is illustrated in Fig. 18.15 (a). Transistors $Q_1(p\text{-}n\text{-}p)$ and $Q_2(n\text{-}p\text{-}n)$ which are directly coupled, constitute the oscillator pair, while Q_3 is the waveshaping transistor. Resistors R_1 and R_2 form a potential divider across V_{CC} supply through the decoupling network R_3, C_6, to provide positive voltage both at the base of Q_1 and collector of Q_2. The voltage at the emitter of Q_1 is developed by capacitor C_1, when it charges towards V_{CC} (+ 20 V), through resistance R_4 and potentiometer R_5 connected in series.

At the instant dc supply is switched on to the circuit, both the transistors are at cut-off, because the base of Q_1 is biased positively and its emitter is at zero potential. However, capacitor C_1 starts charging at once driving emitter of Q_1 positive. When the rising voltage across C_1 offsets the positive voltage at the base of Q_1, the transistor turns on. This makes the base of Q_2 positive which also goes into conduction. The collector current of Q_2 flows through R_1 and the resulting drop across it, lowers the potential at the base of Q_1, thus making it more negative with respect to its emitter. This results in increased current through Q_1 and the regenerative feedback action that follows soon saturates Q_1. When Q_1 is on, its emitter current starts discharging C_1. As soon as the emitter voltage drops sufficiently to remove forward bias on Q_1 it is driven out of conduction. This in turn cuts off Q_2, thereby completing the regenerative cycle. The capacitor C_1 starts charging again towards V_{CC} to repeat the sequence of events explained above.

In the absence of any sync input, Q_1 and Q_2 repeat the on-off cycle at a rate determined by the time constant $C_1(R_4 + R_5)$. This time constant determines frequency of the oscillator. Potentiometer R_5 is the 'hold control' and can be varied to change the frequency. A negative going vertical sync pulse applied at the base of Q_1, through the integrating networks R_9, C_3 and R_{10}, C_4, brings the transistor out of cut-off before it normally would. This synchronizes the oscillator to the applied sync pulses.

The voltage waveform (i) at the emitter of Q_1 (see Fig. 18.15 (b)) is a sawtooth wave developed by the charging and discharging of C_1. Waveform (ii) is the sharp positive pulse developed at the emitter of Q_2 when both the transistors are turned on.

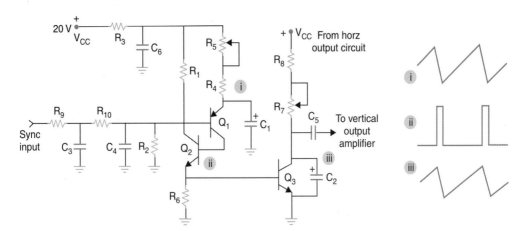

Fig. 18.15 (a). Complementary symmetry relaxation oscillator.   Fig. 18.15 (b). Wave shapes at the emitters of Q_1, Q_2 and collector of Q_3.

Wave Shaping

The wave shaping transistor Q_3 is normally biased to cut-off. It is triggered 'on' by the positive pulse developed at the emitter of Q_2 and directly coupled to its base. Capacitor C_2 connected across the collector and emitter of Q_3 charges during the 'off' period of Q_3 and discharges when the transistor turns on for a short interval of time. The resulting sawtooth voltage across C_2 (waveshape(iii)) is coupled through C_5 to the vertical output stage of the receiver.

The rate at which C_2 charges during the 'off' interval of Q_3 is determined by resistors R_7 and R_8. R_7 is a potentiometer that can be varied to control the amplitude of the vertical sweep and through this the height of the picture.

The voltage towards which C_2 charges is derived from a rectifier in the horizontal output circuit. Any change, in the horizontal deflection output voltage, will also alter the dc voltage fed to this circuit and affect the amplitude of the sawtooth voltage. Thus, if horizontal size of the picture changes, vertical size also changes proportionately, preserving the ratio of height to width of the picture.

18.6   SINE-WAVE DEFLECTION OSCILLATORS

High frequency sine-wave oscillators are more stable in their operation as compared to corresponding relaxation oscillators. Since the horizontal sweep frequency is quite high such oscillators are often used in horizontal deflection circuits. For such an application the oscillator is overdriven so that the tube or transistor acts like a switch and allows a sawtooth forming capacitor to charge and discharge. The frequency of the oscillator is controlled by a reactance tube or transistor placed between the AFC circuit and oscillator.

Reactance Tube Sweep Generator

A typical circuit of a vacuum tube horizontal deflection oscillator cum sweep generator is shown in Fig. 18.16. It employs a tuned grid configuration where mutual coupling between coils L_1 and L_2 can be varied to change the oscillator frequency. The desired frequency is 15625 Hz. The screen grid of pentode V_2 acts as plate for oscillator action. This isolates the oscillator circuit from large plate voltage variations which occur on account of switching action of the tube.

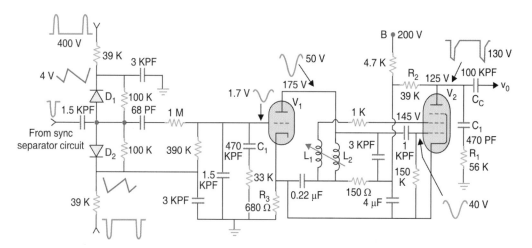

Fig. 18.16. Reactance controlled horizontal sweep generator.

The oscillator uses self-bias and develops about – 15 V at the control grid. For a pentode, this voltage is enough for class C operation. The oscillator is overdriven so that plate current flows in short-duration pulses only during extreme positive peaks of the grid voltage, The capacitor C_1 charges towards B_+ to develop trace period when the tube is off. The retrace is formed when the tube conducts for a short duration. The high (39 K) plate load resistance R_2 ensures a large (130 V P-P) sweep voltage.

Frequency Control

The reactance tube V_1 shunts the oscillator tank circuit and appears as a capacitive reactance. As explained in Chapter 7, Such a behaviour can be simulated by a capacitive feedback between plate and control grid of the tube. The equivalent capacitance across the output terminals of the tube is proportional to mutual conductance ($C_{eq} = g_m \times RC$) of the triode. In the circuit under consideration, the bias of the reactance tube is the combined result of AFC derived grid voltage and voltage drop across R_3, the cathode resistor. And drift in the oscillator frequency results in a change in the dc control voltage. This in turn shifts the operating point to change g_m of the reactance tube. The consequent change in the simulated equivalent capacitance C_{eq}, forces the oscillator to correct its frequency. The continuous feedback action results in a very stable operation of the oscillator.

Reactance Transistor Sweep Generator

Figure 18.17 shows a transistor sine-wave oscillator whose frequency is controlled by a reactance transistor. In this circuit a Hartley oscillator is used. It performs the same functions as its vacuum tube counterpart discussed earlier. The output of the sine-wave transistor oscillator is usually fed to the horizontal output circuit through a driver to provide isolation between the output stage and oscillator circuit.

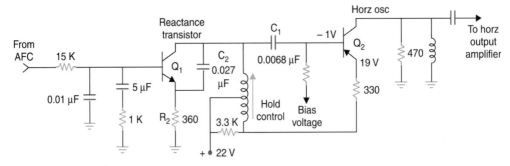

Fig. 18.17. Transistor sine-wave horizontal deflection oscillator and its associated reactance transistor.

As shown in the circuit diagram a reactance transistor (Q_1) is used between the AFC output and horizontal oscillator. It acts like an inductor instead of a capacitor, In order to simulate such a behaviour the phase shift network C_2, R_2 couples some of the oscillator's voltage into the emitter of the reactance transistor and at the same time shifts its phase by 90°. Since the feedback is returned to the emitter, transistor Q_1 can be considered to be a common base amplifier. As such, the emitter voltage and collector current are 180° out of phase. Thus the transistor current lags behind the collector voltage by 90°. Therefore, the oscillator 'sees' the reactance transistor as if an inductor has been connected across its tank circuit.

Any change in the oscillator frequency is sensed by the AFC circuit which couples a proportionate dc error voltage into the base of the reactance transistor. This shifts the operating point of the transistor in the same way as AGC bias does to control gain of tuned amplifiers.

Any decrease in oscillator frequency results in a positive dc control voltage thereby increasing forward bias of the reactance transistor. The resulting increase in collector current shifts the operating point where the transistor gain is lower. In turn, this acts to decrease the effective reactance at the output terminals of the transistor. This causes the total inductance shunting the oscillator tuned circuit to decrease. As a result, frequency of the oscillator is increased to return to its correct value. The reverse would occur if the bias were to decrease for any increase in the oscillator frequency.

REVIEW QUESTIONS

1. Sketch and label the current waveforms that must flow in the deflection yoke coils to produce a full rester. Explain the basic principle of generating such waveforms.
2. Why is a trapezoidal voltage waveform necessary to drive the vertical deflection coils ? What is the effect of source impedance and frequency on the shape of the driving voltage waveform ? How is the basic sawtooth voltage modified to obtain the desired driving voltage waveform ?

3. Draw the circuit diagram of a blocking oscillator-cum-waveshaper which employs a single triode and discuss its operation. Explain the operation of 'hold control' and 'height control' as provided in the circuit drawn by you.

4. A blocking oscillator that employs a transistor is shown in Fig. 18.9. Explain how the circuit operates to develop a sawtooth voltage. In particular, explain the operation of hold and height controls. Why are diodes D_1 and D_2 provided in the feedback transformer circuit ?

5. Explain with suitable waveforms how the frequency of the blocking oscillator is controlled with the help of sync information. Why is the free running frequency of the oscillator kept somewhat lower than the desired frequency ? What will happen if the uncontrolled frequency is higher than the correct frequency ?

6. Draw the basic circuit of a multivibrator employing transistors. How are the feedback networks designed to obtain correct trace and retrace periods ? Explain how the sync pulses control the frequency of such an oscillator.

7. The circuit of a complementary-symmetry relaxation oscillator is given in Fig. 18.15 (a). Describe briefly how the circuit operates to generate sawtooth output voltage. Discuss with suitable waveforms how the sync pulses keep the oscillator in synchronism with corresponding oscillator at the transmitter.

8. Why is a sine-wave oscillator preferred in horizontal deflection circuits ? Explain with a circuit diagram how a reactance tube/transistor connected between the AFC circuit and oscillator operates to maintain a constant frequency.

Chapter 18

19

Vertical Deflection Circuits

The output from the vertical and horizontal oscillators is too low to drive the deflection coils to generate a full raster. Therefore, amplifiers are used to produce enough power to fully drive the coils in the deflection yoke. The horizontal oscillator drives the horizontal output stage to produce horizontal scanning and the vertical oscillator drives the vertical output stage to produce vertical deflection. These two motions occur simultaneously to produce raster on the screen.

19.1 REQUIREMENTS OF THE VERTICAL DEFLECTION STAGE

The purpose of a field-scan output amplifier is to convert the input sawtooth or trapezoidal voltage waveform into a sawtooth current in the field deflection coils, of sufficient magnitude to scan the face of the picture tube. There are many different vertical-output circuit configurations in use that employ all tubes, combined tubes and transistors (hybrid) and only transistors. However, all circuits operate under class 'A' or 'AB' condition to ensure linear operation.

The limitations of inductive coupling that is employed between the output stage and yoke coils, together with the imperfection of active devices, make it necessary to modify the output voltage waveform to achieve linear deflection. This necessitates the use of negative feedback and other waveshaping techniques. Furthermore, transistor circuits need somewhat different approach than the one employing vacuum tubes. All this tends to make the deflection circuits somewhat complex. However, this problem is solved by first discussing various sections of the circuit separately and then putting them together to study the complete set-up.

(a) Vertical Yoke Drive

A voltage stepdown transformer provides an efficient means of coupling the output stage to the deflection coils. Figure 19.1 (a) shows such an arrangement that employs a vacuum tube as the vertical sweep amplifier. The output transformer matches the relatively high impedance of the tube to the low impedance deflection coils. The impedance of the coils is mainly resistive because of the low frequency spectrum that constitutes the 50 Hz trapezoidal sweep voltage. This makes the design of the output circuit basically the same as that of an audio output stage to match a loudspeaker.

In order to amplify the 50 Hz sawtooth with little distortion the frequency response of the amplifier must extend down to 1 Hz. The turns ratio of the output transformer is chosen to match the scanning coil impedance to the optimum output impedance of the tube. This is in the range of 30 : 1 to 6 : 1 depending upon the impedance of the coils. Typical value of primary inductance is around 45 mH with a dc resistance of the order of 40 ohms. The resistance of the secondary winding is quite low and varies between 5 and 10 ohms.

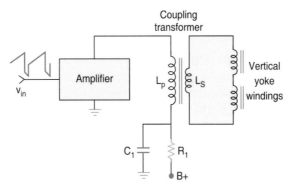

Fig. 19.1 (a). Vertical output stage employing an isolated primary and secondary step-down transformer.

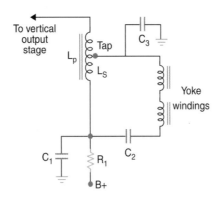

Fig. 19.1 (b). Vertical output circuit employing an autotransformer.

In many circuit designs an autotransformer coupling (Fig. 19.1 (b)) is employed because of its higher efficiency. The position of the tap determines the effective turns ratio of the transformer. If all other conditions are indentical, for the same input power an autotransformer gives a greater output power. The primary and secondary ac currents flow in opposite directions through a common portion of the autotransformer winding and thus tend to cancel each other. Since the net current flowing through these windings is smaller, power losses are less as compared to an isolated winding transformer. This means that for the same power ratings a thinner wire can be used in an autotransformer. In practice this results in a smaller and less expensive transformer.

A shown in Fig. 19.1 (b) a coupling capacitor (C_2) is used to prevent any dc current flow in the deflection coils which would otherwise shift centering of the electron beam. The capacitor C_3 connected between secondary tap and ground is effectively in parallel with the secondary circuit. It improves frequency response of the transformer by speeding up collapse of the induced field during retrace.

Chapter 19

The self-induced voltage in the secondary winding during retrace is stepped up because of the large turns ratio between the primary and secondary. Therefore, the transformer winding must be insulated to withstand induced voltage peaks. As shown in Figs. 19.1 (a) and (b) the network R_1, C_1 provides decoupling, to prevent any undesired feedback from output stage to any other stage through the common dc source.

In vertical output stages that employ transistors, an output transformer is not necessary, because power transistors provide large output current at a relatively low impedance level, and thus can be coupled through a capacitor directly to the yoke coils. The circuit arrangement that is often used is shown in Fig. 19.2. The choke shown in the collector circuit isolates the collector signal path from the power supply. There are some solid state vertical output circuits which use an isolation transformer with a unity turns ratio. This arrangement provides ac driving current to the yoke without the need for a large electrolytic coupling capacitor. A VDR (voltage dependent resistance) is often used in the output circuit across the choke (Fig. 19.2) to protect the output transistor against voltage transients in the collector circuit.

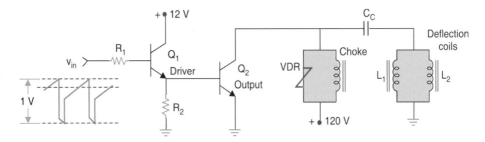

Fig. 19.2. Basic circuit of a transistor vertical amplifier.

(b) Vertical Deflection Coils

The magnetic field for vertical deflection is developed by two coils mounted 180° apart on the neck of the picture tube. Physically these coils are a part of the yoke assembly that also contains horizontal windings. During the vertical deflection period the flux generated by the sawtooth current in the windings, moves the electron beam down and then returns it to the top for a quick retrace. The vertical yoke inductance and resistance are kept as low as possible to accomplish higher deflection efficiency.

Most field scanning coils are designed with an inductance that varies between 3 mH and 50 mH with a resistance of 4 to 50 ohms. The impedance and number of turns of the coils are dependent on the required deflection angle. Wider deflection angle tubes require a stronger magnetic field to cover the entire raster. The peak-to-peak sawtooth current needed by different vertical deflecting coils varies between 300 mA to 2.5 Amp. Typically, a short neck 36 cm picture tube requires a 45 mH deflection coil and a peak-to-peak driving current of 400 mA.

The two halves of the deflecting coils are normally connected in series across the secondary of the drive transformer in vacuum tube circuits. However, in transistor circuits these are connected in parallel to provide impedance match for maximum efficiency.

(c) Vertical Linearity

The patterns of Fig. 19.3 illustrate the effect of three different current waveforms on the reproduced raster. When the sawtooth wave is linear (Fig. 19.3 (*a*)), the vertical scan allows equal spacing between horizontal lines and there is no distortion. In Fig. 19.3 (*b*) the sawtooth wave tends to become flat towards its close, hence the magnetic field increases at a reduced rate with the result that the horizontal lines crowd together at the bottom of the raster. In the third pattern (Fig. 19.3 (*c*)) the horizontal lines are crowded at the top of the raster because of the slow rate of change of current at the beginning of the sawtooth wave. Non-linear vertical scanning is more noticeable than horizontal non-linearity. Crowding of horizontal lines in any region of the raster makes the objects look flatter in that part of the picture. However, when the scanning rates becomes faster than normal in any portion of the raster the corresponding picture area appears elongated. Non-linearity may be caused due to any of the following reasons.

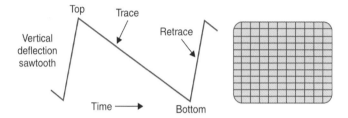

Fig. 19.3 (a). Linear sawtooth causing equal spacing between the lines on the raster.

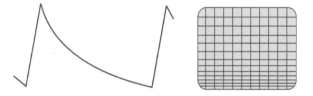

Fig. 19.3 (b). Non-linear sawtooth causing crowding of lines at the bottom of the raster.

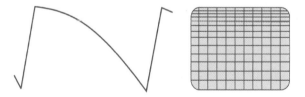

Fig. 19.3 (c). Non-linear sawtooth causing crowding of lines at the top of the raster.

(*i*) A short time-constant sawtooth forming RC circuit develops a non-linear sweep because too much of the exponential charging curve forms part of the trace period. Similar nonlinear compression at the bottom of the raster may be caused by positive peak clipping of the trapezoidal voltage driving the grid/base of the vertical output stage. This is generally the result of incorrect tube/transistor bias. However, such shortcomings can be easily avoided by a careful design of the sawtooth forming circuit and appropriate choice of the device's operating bias.

(*ii*) The magnetizing current of the coupling transformer will also cause nonlinear distortion. This can be made quite small if the transformer has a high primary inductance. This, however, leads to a heavier and costlier output transformer.

Chapter 19

The solution to this problems, which is often employed, is to use a transformer with not-too-high a primary inductance but modify the input voltage to the amplifier for optimum results.

(*iii*) Another aspect, which needs attention, is the attenuation caused at low frequencies by the finite primary inductance of the output transformer, that effectively appears in parallel with the deflecting coils. This is corrected by applying negative feedback from the output circuit to input of the amplifier through a frequency selective network.

(*iv*) Another problem is the pincushion distortion. Since magnetic deflection of the electron beam is along the path of an arc, with point of deflection as its centre, the screen should ideally have corresponding curvature for linear deflection. However, in flat screen picture tubes it is not so and distances at the four corners are greater as compared to central portion of the face plate. Therefore electron beam travels farther at the corners causing more deflection at the edges. This results in a stretching effect where top, bottom, left and right edges of the raster tend to bow inwards towards the centre of the screen. The result, as illustrated in Fig. 19.4, is a pincushion like raster. Such a distortion is more severe with large screen picture tubes having deflection angles of 90° or more.

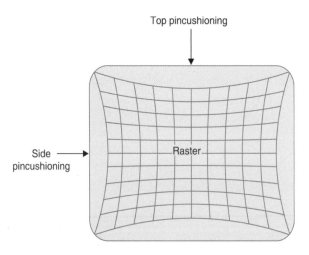

Fig. 19.4. Pincushion distortion of the raster.

In black and white receivers, pincushion distortion is eliminated by suitable design of the yoke and the use of small permanent magnets mounted close to the yoke. These magnets are so positioned that they stretch the raster along the sides of the picture and thus compensate for the pincushion distortion. In colour picture tubes the deflection current in the yoke is modified by special pincushion correction circuits.*

In output stages, (see Fig. 19.5) which use power triodes, the linearity is often controlled by shifting the operating point on the transfer characteristic, by varying a part of the cathode bias resistance. The variable resistance then performs as *Linearity Control*.

Transistor vertical output stages are generally similar to their vacuum tube counterparts. Since the input impedance of a power transistor is only the order of a few hundred ohms, it

*This is explained along with dynamic convergence circuits in chapters devoted to colour television.

becomes necessary to drive the output stage from an emitter follower (driver) so as not to place too high a load on the sawtooth generator. In high power transistor circuits some form of thermal stabilization is also provided for stable operation. This ensures linear operation by maintaining the operating point at the set optimum value. Negative feedback is also used to maintain a frequency response that is flat down to 1 Hz.

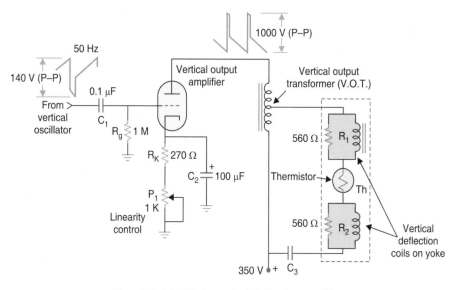

Fig. 19.5 (a). Triode vertical deflection amplifier.

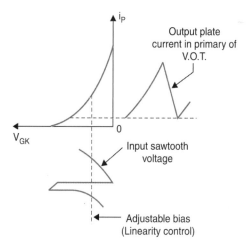

Fig. 19.5 (b). Cancellation of non-linear curvature of the sawtooth wave by optimum choice of the operating point on the output characteristics of the tube.

Whenever possible direct coupling is employed between the driver and output stage to avoid use of a large coupling capacitor. If necessary, linearity correction is also employed by modifying the waveshaping network. This is fully explained in a later section of the chapter while describing a typical transistor output circuit.

Chapter 19

(d) Suppression of Undesired Oscillations

During retrace time when the current through vertical deflecting coils suddenly drops to zero, energy stored in the collapsing magnetic field sets up high frequency oscillations. The frequency of 'ringing' thus produced, depends on the inductance and the distributed capacitance of the deflecting coils. The effect of such oscillations, if not suppressed, would be to pull the beam upwards and downwards alternately instead of the smooth motion downwards. This results in a narrow white strip at the top of the screen. The strip then appears more bright than the rest of the raster. To prevent ringing two shunt damping resistances are connected across the two halves of the vertical scanning coils. Figure 19.6 (*a*) illustrates the ringing effect on the deflection current and Fig. 19.5 (*a*) the use of damping resistors R_1 and R_2 across the vertical scanning coils. It may be noted that the output from the amplifier contains considerable power and the damping resistors do not attenuate the signal appreciably.

(e) Height of the Raster

To fill the entire raster from top to bottom of the screen a definite amplitude of sawtooth current must flow into the deflecting coils. The output amplifier is designed to meet this requirement. In addition a control is provided at the input of the amplifier to finally adjust the height of the picture. This control is known as 'height control'. In some designs the linearity control resistor when provided in the cathode (or emitter) lead of the amplifying device is left unbypassed (by a capacitor) and also acts as the height control. A change in the linearity control resistance varies gain of the stage which in turn controls the height of the raster. The effects of either insufficient or excessive drive are illustrated in Fig. 19.6 (*b*) which clearly indicate the need for a 'height control'.

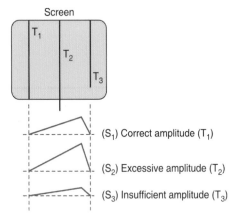

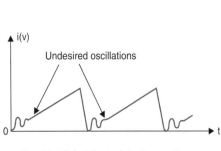

Fig. 19.6 (a). Effect of ringing on the vertical deflection current.

Fig. 19.6 (b). Effect of deflection current amplitude on the height of the raster.

Another factor which affects the height of the picture is the increase in resistance of the deflection coils when current flows through them. To counteract this, a thermistor (see Fig. 19.5) is added between the coils in the yoke to equalize the resistance under high heating conditions. When temperature of the windings rises, the thermistor also gets heated and its resistance decreases sufficiently to compensate for the increase in winding resistance. This maintains a constant deflection current to prevent any change in the height of the raster.

(f) Vertical Rolling of the Picture

The rolling of the picture upwards or downwards on the screen occurs on account of incorrect vertical scanning frequency. This is illustrated in Fig. 19.7 for the case when the vertical oscillator frequency is higher than 50 Hz. It may be noted that relative timings of vertical blanking in the composite video signal and vertical retrace in the sawtooth deflection current are not the same, with the result that vertical blanking occurs during trace time instead of during retrace time. For a vertical frequency that is higher than 50 Hz, each sawtooth advances into trace time for succeeding blanking pulses. As a result the black bar produced across the screen, by the vertical blanking pulse, appears lower and lower down the screen for successive cycles. The opposite effect takes place when the oscillator frequency is lower than 50 Hz. The picture rolls at a faster rate if the oscillator frequency deviates much from 50 Hz. However, when the sync frequency and sawtooth frequency are the same, every vertical retrace occurs within the blanking time. The blanking bars are then produced at the top and bottom edges of the raster and are not visible. Thus, when the vertical oscillator is locked by sync pulses each frame is reproduced over the previous one, and the picture holds still.

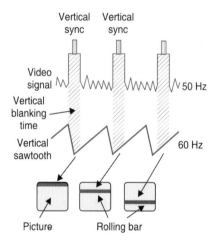

Fig. 19.7. Vertical rolling when the deflection frequency is more than 50 Hz. Note that horizontal blanking details have been omitted for clarity.

As explained in the previous chapter, the oscillator frequency is controlled by the 'hold control'. Since no elaborate sync processing circuitry is provided for the vertical sync pulses, the vertical hold control is provided on the front panel of all TV receivers, for resetting the frequency, when picture rolls continuously up or down.

The vertical jumping of the picture that sometimes occurs should not be confused with vertical rolling. Vertical jumping occurs on account of unequal sync pulse amplitudes for odd and even fields. Occasional jumping is caused by wrong triggering of the vertical oscillator by a stray noise pulse.

(g) Internal Vertical Blanking

The voltage pulses produced during retrace intervals, in the vertical output circuit are coupled to the picture tube to provide additional blanking during vertical retrace time. This is in addition to the blanking voltage at the cathode or grid circuit of the picture tube, which is part of the composite video signal. In receivers, that do not fully retain dc component of the video signal at the picture tube, the blanking pulses fail to reach the required level to cut-off the beam, when vertical retrace is taking place. Thus the horizontal lines that get traced during this period, appear on the screen. However, the use of blanking pulses ensures complete blanking during retrace for any setting of the brightness control, with or without the dc component.

All receivers use a vertical retrace suppression circuit, to turn-off the electron beam, until it is back at the top of the screen to begin the next vertical trace. Figure 19.8 shows two different circuit arrangements, one for obtaining positive pulses, that are suitable for injecting in the cathode circuit and the other for obtaining negative pulses to be applied at the grid of picture tube. The amplitude of these pulses varies between 50 to 150 V in different circuits.

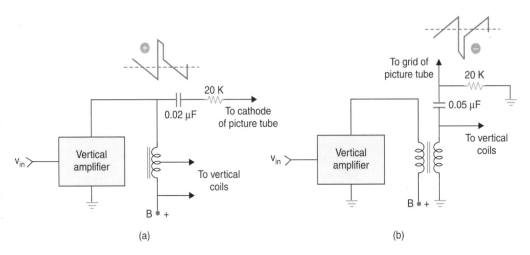

Fig. 19.8. Circuits for obtaining internal vertical blanking pulses
(a) positive retrace pulses (b) negative retrace pulses.

In transistor circuits, where such high amplitudes cannot be obtained, the low amplitude pulses as obtained from the output stage are fed into the video amplifier, which amplifies them before feeding to the picture tube circuit.

In the circuit shown in Fig. 19.8 (a) the pulses are obtained at the primary of the output circuit to yield positive going pulses, whereas in the circuit of Fig. 19.8 (b) negative going pulses are obtained from the secondary of the same output transformer.

Based on the above discussion about the problems of vertical output stage and the techniques used to overcome them, it is reasonable to assume that a large number of circuit arrangements are possible to achieve linear vertical deflection. However, a few typical circuits are discussed to broadly identify various configurations that have found wide acceptance in television receiver practice.

19.2   VACUUM TUBE VERTICAL DEFLECTION STAGE

Figure 19.9 is the circuit of a vertical output stage which is widely used in vacuum tube receiver. As is the common practice, a single tube (triodepentode), serves both as oscillator and output stage. While the pentode section performs as vertical output stage, both triode and pentode (V_1 and V_2) operate together as an unsymmetrical free-running plate coupled multivibrator. Positive going pulses which are obtained from winding L_{S2} on the output transformer are fed back via R_1 and C_1 to the grid of V_1 to provide multivibrator action. The pentode conducts for a longer period to supply a sawtooth current to the deflection coils. During this period the triode is cut-off and a sawtooth voltage builds up at its plate due to C_2 charging through R_2, P_2 and R_3 from B_{++} supply. The potentiometer P_2 (1 M) can be varied to adjust the amplitude of sawtooth and hence serves as height control.

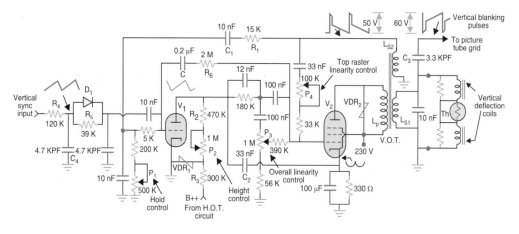

Fig. 19.9. Vacuum tube vertical output stage driven by a multivibrator sweep circuit.

For improved linearity and constant aspect ratio the dc voltage for charging C_2 is obtained from the horizontal output circuit. The VDR_1 connected across R_3 and ground stabilizes this dc source. Any change in width of the picture will indirectly affect B_{++} supply and hence vary amplitude of the sawtooth voltage. The resultant change in drive voltage to the vertical output stage will vary height of the picture, thereby maintaining a constant aspect ratio.

The setting of potentiometer P_3 makes it possible to alter the shape of sawtooth voltage applied to the grid of V_2. This is done to obtain a certain curvature in the sawtooth current that flows through the deflection coils for improving overall linearity. In addition, setting of potentiometer P_4 (100 K) in the negative feedback path from winding LS_2 to the grid of V_2 (pentode) affects the form of drive voltage just after flyback and hence serves as top linearity adjustment. The network R_6 (2 M), C_5 (0.2 µf) provides a small negative feedback from grid of V_2 to the input circuit of V_1 to ensure stable operation of the circuit.

The vertical sync pulses are integrated by network R_4, C_4 which constitutes a low-pass filter. The output across C_4 is passed through the parallel combination of diode D_1 and resistance R_5 to obtain sharp positive going pulses. These are fed to the grid of V_1 to synchronize the frequency of the multivibrator.

The vertical blanking pulses are taken from the secondary of the output transformer through capacitor C_3. The voltage dependent resistance V_{DR2} provides protection to the output tube and transformer by suppressing high voltage transients that develop across the primary winding during retrace periods.

Chapter 19

19.3 TRANSISTOR MULTIVIBRATOR DRIVEN VERTICAL OUTPUT STAGE

In this circuit (see Fig. 19.10) three transistors Q_1, Q_2 and Q_3 form a collector coupled multivibrator. The output of Q_1 is fed to the base of Q_3 through emitter follower Q_2 and output of Q_3 is coupled back to the input of Q_1 through network C_7, R_7 to complete positive feedback loop for multivibrator action. The emitter follower Q_2 provides impedance match between Q_1 and Q_3, effectively isolates sawtooth forming capacitors C_1 and C_2 from the low input impedance of the output transistor, and provides increased current drive to the base of Q_3 to obtain full vertical deflection. The resistance R_8 and capacitor C_6 form a filter to bypass any pickup by vertical coils from the adjoining horizontal deflection coils.

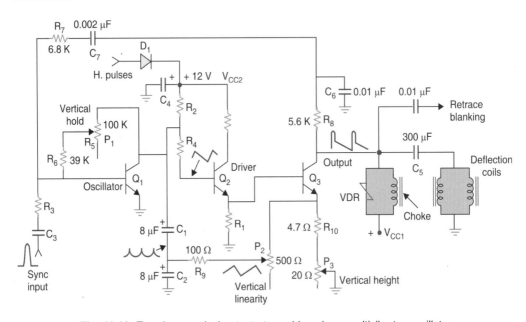

Fig. 19.10. Transistor vertical output stage driven by a multivibrator oscillator.

The capacitors C_1 and C_2 charge towards V_{CC2} to generate the trace portion of the sawtooth wave. Initially dc voltage at the collector of Q_1 is low and it stays at cut-off. As the capacitor C_1 charges and reaches a certain value, Q_1 starts conducting. It is reinforced by positive feedback from Q_3 to the base of Q_1. This action being regenerative, Q_1 instantly goes to saturation and Q_3 is cut-off. The series capacitors C_1 and C_2 then quickly discharge to almost zero voltage through the collector-emitter path of Q_1 thus generating retrace period. At this point Q_1 again returns to cut-off because of disappearance of dc voltage, both at its collector and base.

A portion of the sawtooth wave at the emitter of Q_3 is fed through potentiometer P_2 and resistance R_9 to the junction of C_1 and C_2, where capacitor C_2 integrates it to form a parabolic wave. Since this is in series with the voltage across C_1, the shape of the sawtooth wave fed to the base of Q_2 and through that to the base of Q_3 is also modified. This is done to improve linearity of the sawtooth current to the yoke coils. Potentiometer P_2 thus acts as linearity control. Similarly potentiometer P_3 is adjusted to control negative feedback in the output stage and acts as vertical height control.

The base current to transistor Q_1 is adjusted by potentiometer P_1, which in turn determines the instant when Q_1 will start conducting in each cycle. In this way potentiometer P_1 controls frequency of the multivibrator and operates as 'hold control'.

The voltage V_{CC2} for charging C_1 and C_2 is obtained from the dc voltage, that develops across C_4 by horizontal flyback pulses coupled through diode D_1 to charge it. This capacitor is sufficiently large to provide a steady dc source of about 12 volts. The reason for using horizontal flyback pulses to provide dc voltage is to make the vertical output dependent on the horizontal scanning amplitude. If excessive load current reduces horizontal scanning width, the vertical height will also decrease thus maintaining correct aspect ratio of the raster.

Synchronization is achieved by feeding positive sync pulses at the base of Q_1 through C_3, R_3. The circuit has very good frequency response because of direct coupling between Q_1 and Q_3. However, the value of coupling capacitor C_5 limits the response at very low frequencies.

19.4 BLOCKING OSCILLATOR DRIVEN OUTPUT STAGE

Figure 19.11 is the circuit of a transistorized vertical sweep circuit driven by a blocking oscillator. The sawtooth forming network is in the emitter lead of Q_1, the oscillator transistor. The discharge of C_1 during long cut-off periods of Q_1, provides the basic sweep output. Its amplitude

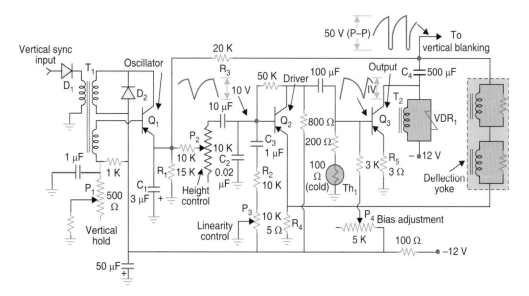

Fig. 19.11. Transistor blocking oscillator and vertical amplifier.

is set by P_2 (10 K Pot), to adjust height of the picture. The sweep voltage is shaped by network C_2, C_3, R_2 and P_3. Potentiometer P_3 is used as linearity control. Linearity is also improved by negative feedback (i) from the output circuit to emitter of Q_1 through R_3, and (ii) by passing the deflection yoke current through R_4, the emitter resistance of driver transistor Q_2. A thermistor (Th$_1$) connected in the base circuit of output power transistor Q_3 is placed on the heat sink as a check against any possibility of thermal runway. Any undue increase in collector current of Q_3 will cause more heating. The consequent increase in temperature of the heat sink will lower resistance of the thermistor. In turn, the reduced emitter-base drive to Q_3 will lower its collector current. Transistor

Q_3 is also stabilized by leaving its emitter resistance R_5 unbypassed. Potentiometer P_4 is varied for optimum bias, more so, when it becomes necessary to replace Q_3. The VDR_1 connected across T_2 keeps the amplitude of output waveform within narrow limits. The sweep output is fed to the deflection coils (connected in series) through C_4, a large electrolytic capacitor. The induced sharp pulses (typically 50 V) during retrace serve as vertical blanking pulses.

The oscillator frequency is set by varying P_1 (hold control) to be slightly less than 50 Hz. Sync pulses, which are injected at the base of Q_1 through T_1, synchronize the frequency with that at the transmitter. In practice, the vertical hold control is varied until the oscillator frequency locks with the incoming sync pulses to provide a steady picture.

19.5   TRANSFORMERLESS OUTPUT CIRCUIT

The high wattage transistor needed for the output stage can be replaced by two medium power transistors in a suitable push-pull configuration as is the common practice in audio amplifiers. One such method uses p-n-p and n-p-n transistors in a complementary symmetry as illustrated by the basic circuit of Fig. 19.12. The output impedance provided by the complementary pair is quite low and so there is no need for an output transformer. This results in higher efficiency and improved performance because the output transformer is often heavy, expensive and introduces frequency distortion.

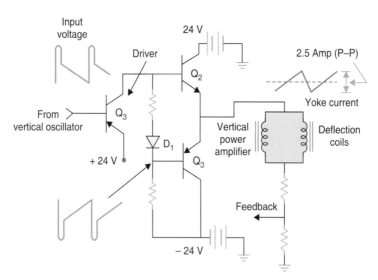

Fig. 19.12. Transformerless vertical output amplifier.

The output transistors Q_2 and Q_3 operate class B and are alternately driven into conduction by a common trapezoidal input signal. When Q_2 is on and Q_3 is off, current flows through the yoke from the positive 24 V supply. On alternate half of the input signal when Q_3 is on and Q_2 is off, current flows in the opposite direction from the negative 24 V supply. This amounts to an ac current flow through the yoke. Diode D_1 is forward biased and voltage drop across it provides suitable bias to Q_2 and Q_3 thereby preventing any crossover distortion. The conduction of D_1 also ties the bases of Q_2 and Q_3 allowing signal output from the driver (Q_1) to feed both of them simultaneously.

19.6 VERTICAL SWEEP MODULE

Figure 19.13 is the block schematic diagram of a typical vertical sweep module. It employs nine transistors and four diodes besides a large number of capacitors and resistors. Various blocks are suitably labelled.

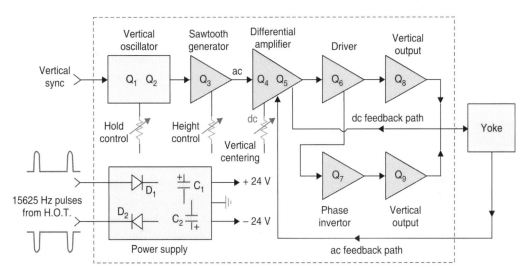

Fig. 19.13. Block diagram of a vertical sweep module.

Oscillator

A explained in the previous chapter, transistors Q_1 and Q_2 operate as complementary symmetry oscillator and provide output at 50 Hz. Hold control forms part of this circuit. Its output is waveshaped by transistor Q_3 and the associated capacitor, to provide a sawtooth voltage. The wave shaping circuit has provision for varying amplitude of the sawtooth voltage and this is the height control.

Differential Amplifier

Transistors Q_4, Q_5 constitute a differential amplifier with two ac and two dc inputs. One ac input is the sawtooth voltage developed by Q_3 and the other is the feedback signal from vertical windings of the yoke. Such an arrangement acts to eliminate, by negative feedback, any distortion introduced by the output stages, ensuring a linear vertical sweep. It also makes the use of linearity control and consequent adjustments unnecessary.

The single ended output of the differential amplifier (transistors Q_4, Q_5) is direct coupled to the output stage. The two dc inputs to the differential amplifier determine the output level which in turn controls vertical (dc) centering of the beam. One of the two dc inputs is a clamped dc voltage that holds the base of Q_4 at near zero voltage. However, the second dc input (at the base of Q_5) is made dependent by feedback on the dc output voltage at the point where deflection windings are connected. The reference dc voltage is obtained from the two dc supplies through a potentiometer which is initially varied to obtain correct centering of the beam.

Driver, Phase Invertor and Output Amplifier

The driver transistor Q_6 couples output from the differential amplifier direct to the output transistor Q_8 and via phase invertor Q_7 to the other output transistor Q_9. Transistors Q_8 and Q_9 are connected as quasi-complementary pair and conduct alternately to provide the total vertical scan.

DC Supply

The power supply block has two rectifiers D_1 and D_2 with associated filter capacitors C_1 and C_2. It provides both positive and negative dc supply sources for the various blocks. The input for dc supplies is derived from the receiver's horizontal scanning circuit. Any change in horizontal scanning will be reflected as a change in dc voltage for the vertical module thus maintaining a fixed aspect ratio of the picture.

19.7 THE MILLER DEFLECTION CIRCUIT

Linearity requirements of colour receivers are very rigorous and need a careful design of the vertical stage. While circuits described in the previous two sections are used both in monochrome and colour receivers, a new circuit known as Miller sweep circuit has found wide acceptance in recent colour receiver designs. This circuit is capable of generating a sawtooth voltage of exceptionally good linearity. As a result, no linearity control is necessary. The basic sweep generation circuit makes use of the Miller effect to enormously increase effective size of the sawtooth forming capacitor.

The simulation of such a large capacitance in the ramp forming circuit can be illustrated by the circuit arrangement shown in Fig. 19.14. The basic amplifier used in this circuit has a very high gain ($|A| \approx 10^6$), inverted output and a very high input impedance. Because of large gain and heavy feedback provided by capacitor C_M the net voltage v_i at the input terminals of the amplifier is nearly equal to zero and hardly any current flows across the terminals marked X and Z. Thus X and Z are virtually a short from the circuit analysis point of view.

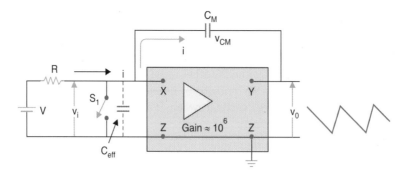

Fig. 19.14. Basic Miller sweep circuit.

The sawtooth forming current i that flows through $R = \dfrac{V - v_i}{R} \approx \dfrac{V}{R}$, $(v_i \approx 0)$. Since practically no current flows between X and Z, the current i completes its path through C_M and charges it.

Therefore i is also equal to $\dfrac{v_i - v_0}{X_{C_M}} \approx \dfrac{-v_0}{X_{C_M}}$, where X_{C_M} is the reactance of capacitor C_M. The current i being equal to V/R is independent of v_0, the effective voltage across C_M and so remains constant to charge the capacitor thus developing a linear ramp across it. The admittance across

the terminals X, $Z = \dfrac{i}{v_i} = \dfrac{-v_0}{v_i} \times \dfrac{1}{X_{C_M}} \left(\text{since } i = \dfrac{-v_0}{X_{C_M}} \right)$

$$= 2\pi f A C_M \text{ because } \left(\dfrac{v_0}{v_i} = -A \text{ and } X_{C_M} = \dfrac{1}{2\pi f C_M} \right)$$

Thus effective value of the capacitance in series with R is gain times the value of C_M. With $|A| \approx 10^6$, the effective capacitance (C_{eff}) will appear to be one million times the external sawtooth forming capacitor C_M. Consequently the charging time constant of the sawtooth forming RC circuit becomes very large and only a small portion of the RC charging curve is used to form trace portion of the sawtooth. Switch S_1 (see Fig. 19.14) can be opened and closed at appropriate moments to generate a sawtooth voltage at the output. In actual deflection circuits a transistor switch which is a part of the oscillator circuit controls the charge and discharge periods. In this manner a sawtooth of excellent linearity is obtained which is later amplified to the desired level. Note that the ramp voltage that appears at the output terminals of the amplifier increases in negative direction because of 180° phase reversal between input and output of the basic amplifier.

Miller Vertical Sweep Circuit

Figure 19.15 is a simplified schematic diagram of a Miller vertical defection circuit. The circuit employs a complimentary symmetry power amplifier to drive the yoke. The waveshaping circuit has a high gain amplifier; a driver circuit and an oscillator switching transistor. Because of overall positive feedback the circuit behaves like a relaxation type of free running oscillator. Transistor Q_1 switches at the vertical frequency rate to provide trace and retrace periods.

The circuit employs four feedback paths to accomplish all the functions. Feedback path A effectively increases the value of Miller capacitor C_M thus ensuring excellent sweep linearity. Feedback path B provides positive feedback needed to convert the high gain amplifier into a sawtooth generating multivibrator. C_M charges through Q_1 but discharges through Q_2. The adjustments of this part of the circuit include height control, and hold control (frequency control) but there is no linearity control. Feedback path C is used to improve frequency stability of the oscillator circuit.

While the Miller sweep circuit provides a very linear sawtooth but the curvature of the tube face plate and its rectangular construction (pincushion effect) necessitate modification of the current that flows through the deflection windings. Therefore feedback path D is used to introduce 'S' shaping or correction to the yoke current to compensate for the stretch at the top and bottom of the raster. 'S' compensation is also achieved by generating a parabola of current and combining it with the sweep output. The modified yoke current then provides linear deflection despite curvature of the face plate. In addition, a pincushion correction circuit is connected in series with the yoke to cause necessary change in the current waveshape.

Chapter 19

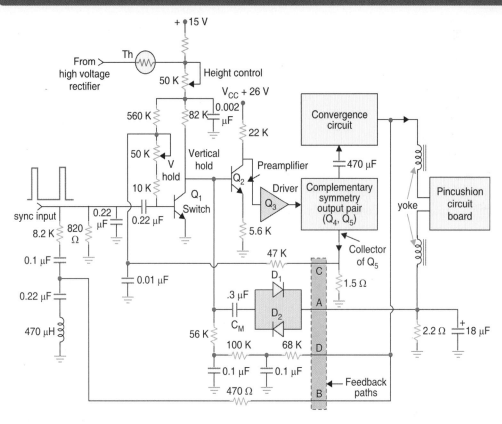

Fig. 19.15. Simplified schematic diagram of a Miller vertical deflection circuit.

19.8 INTEGRATED CIRCUIT FOR THE VERTICAL SYSTEM

Several dedicated ICs are now available which perform all the functions necessary for obtaining linear vertical deflection of the beam. One such IC is BEL 1044, the schematic diagram of which is given in Fig. 19.16. It is a silicon monolithic integrated circuit containing all stages necessary for vertical deflection in black and white television receivers. It performs the functions of a linear sawtooth generator, a geometric 'S' correction circuit, a flyback booster and an output amplifier. This integrated circuit requires only a few external components to complete the total vertical deflection system. Such components and controls are suitably labelled in the figure.

The BEL 1044 is supplied in a 16-lead dual-in-line plastic package, with an integral bent-down wing-tab heat sink, intended for direct printed circuit board insertion. The device is capable of supplying deflection current up to 1.5 A p-p and positive blanking pulses of 20 V amplitude. The block which needs special mention is the flyback booster circuit. This ensures development of spiked trapezoidal voltage of amplitude almost equal to the supply voltage. The charge stored in the 100 µF capacitor supplies necessary power to the circuit during sharp flyback periods. Note that the diode (BY 125) stays reverse biased during flyback intervals.

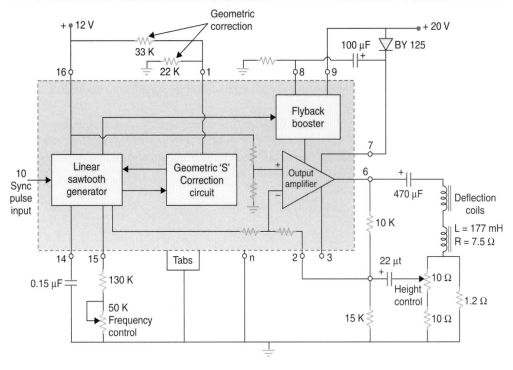

Fig. 19.16. Functional block diagram of the vertical system employing BEL 1044IC.

REVIEW QUESTIONS

1. A well designed vertical scanning section of the television receiver must include the following provisions:

    (i) adjustment for vertical linearity,

    (ii) means for suppressing undesired oscillations in the yoke current,

    (iii) constant height of the raster,

    (iv) correct aspect ratio.

    Explain either with the help of separate circuits or by drawing complete circuit of a typical vertical deflection section, the techniques employed to meet the above requirements.

2. Draw simplified circuit diagram of a vertical deflection amplifier employing transistors and explain its operation. Indicate how positive retrace pulse can be obtained from the output transformer of the circuit drawn by you.

3. The circuit of a commonly used vertical deflection section, that employs tubes is shown in Fig. 19.9. Explain in brief, the operation of this circuit and the need for various controls provided in different parts of the circuit.

4. The circuit of a typical multivibrator controlled vertical output stage that employs a somewhat different method for generating sweep voltage and linearity control is shown in Fig. 19.10. Describe briefly all the essential features of this deflection circuit and also explain how vertical rolling of the picture is prevented.

Chapter 19

5. Specially designed modules for the vertical stage are often used in TV receivers. Draw schematic block diagram of one such module and explain the operation of each section. Why is the dc supply source for the module often made dependent on output of the horizontal scanning circuit?

6. What do you understand by Miller effect ? Explain with a suitable circuit diagram how a large capacitance is simulated at the input of a high gain inverting amplifier for developing a linear sweep.

7. Draw schematic diagram of a typical Miller vertical deflection circuit and explain how a highly linear sweep output is developed by employing suitable positive and negative feedbacks.

8. Draw functional diagram of the vertical system IC-BEL 1044 and explain how various sections of the IC function to develop deflection sweep output.

20

Horizontal Deflection Circuits

The horizontal output stage is a power amplifier which feeds horizontal deflection windings in the yoke to deflect the beam across the width of the picture tube screen. Though the function of this stage is similar to the vertical output stage, its design and operation are very much different on account of the higher deflection frequency, very short retrace period and higher efficiency.

Since the horizontal output load is primarily inductive*, the deflection current induces very high voltage pulses in the output transformer while falling sharply during the retrace period. This energy which is in the form of high voltage pulses is rectified to produce a high voltage dc source (EHT) for the final anodes of the picture tube. Furthermore, the energy associated with the high voltage pulses tends to setup high frequency oscillations in the output circuit. To suppress this high frequency ringing, a diode damper is used instead of resistors employed in the vertical system. The diode while conducting, charges a capacitor, the voltage across which adds to the existing B_+ supply to produce a higher dc source voltage commonly known as boosted B_+ supply. In this manner, part of the energy from the deflection coils is recycled to develop a voltage source for the horizontal stage and other circuits.

Another interesting feature of the horizontal output circuit is that the unidirectional damper diode current which also flows through the deflection coils is used to complete a part of the horizontal trace. This results in improved efficiency because the amplifier need not supply deflection current for the entire cycle.

20.1 HORIZONTAL OUTPUT STAGE

A schematic block diagram of the horizontal output stage is shown in Fig. 20.1. The main sections of the stage are (a) output amplifier V_1, (b) output transformer T_1 (commonly known as flyback transformer), (c) damper diode D_1, (d) high voltage rectifier D_2 and (e) deflection windings in the yoke.

The amplifier operates with self-bias and is designed to perform as a switch which closer for a fixed period during each horizontal deflection cycle to supply power to the output transformer. The transformer's main functions are to provide impedance match between the output

---

* L/R for the vertical deflection yoke is about three times the L/R for the horizontal deflection yoke, but the horizontal scanning frequency is about 300 times the vertical scanning frequency. That is why the horizontal scanning load looks essentially inductive whereas the vertical scanning load looks effectively resistive.

stage and deflection windings and to act as a step-up high voltage transformer for D_2, the EHT rectifier. The damper diode D_1 conducts soon after the flyback stroke to suppress shock excited oscillations. While doing so the damper also causes horizontal scan of the beam on the left-half of the raster.

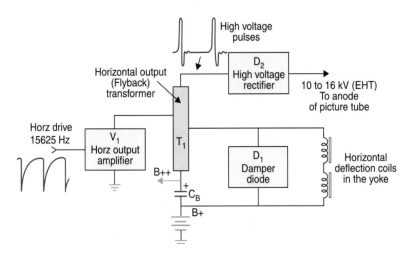

Fig. 20.1. *Block diagram of the horizontal output stage.*

20.2 EQUIVALENT CIRCUIT

The various functions which the horizontal stage performs make its design and operation quite complex. Therefore, it is convenient to first consider a basic equivalent circuit of the stage to help understand its functions and sequence of operations.

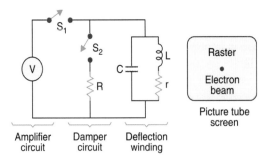

Fig. 20.2. *Simplified equivalent circuit of the horizontal output stage. With no excitation, the electron beam is in the centre of the picture tube screen.*

The circuit shown in Fig. 20.2 is a simplified equivalent circuit of the deflection part of the horizontal output stage where the amplifier* has been simulated by a constant voltage source V

---

* For a typical horizontal deflection coil $L = 2$ mH, $r = 4\ \Omega$, I (p-p) = 3 A, trace period $(t_s) = 54\ \mu S$ and retrace or flyback time $(t_f) = 10\ \mu S$. $\setminus V_L = L\dfrac{di}{dt} = 3$ mH $\times \dfrac{3}{54.\mu S} = 110$ V and corresponding voltage drop $v_r = I \times r = 3 \times 4 = 12$ V. Since $v_r \ll v_L$, a constant voltage V applied across the deflecting coils causes a near linear sweep current through it.

together with switch S_1 to control the flow of current to the deflection coils. The inductor L and the small series resistance r represent lumped inductance and resistance of the flyback transformer and deflection windings. The capacitor C in parallel with this circuit accounts for distributed and stray capacitance of the output stage. The damper circuit is represented by switch S_2 and a resistance R in series with it. The switch S_2 represents damper diode D_1 which when conducts charges C_B through R.

The sequence of operations which cause deflection of the beam across the screen are illustrated in Fig. 20.3 by various circuits and associated waveforms. The location of the beam on the screen is also shown in each case. The direction of beam deflection has been arbitrarily chosen to be from left to right for the direction of current flow from the voltage source.

When switches S_1, S_2 are open, *i.e.*, when both the amplifier and damper diode are not conducting, no current flows through L and r (see Fig. 20.2), and the beam is at the centre of the screen. At t_0 (Fig. 20.3(a)) when S_1 closes (amplifier is turned on) current i_L flows and rises linearly to deflect the beam towards right side of the raster. The beam reaches the edge of the screen when i_L attains a value equal to I at instant t_1. Since the coil resistance is very small, the voltage v_L' across the coil $\left(i_L r + L \dfrac{di}{dt} \right)$ is nearly constant during this period. Its polarity is positive and magnitude small because of the relatively slow rate of rise of current. At instant t_1, the sharp negative spike of the input signal turns off the amplifier, *i.e.*, switch S_1, opens. With S_1, S_2 both open (Fig. 20.3 (b)) the C-L-r circuit gets into free oscillations with a period T, very nearly equal to $2\pi\sqrt{LC}$ seconds (since r is small). The initial conditions are such that i_L continues to grow further until $v_c = 0$ ($t = t_2$). The entire circuit energy, $\frac{1}{2}L\,(I_{max})^2$ is now in the inductance. T/4 seconds later ($t = t_3$) this energy transfers to C (less a very small loss in r) so that $\frac{1}{2}L\,(I_{max})^2 =$

$\frac{1}{2}C(v_{c\,max})^2$ or $v_{c\,max} = I_{max}\sqrt{\dfrac{L}{C}}$. The capacitance C, being only stray capacitances of the circuit, is very small (a few hundred pico-farads atmost). Therefore $V^*_{C\,max}$ is usually very high, of the order of a few thousands of volts.

After another T/4 seconds ($t = t_4$), once more $V_C = 0$ and $i_L = -I_{max}$. In fact the current varies in a co**-sinusoidal manner from its positive peak to its negative peak in a time equal to T/2. If the switches (S_1, S_2) remained open this process of energy exchange would go on for many cycles, the successive values of $V_{C\,max}$ and I_{max} decaying slowly until the whole energy was dissipated in the resistance r. This process is illustrated in Fig. 20.3 (b) with an expanded scale.

---

* $\frac{1}{2}L\,(I_{max})^2 = \frac{1}{2}C(V_{C\,max})^2$

$\therefore$ $\qquad\qquad\qquad\qquad V_{C\,max} = \sqrt{\dfrac{L}{C}}\,I_{max}$

Assuming C, the distributed capacitance = 750 pF

$\qquad\qquad\qquad\qquad V_{C\,max} = 6\ kV$

** $i_L = I_m \cos \omega_0 t$, where $\omega_0 = \dfrac{1}{\sqrt{LC}}$

and $v_L = V_m \sin \omega_0 t$, where $V_m = \omega_0 L I_{max}$.

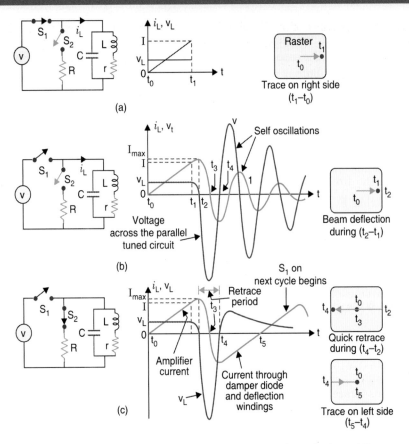

Fig. 20.3. Sequence of operations during scanning of one horizontal line.
(a) amplifier turns on but damper is open (b) amplifier turns off and damper
continues to be open (c) amplifier continues to be off but damper diode turns on.

However, continued oscillations beyond the first half-cycle are not desirable because the continuing oscillatory current in the coil will shift the beam back and forth at the left side of the raster instead of allowing it to trace the next horizontal line at its proper place. Therefore, in order to suppress free oscillations beyond its first half cycle, switch S_2 closes at instant t_4 to dissipate the energy stored in the L, C, r circuit. In fact polarity of the oscillatory voltage changes at t_4 to become positive and thus forward biases the damper diode D_1 (closes switch S_2). With R shunted across L, C, the combination becomes non-oscillatory and the current i_L and voltage v_L then decrease to zero with a time constant approximately equal to L/R where R represents the resistance of the boosted B_+ circuit in series with S_2. The resulting current and voltage waveforms are shown in Fig. 20.3 (c).

L and C are so chosen that the periods of self-oscillations T is twice the horizontal retrace time. Thus as the coil current decreases from positive maximum to zero and reverses to attain its maximum negative value in time $T/2$, the beam gets deflected to the left edge of the raster to complete a fast retrace. The resonant circuit is set at about 60 KHz to provide the desired flyback time of about 8 μS $\left(\dfrac{10^6}{2 \times 6 \times 10^4} = 8\ \mu S \right)$.

Further, the impedance of the B_{++} circuit is so set, or in the circuit under consideration R is so chosen that the current decays to zero in time $(t_5 - t_4)$ which is approximately equal to half the trace period. The decaying current from its maximum negative value to zero deflects the beam to the centre of the screen without any supply of energy from the voltage source. This amounts to recovery of energy, because if not utilized this would not only be wasted but also cause undesired oscillations spoiling the raster.

The switch S_1 (amplifier) closes at t_5 to again connect the voltage source to the LC circuit and just then S_2 (damper) opens to cut-off the damping circuit. The current then smoothly rises to take over deflection of the beam to the right of the raster. This is illustrated in Fig. 20.3 (c). The cycle described above repeats line after line to trace the entire raster.

20.3 HORIZONTAL AMPLIFIER CONFIGURATIONS

Horizontal output amplifiers may employ (i) vacuum tubes, (ii) transistors and (iii) silicon controlled rectifiers. While the end result is same, the mode of operation is somewhat different in each case. Therefore, horizontal output stages employing different devices are described separately. It may also be noted that deflection circuits in colour receivers differ somewhat from corresponding circuits in black and white receivers. However, initially, the discussion is confined to monochrome receivers only.

20.4 VACUUM TUBE HORIZONTAL DEFLECTION CIRCUIT

Fig. 20.4 is the circuit diagram of an earlier version of horizontal output stage used in monochrome receivers. It employs an autotransformer between the amplifier and deflection coils. Each section of the stage is described separately.

(a) Output Amplifier

The amplifier employs a beam power tube having high voltage and power ratings. It is driven by a trapezoidal voltage of approximately 80 V (P-P) which is obtained from the horizontal deflection oscillator. In some circuits a sawtooth voltage that has little or no negative going spike is used to drive the amplifier. In any case, on application of the input signal, grid current flows to develop self-bias. The time constant of the coupling network C_1, R_1 is so chosen that for approximately half of the cycle the input signal drives the grid voltage less negative than cut-off. This results in a linear rise of plate current (see Fig. 20.4) to cause part deflection of the electron beam. With no input drive present, the bias would be zero because there is no provision for self-bias in the

---

* $\frac{1}{2} L (I_{max})^2 = \frac{1}{2} C (V_{C\,max})^2$

∴   ** $V_{C\,max} = \sqrt{\frac{L}{C}}\, I_{max}$

Assuming C, the distributed capacitance = 750 pF

$$V_{C\,max} = 6 \text{ kV}$$

** $i_L = I_m \cos \omega_0 t$, where $\omega_0 = \sqrt{\frac{1}{LC}}$

and $v_L = V_m \sin \omega_0 t$, where $V_m = \omega_0 L I_{max}$.

Chapter 20

cathode circuit. Therefore, the output tube (V_1) should not be operated without grid drive because in the absence of any protective bias the tube will draw excessive current. A slow-blow fractional ampere fuse 'F' is provided in the cathode circuit to protect the tube and transformer against excessive current. The screen grid has the usual decoupling network and a provision to control screen grid voltage. The variation of R_2 changes plate current and thus acts as width control. This is fully explained later along with other methods of width control.

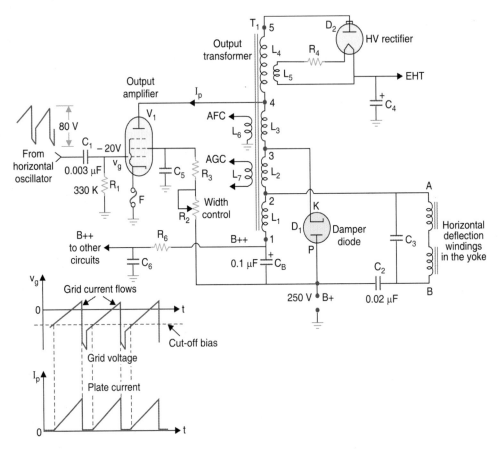

Fig. 20.4. Circuit diagram of a horizontal output stage employing vacuum tubes.

(b) Output Transformer

The circuit employs an autotransformer because of its higher efficiency as compared to one having isolated primary and secondary windings. It is wound on a ferrite core to minimize losses. As shown in Fig. 20.5(a) a small air-gap is provided between the 'U' and 'I' sections of the core to prevent magnetic saturation due to heavy plate current that flows through a section of the winding. The EHT winding and other sections of the autotransformer winding are wound on opposite limbs of the core to obtain higher leakage inductance necessary for third harmonic tuning.

In Fig. 20.4 the winding between 1 and 4 is the primary (L_p) for plate current of the amplifier. This includes L_1, L_2, L_3 and has an inductance of about 100 mH with a winding resistance of about 20 ohms. Winding L_1 is the secondary, to step-down voltage for the deflection coils. Since there is no isolated secondary winding for polarity inversion, all the taps on the auto-transformer have voltages of the same polarity as the plate voltage on the amplifier. The drop in plate current induces voltage of positive polarity across the plate coil for retrace. The winding L_4 at the top steps up the primary voltage to supply ac input to the high voltage rectifier. For a vacuum tube rectifier, its filament power is taken through a small winding L_5 on the same transformer L_4 is a separate winding with a large number of turns of fine wire. It has an inductance of the order of 650 mH. The amount of step-up is limited by the fact that any increase in inductance reduces resonant frequency of the output circuit.

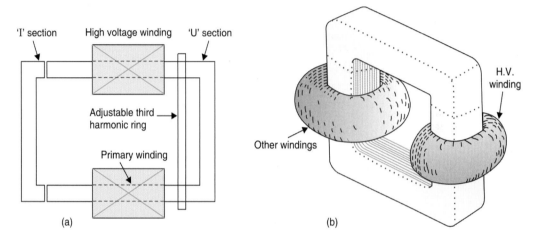

Fig. 20.5. Horizontal output transformer (a) constructional details (b) pictorial view.

Beam deflection. As shown in the waveforms drawn along Fig. 20.4, plate current starts flowing just when the drive signal crosses cut-off bias of the tube. The sawtooth rise of current in the primary circuit induces a negative going rectangular voltage across the secondary winding. This causes a flow of sawtooth current in the deflection windings from terminal B to A to deflect the beam from centre to right side of the raster. Note that the direction of current flow in the yoke windings is opposite to that assumed in the equivalent circuit. However, this does not interfere with the deflection of beam towards the left side of screen because the yoke is so wound that the resulting magnetic field has the correct polarity for causing deflection towards right edge of the raster.

The deflection current is alternating in nature and is coupled to the yoke through capacitor C_2 which blocks the flow of any dc in the deflection windings. This (dc) if allowed will shift the beam from its central location. In some circuits where the blocking capacitor is not used, other techniques are used to bring the beam back to the centre.

Horizontal retrace. Figure 20.6 illustrates input voltage, deflection current and yoke voltage waveforms. As shown, the grid voltage drops sharply to a value much below cut-off soon after the peak of the sawtooth part of the drive voltage is reached. This cuts off the output tube instantly and V_1 ceases to sustain current in the output circuit. The tube remains in cut-off until the grid voltage again rises sharply to permit conduction during the next cycle. The inductance of the output transformer and deflection coils together with their self capacitance act as an L, C circuit that can oscillate at its resonance frequency when excited from a suitable source of energy. When

Chapter 20

plate current drops to zero, the sudden collapse of magnetic field in the deflection circuit generates a high value of induced voltage (typically 6 KV). The energy which thus becomes available in the deflection circuit sets the L, C circuit into ringing at its natural resonant frequency. As explained earlier with the help of an equivalent circuit, the first half cycle of free oscillations results in a quick flyback of the electron beam. It is often necessary to add an additional capacitor (C_3 in Fig. 20.4) to make the resonant frequency nearly equal to 60 KHz for obtaining the desired retrace period.

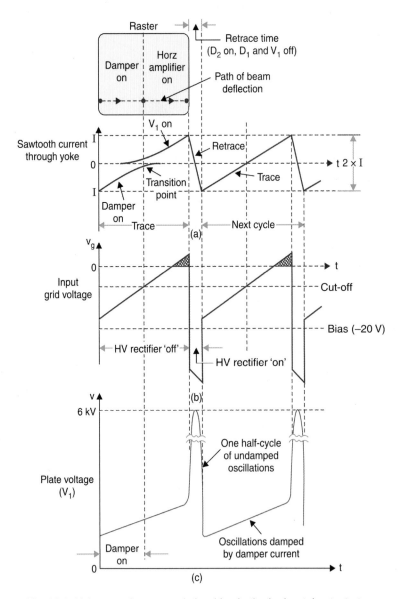

Fig. 20.6. *Voltage and current relationships in the horizontal output stage.*

During retrace the polarity of induced voltage across D_1 is such that it remains cut-off along with the output tube. This is necessary because any flow of current either in V_1 or D_1 will induce a counter voltage that will oppose the current to fall at a fast rate and thus lengthen the retrace period. Thus the first half-cycle of oscillations continues undamped for a fast flyback. If the oscillations are not suppressed after this, they would cause ripples on the scanning current waveform. This results in white vertical bars on the left side of the raster. The bars are on the left side of the raster because the oscillations occur immediately after flyback and these are white because the electron beam scans these areas several times during each horizontal line traced.

(c) Damper Diode

As soon as the polarity of the oscillatory voltage (Fig. 20.6 (c)) reverses, the damper diode (D_1) is forward biased and it conducts to dissipate energy stored in the deflection circuit. The diode, while suppressing undesired self-oscillations after the first half cycle, charges C_B and causes deflection of the beam (see Fig. 20.6) from the left edge of the screen to its centre. This is best explained by considering the two effects separately.

Boosted B_+ supply. The part of circuit in Fig. 20.4 which is associated with the production of B_+ boost is redrawn in Fig. 20.7 (a). Initially when the flyback transformer is not energized the damper diode conducts to supply B_+ voltage to the plate of V_1. However, when the circuit is

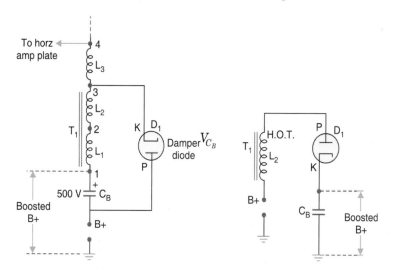

Fig. 20.7 (a). *Boosted B_+ circuit with an autotransformer in the output circuit.*

Fig. 20.7 (b). *Boosted B_+ circuit with a H.O.T. having isolated primary and secondary windings.*

functioning, the diode stays reverse biased most of the time because of high voltage $\left(V_{C_B}\right)$ across capacitor C_B. Immediately after retrace when induced voltage reverses polarity its magnitude is higher than   and so the damper diode is turned on. A part of the damper current completes its path through the boost capacitor C_B and charges it to a still higher potential. In practice the voltage that develops across the capacitor varies between 200 to 600 V. In the circuit under consideration the damper has been tapped up on the autotransformer winding and the voltage that builds up across C_B is about 500 volts. As illustrated in Fig. 20.7 (a) this voltage is in series with the B_+ supply. Therefore the potential difference between the positive side of C_B and chassis ground is around 800 V and is known as boosted B_+ supply. This source is used as a dc supply for the horizontal output amplifier and several other circuits. While supplying current the booster capacitor partially discharges during the time the damper is off. However, this is replenished when the damper conducts during the next cycle. The partial charge and discharge of capacitor C_B at the rate of horizontal scan results in a ripple voltage at this frequency. To remove this ripple a filter circuit, (R_6, C_6 in Fig. 20.4) is often provided.

Figure 20.7 (b) is a boosted B_+ circuit of a horizontal deflection stage employing an output transformer with isolated primary and secondary windings. While the direction of induced voltage can be independently decided but in the circuit under consideration the polarity of the induced voltage in the secondary is opposite to that in the primary. As such, the anode of the damper diode is connected to the upper end of the secondary winding and B_+ supply is connected at its lower end. The diode, initially conducts to charge C_B to the B_+ voltage. However, when the transformer is energized the ac voltage across L_S, soon after flyback, forward biases D_1 which conducts to charge the boost capacitor to a still higher voltage. Thus the entire boosted B_+ voltage appears across the boost capacitor.

The generation of B_{++} voltage explains the principle of energy recovery, where a major part of the energy stored in the coils at maximum deflection is recycled. This improves the efficiency of the horizontal output circuit.

Reaction Scanning. The damper diode D_1 conducts and charges capacitor C_B as explained above. As shown in Fig. 20.8, a part of the diode current also flows through the deflection windings. As C_B charges, the damper diode current decreases. Finally it becomes zero when the voltage across C_B biases the damper diode out of conduction. The diode current that flows through the yoke coils is made to decrease linearly by careful circuit design and this deflects the beam towards the centre. Note that direction of current flow is opposite to that due to plate current of V_1. This is illustrated in Fig. 20.6 (a).

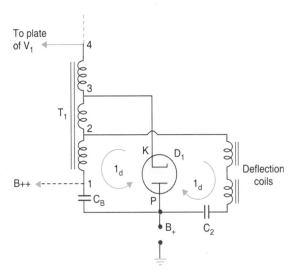

Fig. 20.8. *Reaction scanning by the damper current.*

The circuit is so designed that before the damper current declines to zero, output tube V_1 starts conducting and the resulting current then completes the trace to the right. As shown in the waveform (Fig. 20.6 (a)) the amplifier and damper currents combine to cause horizontal deflection of the picture tube beam. In fact the declining diode current and the rising plate current add to provide a sawtooth current for linear horizontal deflection. This is known as 'Reaction Scanning'.

(d) Generation of High Voltage

During flyback periods the magnitude of induced voltage across the primary (windings L_1 to L_3 in Fig. 20.4) coil is very high. In typical vacuum tube circuits it varies between 3 to 6 KV. By placing a suitably wound coil (L_4) in the magnetic circuit and connecting it to the primary winding in a series aiding fashion, an additional voltage of about 6 to 8 KV is developed by autotransformer action. The corresponding circuit diagram and associated waveshapes are shown in Fig. 20.9. This circuit arrangement results in a total voltage of 9 to 14 KV across two ends of the combined winding. Further, by suitably controlling the distributed parameters of the primary and secondary circuits and utilizing the technique of third harmonic tuning, (discussed in a subsequent section), the total voltage can be enhanced by about 10 to 15 percent. Thus a pulse of the order of 11 to 16 KV is generated. This is fed to the high voltage rectifier D_2 which conducts to charge the filter capacitor to provide EHT supply. It may be noted that the high voltage rectifier conducts for a short time during the flyback period because D_2 is reverse biased at other times on account of 11 to 16 KV dc at its cathode (filament). The conduction of D_2 partially dicharges C_B but it is soon made up when D_1 conducts. Note that while D_2 conducts both V_1 and D_1 remain cut-off to avoid any loading.

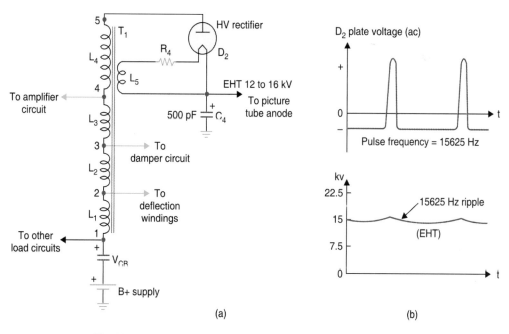

Fig. 20.9. *High voltage (EHT) generation (a) circuit (b) waveshapes.*

(e) Monochrome Yoke

Since the current in the scanning coils deflects the electron beam, the yoke is rated in terms of deflection angle in addition to other essential parameters necessary for generating the required magnetic field intensity. Small screen tubes employ yokes having a deflection angle of 90°. However, the broad screen tubes are provided with 114° deflection angle yokes. Deflection coils suitable for 31 to 61 cm picture tubes are available.

The deflection yoke consists of two sets of coils, one for horizontal deflection and the other for vertical deflection. The coils are independently wound in two halves. These are placed on the yoke at right angles to each other in alternate quadrature segments. The yoke assembly is slipped on to the neck of the picture tube and is oriented to produce a raster parallel to the natural X and Y axes.

In earlier deflection coils additional capacitors were used to balance the distributed capacitance of the two halves and their capacitance with respect to chassis earth. This is necessary if each half coil is to produce equal retrace period for eliminating the possibility of multiple harmonics. However, in the latest designs of scanning coils lumped components are no longer necessary. The two halves of each coils are so wound that they provide the exact desired balance necessary for undistorted scanning. The ratio of inductance to resistance (L/R = time constant) is kept low to achieve minimum permissible limits of distortion. The flared portion of the coils may not exactly match the slope on the neck of the picture tube and, therefore, a pull-back of about 3 mm is provided to permit adjustment of the scanning coils.

Circuit efficiency. The peak energy stored in the inductance L of the deflecting coils is $\frac{1}{2}L(I_{max})^2$ where I_{max} is the peak current. If the damper diode technique is not used, this energy is dissipated in a resistor during each cycle. The power lost is equal to $\frac{1}{8}LI_y^2 f_h$ where I_y is the peak-to-peak amplitude of the deflection current and f_h is the horizontal scanning frequency. For a 61 cm picture tube yoke $L = 2$ mH, and $I_y = 3$ amp. The power consumed is thus about 35 watts. This value is about 25 percent of the total power taken by the television receiver. By the simple technique of replacing the damper resistor R by an efficiency diode the power loss is cut to about one quarter of this value for the same deflection. As illustrated in the waveforms drawn in Fig. 20.6 (a) the total sweep corresponds to a current of $2I$ and yet the peak energy stored in the inductor is $\frac{1}{2}LI^2$ and not $\frac{1}{2}L(2I)^2$. This observation verifies the statement made above that the power loss in the magnetic coils can be cut to one quarter of its previous value by the principle of energy recovery.

The efficiency of the stage is further improved by storing a part of the excess energy in a capacitor (C_B) and feeding it to auxiliary circuits in the receiver.

The ingenious way the flyback pulses are utilized, to produce EHT supply, results in overall economy, because, if the same voltage were obtained from the supply mains, it would require a very bulky and expensive step-up transformer and filters.

The use of first half-cycle of the sinusoidal oscillation is yet another important feature of electronic circuit design which results in such an inexpensive method of obtaining flyback period of a few micro-seconds. All these innovations have resulted in very efficient horizontal output circuits that form part of all present day television receivers.

Chapter 20

20.5 SEQUENCE OF OPERATIONS

It is now obvious that efficient design of the horizontal output circuit and the various auxiliary functions that it performs make its operation quite complex. It is thus desirable to summarize the sequence of operations for a clear understanding of the functioning of horizontal output stage. These are:

(a) Self-bias is used to control conduction of the output tube and is so adjusted that the tube conducts only during a part of the input voltage swing to supply sawtooth current. The rising plate current provides a little more than half of the total rater scanning on the right side of the screen. The damper diode remains reverse biased during this interval. The high voltage rectifier also does not conduct because of the large dc voltage at its cathode.

(b) The output tube suddenly goes to cut-off because of sharp fall of the grid signal voltage. This shock-excites the secondary circuit and it sets into oscillations. The onset of oscillations induces high voltage pulses which make the anode of the high voltage rectifier highly positive. It then conducts during the short pulse periods to provide EHT. Note that the damper diode and output tube continue to be in cut-off during this interval.

(c) The first half-cycle of oscillation is allowed to continue undamped for a fast flyback that brings the picture tube beam to the left side of the raster. During this half-cycle the oscillatory voltage has a polarity which makes the damper diode plate negative. Therefore, both the output tube and damper remain in cut-off while the retrace takes place. Note that the high voltage rectifier goes off soon after the sharp high voltage pulse occurs.

(d) After the first half-cycle of oscillation is over the voltage wave polarity reverses to make the damper diode plate positive. It then conducts (i) to damp oscillations, (ii) to charge the boost capacitor to provide boosted B_+ supply and (iii) to deflect the beam towards the centre of the screen by 'reaction scanning'. The damper capacitor (C_B) which is in series with the damper diode continues to build up charge and finally biases the damper out of conduction.

(e) Before the damper current declines to zero the output tube once again starts conducting to continue the trace to right side of the raster.

20.6 HORIZONTAL AMPLIFIER CONTROLS

A horizontal output amplifier of a monochrome receiver may have a drive control, a width control and a linearity or efficiency control.

(a) Drive Control

This control can be used to adjust the width of the raster by varying peak-to-peak amplitude of the drive voltage to the amplifier. A capacitive or resistive potentiometer is provided in the grid circuit to control the magnitude of input voltage. In modern receivers such a control is not provided and instead width of the picture is adjusted by controlling current in the deflection coils. However, optimum drive voltage is obtained by adjusting output voltage of the horizontal oscillator.

Chapter 20

(b) Width Control

Figure 20.10 (a) illustrates five possible methods of controlling width of the picture. Each control has been suitably numbered for convenience of explanation.

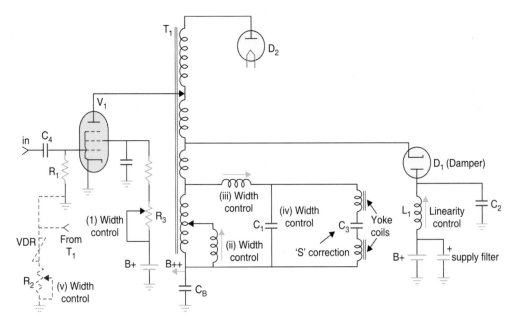

Fig. 20.10 (a). *Horizontal output amplifier controls. Note that in any receiver only one method of width control is used.*

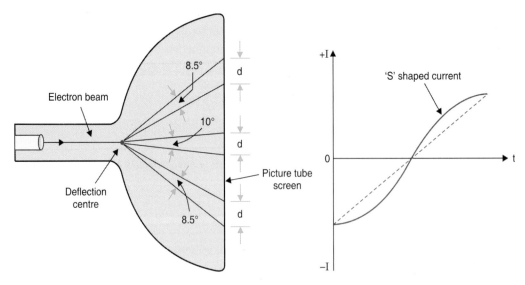

Fig. 20.10 (b). *Effect of flat screen on beam deflection.*

Fig. 20.10 (c). *Modified deflection current.*

(*i*) This control operates by varying screen grid voltage of the amplifier tube. Any increase in screen voltage causes an increase in plate current. This means an increase in deflection energy which is converted into an increase in horizontal deflection. Reducing the screen voltage will have the opposite effect.

(*ii*) In this method a small variable coil (10 to 40 mH) is placed across part of the flyback transformer. It acts as an additional load on the horizontal output stage and absorbs some energy. A reduction in its inductance causes higher current flow through the coil and thus less current flows through the yoke. This results in reduction of picture width.

(*iii*) A variation of the above method is to place a coil in series with the deflection windings. In this case, the width control limits the current that can flow through the yoke. Shorting it will result in maximum deflection. Note that using a coil rather than a resistor as a width control has the advantage of controlling the yoke current without any appreciable loss of power.

(*iv*) In this method of width control, a small capacitor C_1 (50 to 150 pF) is connected across the deflection coils. This lowers the resonance frequency of self oscillations. A lower frequency increases retrace time which in turn causes the high voltage to decrease. Any reduction in the picture tube anode voltage lowers the beam velocity. Thus the beam stays for a longer period under the influence of the deflection field and more deflection is caused. The value of C_1 is chosen for optimum deflection.

(*v*) Another method of width control, part circuit of which is shown in Fig. 20.10 (*a*) by dotted chain lines, accrues from the EHT and raster width stabilization circuit which is often provided in modern receivers. A fraction of the high voltage pulse (see Fig. 20.11) which develops in the flyback transformer is fed to a series circuit consisting of a VDR and a potentiometer (R_2). The VDR rectifies the unsymmetrical pulse voltage. The resulting current develops voltage drops across it and resistor R_2. The amplifier input circuit is modified to include these voltage drops in the self-bias circuit. The negative voltage drop across the VDR adds to the self-bias voltage and counteracts any increase in the pulse voltage. However, drop across the potentiometer which is in opposition to the self-bias voltage can be varied for optimum output circuit current which is turn controls the width of the raster. Thus R_2 can be used as a width control. The stabilizing action of the circuit is explained in Section 20.9.

(c) Linearity Control

Some non-linearity can be caused in the centre of the screen where the deflecting damper current drops and tube current takes over. As shown in Fig. 20.10 (*a*), one method of correcting this consists of a variable inductor connected in series with the damper diode. The coil L_1 forms a parallel tuned circuit with C_2 to be resonant at the horizontal scanning frequency. The switch like operation of the damper (D_1) shock excites the tuned circuit into ringing at the same frequency. The self-oscillatory voltage at the plate of D_1 is modified when damper conducts. It attains the shape of a parabolic ripple voltage of magnitude 30 to 50 volts. The phase and timing of this ac (15625 Hz) voltage controls the conduction time of the damper. Adjusting the coil inductance changes the phase of the anode voltage which in turn changes damper conduction. Thus by properly adjusting the coil inductance a smooth transition from damper current to amplifier current can be obtained so as to have linear deflection of the beam from left edge of the screen to its right side end. In fact when the coil is properly adjusted for linear deflection, dc current of the horizontal output stage will be minimum. This improves efficiency of the stage. Therefore, this coil is also known as efficiency coil and the control as efficiency control.

Chapter 20

Yoke coils of present day monochrome receivers are so well designed that they do not need any linearity adjustment. However, in colour receivers, on account of other complexities adjustment of the efficiency coil is very important and such a control is always provided.

Nonlinearity is also caused due to resistance of deflection coils and on-resistance of the switching device which is not zero. During scans, the voltage available to the yoke inductance decreases as the deflection current rises because of increasing voltage drops across these resistances. This causes appreciable nonlinearity and a correction becomes necessary. This is provided by saturable core reactor (LC) connected in series with the deflection coils (see Fig. 20.11). The saturable reactor (also called linearity coil) consists of a small coil wound on a ferrite core. When the scan current is low, core does not saturate and hence voltage drop across the coil is maximum. However, as the deflection current increases, the ferrite core saturates and drop across it is reduced. Thus the decreasing voltage drop across the saturable reactor compensates for the increasing drop in the deflection windings and the driving device's forward resistance. This ensures an almost constant voltage source across the coil inductance and a sawtooth current flows to cause linear deflection.

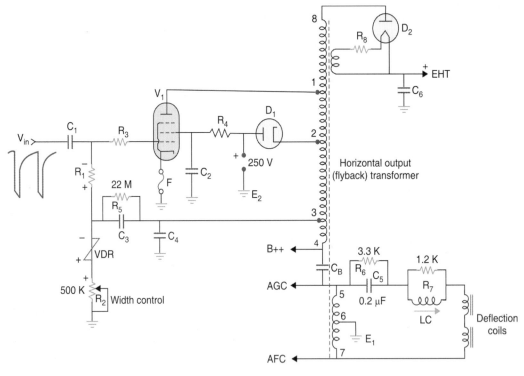

Fig. 20.11. *Improved line (horizontal) output circuit.*

20.7 'S' CORRECTION

A linear variation of current in the deflection coils will cause linear displacement of the spot on the picture tube screen if the face plate has a suitable curvature. However, in modern wide angle picture tubes which have nearly a flat screen, a uniform angular motion of the beam will result in nonlinear deflection. As the deflection angle increases the electron beam has to travel farther to reach the face of the picture tube. Thus more displacement occurs for the same angle of deflection. As shown in Fig. 20.10 (b) a deflection angle of 10° is necessary to scan a distance d at the centre of the screen, whereas an angle of 8.5° is necessary for the same deflection at the edges. Therefore, to produce linear deflection the rate of change of deflection should decrease as the beam approaches edges of the screen. Hence, in order to produce an undistorted picture display, the rate of change of scanning current must be reduced as the deflection angle increases. This means that nearly an 'S' shaped current should flow into the deflection windings. This is illustrated in Fig. 24.10 (c). A capacitor in series with the yoke coils (C_3 in Fig. 20.10(a)) provides the required S-correction. A parabolic voltage develops across the capacitor when the sawtooth deflection current tends to flow through it. This modifies the sawtooth current in such a way that a faster rate of deflection occurs while the beam is at central portion of the screen than when it approaches the edges. Thus the effect of a flat screen is counteracted and linear deflection occurs across the entire width of the faster.

20.8   IMPROVED LINE OUTPUT CIRCUIT

Modern horizontal output circuits employ feedback techniques to achieve linear deflection and make the raster size almost independent of mains voltage variations. Figure 20.11 is the schematic diagram of such a circuit where V_1 is the output tube, D_1 the damper or efficiency diode and D_2 the EHT rectifier. It may be noted that the damper diode is in the primary circuit of the autotransformer.

Circuit Operation

The circuit employes self-bias technique to control conduction and cut-off periods of the output tube. The resistor R_3 prevents excessive grid current flow during positive peaks of the trapezoidal drive voltage. In each cycle, when V_1 conducts, the linearly rising current in winding 1, 2 develops a rectangular voltage across the secondary (5, 7) winding. The resulting sawtooth current through the yoke coils deflects the beam from centre to right side of the raster.

As soon as the grid drive voltage goes negative, tube is cut-off and current flow ceases both in V_1 and D_1. The standing current in the primary winding shock excites the associated L, C circuit. The resulting variation of the self-oscillatory current from its positive to negative peak during the first half cycle of oscillation deflects the beam to the left side of the raster thus providing a fast flyback. The associated induced voltage keeps the cathode of D_1 highly negative during this period and no current flows through V_1 and D_1 during flyback. As soon as the second half cycle of self oscillation begins, cathode of D_1 becomes negative and damper diode conducts via winding 2, 4, capacitor C_B, winding 5, 6, ground points E_1, E_2 and B_+ supply. The damper current induces a voltage in winding 5, 6, which circulates a sawtooth current in the yoke to deflect the beam from left to middle of the screen.

The current that flows through C_B charges it with the polarity marked across it. As soon as potential at the cathode of D_1 becomes equal to B_+ supply, V_1 takes over and conducts again to deflect the beam from middle to right side of the raster. This sequence continues till the boost capacitor C_B is charged to the maximum possible value. Once C_B is fully charged it becomes the supply source for V_1. However, B_+ voltage source continues to supply make-up current during each cycle to replenish part energy lost by C_B while feeding V_1 and other auxiliary circuits.

Chapter 20

The capacitor C_5 in the secondary circuit is for 'S' correction while the linearity coil L_C compensates for any nonlinearity introduced by resistive voltage drops in the output circuit and supply source. Resistors R_6 and R_7 connected across C_5 and L_C respectively suppress any self-oscillations when the circuit is shock excited by deflection current flow.

The voltage induced in winding 1, 8, during flyback together with the voltages across other sections of the autotransformer becomes the high voltage ac source which is rectified by diode D_2 to provide EHT supply. Many line output circuits employ a stack of selenium rectifiers (for example TUS 15) instead of a vacuum tube rectifier.

Third Harmonic Tuning

In the equivalent circuit shown in Fig. 20.12 (a) L_1 is the effective primary inductance and C_1 the primary capacitance. The inductance L_K represents leakage inductance of the EHT winding when referred to the primary side. Similarly C_K is the effective capacitance across L_K while C_2 accounts for the capacitance between EHT winding and ground. The leakage inductance is deliberately made large so that it resonates with the stray capacitance at about third harmonic (2.8 times) of the flyback frequency. The correct tuning is obtained by varying location of the third harmonic ring (see Fig. 20.5) between the two winding and observing voltage waveshape on a cathode ray oscilloscope. When tuned properly, the phase of the third harmonic voltage cancels part of peak voltage at the plate of V_1. At the same time, phase of the third harmonic ringing tends to increase the amplitude of high voltage pulse delivered to the EHT rectifier. The two effects are illustrated in Figs. 20.12(b) and (c). While the reduction in peak voltage by about 10% at the plate of output tube permits the use of a tube having lower voltage breakdown ratings, the enhanced high voltage pulse yield 15 to 20% higher EHT voltage. All this results in increased economy and more efficient operation.

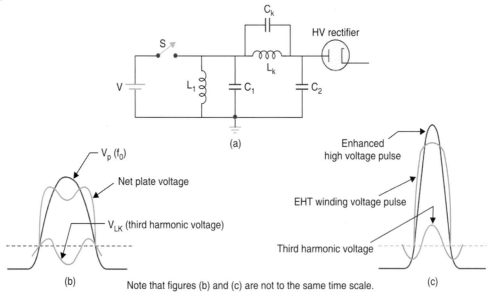

(a)

(b)

(c)

Note that figures (b) and (c) are not to the same time scale.

Fig. 20.12. *EHT winding equivalent circuit and third harmonic tuning.*
(a) Equivalent circuit with HV (EHT) circuit referred to the primary side
(b) Effect of third harmonic tuning on the plate voltage (c) Increased HV
pulse amplitude due to in-phase third harmonic voltage.

20.9   OUTPUT CIRCUIT STABILIZATION

Output tube's operating conditions influence EHT supply and raster width. This is stabilized by regulating the plate current through a feedback loop from the output circuit. A component of the grid bias voltage is obtained by rectifying a part of the pulse voltage which develops across the output transformer. The feedback circuit in Fig. 20.11 includes a VDR and a potentiometer.

A VDR is a semiconductor the resistance of which decreases as the voltage across its terminals increases. It has symmetrical nonlinear V-I characteristics (see Fig. 20.13) and therefore allows equal current flow in either direction if a symmetrical voltage is impressed across it. However, if an unsymmetrical voltage of sufficient amplitude is applied, rectification takes place. This implies that unequal current flows in forward and reverse directions. Thus a net voltage develops across the VDR and its value depends on the magnitude of the impressed voltage.

In the circuit under consideration (Fig. 20.11) a direct current flows through VDR and R_2 because of dc voltage across the boost capacitor. On this direct component is superimposed a pulsating current which is derived via C_3 from the flyback pulses. This is illustrated in Fig. 20.13 where the voltage and current waveshapes have been assumed to be rectangular for the sake of simplicity. As indicated in Fig. 20.13 the working point 'w' is determined by the mean value of this current. The voltage across the VDR fluctuates and its mean value is negative. This actually means that voltage drop across the VDR decreases because the magnitude of reverse current is more than the forward current.

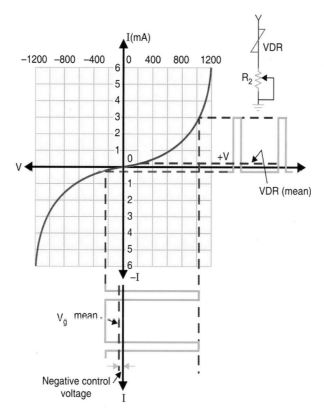

Fig. 20.13. *VDR characteristics and development of control voltage.*

The pulsating negative voltage is filtered by the RC network formed by grid-leak resistor R_1 and capacitance of the oscillator drive circuit. Thus an additional dc voltage proportional to the pulse amplitude becomes part of the amplifier grid bias. The potentiometer R_2 is set for optimum bias so that both the EHT voltage and raster width are of correct values. It due to any reason the EHT voltage rises, it is obvious that the magnitude of the negative voltage across the VDR will increase. However, this cannot happen in practice because an increase in the negative control voltage reduces EHT pulse voltage on account of reduction in tube plate current. Similarly any decrease in EHT pulse will be counteracted by a decrease of the negative control voltage. Thus any tendency for increase in raster width or EHT voltage which may occur due to mains voltage fluctuations or any other reason is prevented by adding or subtracting a control voltage to the self-bias circuit.

Width Control

The steady current that flows through the VDR and R_2 fixes the threshhold value for proper functioning of the VDR. Since the voltage drops across the VDR and R_2 are opposite in polarity, the potentiometer R_2 can be varied to control the operating point of V_1 for optimum deflection current. Thus R_2 acts as a width control.

The capacitor C_4 is chosen to obtain the correct flyback period (frequency to self-oscillations) and thus the required EHT voltage.

The boosted B_{++} supply which is developed in the line output stage feeds the vertical (frame) output circuit in addition to the first and second anodes of the picture tube. The feedback circuit tries to maintain the boosted B_{++} supply at a constant value. However, if any small changes occur, they affect the horizontal and vertical output stages equally thus maintaining correct aspect ratio of the reproduced picture.

Auxiliary Windings

Instead of providing separate windings, AGC and AFC circuits are fed from tappings on the autotransformer. The AFC circuit receives voltage that develops across tap 7 and ground. Similarly the AGC circuit is connected across tap 5 and ground. The connections can be interchanged between taps 5 and 7 depending on the polarity of the pulses needed at the AFC and AGC circuits.

20.10 TRANSISTOR HORIZONTAL OUTPUT CIRCUITS

The two major disadvantages of a vacuum tube horizontal output stage are (i) lot of power is wasted in the tube's heater circuit and (ii) a vacuum tube has a relatively high plate resistance that necessitates the use of a matching transformer between the tube and yoke coils. In a transistor circuit such problems are not there. However, the output transistor while having a high current rating must be able to switch at fast speeds. This is difficult to obtain in the same transistor and proved to be the major obstacle in transistorizing the horizontal output stage. While transistors are now available which can switch a peak power of 200 watts and have an average power rating of 20 to 50 watts but their breakdown voltage capability $V_{CE(max)}$ and $V_{CB(max)}$ is limited to about 1500 volts.

Output Transformer

The output transformer of a transistor line output stage has a somewhat different role than that used in a corresponding vacuum tube circuit. In tube circuits the output transformer is primarily responsible for obtaining rapid retrace because of its self-oscillations. In addition, it is used for providing impedance match to the yoke windings and high voltage pulses to the EHT rectifier. In transistor circuits, retrace time is determined by the resonant circuit formed by effective inductance of the output circuit and an external capacitor connected in the collector circuit of the transistor. This capacitor is known as the flyback capacitor. The output transformer is instead tuned with its distributed capacitance to about third harmonic of the yoke oscillations. This as explained earlier reduces collector to emitter voltage and increases amplitude of high voltage pulses.

Boosted B$_+$ Voltage

In transistor circuits the high dc voltage is developed in a different way. While the circuit has a damper diode but there is no boost capacitor. Instead, the B_+ needs of transistor circuits are met from taps on the output transformer. The dc pulses from the taps are rectified and filtered to obtain different dc voltage.

Driver Stage

In vacuum tube circuits the output trapezoid of the horizontal deflection oscillator is a sufficient amplitude to drive the output stage directly. In transistor circuits a driver stage is provided between the oscillator and output stage. This stage is a buffer amplifier that isolates the oscillator from the output stage and has sufficient gain to meet the input power requirements of the horizontal output stage. The drive power is usually large. For example, during the interval when output transistor is on, the base current may exceed 400 mA. Note that because of low impedance values in transistor circuits, the drive voltage is usually an asymmetrical rectangular wave instead of a trapezoid.

20.11   TRANSISTOR LINE OUTPUT STAGE

Figure 20.14 is the circuit of a basic line (horizontal) output stage. The corresponding input and output signal waveforms are illustrated in Fig. 20.15. The damper diode D_1 and flyback capacitor C_1 are connected directly across the collector and emitter of the transistor. The value of the flyback capacitor C_1 is so chosen that it forms a resonant circuit with the equivalent inductance L of the deflection circuit at a frequency whose period is twice the retrace interval. During the period marked T_1 in Fig. 20.15(c) the input signal forward biases the output transistor into full conduction and the resulting current through the yoke coil deflects the beam to the right of the raster. The damper diode remains reverse biased by the V_{CC} supply during this interval. At instant t_1 the sharp negative pulse at the base of Q_1 turns it off and the current sharply declines to initiate retrace. The beam returns to the left of the raster during the first half cycle of self-oscillations. The self-induced voltage across the coil keeps the damper diode back-biased during this (T_2) interval.

The polarity of the induced voltage reverses at instant t_3 and forward biases D_1 which then conducts to deflect the beam to the centre of the screen. The energy stored in the coil is thus expended and any subsequent oscillations are suppressed. The transistor Q_1 continues to be off for most of the time interval T_3, i.e., $(t_4 - t_3)$. A little before the instant t_4 (see Fig. 20.15(a)) the sharp rise of base voltage forward biases Q_1 and it conducts heavily to continue deflection of the beam to the right of the raster. The diode D_1 immediately ceases to conduct because of the disappearance of self-induced voltage in the yoke. The V_{CC} supply, however, keeps it off during this interval and thus the cycle repeats to scan the entire raster.

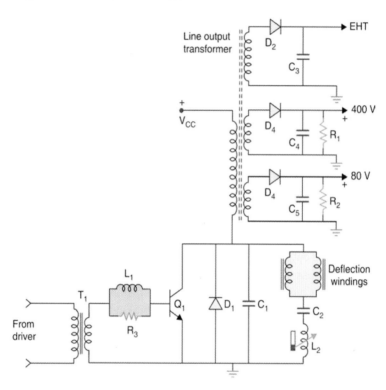

Fig. 20.14. *Transistor line output circuit.*

The capacitor C_2 is for 'S' correction. The correction occurs because of the voltage which develops across it when sawtooth current flows through the coils. This capacitor also blocks dc to the yoke coils which would otherwise cause heating of the coils due to dc resistance of the windings. The high voltage pulses are rectified to provide EHT supply. A stack of silicon diodes is usually used as a high voltage rectifier.

As explained earlier, unlike tube circuits, the efficiency (damper) diode does not provide any boosted B_{++} supply. However, high voltage requirements for the picture tube anodes and other auxiliary circuits are obtained by transforming flyback pulses and rectifying them. As shown in Fig. 20.14 suitable ac pulse sources are rectified by diodes D_3 and D_4 to produce dc voltages of 400 V and 80 V respectively. The capacitors C_4 and C_5 are the filter capacitors to remove 15625 Hz ripple from the rectified outputs. Resistors R_1 and R_2 act as bleeder resistors to discharge the capacitors when the receiver is switched off.

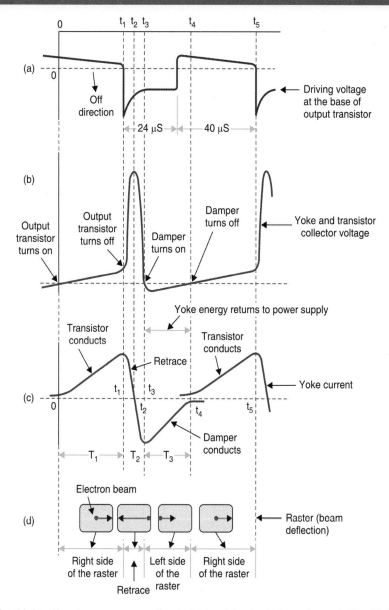

Fig. 20.15. *Waveforms corresponding to horizontal output stage shown in Fig. 20.14.*

20.12 HORIZONTAL COMBINATION IC CA 920

The 'BEL' CA 920 is a 16-pin dual-in-line plastic package, line (horizontal) oscillator combination IC which contains all the small signal stages for horizontal deflection in television receivers. The various functions performed by the device are (*i*) sync pulse separation with optional noise gating, (*ii*) current controlled line oscillator, (*iii*) phase comparison between sync pulses and oscillator output, (*iv*) high synchronizing noise immunity and (*v*) phase comparison between the oscillator waveform and middle of the line flyback pulse for automatic correction of switching delays in the horizontal driving and output stages.

Chapter 20

Phase comparison can take place either between the oscillator signal and sync pulse or between the sync pulse and line flyback signal. With phase comparison between sync pulse and line flyback signal, the various inherent phase shifts that are associated with the oscillator, line driver and line output stages are automatically compensated for. The disadvantage, however, of this method is that the amplitude, shape and duration of the line flyback pulse are dependent on picture tube beam current and other dynamic variations. Hence design on the filtering circuits cannot be optimum both for good synchronizing and noise immunity. If, however, phase comparison between sync pulse and oscillator signal is used, the controlled phase position is independent of the amplitude, shape and duration of the line flyback pulse. The difficulty in this case is that the static and dynamic changes in the switching times of the output stage may now appear and suitable measures have to be taken for their correction. Simple discrete circuit for flywheel synchronization will not be able to provide a fully satisfactory solution because of the contradictory requirements involved. By incorporating two control loops, one comparing the sync pulse and the oscillator signal and the other the oscillator signal and the line flyback pulse, the problem is comprehensively solved in this IC (CA 920).

Typical Application Circuit

A typical drive circuit using CA 920 for horizontal deflection in TV receivers is shown in Fig. 20.16. Composite video signal having peak to peak amplitude of 1 to 7 volts with positive going sync pulses is applied at the input. The low-pass filter R_1, C_1 with cut-off frequency of about 500 KHz filters out noise present without introducing excessive delay to the signal. Before this video signal is applied to terminal (8) of the IC, proper clamping should be done so that only the sync tips drive the sync separator stage. This is achieved by the circuit consisting of R_2, R_3, C_2 and C_3. R_2, C_2 along with base-emitter junction of the input transistor within the IC, clamps the video signal while R_3, C_3 combination ensures that only since tips drive current into terminal (8) irrespective of video signal amplitude. A good set of values for R_2, R_3, C_2 and C_3 are 1.5 M, 10 K, 100 KPR and 1 KPF respectively. Impulsive noise present in the composite video is separated from the signal by the circuit D_1, R_7, to R_{11}, C_6 and C_7 and applied to terminal (9) of the IC. This voltage inhibits the sync separator stage under impulsive noise conditions thus making pin (9), the noise gating terminal of the IC. The composite sync at terminal (7) is processed into vertical sync by means of the integrator R_4, C_4. The time constant of this integrator is chosen to be around 100 μs in order to obtain as large an amplitude as possible for the vertical sync pulses and for maximum rejection of horizontal sync pulses. The horizontal sync pulses are separated by the differentiator R_5, R_6, C_5 combination. This circuit provides voltage division, and has a time constant of approximately 8 μs.

A value of 10 KPF for C_{osc} and 2.7 K for R_{osc} are chosen to be connected at terminals (14) and (15) respectively to determine free running frequency of the oscillator and to obtain optimum transconductance for the oscillator. The free running frequency can be adjusted by the resistor network R_{12}, R_{13}, R_{14} and R_H. The capacitors C_8 and C_9 act as low-pass filter in this control loop. The control current I_{12} is converted into a dc voltage by this filter. This dc voltage drives a controlling current determined by R_{16} connected between terminals (12) and (15). A value of 33 K for R_{16} is optimum in order to achieve equal pull-in and hold ranges.

The flyback pulse derived from the line output transformer is fed at terminal (5) of the IC through a suitable resistor to provide approximately 1 mA of current. The control current I_4 at terminal (4) of phase comparator is converted into a dc voltage by R_{18}, C_{11} combination and is connected to pin (3) to affect the controlled phase shifter. Since this phase comparator suffers no interference from noise signals, it is possible to use a small time-constant for the filtering of current I_4 and thus to achieve a short correction time. A value of 22 KPF is selected for C_{11} keeping in view a short correction time and good stability.

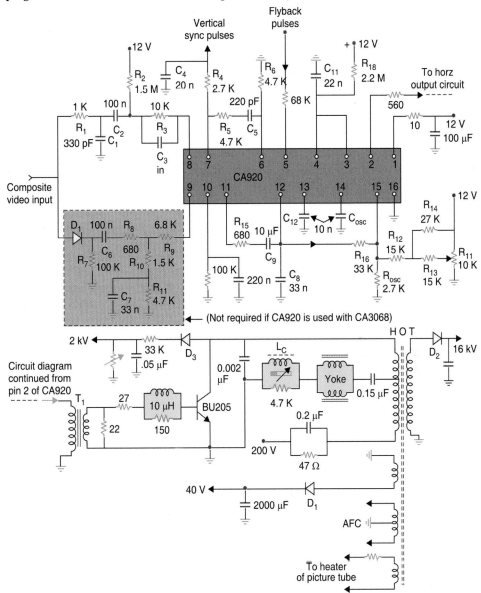

Fig. 20.16. Typical circuit of the horizontal (line) output stage employing
BEL CA920 in conjunction with power output transistor BU205 and associated circuitry.

The output, available at terminal (2) of the IC can drive both transistor and thyristor (SCR) output stages. The 12 volts power supply is fed at pin (1). This voltage is derived from a zener regulator. A 10 ohm resistance is connected in series with pin (1) to limit the supply current in case of increased supply voltage. The total supply current drawn by the entire circuit is about 25 mA at 12 volts. This circuit when used in conjunction with a typical line output stage using transistors BD 115, BU 205 and line output transformer AT 2048/12 provides equally large pull-in and hold ranges of about ± 4.5 percent.

The output circuit shown in Fig. 20.16 in conjunction with this IC employs a BU 205 power transistor and the associated output transformer. This set-up is suitable for driving the yoke of a 51 cm picture tube. The stage operates from a 200 V dc source obtained from a regulated power supply. Besides providing beam deflection and other outputs, the circuit produces an auxiliary low voltage of 40 V required for audio, and RF-IF stages. The stage also provides heating power to the picture tube filament circuit.

20.13 HORIZONTAL DEFLECTION CIRCUITS IN COLOUR RECEIVERS

While the basic mode of operation is the same, horizontal deflection circuits in colour receivers are somewhat different than those provided in black and white receivers. The following are the main distinguishing features of colour receiver line circuits.

(a) A colour receiver picture tube may require as high as 25-KV at its final anode but such a requirement seldom exceeds 18 KV in a monochrome receiver. In addition, many colour receiver picture tubes need another high dc source of about 6 KV for electrostatic focusing of the electron beam. Since these dc supplies are developed in the horizontal output circuit, colour receivers employ special high voltage rectifier circuits to generate them.

High Voltage Tripler Circuit

To generate a very high voltage dc source a voltage-tripler circuit is often used. This arrangement minimizes insulation requirements of high-voltage transformers. Figure 20.17 shows the schematic of such a tripler circuit. The entire circuit is potted and can only be replaced as a unit in the event of failure. Though there are five rectifiers in the module, two are used to couple the rectified dc from one rectifier to another. The other three rectifiers and associated capacitors make up the voltage tripler. Rectifiers instead of resistors are used to couple the dc voltages because there would be loss of both voltage and power if resistors were used.

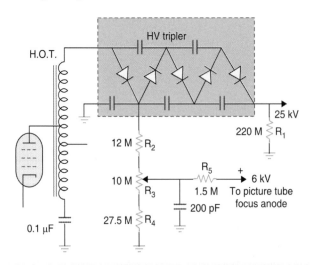

Fig. 20.17. High voltage tripler circuit.

R_1 and R_2 are special high voltage resistors. R_1 (220 M) is a bleeder that discharges the high voltage picture tube capacitor after the receiver is switched off. The other high-voltage resistor R_2 is connected to a tap on the tripler circuit. This together with potentiometer R_3 and resistor R_4 forms a voltage divider network to supply high voltage dc to the focus electrode of the picture tube.

All the high-valued resistors can be adversely affected by handling. Leakage paths caused by fingerprints may result in a dramatic drop in resistance. Therefore, these parts should be handled as little as possible, and when necessary should be held by their leads. If dust or dirt accumulates on them, it should be removed by washing them with alcohol or any other 'Freon' type cleaning solvent.

(b) The electron beam current of a black and white picture tube generally does not exceed 500 μA. The beam current of a colour picture tube may be two to three times this value. At maximum brightness a colour picture tube may dissipate as much as 25 W. The internal resistance of the high voltage power supply in the line output circuit is usually large and variation of brightness results in blooming and poor focus unless suitable regulator circuits are employed. Since the beam current of colour picture tubes is very large such effects are more pronounced in colour receivers.

High Voltage Regulation

In an attempt to keep the picture tube anode voltage almost constant under varying brightness conditions many monochrome and all colour receivers employ special high voltage regulator circuits. One such arrangement which is used in vacuum tube circuits was discussed in Section 20.9 while explaining an improved horizontal output circuit.

Solid state colour receivers do not usually employ direct high voltage regulators, because, compared to vacuum tubes, transistor line output circuits have very low impedance levels. As such, changes in high voltage load current do not cause large changes in dc output voltages. Nevertheless, most transistor colour receivers do incorporate some form of high voltage regulation. It may be noted that it is not practicable to regulate the bias of an output transistor as is often done in tube circuits because the transistor acts more like a switch than an amplifier. Therefore, regulation of sweep and high voltage in transistor colour receivers is often done by regulating B_+ supply voltage that feeds the horizontal output stage. Such regulator circuits are described in Chapter 24 which is devoted to receiver power supplies.

(c) In order to reduce X-ray radiation from a television receiver under breakdown conditions, special hold-down circuits are used to automatically disable the rectifier when a fault occurs which would tend to increase excessively the generation of EHT supply. While most monochrome receivers do not include such circuits but a hold-down circuit is a must in colour receivers where the EHT voltage is very high.

Hold-down Circuit

An example of a high voltage hold-down circuit used in transistor colour receiver circuits is shown in Fig. 20.18. The circuit is so designed that an increase in high voltage above 30 KV causes an SCR to conduct and ground the horizontal oscillator supply voltage. This causes the horizontal oscillator to cease functioning resulting in a loss of both, the high voltage and the raster.

Chapter 20

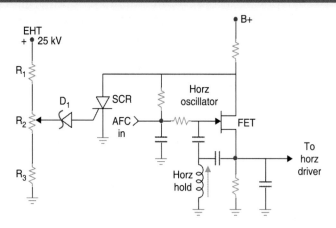

Fig. 20.18. High voltage hold-down circuit.

The gate of the SCR is fed from a voltage divider which is placed directly across the EHT supply. The zener diode (D_1) placed in series with the gate circuit ensures that the SCR will turn on precisely at the desired voltage. As soon as the voltage exceeds 30 KV, zener break-down occurs and a positive gate voltage turns on the SCR. The potentiometer R_2 is set for optimum operation of the circuit. The SCR will remain on, until either the receiver is turned off or the cause of excessive high voltage is removed.

(d) As shown in Fig. 20.19 colour receivers require dynamic convergence and anti-pincushion distortion circuits. These are in part activated by pulses obtained from the horizontal output amplifier. Such circuits are not required in monochrome receivers.

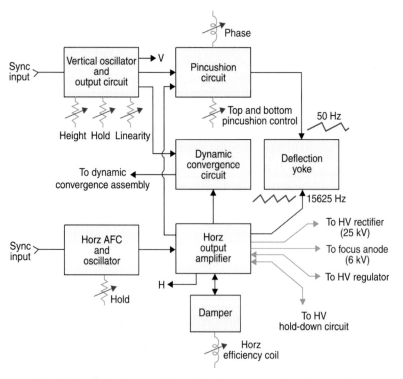

Fig. 20.19. Block diagram of the deflection system of a colour receiver.

(e) Since the sawtooth deflection current requirements in colour receivers are much higher than monochrome receivers, horizontal output amplifiers of colour receivers need a drive voltage which is almost double than that needed to drive corresponding amplifiers in monochrome receivers. As a consequence line output amplifiers of colour receivers consume nearly double power than similar monochrome receiver circuits.

20.14 SCR HORIZONTAL OUTPUT CIRCUIT

In some colour receivers SCRs (silicon controlled rectifiers) are used to produce a sawtooth yoke current in horizontal output circuits. Such a circuit employs a pair of SCRs in conjunction with two special diodes and a number of resonant circuits to form a system of switches which generate the sawtooth deflection current.

An SCR deflection circuit is operated directly from the rectified ac line voltage. It does not need a matching transformer and the purpose of fly-back transformer is to develop high voltages pulses for EHT and other circuits. It also provides, like other line circuits, pincushion correction, convergence, high voltage regulation and high voltage hold-down. However, SCR deflection circuits need exact component values and the tuning of various resonant circuits is quite critical.

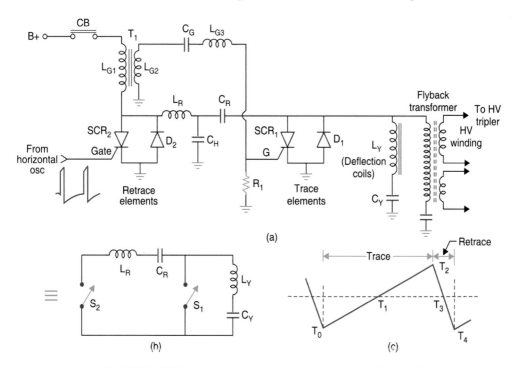

Fig. 20.20. SCR horizontal output stage (a) basic schematic circuit (b) simplified equivalent circuit (c) trace and retrace intervals.

The resonant circuits formed by various combinations of coils and capacitors in the circuit (Fig. 20.20(a)) are allowed to oscillate alternately and later turned off at appropriate moments by the switching action of silicon controlled rectifiers and associated diodes. Since a large number of resonant circuits, each tuned to a different frequency, are used to obtain different segments of the sawtooth current, the circuit operation is quite complicated. Therefore the explanation that follows has been greatly simplified for a better understanding of the basic principle of its operation.

Circuit Components

In Fig. 20.20(a) two SCR units are used in the horizontal output circuit. SCR_1 conducts for trace periods while SCR_2 turns on during retrace periods. Switching of SCR_2 for retrace is called the commutating operation. Each SCR has a special fast recovery diode with inverse parallel connections. The purpose is to allow current in both directions from cathode to anode in either the SCR or its diode, depending on which anode is positive. When the diode conducts its low resistance drops the SCR anode voltage to the cur-off level. During trace time SCR_1 is on and SCR_2 is off. In effect, SCR_1 and D_1 together comprise a controlled single pole-single throw switch (S_1) while SCR_2 and D_2 form another similar switch (S_2) as shown in Fig. 20.20(b). The components L_R, C_R, C_H, and C_Y supply the necessary energy storage and timing functions. The inductance L_{G1} gives a charge path for C_R and C_H from B_+, thereby providing a means to 'recharge' the system from the power supply. The secondary winding (L_{G2}) on transformer T_1 supplies necessary gating current to SCR_1. The capacitor C_H controls the retrace time because it is charged to B_+ voltage through L_R. Pulses from the horizontal oscillator at 15625 Hz turn on the retrace SCR at appropriate moments. The conduction in SCR_2 cuts-off the retrace silicon controlled rectifier SCR_1. Thus the trace and retrace periods are controlled and synchronized with the horizontal oscillator.

Circuits Operation

Figure 20.20(b) is the simplified equivalent circuit of the SCR deflection circuit. The waveforms shown in Fig. 20.20(c) illustrate approximate time intervals during which switches S_1 and S_2 (trace and retrace SCR-diode combinations) close or open to circulate sawtooth current in the deflection windings.

If the circuit has been functioning for many cycles and retrace has just been completed, the yoke (L_Y) magnetic field is at its maximum value and the electron beam is on the left side of the raster. This corresponds to time T_0 in Fig. 20.20(c) where the induced current through L_Y is at its peak negative value.

Trace Time

Referring to Fig. 20.20(b), during the first half of trace time (T_0 to T_1) switch S_1 (SCR_1 and D_1 combination) is closed causing collapse of the magnetic field around the yoke inductance (L_Y). The resulting current deflects the beam to approximately middle of the screen and charges the capacitor C_Y. During the second half of the trace time interval (T_1 to T_2) the current in the yoke circuit reverses because the capacitor C_Y discharges back into the yoke inductance L_y. This current causes the picture tube beam to complete the trace. Note that SCR_1 and D_1 combination (S_1) continues to provide current conduction path during the entire trace interval when current changes from negative to its positive maximum value.

Retrace Initiation

The forward scan stops at T_2 because a pulse applied to SCR_2 from the horizontal oscillator triggers it into conduction (S_2 closes). Both S_1 and S_2 remain closed for a very short interval. The charge previously stored on C_R is released into the commutating circuit comprising of L_R, C_R and

C_H. Thus the current through SCR$_1$ rapidly falls to zero causing the turning off of switch S_1 and the initiation of flyback. As stated earlier the frequency of various resonant circuits is so chosen that trace and retrace intervals are correctly obtained.

Retrace

With S_1 open and S_2 closed current flows through a series resonant circuit comprising of L_R, C_R, L_Y and C_Y. The natural resonant frequency of this circuit is much higher than that of the yoke circuit L_Y and C_Y because the value of capacitor C_R is much smaller than that of C_Y. As a result a fast flyback becomes possible. The retrace current comes to zero (T_2 to T_3) at the midway point of the flyback interval, and just then SCR$_2$ stops conduction. However, diode D_2 now becomes forward biased due to reversal of polarity of the oscillatory voltage. Thus S_2 continues to be closed and a retrace current flows in the opposite direction to complete the flyback at instant T_4. This action (current flow) effectively transfer the energy stored in C_R back to the inductance L_Y of the horizontal deflection coils. At time T_4 the diode D_1 becomes forward biased due to reversal of voltage polarity (S_1 closes) and shortly afterwards S_2 opens. The field about the yoke inductance (L_Y) starts to collapse and the resulting current again commences the next trace period. The cycle repeats to provide horizontal scanning at the rate of deflection frequency.

Note that during flyback the primary of T_1 is connected between B_+ source and ground via SCR$_2$ and D_2. When D_2 stops conduction, L_{GI} is disconnected from ground and C_R charges through this winding from the B_+ supply to replenish energy in the deflection coil circuit. It may also be observed (see Fig. 20.20(a)) that the voltage developed across L_{GI} (primary of T_1) during the charging of C_R is coupled to the gate of SCR$_1$ through the secondary of T_1 and waveshaping network comprising of C_G, L_{G3} and R_1. In turn, the gate waveform that is generated possesses adequate amplitude to enable SCR$_2$ to conduct while its anode is sufficiently positive.

REVIEW QUESTIONS

1. Why is the design of the horizontal output stage very much different than that of the vertical stage, while their functions are very similar to each other ? What are the special features which make the operation of the line output stage very efficient ?

2. Explain fully with a suitable functional diagram and waveshapes, how the first half-cycle of self-oscillations in the output circuit is used to obtain a fast retrace of the beam. How is the continuation of ringing suppressed ?

3. Show how the energy stored into the deflection windings is recycled to generate boosted B_{++} supply.

4. Explain with a circuit diagram how the high voltage pulses, induced in the output transformer windings, are used to generate EHT supply.

5. What do you understand by reaction scanning ? Explain step by step the sequence of operations in the line output stage for reaction scanning. Draw suitable wave-shapes and diagrams to illustrate your answer.

6. Explain the functions of linearity and width controls normally provided in the horizontal output stage. Describe their operation by drawing relevant portions of the output circuit.

Chapter 20

7. With reference to Fig. 20.11 explain how the feedback circuit maintains a constant EHT supply and width of the picture through the voltage drop across the VDR. What is the function of potentiometer R_2 ?

8. What do you understand by third harmonic tuning in the line output transformer ? Justify that is results in lesser peak power from the output tube and more ac voltage from the EHT rectifier.

9. In what respects a transistor line output stage is different from a corresponding vacuum tube circuit ? Why is a driver stage necessary in transistor output circuits ? Why is a separate flyback capacitor provided in such circuits ?

10. Draw the circuit diagram of a typical horizontal output stage which employs a transistor and explain how it functions to provide linear horizontal scanning. What is 'S' correction and how is this applied ?

11. Discuss briefly the main features of the horizontal combination IC, CA 920. With reference to Fig. 20.16 explain the circuit operation of this IC.

12. Describe fully the distinguishing features of horizontal deflection circuits of colour receivers. Why is the deflection current very high in colour receivers ?

13. Draw the circuit of a high voltage tripler and explain how EHT voltage of the order of 25 KV is developed. Why is it necessary to handle high valued resistors very carefully ?

14. Why is a hold-down circuit necessary in colour TV receivers ? Draw the circuit diagram of such a circuit and explain how it operates to disable the horizontal oscillator when EHT exceeds the prescribed upper limit.

15. Draw basic circuit of a horizontal output stage employing SCRs. Explain how the circuit functions to provide linear trace and retrace periods. What are the merits and demerits of such deflections circuits ?

Chapter 20

21

Sound System

A television signal transmitted from a broadcast station consists of an amplitude modulated RF picture carrier and a separate frequency modulated RF sound carrier both within the 7 MHz channel bandwidth alloted to the station. This signal is intercepted by the receiver antenna and coupled through a transmission line to the RF section in the tuner. The block diagram of Fig. 21.1 shows various stages through which the sound signal passes in a monochrome and a colour receiver, before it is finally delivered to the loudspeaker.

In the tuner FM sound signal is amplified along with the picture signal and then transferred to the sound IF frequency from the channel carrier frequency. The sound IF signal together with its sidebands is passed through the IF amplifiers without much amplification. At the video detector it is made to beat with the strong picture IF carrier frequency to transfer the modulating audio signal to yet another sound IF frequency which is equal to the difference of the two intermediate frequencies (38.9 HHz—33.4 MHz = 5.5 MHz). This amounts to heterodyning the sound signal twice. Since the 5.5 MHz sound carrier is obtained by mixing the two IF carrier frequencies, this method of processing the sound signal is known as 'Inter-Carrier Sound System'.

The 5.5 MHz FM sound signal thus obtained is coupled from either video detector or video amplifier to the sound IF amplifier. The sound IF signal, together with its sidebands is amplified to a level of several volts for the FM detector. This stage converts FM modulation to equivalent amplitude modulation and then detects it to recover the audio output. This is amplified in the audio section to have enough output to drive the loudspeaker.

21.1 SOUND SIGNAL SEPARATION

The output from the last IF stage consists of picture IF (38.9 MHz) amplitude modulated with the composite video signal and sound IF (33.4 MHz) frequency modulated with the audio signal. The two intermediate frequencies with their sidebands occupy a frequency spectrum of about 6.75 MHz. This signal is coupled to the video detector (see Fig. 11.6) which employs a diode and rectifies the centre zero modulated signal to produce a dc component, sum and difference frequencies of the various constituents of the input signal and their harmonics. The resultant band of frequencies is passed through a low-pass filter that forms part of the video detector. The cut-off frequency of the filter configuration is set at 5.75 MHz. Thus all the sum components and higher order harmonics which lie above the cut-off frequency (5.75 MHz) are filtered out. The output then consists of two distinct components. The band the lies between 0 and 5 MHz is the demodulated video signal and is fed to the video amplifier. The second component that lies around 5.5 MHz is formed as the difference of the picture IF and sound IF with its sidebands *i.e.*, 38.9 — (33.4 ± sidebands) = 5.5 MHz ± sidebands.

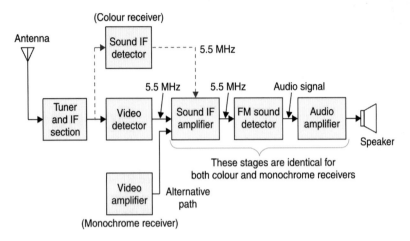

Fig. 21.1. *Sound signal path in monochrome and colour receivers.*

In fact the strong picture signal intermediate frequency acts as the carrier frequency to a relatively low amplitude sound signal intermediate frequency and its sidebands to provide another frequency translation to the sound signal. This shifts the sound modulation around a much lower carrier frequency of 5.5 MHz which is the difference of the two intermediate frequencies. This technique of separating the sound signal is known as the intercarrier sound system. The output has some amplitude variations. If these variations are large, it becomes difficult to remove them at the limiter stage, with the result that the audio output is severely distorted. This can be minimized by keeping the sound carrier signal amplitude very low. This explains why the sound IF which lies on the left skirt of the overall IF response is attenuated by about 20 db in comparison with the picture IF frequency. In fact this is the pivotal point for making the intercarrier sound system a success. Accordingly, the sound carrier together with its sidebands is given only 5 percent gain, in comparison with the video signal, to obtain a beat note at the detector, which contains only the frequency modulation of the original sound IF carrier and practically nothing of the video modulation.

Another factor that needs attention is the buzzing sound that results from the 50 Hz vertical sync signal interference with the 5.5 MHz sound signal. This is known as intercarrier buzz and is the result of cross-modulation between the two signals. Similarly, the 15625 Hz horizontal sync can result in a hissing sound in the loudspeaker. Here again the remedy lies in keeping the magnitude of the sound signal much lower than the picture signal to keep the beat note amplitude low. This, together with the limiting that is provided after the sound IF stage, eliminates practically all amplitude variations in the FM sound signal and normally no intercarrier buzz or hiss is audible in the loudspeaker.

21.2 SOUND TAKE-OFF CIRCUITS

The usual sound take-off point in a monochrome solid state receiver is at the output of video driver stage. In Fig. 21.2 the video driver is connected as a phase-splitter where the emitter circuit feeds the video output amplifier and the tuned transformer T_1 in the collector circuit separates sound IF from the composite video signal. Since the emitter circuit contains full composite signal, including the 5.5 MHz sound IF and its sidebands, a shunt trap (L_1, C_1) tuned to 5.5 MHz is used to eliminate this source of interereference. The transformer T_1 is tuned to 5.5 MHz and delivers sound signal to the base of Q_2, the sound IF amplifier. This transformer is tapped to provide proper impedance match and to maintain a high 'Q' of the tuned circuit.

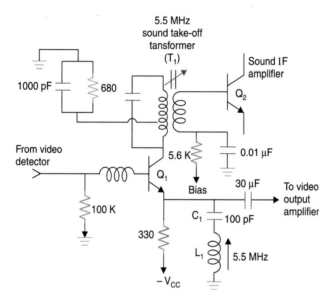

Fig. 21.2. Sound take-off circuit in a monochrome receiver.

A typical sound take-off and detector circuit used in colour receivers is shown in Fig. 21.3. It can be seen that the sound take-off point is located before the video detector to minimize the 1.07 MHz interference that would be produced if the sound IF (5.5 MHz) is permitted to beat

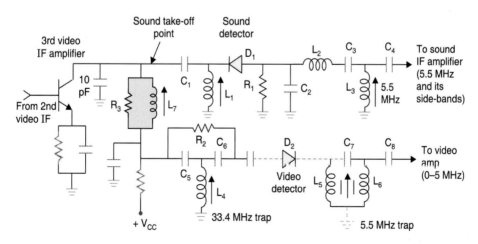

Fig. 21.3. Sound take-off and detector circuit in a colour receiver.

with the 4.43 MHz colour subcarrier. Conversion of sound IF to intercarrier sound (5.5 MHz) is accomplished by the diode D_1 which acts like a series detector. The input to the detector is via a series resonant circuit C_1, L_1. This circuit is tuned to about 36 MHz and is broad enough to include both picture and sound IFs. R_1 and C_2 are the load resistor and IF filter. C_3 and L_3 constitute a filter that separates the sound IF from the composite video that also appears at the output of the sound IF detector. The inductor L_2 blocks the video IF and its harmonics.

Since the sound take-off is before the video detector, greater attenuation of the 5.5 MHz sound IF is possible in the video detector circuit. This is done (see Fig. 21.3) by bridge type trap circuits located before and after the detector. The trap circuit located before the video detector (D_2) is tuned to 33.4 MHz and the one after the detector is tuned to 5.5 MHz. Thus signals at these frequencies are highly attenuated and do not reach the cathode of the picture tube.

21.3 INTER-CARRIER SOUND IF AMPLIFIER

The amplitude of the inter-carrier sound signal at the output of the video detector is very low and so at least two stages of sound IF amplification are provided before feeding it to the FM detector. This second IF stage is also used as a limiter, if necessary. Each IF stage is a tuned amplifier with a centre frequency of 5.5 MHz and a bandwidth of over 150 KHz to provide full gain to the FM sidebands. The desired bandwidth, though large, is easily attained because it is a relatively small percentage ($\approx$ 2%) of the intermediate frequency. This is similar to a radio receiver IF amplifier where the necessary bandwidth of 10 KMz is also nearly 2% of the 455 KHz IF frequency. However, the design and tuning of a TV sound IF stage is critical since in FM, narrow IF bandwidth causes amplitude distortion on loud signals because of insufficient response for maximum frequency deviation. This effect corresponds to clipping and limiting in an audio amplifier which can make the sound unintelligible. In AM, however, the loss of high frequencies that results on account of insufficient IF bandwidth is not too obvious in the reproduced sound.

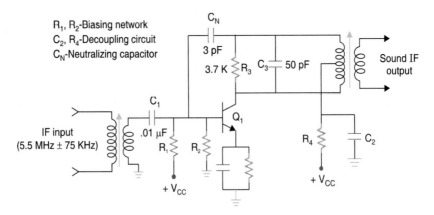

Fig. 21.4. Circuit diagram of a typical sound IF amplifier.

Figure 21.4 shows the circuit diagram of a typical transistor IF amplifier. The signal from the video detector is coupled through an impedance matching transformer, the secondary of which is broadly tuned to the intercarrier sound IF frequency. The output transformer has several extra turns on the primary winding to provide neutralization through a very small capcitor. Without negative feedback the stage will have a tendency to oscillate which can cause undesired high pitched sound from the speaker.

21.4 AM LIMITING

The frequency modulated sound signal has some amplitude variations, both because of heat formation at the video detector and due to somewhat unequal amplification at the RF and IF stages of the receiver. AM interference can be eliminated by limiting amplitude variations. Amplitude modulation rejection can be accomplished either by using the last IF stage as an FM limiter or by using an FM detector that does not respond to amplitude variations. FM discriminator circuits require a limiter for removing AM if present from a FM signal before detection. However, the ratio-detector and the quadrature-grid detector donot need a limiting stage because they are insensitive to amplitude variations in the FM signal.

The Limiter

The limiter is similar to the preceding IF stage but its dynamic range is kept narrow to achieve the desired clipping action. The dynamic range is reduced by using self-bias at the input and causing early saturation by reducing dc supply to the output circuit. Figure 21.5 (a) shows a typical FET amplitude limiter. Note the use of R_g, C_g for self-bias and R_1 for reducing the drain voltage. The stage operates as a class C amplifier. With varying signal amplitudes, the bias automatically adjusts itself to a value that allows just the positive tips of the signal to drive the gate positive and cause gate current flow. This maintains necessary self-bias to cause limiting.

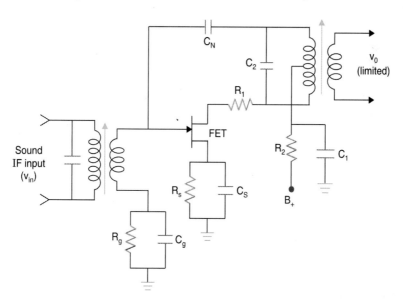

Fig. 21.5 (a). Amplitude limiter circuit employing an FET.

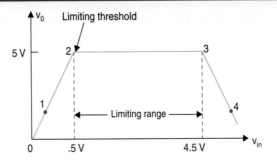

Fig. 21.5 (b). Typical limiter response characteristics.

Figure 21.5 (*b*) shows typical response characteristics of the amplitude limiter. It indicates clearly that limiting takes place only for a certain range of input voltages, outside which output varies with input. When the input voltage is too low (range 1 to 2) the stage behaves like a class A amplifier and no limiting takes place. Once the signal amplitude (peak-to-peak) exceeds the dynamic range, limiting action commences to deliver a constant output voltage. In the limiting region (range 2 to 3), as the input voltage increases the output current flows for a somewhat shorter portion of the input cycle to maintain a constant output voltage. Note that, though the current flows in short pulses (class *C* operation), the output voltage is sinusoidal. This is due to the flywheel action of the output tuned (tank) circuit. However, when the input voltage increases sufficiently (range 3 to 4), the angle of output current flow is reduced so much that less power is fed to the output circuit and the output voltage is reduced.

Thus to ensure proper limiting action the input signal amplitude should remain within the limiting region. In TV receivers this is ensured by AGC action.

Most present day receivers employ transistors in the sound section and use a ratio-detector which does not need a limiter. However, the last IF stage is designed to limit large amplitude variations, if any, and also to provide reasonable gain to the FM sound signal.

21.5   FM DETECTION

FM detection is carried out in two steps. The frequency-modulated IF signal of constant amplitude is first converted into a proportionate voltage that is both frequency and amplitude modulated. This latter voltage is then applied to a detector arrangement which detects amplitude changes but ignores frequency variations.

Slope Detection

Amplitude variations in an FM signal can be provided by using one of the sloping sides of a tuned circuit. Consider a frequency-modulated signal fed to a tuned amplifier (Fig. 21.6 (*a*)) whose resonant frequency is on one side of the centre frequency of the FM signal. As shown in Fig. 21.6 (*b*), the circuit is detuned to bring the centre frequency (5.5 MHz) to point *A* on the selectivity curve. The input to the amplifier is shown to have a frequency variation for a given audio modulation. As the input signal frequency shifts, the frequency deviations of the carrier signal are converted into proportionate amplitude variations (see Fig. 21.6 (*b*)). These AM variations result from the unequal IF gain above and below the carrier frequency. Thus the IF output varies in amplitude at the audio rate, in addition to its continuously changing frequency. The resulting AM signal can be coupled to a diode detector to recover the audio voltage.

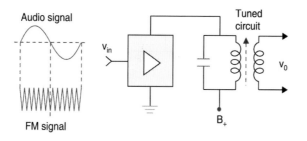

Fig. 21.6 (a). *Tuned amplifier.*

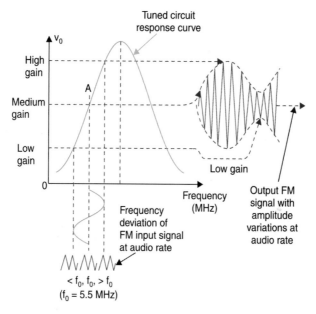

Fig. 21.6 (b). *Slope detection of an FM signal.*

Transformer Action

While the slope detector described above illustrates the principle of FM detection, it is both inefficient and linear over a very limited frequency range. In practical FM detectors, tuned circuits and other coupling techniques are used to transform frequency deviations of the incoming carrier into corresponding amplitude variations. In order to fully understand this transformation, it would be necessary to examine the phase relationships in a tuned transformer as the frequency of the signal applied to it changes.

As shown in Fig. 21.7 (a), the primary is a parallel resonant circuit but the secondary behaves like a series resonant circuit because the voltage is induced in series with the secondary as a result of the primary current. The magnitude of the induced voltage (e induced) is equal to $\pm$ $j\omega MI_p$ where M is the mutual inductance between the two windings. The sign of the induced voltage depends on the direction of winding. It is simpler to assume connections giving negative mutual-inductance and so the phasor diagrams of Fig. 21.7 (b) and subsequent circuits correspond to this assumption.

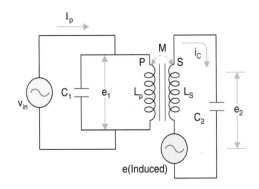

Fig. 21.7 (a). *Primary-secondary relationships in a tuned circuit.*

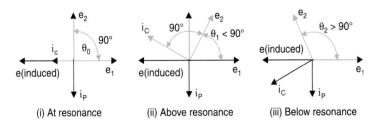

Fig. 21.7 (b). *Phase relationships in a tuned circuit.*

Three cases will be examine: (*i*) at resonance, (*ii*) above resonance and (*iii*) below resonance.

   (*i*) *At resonance.* The induced voltage (*e* induced) 'sees' a resistive circuit at resonance. Therefore, the secondary current (i_c) is in phase with the voltage. This current while flowing into the secondary tuned circuit produces a voltage drop (e_c) across C_2 which lags the current by 90°. The voltage across the capacitor is also the voltage across the secondary winding and is not the same as the induced voltage. The relations between the primary and secondary voltages are illustrated in Fig. 21.7 (b). Notice that the output secondary voltage (e_c) for case (*i*) is 90° out of phase with the voltage across the primary.

   (*ii*) *Above resonance.* If the frequency of operation is increased above resonance (above 5.5 MHz in FM sound signal) the secondary series circuit impedance will appear to be inductive. As shown in the corresponding phasor diagram the secondary current (i_c) now lags the induced voltage by some angle that depends on Q of the circuit and the amount of input signal frequency deviation. Since the voltage across the capacitor C_2 lags the current by 90° the phase angle θ_1 between the primary voltage and the secondary terminal voltage is now less than 90°.

   (*iii*) *Below resonance.* When the input signal deviates below resonance, the secondary circuit becomes capacitive. The secondary current now leads the induced voltage (see phasor diagram for input frequency below resonance) by an angle which depends on the extent of frequency deviation and Q of the circuit. Therefore, the secondary terminal voltage (e_2) and the primary voltage (e_1) are more than 90° apart.

Thus it is evident from the phasor diagrams that a change in frequency will cause a change in phase between the primary and secondary voltages. If these voltages are suitably added together vectorially, as is done in practical discriminators, their vector sum will be seen to change in amplitude. This explains the technique of generating amplitude variations in the FM signal before feeding it to the diode detectors for demodulation.

21.6　FM SOUND DETECTORS

Earlier television receivers used specially designed multifunction vacuum tubes in FM sound detectors. However, tubes were soon replaced by solid state circuitry. In fact FM detector was one of the first receiver circuit block to be made into an integrated circuit. Since that time a number of different sound ICs have been developed. One of the aims of this development has been to cut down or eliminate the need for tuned circuit alignment. The other factors which brought about quick IC development have been system performance, cost and reduced circuit complexity. In general FM sound detectors may be classified as under:

1. Discriminator

2. Ratio detector

3. Quadrature detector

4. Differential peak FM detector

5. Phase-locked loop FM detector

One of the earlier sucessful detectors was the Foster-Seely discriminator. Since a discriminator must be preceded by a good limiter, it was soon replaced by a ratio detector which has an inbuilt limiter. The ratio detector was commonly used in hybrid television receivers till the advent of ICs for the sound section. Earlier ICs employed ratio or quadrature detectors for demodulating the FM sound signal. However, recent sound ICs employ either a differential peak or a phase-locked loop detector because they need least tuned circuit alignment. When solid state devices were not available specially designed tubes (gated-beam) were developed which perform as limiter, detector and first audio amplifier, all in one glass envelope. However, their use is now limited to the old 'all tube' receivers which are still in operation.

1. The Foster Seeley Discriminator

As already explained detection of the frequency modulated signal is carried out by first modifying the frequency spectrum of the wave in such a manner that its envelope fluctuates in accordance with the frequency deviations. In the Foster-Seeley discriminator (see Fig. 21.8 (a)) this is carried out by the vector addition of the instantaneous voltages that develop across the input transformer. Since the magnitude of the resultant voltage depends on the instantaneous phase angle between the primary and secondary voltages this detector is also known as 'phase-shift' discriminator. The resulting amplitude-modulated wave is then applied to the diodes D_1 and D_2, that form part of the balanced AM detector, to recover the audio modulating voltage.

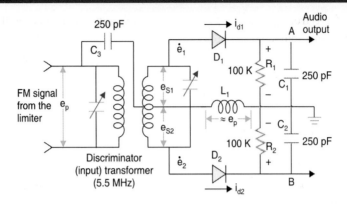

Fig. 21.8 (a). *Foster Seely discriminator.*

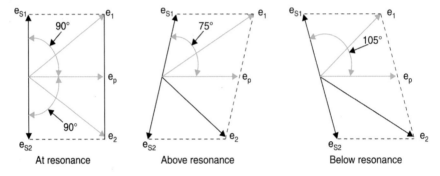

At resonance   Above resonance   Below resonance

Fig. 21.8 (b). *Discriminator transformer voltages.*

The action of the discriminator may be explained as follows:

The input transformer is double-tuned with both primary and secondary resonant at the centre frequency of 5.5 MHz. In addition to the inductive coupling between the two tuned circuits, the primary voltage e_p, is also coupled through capacitor C_3 (250 pF) at the centre of the secondary winding. The centre tap of the secondary winding is also connected to the common points of R_1, R_2 and C_1, C_2 through L_1, a radio frequency choke. Thus the circuit composed of C_3 and L_1, with the other end of L_1 grounded, comes effectively across the primary winding. At and around 5.5 MHz the reactance of L_1 greatly exceeds that of C_3. Therefore, voltage across L_1 is almost equal to e_p, the applied primary voltage. Accordingly the voltage fed to each diode is the vector sum of the primary voltage and the corresponding half-secondary voltage. As is obvious from the circuit diagram (Fig. 21.8 (a)), the induced voltage across the secondary divides into two parts at the centre-tap and is thus applied in push-pull to the diode plates. However, the primary voltage which effectively appears between the centre-tap and ground is applied in parallel to the two diodes. Therefore, the IF signal voltage for the two diodes is the resultant (vector sum) of the induced secondary voltage applied in push-pull and the primary voltage applied in parallel to the two diodes.

Diode Voltage Phase-Relations. As explained earlier, in a tuned circuit the secondary voltage is 90° out of phase with the primary voltage at resonance. As the applied IF frequency swings above and below the resonant frequency, the phase angle varies around 90°. In the circuit of Fig. 21.8 (a) the secondary voltage is devided into two equal halves e_{s1} and e_{s2} by the centre tap connection. The voltages are of opposite polarity with respect to each other. Thus at the resonant frequency of the tuned-secondary circuit, the secondary voltages e_{s1} and e_{s2} are in quadrature with voltage e_p existing across the primary inductance. When the applied frequency is either higher or lower than the resonant frequency of the secondary, the phase angles of e_{s1} and e_{s2} relative to e_p will differ from 90°. The vector relations and magnitudes of the resultant phasors e_1 and e_2 are illustrated in Fig. 21.8 (b).

At the centre (IF) frequency both e_{s1} and e_{s2} are 90° out of phase with e_p with the result that e_1 is equal to e_2. At some IF signal frequency swing above the centre frequency, e_p is shown 75° out of phase with e_{s1} instead of 90°. This makes e_p closer to e_{s1} and as a reuslt e_1 increases while e_2 decreases. Similarly for some carrier frequency swing below the centre frequency, angle of e_s with e_p increases to 105°, thus making e_2 greater then e_1. The voltages e_1 and e_2 are applied across the diodes D_1 and D_2 respectively. With no modulation, *i.e.*, when IF frequency is equal to the centre frequency, e_1 is equal to e_2 and both the diodes conduct to pass equal currents. Since R_1 is equal to R_2, the net output voltage is zero. However, when the centre frequency swings above resonance to make e_1 greater than e_2, i_{d1} exceeds i_{d2} and the net output voltage is positive. Similarly when the carrier frequency swings below the centre frequency, e_2 exceeds e_1, and i_{d2} becomes greater than i_{d1} to develop a net negative output voltage. Note that the RF choke does not permit any current flow at the IF frequency and all high frequency components of the rectified current complete their path through ground. The rectified dc component, however, complete its path through the RF choke.

The time-constant $R_1C_1 = R_2C_2$ is kept much larger than the period of IF frequency so that voltages across C_1 and C_2 cannot follow high frequency variations. However, it is kept much smaller than the highest audio signal period so that the voltage variations across the two capacitors can follow the audio variations. Note that the voltages across C_1 and C_2 subtract from each other to develop audio voltage across the output terminals A and B.

Discriminator Response. The output voltage is the arithmetic difference of e_1 and e_2 which vary with the instantaneous frequency as shown in Fig. 21.9 (a). Therefore, deviations in the instantaneous frequency away from the carrier frequency cuase the rectified output voltage V_{AB} to vary in accordance with the curve of Fig. 21.9 (b). The slope of this curve shows

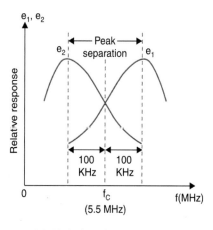

Fig. 21.9 (a). Variation of e_1 and e_2 with frequency.

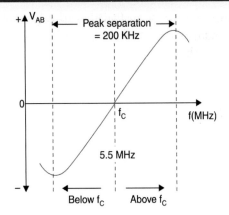

Fig. 21.9 (b). Discriminator ('S' shaped) response curve.

that the rectified output V_{AB} will reproduce with reasonable accuracy, the variations of the instantaneous frequency as long as the operation is confined to the region between the peaks of e_1 and e_2. In the illustration this is shown to be ± 100 KHz with a centre frequency of 5.5 MHz.

2. Ratio Detector

In the Foster-Seeley discriminator any amplitude variations of the input signal give rise to unwanted changes in the resulting output voltage. This makes limiting necessary. It is possible to modify the discriminator circuit to provide limiting, so that the amplitude limiter can be dispensed with. A circuit so modified is called a ratio detector.

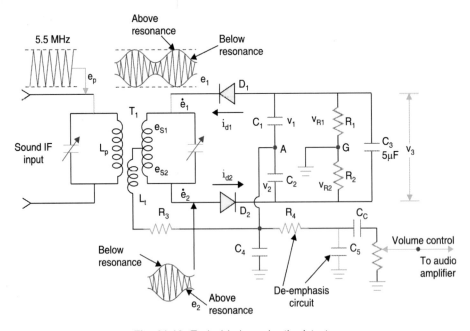

Fig. 21.10. Typical balanced ratio detector.

Balanced Ratio-Detector. The circuit of a balanced ratio detector is shown in Fig. 21.10. Neglecting capacitor C_3 for the moment, this arrangement is seen to differ from the Foster-Seeley discriminator in (i) diode D_1 has been reversed and (ii) the output voltage is obtained between the grounded junction of R_1 and R_2 and A, the common point of capacitors C_1 and C_2.

The input coupling transformer T_1 has the same function as in a phase-shift discriminator. Both the primary and secondary tuned circuits are resonant at the IF (centre) frequency. The secondary is centre-tapped to provide equal voltages of opposite polarity for the diode rectifiers. The primary voltage is applied in parallel to both the diodes by a tertiary winding L_t which is wound on the same core and is connected at the centre-tap of the secondary winding. L_t is wound directly over the primary winding for very close coupling so that the phase of the primary voltage across L_p and L_t is practically the same. This arrangement (with the tertiary winding) of providing primary voltage to the secondary circuit is preferred in order to match the high impedance primary to the relatively low impedance secondary circuit. The resistor R_3 in series with L_t, limits the peak diode current for improved operation.

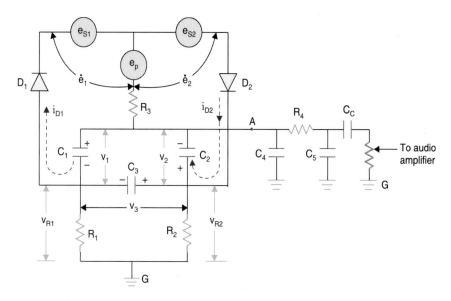

Fig. 21.11. Equivalent circuit of the balanced ratio detector.

Circuit Operation. Consider the equivalent circuit of the ratio detector shown in Fig. 21.11. The primary voltage e_p combines with the secondary voltages e_{s1} and e_{s2} to produce resultant voltages e_1 and e_2 as illustrated by the phasor diagrams of Fig. 21.8 (b). The two series circuits formed by equal resistors R_1, R_2 and equal capacitors C_1, C_2 are connected in parallel across the plate of D_1 and cathode of D_2. The voltages e_1 and e_2 are applied across the diodes D_1 and D_2 respectively. The rectified currents marked i_{D1} and i_{D2} flow through R_3 to charge capacitors C_1 and C_2 with the polarity marked across each capacitor. The voltages v_1 and v_2 which develop across C_1 and C_2 are approximately equal to the peak value of the IF signal applied to each rectifier. The net voltage ($| v_1 | + | v_2 |$) which develops across the two capacitors also appears across the series combination of R_1 and R_2. Since $R_1 = R_2$ and their common point is grounded, the total voltage gets equally divided across the two resistors with the result that $v_{R1} = v_{R2}$

$$= \frac{|v_1| + |v_2|}{2}$$. Therefore, the net output voltage (v_{AG}) at the audio take-off points A and G equals

$$| v_1 | - \frac{|v_1| + |v_2|}{2} = \frac{|v_1| - |v_2|}{2}.$$

At the centre frequency where $e_1 = e_2$, equal voltages are developed across the capacitors making $v_1 = v_2$. This results in zero audio output voltage. When the FM input signal is above centre frequency, e_1 is greater than e_2, with the result that v_1 exceeds v_2. This makes point A more positive producing audio output voltage of positive polarity. Below the centre frequency e_2 is greater than e_1, and thus v_2 exceeds v_1. The result is audio output voltage of negative polarity at point A. The response curve of the ratio detector is also S-shaped and is essentially the same as shown in Fig. 21.9 (b) for the discriminator circuit.

An important characteristic of the ratio detector circuit can be illustrated by considering numerical values of v_1 and v_2. Assume that at the centre frequency, $v_1 = v_2 = 4$ V. As the frequency deviates above its central value, v_1 increases in magnitude and v_2 decreases by the same value. If v_1 becomes 6 V, v_2 reduces to 2 V. This makes point A more positive by 2 volts with respect to ground $\left(\frac{(6-2)}{2} = 2 \text{ V} \right)$. Similarly, when frequency swings below the centre frequency by the same amount, v_1 reduces to 2 V and v_2 rises to 6 V. This makes point A, 2 volts negative with respect to ground $\left(\frac{(6-2)}{2} = -2 \text{ V} \right)$. Thus an audio output voltage with a peak value of 2 volts is generated. This is exactly half the magnitude to the output voltage obtained from a discriminator (Fig. 21.8 (a)) under similar input voltage conditions. This is so, because, in a discriminator the voltages across the capacitors combine to produce the audio output voltage whereas in the detector under discussion, the output arises as a result of variations in the ratio $| v_1 | / | v_2 |$, while the sum $| v_1 | + | v_2 |$ remains substantially constant. In fact it is this behaviour that gives it the name—*ratio detector*.

Stabilizing Voltage. In order to make the ratio detector insensitive to AM interference, the total voltage $v_3 = v_1 + v_2$, must be stabilized so that it cannot vary at the audio frequency rate. This is achieved by connecting a large capacitor C_3 across the series combination of C_1 and C_2. Since C_3 is large it acts as a low impedance load to any change in amplitude that might otherwise occur. For example a momentary increase in the amplitude of v_3 causes a large charging current to flow through the diodes into C_3. This represents power absorbed from the primary and secondary resonance circuits, and so reduces the voltage applied to the diodes. Conversely, if the amplitude of the incoming signal tends to drop below the average amplitude, then C_3 prevents the voltage v_3 from dropping by discharging into R_1 and R_2. It is thus seen that the presence of C_3 prevents amplitude variations which would otherwise occur in the voltage v_3 and likewise in voltages v_1 and v_2.

It may be noted that when a channel is changed and the magnitude of sound IF signal shifts to a new value, the voltage across C_3 changes to another value and maintains itself. A 5 μF capacitor (C_3) is considered adequate for this purpose. The capacitor C_4 is connected across the audio output terminals to bypass higher frequency components. However, the audio voltage (V_{AG})

effectively develops across C_4 and varies in accordance with frequency modulation of the sound carrier.

The audio output is followed by a de-emphasis circuit (R_4, C_5) before feeding into the audio section.

Single Ended Ratio Detector. A simple and economical ratio detector is shown in Fig. 21.12. This is a single ended or an unbalanced version because the two diodes are not equally balanced with respect to ground. A component of the diode currents i_{d1} and i_{d2} flows through the series circuit formed by the two diodes, R_1 and C_1 in parallel and the secondary winding circuit. However, some current flows via C_2 and ground to complete the circuit. Note that a part of the D_1 current (i_{d1}) flows through C_2 towards the ground while the corresponding part of D_2 current (i_{d2}) flows through C_2 in the opposite direction. Thus the two diode currents subtract across C_2 to develop a voltage which varies at the audio rate. This function in the other ratio detectors is carried out by the capacitors connected across the series circuit formed by the diodes. The resistors R_2 and R_3 (1 K) in series with D_1 and D_2 respectively limit the peak diode currents to improve dynamic balance of the diodes at higher signal conditions. One of the series resistances (R_2) is variable and may be adjusted for better balance. The time-constant of R_1 in parallel with C_1 is so large (200 ms) that the circuit neither responds to fast noise amplitude changes nor to the relatively slow changes in amplitude due to spurious amplitude modulation. This as explained ealier provides the desired limiting action. R_4, C_3 is the de-emphasis circuit which also filters out unwanted IF components from the diode currents. Since the output voltage has only one ground terminal no dc level shift is necessary.

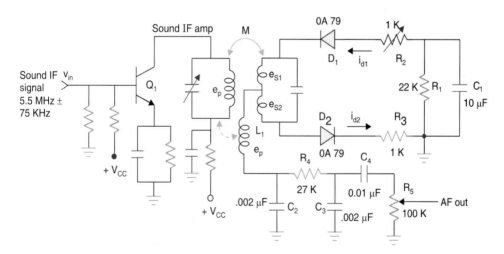

Fig. 21.12. *Single-ended ratio detector.*

3. Quadrature Detector

Quadrature detection is another approach to FM demodulation. This method combines the functions of a limiter, a discriminator and an audio voltage amplifier. A specially designed tube known as gated-beam tube performs all these functions. The transistor version of this method employs several transistors to provide the same functions. Normally this forms a section of an IC specially developed for sound section of the receiver.

Gated-Beam Tube. A dual-control quadrature detector or gated-beam tube is a specially constructed sharp cut-off pentode. The important characteristic of this tube is that the plate current can be sharply cut-off either by control grid or suppressor grid. When the control grid voltage changes from negative to positive the plate current rises sharply from zero to attain its maximum value. The cut-off voltage is usually close to − 2 V. Thus the control (limiter) grid provides limiting action to any amplitude variations which may be present in the incoming FM sound signal.

The electrons passing through the limiter grid are accelerated by the positive potential on the screen grid. The quadrature grid (suppressor grid) also exercises similar control over the plate current to produce sharp cut-off and saturation characteristics. If the quadrature grid is made strongly negative the plate current of the tube is cut-off, on matter how positive the limiter grid is. Thus the two grids act as current control gates and unless both permit conduction simultaneously, no plate current can flow. It is this action which gives the tube its name of gated beam tube. Later circuits used a pentode having similar cut-off characteristics. Such circuits came to be known as quadrature grid detectors.

Circuit Operation. A typical quadrature grid detector employing a pentode is shown in Fig. 21.13. The plate and screen grid are operated at positive potentials in a manner similar to corresponding circuits of conventional tube amplifiers. The FM signal v_L is coupled to the limiter grid (G_1) through a circuit broadly tuned to the intercarrier sound IF frequency. For an input signal over one volt peak-to-peak, the tube will be driven from cut-off to saturation or vice-versa to conform with the positive and negative excursions of the input signal. This results in excellent limiting and almost a square wave beam current is produced in the region beyond the input grid.

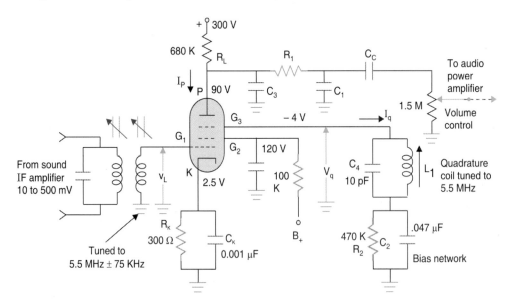

Fig. 21.13. Quadrature grid FM detector circuit.

The quadrature circuit consists of a slug tuned coil shunted by a fixed capacitor forming a parallel resonant circuit. The circuit is tuned to the sound IF centre frequency. It is connected to the quadrature grid G_3, which is biased to about -4 volts.

Since grid 3 is at -4 V and the screen grid at 120 V, the electrons in the space between the two grids are slowed down to form an outer space-charge. This is in addition to normal space charge near the cathode. The outer space-charge serves as a source of electrons for the suppressor or quadrature grid G_3 and plate.

As the input voltage swings from positive to negative, the amount of space charge at the virtual cathode near grid 3 also varies. This varying space charge causes current flow by electrostatic induction, in the circuit connected to the quadrature grid. The induced current lags the input voltage by 90°. The voltage across the tuned circuit also lags by 90° when the input signal frequency is 5.5 MHz to which the quadrature circuit is tuned. This is why G_3 is called the quadrature grid. When the input signal frequency deviates below or above the centre frequency, the phase of quadrature voltage v_q also varies below and above 90°.

Figure 21.14 shows effect of limiter and quadrature grids on the plate current. When the signal swings above centre frequency, v_q lage v_L by more than 90°, because the tuned circuit behaves like a capacitive reactance at a frequency which is higher than its resonant frequency. The additional lag keeps either of the two grids in cut-off most of the time during the intervals of positive swing in the signal. This results in narrow pulses of plate current. Despite the fact that maximum value of plate current is the same, narrower pulses mean a smaller value of average plate current.

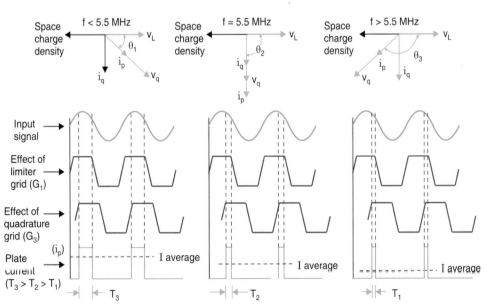

Fig. 21.14. Effect of control and quadrature grids on the plate current of a quadrature grid detector.

Similarly when the signal frequency becomes less than the centre frequency the tuned circuit presents an inductive impedance and v_q lags v_L by less than 90°. The two signals permit longer conduction periods resulting in wider pulses with a higher average value of plate current. These pulses are integrated by the network R_L and C_3. Note that changes in the phase of v_q result from frequency deviations in the FM signal. Similarly variations in the average value of plate current follow amplitude and frequency of the modulating audio signal. Thus the integrated output voltage at the plate of the tube gives the required audio output. C_3 also acts as a plate bypass capacitor for 5.5 MHz signals. R_1, C_1 constitute the de-emphasis network.

Tansistor Quadrature Detector. The transistor version of the quadrature detector employs three identical transistors. The circuit arrangement is shown in Fig. 21.15. Transistors Q_1 and Q_2 perform the same functions as the limiter and quadrature grids in a gated-beam tube. The sound IF signal is fed directly to the base of Q_1, but is given a 90° phase-shift before applying it to the base of Q_2. The phase shift is accomplished by driving the signal through a very small capacitor C_2. The circuit formed by L and C_3 does not produce any further phase-shift at 5.5 MHz because it is tuned to be resonant at this frequency.

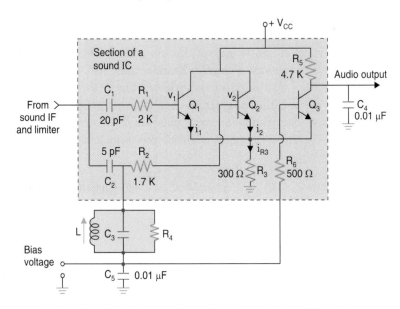

Fig. 21.15. Circuit diagram of a solid state quadrature FM detector.

The sound IF signal attains sufficient amplitude before it is coupled to the detector circuit and the biasing is so set that the two transistors are driven into saturation during the positive swing of the input signal applied to each transistor. The emitters of the two transistors are grounded through a common resistor R_3. Thus the output voltage that develops across R_3 depends on the conduction of both Q_1 and Q_2. Since the voltage drop across R_3 reverse biases the two transistors, the net bias automatically adjusts itself to keep the magnitude of peak current through R_3 at nearly a constant value. However, the duration for which the two transistors conduct together varies and depends on the relative phase angle between the two input signals.

As shown in the waveforms of Fig. 21.16, at the centre frequency where the phase difference between the two voltages is 90°, the current through R_3 flows for a period T_2 to develop an average current I_2. However, when the signal frequency becomes more than the centre frequency, the phase angle between the two input voltages exceeds 90° and current through R_3 flows for a longer period to develop an average current I_1 which is greater than I_2. When the input signal frequency become less than the centre frequency, the relative phase angle reduces to less than 90° and the magnitude of average current through R_3 changes to I_3 which is less than I_2.

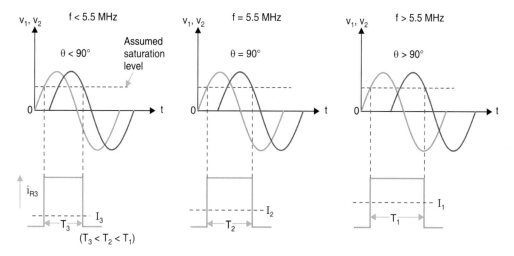

Fig. 21.16. *Effect of frequency deviations on the average output current in a transistor quadrature detector. Note that cut-off bias for both Q_1 and Q_2 has been assumed as zero volt.*

Since the relative phase change between the input voltages to Q_1 and Q_2 results from frequency deviations in the FM signal, the output voltage which develops across R_3 due to variations in the average value of current through it follows the audio modulation. The audio voltage thus developed is in the input circuit of transistor Q_3 where it is amplified before feeding it to the audio section. Note that capacitor C_4 (0.01 μF) by passes all IF components to develop only audio voltage at the collector of Q_3.

It is interesting to note that in a gated-beam tube the limiter grid and quadrature grid act like two gates in series and when the phase angle between v_L and v_q exceeds 90°, the average output current decreases. However, in the transistor circuit, under similar conditions the average output current increases. This is because the two transistors Q_1 and Q_2 act as two gates in parallel to permit current flow when either of the two input signals is positive. Thus there is a 180° phase difference between the output currents of the two quadrature detectors for the same frequency swing. Since the audio signal is symmetrical about its centre zero axis there is no difference in the sound reproduced at the loudspeaker.

4. Differential Peak FM Detector

In this system a new technique known as differential peak detection is used to develop audio voltage from the frequency modulated signal. Figure 21.17 (*a*) is the block diagram of this detector. It forms part of the IC, CA 3065 described in a later section of this chapter. The detector employs only a single tuning coil.

The two peak detectors employ differential amplifier configurations with emitter followers at their inputs. The output circuits of the two detectors have identical RC networks of suitable time constant to provide peak detection of the signals fed to them. The high input impedance of the emitter followers isolates the detectors from the preceding sound IF section.

The external frequency sensitive network comprises of a parallel LC circuit and a series capacitor. The voltage e_1 obtained from the sound IF amplifier and limiter output is fed (actually applied at the input terminal of the IC) both to this frequency selective network and one of the peak detectors.

At the centre frequency ($f_c = 5.5$ MHz) the parallel circuit behaves like an inductor in series with capacitor C_2. This combination provides a slight boost to the signal across C_2 so as to make e_2 equal to e_1. The signal voltage e_2 is applied at the input of the other peak detector. The two peak detectors develop proportionate dc voltages at their output terminals which feed into corresponding input terminals of the differential amplifier. Since e_1 is equal to e_2, the differential output voltage is zero in the absence of any modulation. As the intercarrier frequency swings to become low, the parallel network L_1, C_1 along with C_2 in series, approaches its series resonant frequency. Thus e_2 (across C_2) attains a higher magnitude than e_1. This in turn results in a negative output voltage from the differential amplifier. As the carrier frequency increases above f_c, the parallel network

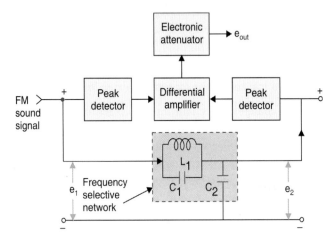

Fig. 21.17 (a). Block diagram of a differential peak detector.

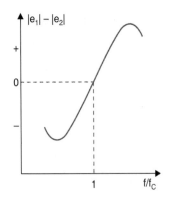

Fig. 21.17 (b). Difference signal as a function of frequency.

(L_1, C_1) approaches its parallel resonant frequency and its impedance rises to make e_2 less than e_1. The differential amplifier output then swings to become positive. An analysis of the frequency response of e_1 and e_2 indicates that $|e_1| - |e_2|$ follows the normal 'S' curve (see Fig. 21.17 (b)) resulting in FM detection. The linearity of this detector is good and total harmonic distortion in the output is less than one percent.

5. Phase-Locked Loop FM Detector

The latest in FM detection is the phase-locked loop detector shown in Fig. 21.18. It may be classed as an inductorless detector requiring no alignment. In this circuit a dc error voltage is developed by a phase detector that compares the phase of the input FM signal with that of a locally generated oscillator of the same frequency. The dc control voltage obtained at the output of low-pass filter is directly related to the degree of phase difference between the two signals. This control voltage which is returned to the voltage controlled oscillator (VCO) after amplification forces the oscillator to change its frequency thereby reducing the phase error. When the free-running frequency of the VCO becomes close to the input signal frequency, the system locks at this frequency and the VCO tracks the input FM signal. Since the dc control voltage will vary in step with the frequency deviations of the FM signal, it represents audio modulation. Thus the output from the dc amplifier is the demodulated audio signal. The IC contains a voltage amplifier (audio amplifier) the output of which can drive an audio power amplifier. The capacitor C_1 is the de-emphasis capacitor connected to the corresponding pin of the monolithic IC package. For a noise-free reception, this FM detector requires a good limiter preceding the phase detector.

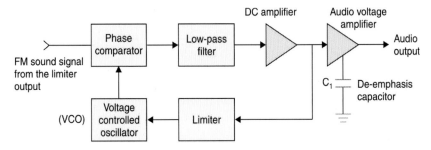

Fig. 21.18. *Schematic block diagram of a phase-looked loop FM detector.*

21.7 SOUND SECTION INTEGRATED CIRCUITS

As already stated early hybrid receivers employed some form of a ratio detector for FM detection and vacuum tubes like ECL 82 (triode-pentode) for the audio section. The tubes were later replaced by transistors and a large number of circuits were developed for high quality sound reproduction. However, as soon as the integrated circuit technology was perfected, a large number of monolythic ICs were developed to perform the functions of sound IF limiter-amplifier, FM detector and audio pre-amplifier. In addition many ICs for the audio section are also now available. Out of the many sound section ICs that have been developed, TAA570, TBA750A and CA3065 are commonly used in modern circuits. All such ICs employ complex circuitry in order to minimize the use of transformer couplings and inductors.

IC-TAA570

This IC employs a four-stage differential amplifier in the sound IF amplifier and limiter section. All the four stages receive their collector voltages through a multiemitter transistor which provides a self-limiting action on the voltage-swings at the collectors. All this results in very good noise suppression and excellent amplitude limiting. The FM detector section is a symmetrical phase detector which is a modified version of the transistor quadrature detector described earlier. The audio driver section is designed to give a maximum gain of about 80 db and can deliver an AF output equal to 1.8 volts for a frequency deviation of 50 KHz.

Figure 21.19 shows necessary details of its external connections and the circuit of the associated audio output stage. This is commonly used in hybrid Television receivers.

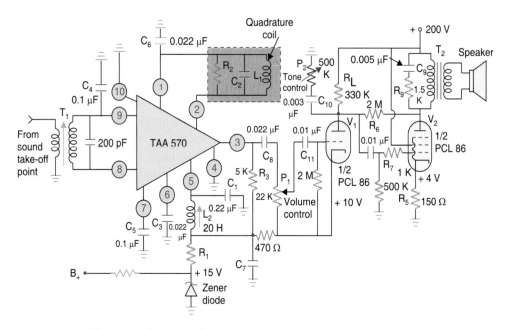

Fig. 21.19. *Sound section of a hybrid receiver employing IC TAA570.*

The sound IF signal from the video detector is coupled through the tuned matching transformer T_1 to the input terminals (pins 8 and 9) of the IC. Stabilized dc supply is fed at pins 3 and 5 through resistor R_1 and decoupling capacitor C_7. L_2 is the choke and C_1 the decoupling capacitor to the dc supply at pin 5.

The components shown within the dotted-chain box consitute the quadrature circuit where L_1, C_2 form a tuned circuit resonant at 5.5 MHz and R_2 is the damping resistance across the quadrature coil. De-emphasis is provided by capacitor C_3 connected at pin 6 of the IC. The capacitors C_4, C_5 and C_6 are bypass capacitors at the various pins of the integrated circuit. R_3 is the load resistor of the audio amplifier transistor and its output is available at pin 3. The output from the IC is coupled to the voltage amplifier tube V_1 (1/2 PCL86) through C_8 and potentiometer P_1 which is the volume control. Potentiometer P_2, that is connected at the plate of V_1 through a capacitor (C_{10}) functions as tone control. The output tube V_2(1/2 PCL86) provides power amplification and the sound output is delivered to the loudspeaker through T_2 the matching output transformer. The series circuit formed by C_9, R_9 and connected across the primary circuit of the output transformer acts as tone correction network. Negative feedback is applied from the plate of V_2 to that of V_1 through R_6 for improved tone control operation. R_5, the cathode resistor of V_2, has been left unbypassed to provide overall gain stability and improved frequency response of the audio amplifier. Resistance R_7 in series with the control grid circuit is for suppression of any parasitic oscillations.

IC-BEL CA 3065

This is a linear IC chip specially designed for the sound section of monochrome television receivers. Its circuit basically consists of (i) a regulated power supply, (ii) an IF amplifier-limiter, (iii) an FM detector, (iv) an electronic attenuator and buffer amplifier and (v) an audio driver. The schematic diagram of this IC is shown in Fig. 21.20.

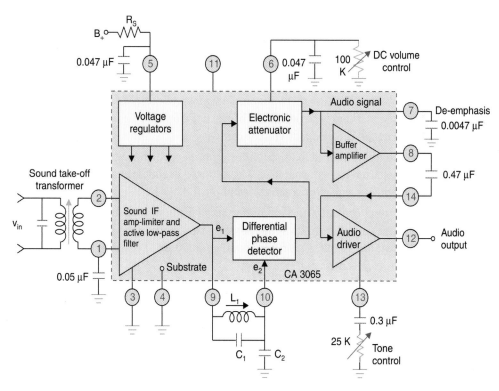

Fig. 21.20. Block diagram of the IC(BEL) CA 3065 in a typical circuit application. Note terminal 5 may be connected to any positive voltage through a suitable resistor (R_s) provided power rating of the IC is not exceeded.

(i) *Regulated power supply.* The input power supply to the IC is regulated through a combination of zener diodes. Separately regulated and decoupled dc supply is fed to different sections of the IC for stable circuit operation.

(ii) *IF amplifier-limiter.* The single ended differential amplifier is the basic configuration for each of the three stages of the amplifier-limiter section of CA 3065. The quiescent currents of different transistors employed in the amplifiers are so adjusted that best possible limiting and AM rejection characteristics are obtained. The limiting amplifier is followed by an active low-pass filter which suppresses harmonics from the signal. The output of this filter feeds the FM detector.

Chapter 21

(*iii*) *FM detector.* A differential peak detector, the working principle of which was explained in an earlier section (see Fig. 21.17), is used for obtaining audio voltage from the frequency modulated signal. As shown in Fig. 21.20 voltage e_1 obtained from the amplifier-limiter is applied at terminal 9 of the IC and thus fed simultaneously to the frequency selective network and one peak detector. The signal voltage e_2 obtained at the output of the frequency selective network (L_1, C_1, and C_2) is applied at the input (pin 10) of the other peak detector. The outputs from the two peak detectors feed into the differential amplifier, the output of which is the demodulated audio signal.

(*iv*) *Electronic attenuator and buffer amplifier.* The output of the differential peak detector is fed to the electronic attenuator which performs the conventional function of the audio volume control. The stage employs a single ended differential cascade amplifier. The gain of this differential amplifier is controlled by varying the quiescent current of the transistors. This is achieved by changing the B_+ voltage to one of its dc inputs, while the other dc input is kept at constant. An external variable resistance connected between terminal 6 of the IC and ground is used to vary the dc potential. Because there is no audio signal present across the variable resistance from terminal 6 to ground, there is no possibility of any noise or hum pick-up in the audio signal. This type of volume control is called a dc volume control.

The output of the attenuator is fed to a buffer amplifier which is an emitter follower and isolates the attenuator from the external load. The detected output is available at terminal 8 of the IC and the de-emphasis capacitor is connected at pin 7.

(*v*) *Audio driver.* The output of the buffer stage feeds into an audio driver which is a simple common emitter amplifier sandwitched between two emitter followers for input isolation and low output impedance. Typical gain of this stage is 20 db. Terminal 13 is available for incorporating suitable tone control network.

21.8   AUDIO OUTPUT STAGE

The output from the IC (CA 3065) can drive a PCL84/PCL86 audio amplifier, a solid state complementary output stage, or a sound IC like the BEL CA810.

A complete sound IF circuit employing CA 3065 along with a complementary transistor output amplifier is shown in Fig. 21.21. The output from the IC at pin 12 is fed to the driver transistor Q_1 through coupling capacitor C_3. Transistor Q_1 (BC 148B) forms a boot-strapped driver stage which feeds the matched output pair Q_2 and Q_3 (AC 187/01, AC 188/01). The output transistors drive the loudspeaker (8 ohms) through the electrolytic capacitor C_4, R_1 and R_2 and thermistor (Th) provide necessary temperature compensation in the biasing circuit of the output pair. The stage operates class AB to avoid cross-over distortion. Typical data of the audio amplifier is as under:

Supply voltage	15 V
No signal current	55 mA
Current drain at 2.5 W	300 mA
Max. power output	2.5 W
Third harmonic distortion at 2.5 W	3%
A.M. rejection	35 db
Speaker impedance	8 Ω

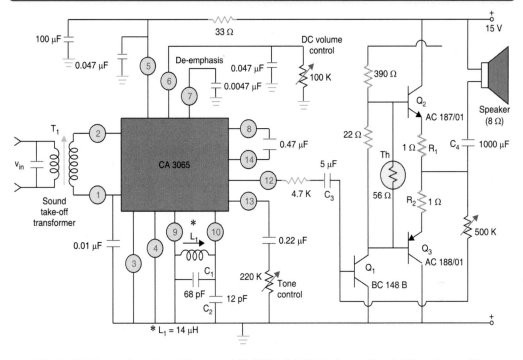

Fig. 21.21. *External circuitry of the sound IC (BEL) CA 3065 along-with a 2.5W audio amplifier.*

Audio IC CA 810

The IC BEL CA 810 has been specially developed for the audio section of television receivers. This audio amplifier as illustrated in Fig. 21.22 is capable of delivering 4 W output power into an 8 ohm load. Good frequency response and high input impedance are some of the salient features of this circuit. Its typical performance characteristics are as under:

(1) Supply voltage	16.0 V
(2) Maximum output power at 10% THD	4.1 W
(3) Input sensitivity at 50 mW output power	9.2 mV
(4) Input sensitivity at maximum output power	70 mV
(5) Zero signal current drain	8.6 mA
(6) Supply current at maximum power output	380 mA
(7) Input impedance	100 KΩ
(8) Frequency response (3 db)	20 Hz to 20 KHz

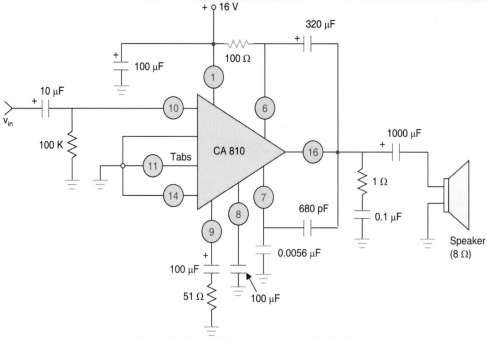

Fig. 21.22. Audio amplifier employing IC (BEL) CA 810.

REVIEW QUESTIONS

1. Explain the mechanism of separating the sound signal in a video detector. What is the pivotal point the makes the intercarrier sound system a success ?

2. Explain with suitable circuits and phasor diagrams the phase relationships between the input and output of a tuned circuit when the input signal frequency is varied below and above resonance.

3. Explain the basic principle of FM detection. Describe with a circuit diagram how FM detection takes place in a Foster-Seeley detector. Why such a circuit must precede by a voltage limiter ?

4. Draw the circuit diagram of any type of ratio detector and explain how the audio signal is detected. How is the limiting action achieved in a ratio detector ?

5. Explain with a circuit diagram and waveshapes, how FM detection takes place in a quadrature grid detector that employs a specially designed pentode. Why it is called quadrature detection ?

6. Fig. 21.15 is the circuit of a quadrature FM detector which employs transistors. Explain how the circuit performs as a limiter, FM detector and audio preamplifier. How is the detection mechanism different from the circuit employing a gated-beam tube ?

7. Explain with a suitable circuit diagram the basic principle of peak FM detection. Explain the manner in which FM signal is made to acquire amplitude variations before feeding it to the differential amplifier.

8. Draw schematic block diagram of a phase-locked loop FM detector and explain how the circuit performs to demodulate the FM signal.

9. The block schematic of the sound IC CA 3056 is given in Fig. 21.20. Explain briefly the function of each block. Why is the electronic attenuator block called a dc volume control ?

10. Draw the circuit diagram of an audio output stage that can be fed from the sound IC AC 3065 and explain how it produces enough audio output power with negligible distortion.

RF Tuner

T he RF amplifier, the local oscillator and the mixer stages form the RF tuning section. This is commonly known as 'Tuner' or 'Front end'. A simplified block diagram of the tuner is shown in Fig. 22.1. It is the same for both monochrome and colour receivers except that automatic frequency tuning is provided in colour receivers only. The function of the tuner is to select a single channel signal from the many signals picked up by the antenna, amplify it and mix it with the CW (continuous wave) output of the local oscillator to convert the channel frequencies to a band around the intermediate frequency of the receiver. It is the local oscillator that tunes in the desired station. Its frequency is unique for each channel, which determines the RF signal frequencies to be received and converted to frequencies in the pass-band of the IF section.

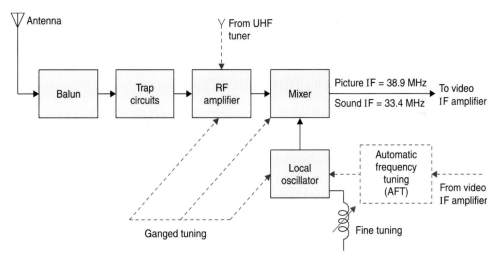

Fig. 22.1. Block diagram of a VHF tuner. Note that AFT is provided in colour receivers only.

22.1 TUNER OPERATION

As shown in the tuner block diagram the function of channel selection is accomplished by simultaneously adjusting the tuned circuits of all the three stages that make up the tuner. This means that three or four tuned circuits must be switched while changing channels. The tuned circuits found in both vacuum tube and transistor tuners are as follows:

(i) Input tuned circuit to the RF amplifier

(*ii*) Output tuned circuit of the RF amplifier

(*iii*) Input tuned circuit to the mixer

(*iv*) Local oscillator tuned circuit

In some tuners the mixer input tuned circuit is left out and thus they have only three tuned circuits. Each tuned circuit consists of a coil and a capacitor. The resonating capacitance consists of distributed capacitance of the circuit plus small fixed ceramic capacitors. The fine tuning control is varied to obtain exact picture and sound intermediate frequencies for which the receiver is designed. The correct setting of the local oscillator frequency is indicated by the best picture obtained on the screen. Physically, the tuner is wired as a sub-assembly on a separate chassis which is mounted on the main receiver chassis. The IF signal from the tuner is coupled to the first picture IF amplifier through a short coaxial cable. The AGC line and B_+ supply are connected from the main chassis to the tuner subchassis. The tuner is enclosed in a compact shielded box and all connections are usually brought out at the top of the tuner via feed through capacitors. In tube-type receivers, the filament power to the tuner tubes is also supplied from the common power supply.

Though the principle of selection in television receivers is the same as that in radio receivers the design of the tuner is very different from the heterodyning section of the radio receiver. This is because of large band width requirements and very high frequencies employed for TV transmission. In fact tuner design techniques differ so much at ultra high frequencies that all true multichannel receivers employ separate tuners for the VHF and UHF channels. We will confine our discussion to channels with frequencies 47 to 223 MHz in the VHF band (30 to 300 MHz). The basic principle of UHF tuners is discussed separately in a later part of this chapter.

22.2 FACTORS AFFECTING TUNER DESIGN

The factors that must be considered before attempting to learn details of the tuner circuitry are:

(a) Choice of IF and Local Oscillator Frequencies

The merits of choosing intermediate frequencies close to 40 MHz were fully explained in Chapter 8. However, it may again be emphasized that such a choice results in (*i*) high image rejection ratio, (*ii*) reduced local oscillator radiation, (*iii*) ease of detection and (*iv*) good selectivity at the IF stages. The local oscillator frequency is kept higher than the channel carrier frequency since this results in a relatively narrow oscillator frequency range. It is then easier to design an oscillator that is stable and delivers almost constant output voltage over its entire frequency range.

Intermediate frequencies have been standardized in accordance with the total bandwidth allotted to each channel in different TV systems. In the 625-B monochrome system, the picture IF = 38.9 MHz and sound IF = 33.4 MHz.

(b) Need for an RF Amplifier Stage

In principle an RF amplifier is not necessary and the signal could be fed directly to the tuned input circuit of the mixer as is the normal practice in radio receivers. However, at very high frequencies the problems of image signals, radiation from the local oscillator through the antenna circuit and conversion loss at the mixer are such that a stage of amplification prior to the mixer is desirable. Another very important purpose of providing one stage of amplification is to feed enough RF signal into the mixer for a clean picture without snow. White speckles called snow are caused by excessive noise signals present in the video signal. The mixer circuit generates most of the noise because of heterodyning. Except in areas very close to the transmitter the signal level is moderate and a low noise RF amplifier stage with a gain of about 25 db becomes necessary to obtain the desired signal-to-noise ratio of 30 : 1. In fringe areas where the signal level is very low an additional RF amplifier known as *booster* is often employed to maintain a reasonable signal-to-noise ratio.

In all tuner designs the gain of the RF amplifier is controlled by AGC voltage to counter any variations in the input signal to the receiver. A signal level of about 400 µV at the input to the receiver is considered adequate for a snow free picture.

(c) Coupling Networks

Parallel tuned networks are used to accomplish the desired selectivity in RF and IF sections of the receiver. For minimum power loss the coupling network should match the output impedance of one stage to the input impedance of the following stage. In circuits employing tubes this does not present any problem because the output and input impedance levels are not very different and the main criterion is to transfer maximum voltage to the next stage. However, in transistor amplifiers, where the input circuit also consume power, the interstage coupling network must be designed to transfer maximum power from the output of one stage to the input of the following stage. The coupling design is further complicated by the fact that in transistors the output and input impedance levels are much different from each other. For example the input impedance of a common base configuration at very high frequencies is only a few tens of ohms, whereas its output impedance is of the order of several tens of kilo-ohms.

22.3 BASIC COUPLING CIRCUITS

Before attempting to discuss specific coupling arrangements it is necessary to have a clear concept of impedance levels, matching and bandwidth of tuned circuits. Consider the basic tuned coupling circuits of Fig. 22.2. In Fig. 22.2(a) the impedance matching is obtained by connecting the input between terminals 2 and 3 instead of 1 and 3. Since the ratio of impedance levels is associated with square of the turns ratio, the input impedance between points 2 and 3 will be one fourth of the impedance between points 1 and 3, if point 2 is at the centre tap of the coil.

Figure 22.2(b) shows another approach to selecting any impedance level where the capacitor C has been split into two separate capacitors C_1 and C_2 to provide a tap. In this case $Z_{2-3} = Z_{1-2} \times \left(\dfrac{C_1}{C_2}\right)^2$. Thus the input impedance Z_{2-3} increases with a decrease in the value of C_2. Note that as in the previous circuit the resonant frequency and quality factor (Q) of the tuned circuit remain the same if $C = (C_1 \times C_2)/(C_1 + C_2)$ is maintained.

Figure 22.2(c) shows the basic transformer coupling configuration where the tuned primary circuit is coupled to a nonresonant secondary. In this case if N_1 and N_2 are the number of turns in the primary and secondary respectively, the required turns ratio for an impedance match between the two sides is $n = N_1/N_2 = \sqrt{r_0 / r_i}$ where r_0 represents the impedance of the driving circuit and r_i that of the load circuit.

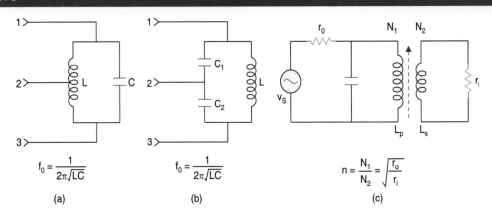

$$f_0 = \frac{1}{2\pi\sqrt{LC}}$$

(a)

$$f_0 = \frac{1}{2\pi\sqrt{LC}}$$

(b)

$$n = \frac{N_1}{N_2} = \sqrt{\frac{r_0}{r_i}}$$

(c)

*Fig. 22.2. Impedance matching in coupling circuits (a) by inductive tapping
(b) by capacitive tapping and (c) by transformer action.*

The use of these concepts is illustrated by a practical example. Assume that a coupling network is to be designed for a high frequency bandpass amplifier using a common emitter configuration both in the input and output circuits. Taking r_0 of transistor $Q_1 = 10$ K ohms and r_i of transistor $Q_2 = 500$ ohms, the calculated value of L_p (primary inductance) and L_s (secondary inductance) for a given bandwidth turns out to be of the order of a few μH and a fraction of a μH respectively. To construct a transformer with a high value of unloaded Q of, say, 100 using a very low primary inductance of a few μH is very difficult. In practice the problem is solved by using a tapped primary winding where the inductance of the primary winding can be made many times the calculated value. An auto-transformer can also be used by choosing appropriate taps both for input and output circuits. Figure 22.3 (a) shows such a configuration.

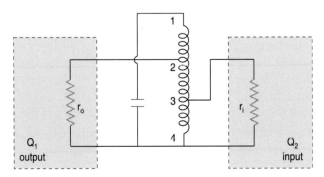

*Fig. 22.3 (a). Interstage network using autotransformer
coupling with tuned primary winding.*

Capacitive Coupling

In many applications difficulty is often encountered in obtaining 'unity' coupling between primary and secondary windings. This problem becomes particularly severe in common-base amplifier configurations operating at higher channel frequencies because the input impedance becomes as low as 20 ohms. In such cases capacitive coupling is often used. Figure 22.3(b) shows such a circuit configuration. Since C_1 is effectively shunted by the low input resistance of the common-base circuit (transistor Q_2) the total capacitance $C_T = C_1 + C_2$, and f_0 (centre frequency)

$= 1/\sqrt{2\pi L_1 C_T}$. To achieve a higher quality factor the value of inductance L_1 can be increased and C_T accordingly decreased. However, this would need connecting the output of transistor Q_1 to a suitable tapping on the coil to maintain the desired impedance match.

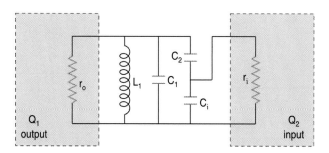

Fig. 22.3 (b). Interstage network employing capacitive coupling using two capacitors.

Double Tuned Networks

A single tuned circuit, employing any one form of the couplings discussed above, is normally used in the input circuit of the RF amplifier to provide a match with the antenna circuit. The output circuit of this amplifier is usually double tuned because most of the second channel rejection of the tuner is effected here. The advantages of double tuned interstage coupling networks (Fig. 22.4 (a)) over single tuned circuits are:

    (i) the frequency response is flatter within the pass-band.

    (ii) the drop in response is sharper immediately adjacent to the ends of the pass-band.

    (iii) attenuation of frequencies not in the pass-band is higher.

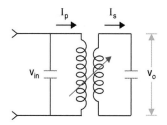

Fig. 22.4 (a). Double tuned circuit.

    The above advantages are clearly seen in the response curves shown in Fig. 22.4 (b). For loose coupling the response curve has a single peak at f_0 like a single tuned circuit but with low amplitude. As the coupling is increased, the output amplitude increases and at a middle value of coupling called 'critical coupling' the secondary has its maximum amplitude and greatest bandwidth without double-peaks. When the coupling is further increased to what is known as overcoupling, two peaks result in the secondary output to broaden the output response. Varying the coupling beyond critical coupling results in greater bandwidth of the coupled circuits that are tuned to the same resonant frequency. Also as shown in the response curve, for an overcoupled case the drop in response is sharper on either edge of the pass-band.

Chapter 22

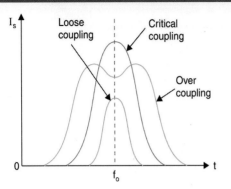

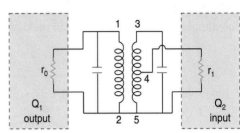

Fig. 22.4 (b). *Effect of coefficient of coupling (k) on the response curve of a double tuned circuit.*

Fig. 22.4 (c). *Double tuned interstage coupling network employing inductive coupling with a tap on the secondary to obtain impedance match and selectivity.*

Various forms of interstage coupling networks employing two tuned circuits are commonly used where both the primary and secondary sides are tuned to the same resonant frequency. Figure 22.4(c) shows a typical circuit where the secondary winding has been tapped to obtain the desired impedance match and selectivity. In all double tuned circuits damping resistances can be connected across both the primary and secondary circuits to get larger bandwidth with more flat top. The use of resistors, however, reduces gain because of decrease in the value of Q of both the circuits.

Impedance Coupling

Another form of coupling, that employs a common impedance Z_m for mutual coupling between two tuned circuits, is shown in Fig. 22.5 (a). This method is also called band-pass coupling because it usually has the bandwidth of an overcoupled double tuned transformer. In Fig. 22.5(b) coil L_m is common to both the tuned circuits with L_1, C_1 and L_2, C_1. As a result, signal voltage across L_m is coupled from one tuned circuit to the other. There is no direct coupling between L_1 and L_2 and both L_1, C and L_2, C_2 are pretuned by means of brass slugs to the same centre frequency. The bandwidth is controlled by the value of L_m. The greater the value of L_m the greater the mutual impedance and greater is the bandwidth. This type of coupling is commonly employed between the RF amplifier and the mixer. Values of L_m and damping resistances are chosen to provide a band pass of approximately 7 MHz.

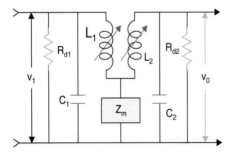

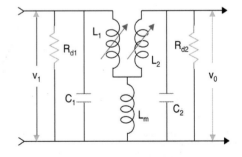

Fig. 22.5 (a). *Mutual coupling through a common impedance Z_m.*

Fig. 22.5 (b). *Inductive (L_m) bottom coupling.*

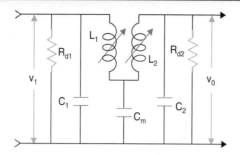

Fig. 22.5 (c). Capacitive (C_m) bottom coupling.

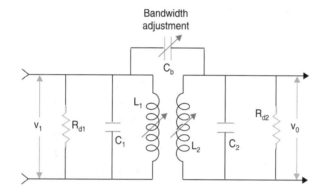

Fig. 22.6. Interstage broadband coupling with adjustable bandwidth.

Another form of this type of coupling is shown in Fig. 22.5 (c) where the coil L_m is replaced by a capacitor C_m. The smaller the value of C_m the greater the mutual impedance and more the bandwidth. The circuit arrangement shown in Fig. 22.6 is yet another method of coupling between stage in order to achieve a broadband response. Here a small capacitor connects the primary and secondary windings. The value of the capacitor is low (10 to 20 PF) and it controls the bandwidth. A damping resistance is sometimes connected across the tuned circuits to adjust the bandwidth.

22.4 TUNER CIRCUIT ARRANGEMENT

For earlier receivers special tubes were developed for use in tuner circuit. The tubes, both pentodes and triodes, are miniature type with reduced interelectrode capacitance and lead inductance. A pentode yields more gain but a triode is preferred because it generates less noise. This is because noise originating in a tube varies directly with the number of positive elements within a tube. To compensate for lower gain available from a triode special dual triodes have been developed for use in a cascode configuration. Special frame grid construction having sturdy supports is employed to reduce microphonic effects. Fine grid wire with close spacing between cathode and control grid results in increased transconductance. Thin grid also results in reduced grid to cathode (C_{gk}) capacitance. By means of grid guide plates, the electron flow is concentrated into a smaller area to reduce noise which otherwise results due to random motion of electrons. The twin triode RF amplifier is followed by triode-pentode mixer to convert channel frequencies into the IF region.

In tuners employing transistors, three transistors are normally employed, one each in RF amplifier, local oscillator and mixer circuits. In almost all tuners, whether tube or transistor version, a single tuned circuit provides coupling between the antenna circuit and RF amplifier and a double-tuned circuit is used between the RF amplifier and mixer.

Neutralization

Unwanted feedback due to junction or interelectrode capacitances is much greater in transistor than in tubes. This needs neutralizing. In a transistor amplifier circuit the signal fed from collector to base via the collector-base junction capacitance may be either regenerative or degenerative depending on the frequency of the signal and the nature of the collector load. The purpose of neutralization is to deliberately feed a signal from the collector to the base which is 180° out of phase with the feedback signal travelling through the collector-to-base junction capacitance. The neutralizing signal cancels the effect of undesired feedback.

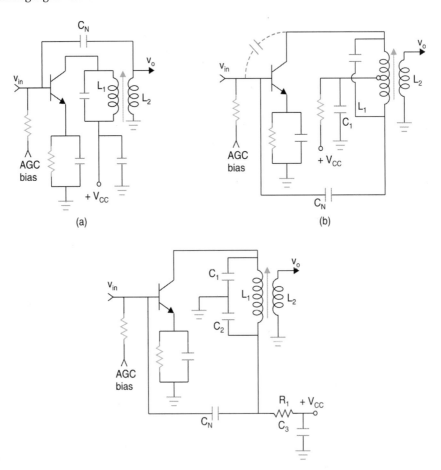

Fig. 22.7. Methods of neutralizing a high frequency amplifier stage. C_N is the neutralizing capacitor (a) feedback voltage from secondary of output transformer (b) feedback from tapped primary coil (c) feedback through capacitive voltage divider (C_1, C_2) across L_1.

Figure 22.7 illustrates three methods of obtaining signal voltage which is opposite in polarity to the collector voltage. The feedback capacitor C_N couples back the neutralizing voltage and blocks any dc voltage between the collector and base. In the circuit of Fig. 22.7(a) feedback voltage is obtained from the secondary winding which is 180° out of phase with respect to the primary. In Fig. 22.7(b) the bottom end of L_1 has signal of opposite polarity with respect to the top. The ground is provided by C_1. Therefore the neutralizing voltage fed back by C_N cancels the internal feedback.

The circuit of Fig. 22.7(c) uses a capacitive voltage divider arrangement for feedback. It may be noted that the junction of C_1 and C_2 is grounded thus providing out of phase voltage at the two ends of the winding with respect to ground. Thus the voltage fed back through C_N is out of phase with the internal feedback voltage. The isolating resistance R_1 is needed in the collector circuit to prevent shorting of the signal through low impedance of the dc supply. This feedback arrangement is often used in RF amplifiers while the previous two circuits find favour in IF amplifiers.

22.5 TYPES OF TUNERS

All TV receivers have separate VHF and UHF tuners. The VHF tuner may be single channel or multichannel to cover all Band I and Band III channels. Normal switching methods of tuned circuit selection are not used because longer leads may cause parasitic oscillations. The VHF tuners are either turret type or rotary-wafer type.

(i) Turret Tuner

The turret or drum type tuner is so named because coils for the various channels are mounted on a slotted drum type structure that is rotated by the channel selector. The tuned circuits are mounted on separate strips for each channel which are clipped into the turret. The coils required are selected by rotating the turret thus maintaining the shortest possible length of connections. Each strip has coils for RF amplifier, mixer and local oscillator for a single channel. The schematic of a VHF turret tuner shown in Fig. 22.8 illustrates the manner in which a coil strip on the drum makes connections with the rest of the circuit.

(ii) Wafer or Incremental Tuner

This tuner employs a wafer switch type construction where a tier of wafer switches permits selection of the proper RF, mixer and oscillator coils. The coils are usually mounted around the outer rim of the switch with a few turns of wire for the lower channels and progressively decreasing for the higher channels. The wafer switch is rotated to connect one set of RF, mixer and oscillator coils for each channel. Figure 22.9 shows schematic of VHF tuner-wafer switches. Note that the coils for RF amplifier, oscillator and mixer stages are on separate wafer switches ganged on a common shaft. The oscillator coils are always on the front section for convenience of adjusting the inductance to set the frequency.

The wafer type tuners are also called incremental tuners because the change of channel is accomplished by a progressive shorting of sections of the total inductance. For channels 2 to 6 individual coils are connected in series and are progressively shorted out. For channels 7 to 11 only a single turn or half a turn is sufficient for tuning.

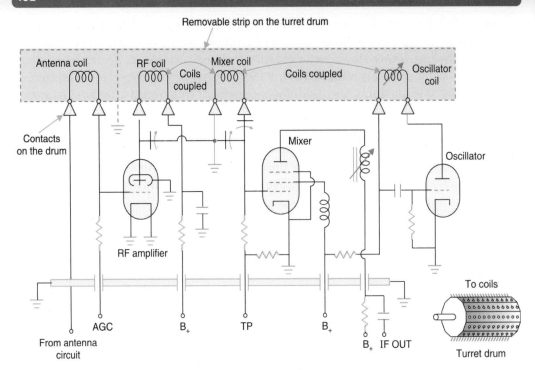

Fig. 22.8. VHF turret tuner.

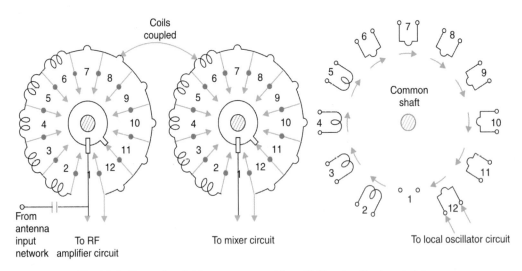

Fig. 22.9. Channel switching arrangement in a VHF wafer (incrimental) tuner.

Channel I is not assigned for TV broadcasting any more and this position on the selector switch is generally used to turn on the UHF tuner. The recent version of tuner wafer construction utilizes a printed circuit wafer. In this method all coils for higher channels are replaced by printed circuit inductances. This method has the advantage of greater reliability, greater uniformity of alignment and lower cost.

Channel tuning. There is a separate adjustment for each channel. Resonance is achieved with distributed capacitance of the coil, stray and trimmer capacitors. For fine tuning the oscillator coil is usually constructed with an aluminium or brass screw as the core. Inductance decreases with an aluminium or brass core because of eddy currents in the metal. The screw is turned by a plastic gear wheel which is engaged when the fine tuning control is pushed-in against the holding springs. The frequency variation thus obtained is enough to tune in the next channel station.

Some receivers employ automatic frequency tuning (AFT) where an automatic frequency control is applied to the local oscillator in the tuner. Such circuits are discussed in a separate chapter devoted to special circuits.

Mechanical detent. Good connections at the switch contacts of the tuner are essential to ensure resetability. Resetability is the ability of a tuner to repeatedly return to exactly the same frequency after being switched. To achieve this the tuner has a wheel with notches for each position. Each notch or detent holds the switch steady for good contact by means of a detent spring. The tuner should be tight to turn as it clicks into position on each channel. To achieve good connection the switch contacts are silver plated for very low contact resistance essential for selectivity in very high frequency circuits.

Tuner subassembly. The tuner with tubes or transistors is a separate unit on a subchassis. It is often situated at some distance from the IF amplifiers to avoid any radiation and interference problems. In tuners special capacitors known as 'feed through' capacitors are used for bypassing on the B_+ and AGC lines into the tuner. As shown in Fig. 22.10 one side of the capacitor is the metal base mounted on the chassis. The two end terminals are both connected internally to the opposite capacitor plate. This construction has very low inductance and prevents radiation of signal from the leads which must go through metal case of the tuner. The end terminals also serve as convenient points for connecting test equipment.

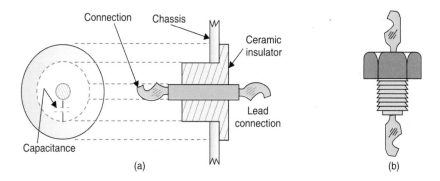

Fig. 22.10. Feedthrough capacitor (a) constructional details (b) pictorial view.

22.6 VARIOUS SECTIONS OF A VHF TUNER

As shown in Fig. 22.11 the tuner consists of the following sub-sections.

(a) Balun-Input Transformer

The tuner must have an input impedance equal to the characteristic impedance of the antenna feeder for maximum signal transfer to the tuner and to avoid reflections on the line. The balun

Chapter 22

matches the twin wire ribbon feeder impedance of 300 ohm to the 75 ohm input impedance of RF amplifier. It consists of two 150 ohm (Z_0) quarter-wave sections of bifilar windings on a ferrite core. As shown in Fig. 22.12, at one end the two quarter-wave lines are connected in parallel to give 75 ohm impedance and at the other end they are connected in series to provide an impedance equal to 300 ohms. If a coaxial cable is used to connect the antenna to RF amplifier, an identical parallel arrangement can be used on both sides of the balun to provide the match. Either side of the balun can be used for input with the opposite side for output.

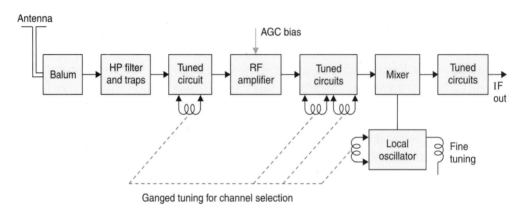

Fig. 22.11. Sub-sections of a VHF tuner in a monochrome receiver.

The capacitors C_1 and C_2 (200 pF), one in each lead, isolate the antenna from the chassis as a safety precaution, in transformerless receivers and to prevent damage due to lightning. The 2 MΩ shunt resistors R_1 and R_2 discharge any static charge accumulated on the capacitors.

(b) HP Filter and Trap Circuits

Figure 22.12 shows various trap circuits and a high-pass (HP) filter that follow the balun. Unwanted spurious signals which lie in the IF band of 32 to 42 MHz are rejected by two trap circuits connected on cither side of the HP filter configuration. The HP filter has a cut-off frequency of 45 MHz and thus rejects all frequencies below channel number 2. These filter

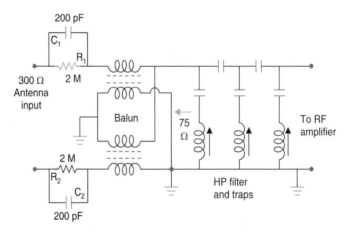

Fig. 22.12. Typical input circuit of a VHF tuner.

sections are essential because low 'Q' RF circuits may not be able to attenuate the adjacent channel frequencies due to their broad bandwidth, particularly when the receiver is tuned to lower channels. It is very difficult to reject such signals once they reach the mixer and IF stages. In some receivers a parallel resonant trap circuit is provided in series with the signal path to reject signals from the commercial FM broadcast band extending from 88 to 108 MHz. The balun and filter circuits are normally placed side by side on separate mounts very close to the tuner.

(c) RF Amplifier

The RF amplifier is designed to provide adequate gain to weak signals. It is essential to maintain a high signal-to-noise ratio at the mixer which generates lot of noise. The equivalent noise voltage at the input of the RF amplifier sets a limit to the minimum signal strength that should be received to avoid excessive snow effects. As already stated RF amplifier besides providing gain to the selected channel signals (i) prevents radiation from the local oscillator and (ii) rejects image frequency signals.

The RF stage is best suited for AGC because the signal amplitude is small and the gain control is most effective with least distortion. Application of AGC is usually delayed to the RF amplifier in order to maintain good S/N ratio for weak signals.

The RF amplifier should have a pass-band (see Fig. 22.13) broad enough to pass the selected channel and allow for small variations in the tuned circuit frequencies. Double tuned circuits adjusted to provide a flat top response are used for coupling the RF amplifier to the mixer stage. The response as shown in Fig. 22.13 has a dip of about 1 db in-between the two peaks.

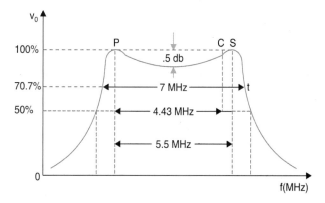

Fig. 22.13. RF amplifier response.

(d) Local Oscillator

The local oscillator generates an unmodulated sinusoidal (CW) voltage that is heterodyned in the mixer with the incoming RF signal to affect its frequency translation to the IF band. The local oscillator frequency is changed whenever a different channel is selected to maintain difference equal to the intermediate frequency between the local oscillator frequency and the tuned channel carrier frequency. The fine tuning control allows the oscillator frequency to be varied over a narrow range.

An important characteristic of the local oscillator used in tuners is its freedom from 'drift'. Drift occurs due to temperature changes, ageing of circuit components, changes in dc voltage and variation of the oscillator load. Special compensation techniques are used to obtain oscillator stability. In monochrome receivers a small amount of drift can be tolerated but in colour receivers even very small frequency changes can result in the loss of colour portion of the signal. Therefore in tuners designed for colour receivers, ATC (automatic tuning control) or AFT (automatic frequency tuning) circuit is used for strict control of the oscillator frequency.

In tuners employing transistors the problem is more serious because transistor operation is highly temperature dependent and sensitive to dc supply fluctuations. In such circuits, in addition to the use of usual compensation techniques and high quality components, a stabilized power supply and special capacitors with zero temperature coefficient are employed.

(e) Mixer Stage

The function of this circuit is to convert the incoming RF signal frequencies from different channels into a common IF pass-band of the receiver. This is achieved in the mixer by heterodyning or beating the local oscillator frequency with the signal obtained from the RF amplifier. The mixer stage, combined with the local oscillator, may be considered as a frequency converter. To produce frequency translation, rectification in the mixer is necessary. This is achieved by keeping the amplitude of the local oscillator signal large and operating the mixer tube or transistor in the non-linear region of its dynamic characteristics. The operation in the nonlinear region results in serve amplitude distortion which in turn generates sum and difference frequencies of all the constituents of the input to the mixer. Higher harmonic products with decreasing amplitudes are also generated. However, the major products at the output of the mixer are the sum and difference frequencies (sidebands) which the input signal produces by beating with the strong local oscillator frequency.

Since the difference between the RF signal and local oscillator signal frequencies is appreciable, the two sideband regions are quite apart from each other. The desired IF band is thus easily separated by a tuned circuit. The double tuned circuit used at the output of the mixer is broadly tuned to select the desired band between 33.15 to 40.15 MHz and reject all other frequencies.

The conversion gain at the mixer stage is defined as the IF signal output divided by the RF signal input. Since the conversion trans-conductance in vacuum tubes is quite low, a gain of only about six is possible. With an RF input of 2 mV an IF output of about 12 mV becomes available. In mixers using transistors a conversion power gain between 15 to 20 db is obtained for most VHF channels.

22.7 ELECTRONIC TUNING

Varactors or varicaps are used for electronic tuning in tuner circuits. Varactor is a special silicon diode, the junction capacitance of which is used for tuning. This capacitance varies inversely with the amount of reverse bias applied across the diode. The resonant frequency of the tuned circuits in which they are connected, is controlled merely by changing the reverse bias across the varactor. Figure 22.14(a) illustrates how a varactor diode D is used in a resonant circuit. The capacitor C_1 is quite large and thus has negligible reactance at resonant frequency of the tuned circuit, but it is needed to block the dc supply voltage. The dc supply through R_1 supplies reverse bias to the diode. R_2 is the isolating resistance and R_3 can be varied to change the reverse bias voltage.

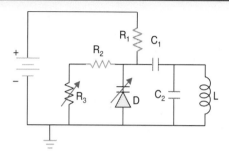

Fig. 22.14 (a). Basic circuit for varactor diode tuning.

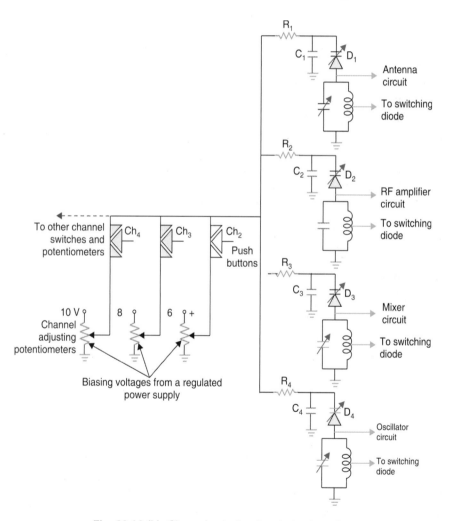

Fig. 22.14 (b). Channel selection by electronic tuning.

Since C_1 is a short for ac, it puts the capacitance of the varactor in parallel with C_2. At higher channels, capacitor C_2 is not necessary and only the varicap is enough to provide necessary capacitance.

Channel Selection

A simplified circuit arrangement of a common method by which various channels can be selected by electronic tuning is shown in Fig. 22.14 (b). In this circuit, selection of dc voltage for reverse biasing the varactor diode in each tuned circuit is done by push buttons. The amount of reverse bias determines the diode capacitance and thereby the resonant frequency of each tuned circuit. All the tuned circuits of a channel are adjusted simultaneously to ensure proper tracking. A separate potentiometer is provided for each channel. Each potentiometer acts as a fine tuning control for the associated channel tuned circuits. The dc tuning voltage is obtained through a voltage regulator for stable operation.

In Fig. 22.14 (b) the resistors R_1, R_2, R_3, and R_4 couple dc voltage to the varactors from potentiometers and isolate one varactor from the other. The capacitors C_1, C_2, C_3 and C_4 are large and provide a short to the RF signals. Therefore, from the ac point of view the varactors, in effect, appear in parallel with the tuned circuits.

Band Switching

The VHF channels 2 to 11 include a wide frequency range and it is too much of a variation for capacitive tuning. Therefore, part of the tuning coil is shorted to reduce its inductance for the high-band VHF channels. A switching diode is used for shorting the tap on the coil to ground. One possible method of doing so in illustrated in Fig. 22.15. The diode D_1 is the switching diode connected through C_1 to a tap on the tuned-circuit inductance L_1. When D_1 is not forward based, it is an open circuit and has no effect on the coil. However, when positive potential is applied at the anode of D_1 by closing the band switch, it conducts. Then its low resistance is equivalent to a short circuit across the lower section of the coil. The capacitor C_1 prevents the diode voltage from shorting to ground through the coil. Thus when the switch is open all the turns in L_1 are used. However, when S_1 is closed only some turns on L_1 are used for the high-band VHF channels 5 to 11.

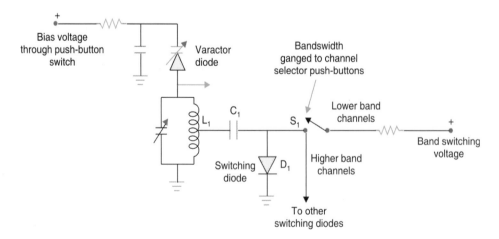

Fig. 22.15. A typical circuit for band changing by a switching diode in electronic tuning.

Only one switching diode is shown in Fig. 22.15. Actually each of the tuned circuits in the tuner has a switching diode connected to a tap on the corresponding coil. The positive band-switching voltage is applied to all the switching diodes. Note that switch S_1 is ganged with the channel selector to apply switching voltage when higher channels are selected. As explained earlier the push buttons continue to apply tuning voltage to the varactors for selecting each of the channels.

22.8   VACUUM TUBE TUNER

The tuner circuit shown in Fig. 22.16 (a) employs a twin-triode (PCC 88) in the RF section and a pentode-triode in the converter section.

RF Amplifier

The two triodes of PCC 88 connected in a cascode configuration constitute the RF amplifier. The grounded cathode stage V_1 drives the grounded grid amplifier V_2. The tube V_1 has AGC bias and is dc coupled to the cathode of V_2 through an inductor. RF input signal is applied through coupling transformer T_1 to the grid of V_1. In its plate lead is the resonant circuit formed by L_k, C_0 and C_{in} which form a pye type filter with a response broad enough to include all VHF channels. The plate load of V_1 is very small (input resistance looking into the cathode of V_2) and hence provides very little gain. Neutralization is affected by balancing the bridge formed by C_1, C_2 (variable), C_{gp} and C_{gk}. Most of the gain (close to that of a pentode) is given by the grounded-grid stage V_2 that needs no neutralization. Since V_1 and V_2 are identical and same dc current flows through them, the plate-cathode voltage for each tube is one-half of the B_+ supply. The voltage divider network formed by R_1 and R_2 provides the necessary grid bias for V_2. The amplified signal in the output circuit of V_2 is coupled by transformer T_2 to the mixer grid. Tuning of all the resonant circuits is obtained by the self input and output capacitances and no physical capacitors are normally provided. RF tuning for each station is changed by switching coils in transformer T_2.

Mixer Stage

The pentode-triode combination in a single glass envelope (PCF 80) functions as a converter. RF signal is coupled through L_1 and oscillator output through L_3 to the grid circuit (L_2) of the pentode. The local oscillator coil L_3 is positioned close to the mixer grid coil (L_2) for inductive coupling. A non-linear operation of V_3 is set with proper biasing obtained through both R_k-C_k in the cathode circuit and the grid-leak bias produced by the injected oscillator voltage. The non-linear operation results in heterodyning action.

Both plate and screen voltage dc supplies have decoupling networks to ensure proper operation of the mixer. Plate circuit of the mixer is tuned to select the desired IF band. It is coupled to the first IF amplifier through the tuned transformer T_3.

Local Oscillator

A modified Colpitt's oscillator (commonly known as ultraudion oscillator) employing the triode section of tube PCC 80 supplies sinusoidal voltage of constant amplitude at the desired frequency for each channel. Special feature of this oscillator is the use of tube capacitances alone for feedback and tuning. As shown in the equivalent circuit (Fig. 22.16(b)) of the tank circuit, the oscillator coil L_3 is connected from plate to grid across C_{gp}. The capacitances C_{pk} and C_{gk} form a capacitive voltage divider across L_3 and the oscillator voltage which develops across C_{gk} is fed back to the grid for sustained oscillations. Total capacitance across L_3 is C_{gp} in parallel with the series combination of C_{gk} and C_{pk}. The self capacitance values very between 2 to 3 pF needing an inductance (L_3) of about 1 μH for tuning channel 4 in the VHF band.

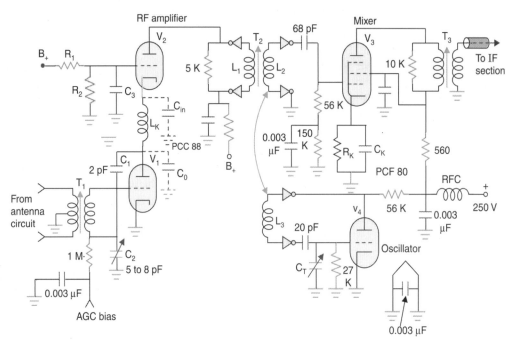

Fig. 22.16 (a). Vacuum tube VHF tuner.

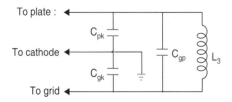

Fig. 22.16 (b). Ultraudion oscillator's equivalent circuit.

22.9   TRANSISTOR TUNERS

With their low power drain, miniature size and reliability, transistor are now universally used in tuners of modern TV receivers. However, the transistors chosen must have the required high frequency performance, provide adequate gain and low noise figure to achieve good overall sensitivity.

While the mixer stage employs a common emitter configuration, both common base and common emitter configurations can be used for the RF amplifier and local oscillator. The grounded base circuit, needs no neutralization, but on account of its very low input impedance, presents matching problems. Its signal handling capacity is also low in comparison with a common emitter arrangement. The grounded emitter configuration provides high overall gain and has a large signal handling capacity. But its parameters change with shift in the operating point due to AGC action and introduce higher cross-modulation. However, this can be reduced by inserting a small unbypassed resistance in the emitter circuit.

A dual gate MOSFET transistor can also be used in the RF section. It has the advantage of good gain, high input impedance, low noise factor and less cross-modulation effects. The dual-gate feature permits the use of simple AGC circuitry requiring very low control power. The BEL 2N300 FET has all the above merits and is employed as an RF amplifier in many tuner designs.

Transistor VHF Tuner

The circuit of a commonly used transistorized VHF tuner is shown in Fig. 22.17. It employs BF 196 for the RF amplifier and mixer stages, and BF 194-B in the local oscillator circuit. All the three transistors operate in grounded emitter configuration. The 75 ohm impedance of the balun on antenna side is matched to the input of Q_1 (RF amplifier) by a tap on coil (L_1) and the impedance transforming network formed by the two capacitors (10 pF and 15 pF) which shunt the coupling coil. Forward AGC voltage is fed at the base of BF196 via 1 KΩ resistance and a decoupling network. The 47 ohm resistance inserted in the base circuit helps to suppress any

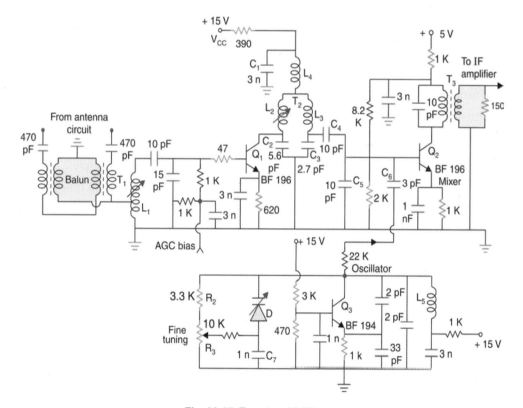

Fig. 22.17. Transistor VHF tuner.

parasitic oscillations. The RF amplifier output circuit is double-tuned and mutually coupled through common inductor L_4. Note that capacitors C_2 and C_3 are effectively in parallel with L_2, L_4 and L_3, L_4 respectively to provide tuning. The cores of L_2 and L_3 are adjusted to obtain a broad tuning with peaks at the picture and sound carrier frequencies.

Chapter 22

The output from the RF amplifier is coupled to the base of mixer transistor Q_2 through the voltage divider and matching network formed by capacitors C_4 and C_5. The oscillator output is also fed at the base of Q_2 through capacitor C_6. The mixer output circuit is double tuned and feeds the first IF amplifier. The 150 ohm resistor across the secondary winding of T_3 is a damping resistance to achieve broad tuning.

In the local oscillator circuit, L_5 in parallel with a 2 pF capacitor constitutes the tank circuit. The oscillator output frequency is adjusted by varying L_5. A varactor diode D is used for fine tuning. As shown in the circuit, it is connected in parallel with the tank circuit. Reverse bias voltage to the diode is fed through the voltage divider formed by R_2 and R_3. The potentiometer R_3 is provided on the front panel of the receiver. It serves as fine tuning control because varying R_3 changes the reverse bias voltage and hence the capacitance across the tank circuit. The capacitor C_7 blocks dc voltage shorting to ground.

22.10   UHF TUNERS

There are two types of UHF tuners. In one type the converter section of the UHF tuner converts the incoming signals from the UHF band (470 to 890 MHz) to the VHF band. Usually the UHF channels are converted to VHF channel numbers 5 or 6 and the VHF tuner, in turn, processes these signals in the usual way. Since the relative positions of video and sound IFs are specified, the converter oscillator of the UHF tuner is tuned to a frequency lower than the incoming UHF signal to take into account the double-heterodyning action involved.

In the second type, the UHF signal is heterodyned with local oscillator output to translate the incoming channel directly to the IF band. The output thus obtained from the UHF tuner is coupled to the VHF tuner, where the RF amplifier and mixer stages operate as IF amplifiers, when the VHF station selector is set to the UHF position. In this position the VHF oscillator is disabled by cutting off its dc supply. The RF tuned circuits are also changed to IF tuned circuits.

In present day tuners, this type is preferred. Figure 22.18 illustrates the block schematic of this circuit. Separate antennas are used for VHF and UHF channels. The UHF tuner does not employ an RF amplifier because the input signal strength being very low is comparable to the noise generated in the RF amplifier. If used it lowers the signal-to-noise ratio and fails to boost the signal level above the noise level. Therefore, the incoming UHF signal is directly coupled to the mixer where a diode is used for heterodyning instead of a transistor or a special tube. The reason for using a diode is two-fold. Firstly, the local oscillator output at ultra high frequencies is too low to cause effective mixing with an active device whereas it is adequate when a diode is employed. Secondly the noise level with a diode is lower than when an active device is used. Since the diode is a non-linear device it can produce beating of the incoming channel frequencies with the local oscillator frequency to generate side-band frequencies.

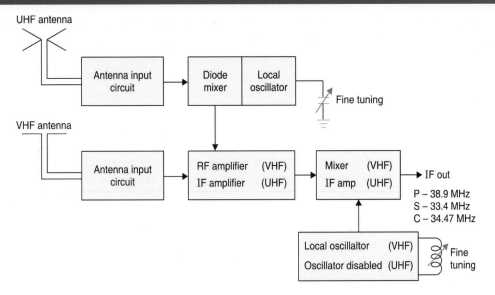

Fig. 22.18. Block diagram of a UHF-VHF tuner.

The weak output from the UHF tuner is coupled to the VHF tuner where both the RF amplifier and mixer stages acts as IF amplifiers to boost the signal to the level normally obtained when a VHF channel is tuned.

In conclusion the various functions performed by a television receiver tuner may be summarized as follows:

1. It selects the desired station and rejects others.
2. It converts all incoming channel frequencies to a common IF band.
3. It provides gain to the weak input signal picked up by the antenna. This improves *S/N* ratio and reduces snow on the picture.
4. It isolates the local oscillator from the feeder circuit to prevent undesired radiation through the antenna.
5. It improves image rejection ratio and thus reduces interference from source operating at frequencies close to the image frequencies of various channels.
6. It prevents spurious pick-ups from sources which operate in the IF band of the receiver.
7. It provides matching between the antenna and input circuits of the receiver, thus eliminating the appearance of ghost images on the screen.
8. It rejects any pick-up from stations operating in the FM band.
9. It has provision to select any channel out of the various allotted in the VHF and UHF bands for TV transmission.
10. It has a fine tuning control for accurate setting of the selected station.

Chapter 22

REVIEW QUESTIONS

1. Describe with a simple block diagram the circuit arrangement of a VHF tuner. Why does an RF amplifier stage always precede the mixer ?

2. Describe different impedance coupling circuits which are often employed between the RF amplifier and mixer. Sketch typical frequency response of the RF amplifier.

3. Why is neutralization applied to high frequency amplifier circuits ? Describe with circuit diagrams the methods which are normally employed for neutralization in transistor amplifier circuits.

4. Describe with suitable sketches different types of mechanical switching arrangements used for channel selection in tuners. What are 'feed through' capacitors and why are they used in a tuner sub-assembly ?

5. Draw detailed block schematic circuit of a VHF tuner and give necessary details of balun and filter circuit configurations provided at the input of the tuner. Why is a fine tuning control provided in all tuners ?

6. Describe briefly the basis design requirements that must be met in the RF amplifier mixer and local oscillator of a tuner for a stable and snow free reproduction of the picture.

7. What is electronic tuning ? Explain with circuit diagrams how different channel can be selected by this method.

8. The circuit diagram of a typical vacuum tube tuner is shown in Fig. 22.16 (a). Explain fully, how the circuit functions to produce frequency translation of all incoming channels to the common IF band. Why a cascode configuration has been employed in the RF section of the tuner ?

9. A commonly used transistor VHF tuner is shown in Fig. 22.17. Explain its working and the function of all the components used in the tuner.

10 Describe with a block schematic diagram the working principle of a UHF tuner. Why is an RF amplifier not used in UHF tuners ?

Video IF Amplifiers

T he tuner selects the desired TV channel and converts its band of frequencies to the intermediate frequency band of the receiver. While the total bandwidth of the channel is 7 MHz, the picture and sound intermediate frequencies are separated by 5.5 MHz. The receiver intermediate frequencies have been fixed at 33.4 MHz for the sound signal and 38.9 MHz for the picture signal. In the PAL colour system, colour subcarrier frequency is located 1.07 MHz before the sound carrier and its IF frequency is 34.47 MHz. Thus the band of frequencies to be handled by the IF sections of both monochrome and colour receivers lies between 33.15 MHz to 40.15 MHz. Practically all the gain and selectivity of the receiver is provided by the IF section.

As shown in the block diagram of Fig. 23.1, in colour receivers the sound take-off point is at the 3rd video IF amplifier. In monochrome receivers, where video and sound detection occurs together, the sound IF is separated at the output of the video detector. The video detector output feeds the video amplifier, sync separator and AGC circuits.

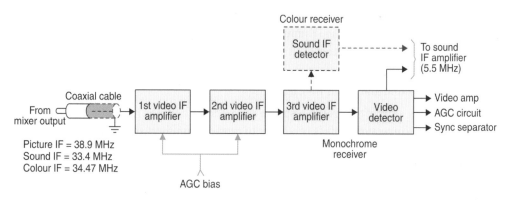

Fig. 23.1. IF section and sound take-off points for monochrome and colour receivers.

23.1 VIDEO IF SECTION

The desired overall response curve of the IF section of a black and white receiver and that of a colour receiver are shown separately in Fig. 23.2. In a monochrome receiver (Fig. 23.2 (a)) the sound IF lies at about 5% of the maximum amplitude because of intercarrier sound detection requirements. However, in colour receivers (see Fig. 23.2 (b)) where the sound take-off is before the video detector, the sound IF lies at almost zero level on the overall IF response curve. This is so because sound signal is no longer required in the video detector and if present can result in sound bars on the reproduced picture. Therefore, an additional trap circuit is provided at the input of video detector to further attenuate the sound IF by about 15 db. This is illustrated in the corresponding response curves.

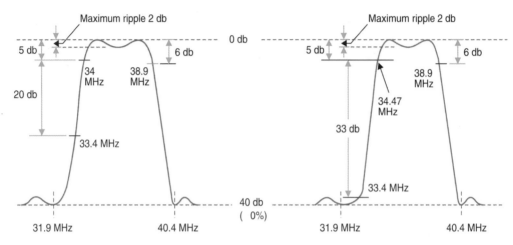

Fig. 23.2 (a). Overall IF response of a monochrome receiver.   Fig. 23.2 (b). Overall IF response of a colour receiver.

The choice of intermediate frequencies and other essential requirements to be met by the IF section were discussed in Chapter 8 which deals with the basic circuit arrangement of a television receiver. Based on that discussion and the need for obtaining overall IF frequency response curves shown in Fig. 23.2, the design requirements of the IF section can be summarized as follows:

(i) The combined gain of the IF amplifiers should be about 8000 to deliver a peak-to-peak signal of 3 to 5 volts at the input of the video detector.

(ii) The stage must provide necessary bandwidth and selectivity for faithful reproduction of picture, sound and colour information.

(iii) The upper skirt of the overall response curve should have the requisite slope for vestigial sideband correction. This is necessary to provide a uniform signal at the video detector over the entire range of video frequencies.

(iv) In monochrome receivers the lower skirt of the IF response should be designed to have a steep slope to reduce the gain at the sound IF to about 5 percent (–26 db) of the maximum gain. This is essential for successful functioning of the intercarrier sound system.

(v) In colour receivers the lower skirt of the IF response should have a slope so as to place the colour IF at about – 5 db level.

(vi) Additional trap circuit should be provided in colour receivers at the input of video detector to attenuate the sound IF by another 14 db.

(vii) The IF section of both monochrome and colour receivers must include rejection filters to suppress adjacent channel interference.

23.2 IF AMPLIFIERS

It is now obvious that the IF section of both monochrome and colour receivers is essentially the same but for small differences in the response curves. Each IF stage is a tuned amplifier. Because of the selective properties of the tuned amplifier the gain is quite large and falls off sharply at frequencies that lie on either side of the passband. In some IF circuit designs a double tuned circuit is used for coupling one stage to the other. This circuit arrangement is shown in Fig. 23.3 where C_{in} and C_{out} are the self shunting capacitances of the input and output circuits respectively. Their values are generally of the order of 10 to 15 pF.

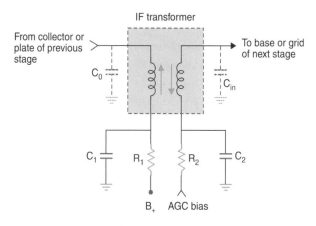

Fig. 23.3. Double tuned transformer of an IF amplifier.

Thus the required values of inductance are between 1 and 2 μH for resonance at about 40 MHz. Usually L_p and L_s each have a slug for tuning the primary and the secondary circuits. Besides the use of double-tuned circuits the quality factors of the tuned circuits are lowered to achieve desired bandwidth. This is affected by connecting damping resistors across the tuned circuits.

In Fig. 23.3, R_1-C_1 and R_2-C_2 are decoupling networks for the dc supply and AGC circuits respectively. It may be noted that C_1 and C_2 are actually parts of the tuned circuits. C_1 returns L_p to C_{out} and C_2 returns L_s to C_{in} to form resonant circuits.

Several tuned amplifier stages are cascaded together to provide a large gain to amplify the weak RF signal received from the mixer. Each stage operates under class 'A' condition for minimum distortion. At least one stage is AGC controlled to change the gain in accordance with changes in the signal picked up by the antenna so as to deliver a signal of almost constant peak to peak amplitude to the video detector. The last IF stage is not fed with AGC bias because here the signal amplitude is quite large and any shift in the operating point by the AGC bias would result in amplitude distortion.

The overall design of the video IF strip becomes quite complex because in addition to gain and bandwidth, matching of tuned circuits, stability, neutralizing and trap circuits are also to be taken into account for proper functioning of the video IF section.

Neutralizing and matching of tuned circuits were discussed at length in the last chapter. The same problems are there in the video IF stages. However, here matching and neutralization are

Chapter 23

less difficult because the operating frequency range is narrow, fixed and much lower than that obtained in tuner circuits.

23.3   VACUUM TUBE AND SOLID-STATE IF AMPLIFIERS

Video IF amplifiers employ miniature pentodes which may not need any neutralizing. Two or three stages in cascade are enough to provide the necessary gain of about 70 db. In vacuum tubes the input and output impedances are fairly high even at very high frequencies and so the quality factor Q, of the coupling networks remains high, thus enabling higher gain per stage. Furthermore, in tube circuits, the input and output impedances are comparable and so this does not present any serious impedance matching problems.

All present day receivers employ transistors and ICs in the IF section. In transistor IF amplifiers, three or four stages of tuned amplifiers become necessary to obtain the desired gain. This is so because the low input impedance of the common emitter configuration at very high frequencies shunts the tuned circuit to lower its Q, and this results in reduced gain per stage. The wide difference in the input and output impedance levels necessitates the use of special coupling techniques in transistor amplifiers. The junction capacitance is also large and makes neutralizing necessary for stable operation.

Integrated circuits have been developed where one chip contains all the three or four IF stages. It also includes the video detector and other allied circuits. The ICs do not, however, contain IF transformers or other coupling impedances. Tiny IF transformers and other circuit components not included in the integrated circuit are mounted around the IC module. These ICs are used in both hybrid and all-solid state receivers.

Amplifier Bandwidth

The overall bandwidth, gain and waveshape of the response curve, close to the ideal (Fig. 23.2) can be achieved in two ways—by synchronous tuning or by stagger tuning of the individual stages. In synchronous tuning the stages are tuned to the same centre frequency and the response of each stage is suitably designed to provide the required overall response. However,

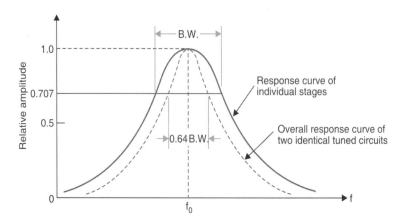

Fig. 23.4. Comparison of response curve of an individual stage with overall response curve of an identical cascaded pair.

when cascaded amplifiers are tuned to the same frequency the overall bandwidth shrinks drastically. The reason is that overall gain equals the product of individual gain values, not their sum. Then the peak values become more peaked while the low-gain frequencies become more attenuated. Figure 23.4 illustrates the response of individual stages and overall response when two identical stages are cascaded. As is obvious from the overall response curve, the peak value become more peaked while the low-gain frequencies are reduced in amplitude. The resultant sharp peaked response with narrow bandwidth is undesirable for the wide passband needed in the video IF amplifier. Thus if synchronous tuning is to be used a large number of IF stages, each designed to have a very wide bandwidth, would be necessary to achieve desired selectivity and gain. This obviously turns out to be very difficult and hence is not much used.

Stagger Tuning

Consider two single tuned amplifiers each with the same bandwidth Δ_1 but with their centre frequencies separated or staggered by an amount equal to the bandwidth (Δ_1). The individual response curves are illustrated in Fig. 23.5 (a), and the overall response of the pair is shown in Fig. 23.5 (b). As a result of stagger tuning the overall bandwidth (to 0.707 point) is $\sqrt{2}$ times the bandwidth of a single stage. It is also seen that staggering has reduced total amplification at the centre frequency (f_0) to half the value obtained before detuning. This is so because at the centre frequency each detuned stage has an amplification of 0.707 of the peak amplification of the individual stage ($0.7 \times 0.7 = 0.5$). Thus the equivalent voltage amplification per stage of the staggered pair is 0.707 times the amplification of the same two stages when used without staggering.

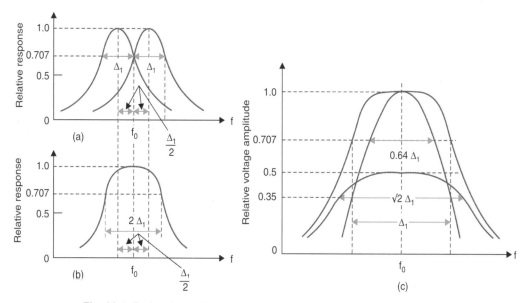

Fig. 23.5. Basic relationships in a stagger-tuned pair and its comparison with single-tuned and double-tuned circuits (a) response of individual stages in a staggered pair (b) overall characteristics of the pair (c) comparison of non-staggered and staggered pairs.

It may be noted that, since amplification curve of the stagger tuned pair has a bandwidth corresponding to that of a critically coupled double-tuned stage, the half-power bandwidth of the staggered pair is $\sqrt{2}$ times as large as the half-power bandwidth ($\Delta_1 = f_0/Q$) of an individual single-tuned stage. Hence the equivalent nominal gain-bandwidth product per stage of a stagger-tuned pair is $0.707 \times \sqrt{2} = 1.00$ times that of the individual single-tuned stages.

Next consider that stagger tuning is retained but the bandwidth of individual tuned circuits is decreased. This is illustrated in Fig. 23.5(c) where overall response of the two stage stagger tuned pair is compared with the corresponding two-stage non-staggered pair having the same resonant circuits. By readjusting circuit Q's and by detuning to give the staggered pair the same bandwidth as that of an individual single-tuned stage, it is possible to obtain a staggered pair which has the same equivalent voltage amplification per stage as that of the single-tuned stage. Thus two-stage staggering systems will have the characteristics shown by the outer curve in Fig. 23.5 (c).

To clarify this further, suppose that the bandwidth of each individual stage is decreased to 0.707 of its original value. To do this 'Q' of the individual circuit is raised to 1.4 times its previous value, which will increase the gain by a factor of 1.4. Now if the stages are staggered by an amount equal to the reduced bandwidth, the overall gain is one-half the product of 1.4 times 1.4, i.e., $1/2(1.4 \times 1.4) = 1$. Thus the overall gain is equal to that obtained earlier when both circuits were tuned to the same frequency. This is the effect of stagger tuning.

Note that in comparison, the two-stage identically tuned system with the same gain per stage, as the staggered pair under consideration, will have a bandwidth of only 0.644 times that of the staggered pair. This is because of the narrowing which results when two identically tuned single-tuned stages are cascaded. This is also illustrated in Fig. 23.5(c).

The concept of stagger tuning is extended to all the three IF stages to get the desired bandwidth. As shown in Fig. 23.6 (a) each circuit is tuned to a different centre frequency. The Q of the coil tuned to the centre frequency of the band is considerably lowered by damping to get a broad response in the middle range of frequencies. The final response of the triplet is shown in Fig. 23.6 (b).

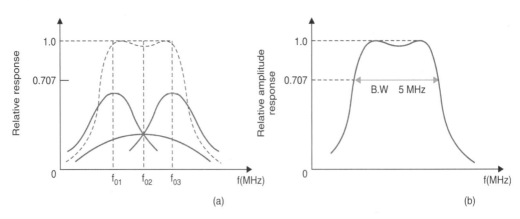

Fig. 23.6. Illustration of stagger—tuning when extended to all the three IF stages.
(a) response of individual stages each tuned to a different centre frequency
(b) final response of the triplet when Q of each stage is readjusted to a suitable value.
Q of f_{o2} (circuit) < Qs of f_{o1} and f_{o3} circuits.

A similar response can be achieved by employing double-tuned circuits in all the three IF stages. However, stagger tuning is preferred since it is (i) less expensive, (ii) easy to tune and (iii) more stable because no two circuits have the same centre frequency and therefore the probability of self-oscillations on account of feedback is low.

Interstage Coupling

The stagger tuning described above is a method of obtaining desired bandwidth and is not a form of coupling. Stagger tuning can be accomplished with either transformer coupling or with impedance coupling. Figure 23.7 (a) is an example of impedance coupling often used in vacuum tube IF stages. AGC bias is fed through R_2 at the control grid. In the absence of any signal AGC voltage drops to zero and the small cathode resistor R_1 prevents the tube from being operated without bias. The plate load consists of a tunable coil. The output signal is taken across this (IF) coil and delivered to the next stage through a capacitor.

A similar coupling arrangement in transistor circuits is shown in Fig. 23.7 (b). However, in this circuit a tap on the coil provides better impedance match to the low input impedance of the transistor employed in the next IF stage.

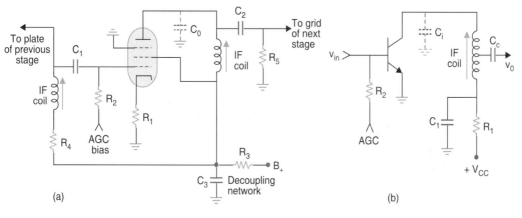

Fig. 23.7. Typical examples of impedance coupling in IF amplifier stages
(a) in vacuum tube circuits (b) in transistor circuits.

Bifilar IF Coils

This type of coil has two windings like a transformer but is not double-tuned. As shown in Fig. 23.8 (a), a bifilar coil is wound with twin conductor wires each insulated from the other. One winding is for the plate or collector and the other winding is for the grid or base of the next stage. Because of unity coupling the output is essentially a single-peaked response, because the double-humped overcoupled response changes into two independent peaks which are widely separated. Only one of the peaks (the lower frequency one) is used for tuned amplifier applications. Since the IF signal is inductively coupled to the next stage, no coupling capacitor is needed. Only the primary is tuned and the coil acts like a single-tuned circuit. Because of these merits it is widely often used in IF stages. The circuit of the coupling network employing a bifilar coil is shown in Fig. 23.8 (b). A tuning slug is moved inside the former on which the coil is wound for tuning to the desired frequency.

Chapter 23

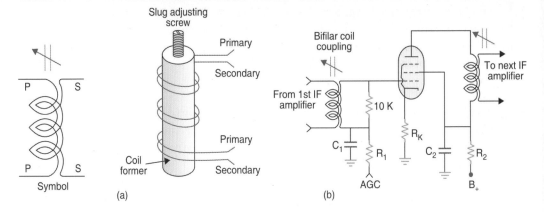

Fig. 23.8. Bifilar IF coil and coupling (a) pictorial view and symbol (b) typical coupling circuit.

Link-coupling

This method is generally used to couple the IF signal from mixer to first IF stage on the main chassis. The circuit arrangement is shown in Fig. 23.9. The link from tuner to the IF stage is a 75 ohms coaxial cable with shielded plugs at both ends to couple the signal at a low impedance level. As a result, the effect of stray capacitance is minimized. Since the shield of the cable is grounded, it prevents any pick-up from interfering signals. The shielded cable also prevents any radiation of the mixer output which otherwise can cause interference in the receiver. As shown in Fig. 23.9, the secondary of the transformer T_1 has just a few turns for a stepdown of voltage and impedance to provide a match to the coaxial cable. Similarly at the other end of the cable a step-up transformer T_2 is provided to raise the signal voltage that is fed to the grid or base of the first IF amplifier stage. Thus the IF signal is effectively coupled from the mixer output to the IF input by means of a low impedance coaxial cable of suitable length without radiation or any stray pick-up. The link coupling generally has the response of an overcoupled stage.

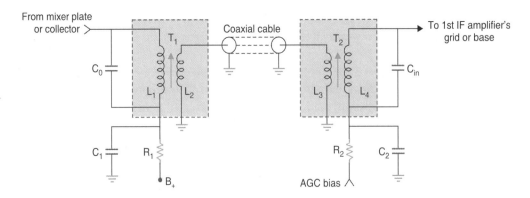

Fig. 23.9. Link coupling from mixer output to the input of 1st IF amplifier on the main chassis.

23.4   VESTIGIAL SIDEBAND CORRECTION

The need for vestigial sideband correction was clearly established in Chapter 4 which deals with channel bandwidth requirements. As explained there, if such a correction is not applied, frequencies between dc to 0.75 MHz will appear twice in the video detector output and thus produce an output that is double of that at higher video frequencies. Also, frequencies that lie between 0.75 to 1.25 MHz will have intermediate values because of the attenuation slope in this region at the transmitting end. The result of this would be that frequencies up to 0.75 MHz will get undue emphasis as compared to higher video frequencies. The extra gain for low video frequencies increases contrast in the reproduced picture that is usually accompanied by excessive smear. Similarly, a relatively lower emphasis of the high video frequencies reduces high frequency detail that is essential to make the picture sharp and clear. This clearly establishes the need for vestigial sideband correction.

In order to equalize the effect of vestigial sideband transmission, the overall IF response is aligned to give the picture carrier approximately 50 percent of maximum response as shown in Fig. 23.2. As a result, sideband frequencies of the double sideband signal are allowed an average response of 50 percent as compared to 100 percent response for the single sideband frequencies. Thus at low frequencies, the output that results from the two sidebands adds up to give an output equal to that at higher video frequencies. The relative output from the video detector will then be the same for all video modulating signals having the same amplitude, irrespective of whether they are transmitted with single or double sidebands. The necessary attenuation slope around the picture IF frequency is provided by suitably choosing the centre frequencies of the IF stages and also by varying their coefficient of coupling. Additional filter (trap) circuits are also used to control the exact shape of the IF section response.

23.5   THE IF SOUND SIGNAL

The picture and sound IF signals are amplified together in the IF section of the receiver. However, the gain around the sound IF frequency is reduced to about 5 percent of the maximum gain, both for successful working of the intercarrier sound system and to avoid appearance of sound bars on the screen. Special rejection filters (trap circuits) are provided normally at the input of the first IF stage, to achieve the desired attenuation slope.

23.6   ADJACENT CHANNEL INTERFERENCE

Several interfering signals are picked up by the receiver antenna along with the signal from the selected channel. Many of these are suppressed in the RF tuner and do not reach the first IF stage. However, some adjacent channel interfering signals are so strong and close to the selected channel, that they pass through the tuner section, and appear at the input of the first IF stage. The major sources of interference are the difference products of the local oscillator frequency beating with the sound carrier of the lower adjacent channel and the picture carrier of the upper adjacent channel. As explained in Chapter 8, these occur at 40.4 MHz and 31.9 MHz, *i.e.*, at the two edges of the IF pass-band. It is essential to reject them at the output of RF tuner, because the IF stages provide a very large gain and it is very difficult to remove such signals once they pass through the IF section. Therefore, trap circuits that provide a rejection of at least 40 db are

Chapter 23

included at the input of the first IF stage to stop their passage to the succeeding IF stages. The effect of such trap circuits is illustrated in the response curve of Fig. 23.2.

IF Wave Traps

A wave trap is a resonant circuit tuned to attenuate a specific frequency. Basically there are five types of trap circuits used at the input of the video IF amplifiers. These are (*a*) series tuned, (*b*) parallel tuned, (*c*) absorption type, (*d*) degeneration type and (*e*) bridge type configurations.

(*a*) *Series trap circuit*—The series trap is a parallel resonant circuit. It offers very high impedance at the rejection frequency to which it is tuned. It is connected in series between two IF stages as shown in Fig. 23.10 (*a*). Since the trap is a sharply tuned-circuit designed to reject one frequency or at most, a narrow band of frequencies, it offers negligible impedance to all other frequencies and thus the desired signal is passed without any attenuation.

(*b*) *Parallel trap circuit*—This trap is a series resonant circuit connected in shunt with the path of the IF signal. Its impedance is very small at the frequency to which it is tuned. It thus acts as a short at the frequency that is to be rejected and bypasses it to ground. Its connection in the IF stage is illustrated in Fig. 23.10 (*b*). It is important that the parallel trap circuit has a very high Q so that is rejects only a narrow range of frequencies.

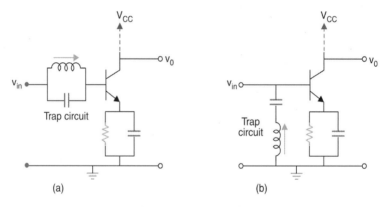

Fig. 23.10. Trap circuits in IF amplifier configurations (a) series (parallel resonant) trap circuit (b) shunt (series resonant) trap circuit.

(*c*) *Absorption trap circuit*—The absorption trap circuit shown in Fig. 23.11 (*a*) is inductively coupled to the plate or collector load inductance of the IF stage. The trap is tuned to the frequency that is to be rejected. As a result, at this frequency maximum current flows within the trap circuit. Since the trap circuit draws energy from the tuned load coil, a sharp decrease in the Q of the tuned load circuit occurs with the result that the stage gain at the undesired frequency is very much reduced. This amounts to a large attenuation at the frequency to which the absorption trap is tuned.

(*d*) *Degenerative trap circuit*—This circuit, illustrated in Fig. 23.11 (*b*), applies a large negative feedback to the IF amplifier at the frequency to be rejected. A parallel L-C circuit is inserted in the cathode or emitter lead of the tube or transistor employed in the IF stage. The circuit is tuned to the frequency that is to be attenuated. Thus, at resonance, it develops a large voltage across it which results in a sharp fall in the gain because of the degenerative action. At all other frequencies the impedance of the trap circuit is very small and hence practically no degenerative feedback occurs at these frequencies.

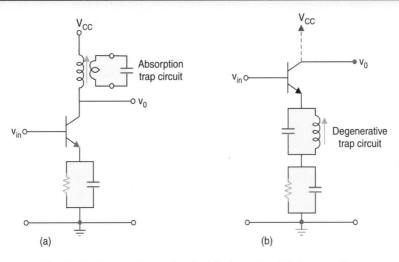

Fig. 23.11. Resonant trap circuits (a) absorption (b) degenerative.

(e) *Bridged 'T' trap circuit*—The circuit configuration of a bridged 'T' filter is shown in Fig. 23.12 (a). The special feature of this circuit is that the value of R is so chosen that it effectively cancels the coil resistance. Then the parallel resonant circuit formed by L, C_1 and C_2 attains a very high Q. The values of L, C_1 and C_2 are suitably chosen to reject the desired frequency. Because of the large effective Q, this network is highly selective. It is put in series with the signal path and the slug in the trap coil is varied for minimum output. An attenuation of the order of 40 db can be easily attained if necessary. The circuit of Fig. 23.12 (b) is another version of this trap circuit. The bridged 'T' trap is most widely used because of its highly selective attenuation characteristics.

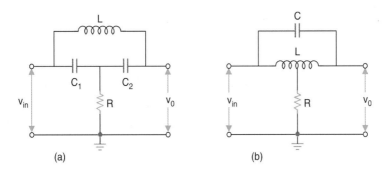

Fig. 23.12. Two different configurations of a bridged 'T' trap circuit.

IF Alignment

It is obvious from the discussion in the preceding sections that several stringent requirements must be met while tuning the IF station of the receiver. This is normally carried out with a sweep generator and the output from the video detector is displayed on a cathode-ray oscilloscope screen for ease of tuning. The sweep generator has markers which are used for locating various frequencies on the displayed pattern for tuning the coils. The detailed procedure for tuning the RF and IF stages of a TV receiver is explained in a subsequent chapter.

Chapter 23

23.7   VIDEO IF AMPLIFIER CIRCUITS

A typical circuit of a monochrome receiver IF stage employing tubes is shown Fig. 23.13. All the three tubes are miniature pentodes having very low interelectrode capacitances. Only the first stage is AGC controlled. Additional bias voltage has been provided through the potentiometer network R_1 and R_2 from the screen voltage dc supply. This ensures operation in the linear region. A part of the cathode resistor of both V_1 and V_2 has been left unbypassed in order to provide negative feedback. This results in overall stability of the stage. However, no such feedback is necessary in the last stage. The dc supply to the plate and screen grid circuits of all the tubes is through decoupling networks in order to avoid any interaction amongst the three sections of the amplifier.

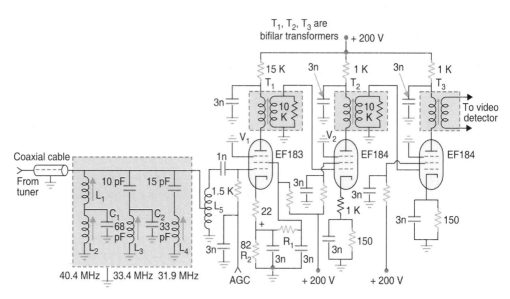

Fig. 23.13. Vacuum tube video IF amplifier.

The coupling between the stages is through bifilar wound coils. The secondary circuits of the coupling transformers T_1 and T_2 are shunted by suitable resistors for obtaining the desired bandwidth. The overall voltage gain can be varied between 6000 to 10,000 by the AGC bias. At the input of the first stage, coil L_5 resonates with the input capacitance of V_1 to couple the signal from the tuner to its control grid. The tapping on this coil is for impedance match between the coaxial cable and input circuit of the first IF amplifier.

A three section filter is connected at the input of IF section to prevent entry of interfering signals into the amplifier. The first section is tuned to reject interfering signal from the upper adjacent channel. This section is a modified form of the parallel trap circuit. Coil L_1 and capacitor C_1 are tuned to 40.4 MHz to reject this frequency. However, coil L_2 that forms a parallel resonant circuit with C_1 is tuned to a frequency somewhat higher than the interfering signal frequency. This ensures a steep rise in impedance just after a sharp fall at 40.4 MHz. This also helps to shape the upper skirt of the IF response curve and no separate trap circuit has been used to lower the gain at the picture IF for vestigial sideband correction. The second filter section attenuates the IF sound signal (33.4 MHz) to the desired level. The third section is a series trap circuit tuned to reject the lower channel interfering frequency of 31.9 MHz. The entire filter section and coupling transformers are shielded to avoid any radiation and undue coupling with each other.

IF Section Employing Transistors

The circuit shown in Fig. 23.14 is that of a three-section four-transistor video IF amplifier. Each one of the transistor has a very high cut-off frequency and a relatively low junction capacitance.

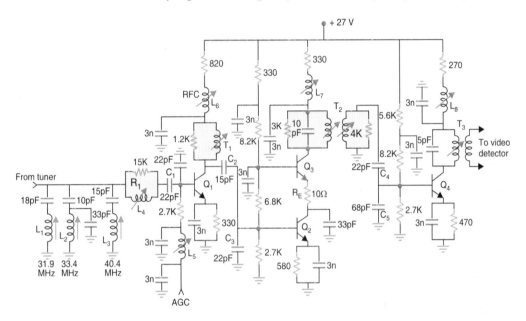

Fig. 23.14. Video IF amplifier with cascoded interstage.

At the input of the first IF amplifier stage a three section filter is provided. The first section is a series trap that blocks the lower channel interfering frequency. Similarly, the third section filters out the upper channel beat frequency. Section two of the filter unit is designed to attenuate the sound IF frequency to about 20 db below the picture IF level. Coil L_4 is for frequency compensation against the low input impedance of the common emitter configuration. This coil is shunted by a resistor (R_1) to damp any oscillations.

Coils L_5 to L_8 are IF chokes inserted in the dc supply lines and the AGC circuit. The signal frequencies are thus contained within the respective stages and returned to ground through separate decoupling capacitors. Besides this, the biasing circuits and the collector supply points are provided with separate decoupling networks. Hence stable operation is ensured without any possibility of self oscillations due to mutual coupling paths. The IF signal is coupled to the base of transistor Q_1 through the coupling capacitor C_1. AGC is fed only to the first amplifier and the remaining two stages employ self biasing. The collector load is a single tuned circuit set at a suitable frequency for stagger-tuning. The output from the load circuit (T_1) is coupled to the base of Q_2 through capacitors C_2 and C_3 which form a voltage divider in addition to providing impedance match between the two stages.

The second stage combines a grounded emitter stage (Q_2) that drives a grounded base (Q_3) amplifier. The output of Q_2 is dc coupled to the emitter of Q_3. The two transistors thus combine to form a cascode pair which gives somewhat large gain compared to a single stage with large bandpass characteristics. The transistor Q_2 is not neutralized because it has very low collector load impedance. R_E is a 10 ohms resistor inserted in the emitter lead of Q_3 for thermal stabilization. Most of the gain is in the grounded-base stage Q_3, which needs no neutralization. The primary winding of transformer T_2 is a tuned circuit connected at the collector of Q_3. The impedance match to the next stage is achieved through the secondary winding of T_2 and capacitors C_4 and C_5. The third IF amplifier has again a common emitter configuration employing a single coil tuned load. The secondary is wound for impedance matching and feeds into the cathode of the diode that is a part of the video detector.

IF Amplifiers Using FETs

Some receiver designs employ dual-gate MOSFETs in the IF section. With their high input and output impedances, matching problems between succeeding stages are greatly simplified. Normally, three stages in cascade provide the desired gain and selectivity. One gate of the MOSFET receives the signal while the other gate is fed with bias from AGC line. With rapid advances in IC technology, MOSFETs and transistors have been replaced with IC modules in the IF section.

23.8   IF SUB-SYSTEMS EMPLOYING ICs

Many dedicated ICs have been developed which include AGC, video detector, and sound conversion circuits besides a multisection IF amplifier. Some of such ICs also include provisions for processing chroma (colour) signal and part AFT (automatic frequency tuning) control circuits.

Initially when a synchronous demodulator IC (TCA 540) became available it was widely used in the IF section of solid state receivers in conjunction with IF amplifiers employing transistors BF 167 and BF 173. However, such circuits have now been superseded by monolithic ICs which incorporate many other functions besides video IF amplification. BEL CA 3068 is one such IC the description of which follows.

Video IF Sub-system CA 3068

The BEL CA 3068 is a 20 lead quad-in-line (plastic package) IC specially designed to perform as a complete IF sub-system in monochrome and colour receivers. This package includes a wrap around shield that serves to minimize interlead capacitance. The IC and other associated components are housed in a modular box and all interconnections to it are made via feed through capacitors or insulated lead throughs. The module is soldered on the main printed circuit board (PCB).

The main sections of IC CA 3068 are:

   (i) High gain wide-band IF amplifiers (overall gain 87 db)

   (ii) Keyed AGC with noise immunity circuits

  (iii) Delayed AGC for the tuner

  (iv) Video detector

   (v) Video preamplifier

(*vi*) Intercarrier sound detector

(*vii*) Sound IF amplifier

(*viii*) Zener regulated dc reference source

As shown in the functional diagram of this IC (Fig. 23.15), the IF signal from the tuner is fed through terminal 6 to a cascode high frequency wideband amplifier, Negative AGC voltage is used to control the biasing of the amplifier. It is designed to have a reverse AGC control over a wide range (55 db) with excellent operational stability. The output of the cascode amplifier feeds into the 1st IF amplifier which is followed by another multistage wide-band amplifier. The combined gain of these amplifiers is enough to provide an overall gain of about 87 db. The bandwidth is controlled by external tuned circuits. The IF output is available at terminal 14 of the IC for automatic frequency tuning (AFT) circuit in the tuner.

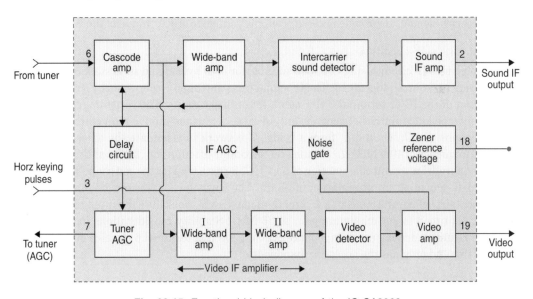

Fig. 23.15. Functional block diagram of the IC CA3068.

The video amplifier output is fed to the AGC circuit via a noise gate. The IF AGC circuit also receives horizontal keying pulses through terminal 3 of the IC and develops a negative AGC voltage (terminal 4) proportional to the input signal carrier strength. The AGC circuit is designed to operate in two modes. Gain reduction up to 40 db is controlled in the first mode, while a further gain reduction of 15 db is obtained in the 2nd mode. This results in precise gain control and stable operation.

The AGC output also feeds the tuner AGC delay circuit. The delay is set by external components and dc supply connected at terminal 8. The biasing network of the delay circuit is so designed that a crossover occurs at a predetermined RF input to the receiver. The reverse AGC gets applied to the tuner after a desired voltage delay. If necessary, reverse AGC available at terminal 7 is inverted through an external discrete transistor to get a forward AGC in case the tuner employs *n-p-n* bipolar transistor. A forward swing of 2 to 7 volts is thus achieved.

A feedback loop in the 2nd video IF amplifier provides low output impedance for driving the video detector thus eliminating the need for a conventional tuned circuit at its input. The video detector is employs the usual peak detector circuit. At its output is an active low-pass filter to remove IF and higher frequency components. The video detector output is fed to the video preamplifier through a differential pair, the output of which maintains correct black level despite input signal variations. The amplifier output is in Darlington configuration and delivers video output (terminal 19) at a low impedance level for ease of coupling to the discrete high voltage video amplifier.

Composite video IF signal from the cascode amplifier is also fed at terminal 12 to a wide-band amplifier the output of which is coupled to the intercarrier sound detector. The resulting 5.5 MHz FM sound intercarrier signal is available at terminal 2 of the IC after amplification by the sound IF amplifier. Frequency selective feedback is provided in this amplifier to obtain channel bandpass characteristics with a peak at 5.5 MHz.

Each stage of the IC is fed through an inbuilt dc voltage regulator. In addition to this, an isolated zener reference dc voltage is available at terminal 18 of the IC. It can be used with an external series element to provide necessary supply voltage to any terminal of the IC.

Practical Video IF Circuit Using CA 3068

A complete video IF circuit using CA 3068 suitable for both hybrid and solid state monochrome TV receivers is shown in Fig. 23.16. The IF signal from the tuner is connected by a coaxial cable to the input of the IC at terminal 6 through a set of trap circuits. The resistors R_1, R_2 and R_3 provide an impedance match between the output of the tuner and input circuit of the IC to avoid any reflections. R_2, C_1, C_2 and L_1 form a bridge T network to attenuate the adjacent channel sound IF frequency of 40.4 MHz. L_1 is tuned for maximum attenuation (35 db) at this frequency. The series parallel resonant combination of L_2, C_4, C_3 forms a trap for the self sound IF frequency of 33.4 MHz providing an attenuation of 38 db. The capacitor C_5 transforms the low source impedance (40 Ω at point A) to 700 Ω. L_3 and C_7 along with R_4 constitute the input band shaping circuit. L_3 is tuned to get maximum gain at 36 MHz. Stability considerations in the cascode amplifier require a driving point impedance of about 500 Ω at terminal 6. A value of 2.7 K for R_4 achieves this and provides the necessary bandwidth.

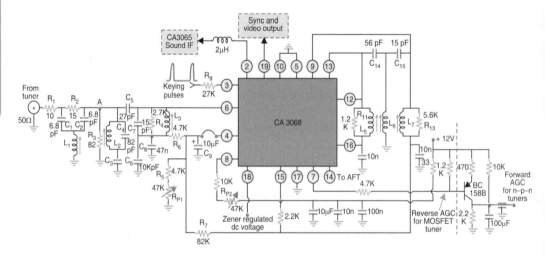

Fig. 23.16. Schematic circuit of a video IF sub-system using IC CA3068.

The network C_8, C_9, R_6 and C_6 smoothens the AGC voltage available at terminal (4). Capacitor C_9 is charged through R_7, R_5 combination. An isolated low dc voltage at terminal (16), derived from the main supply to the IC, is used to feed the required bias at terminals (13) and (12). The series combination of 10 KΩ resistor and R_{p2} sets the delay for the tuner AGC voltage, available at terminal 7 of the IC. The value of R_9 is selected to keep the keying current to about 0.8 mA. At this current the voltage at terminal (3) is around 8 volts.

The required IF response is obtained, through the interstage coupling network, consisting of $R_{11}, R_{13}, L_5, L_6, L_7, C_{14}$ and C_{15} and the single tuned circuit at the input consisting of L_3, C_7 and R_4. In the interstage coupling network, a terminating resistance R_{13} of about 5.6 K is necessary at terminal (9) in order to ensure good stability of the cascode amplifier. Similarly R_{11} is needed to present a driving source impedance of about 1 K at terminals 12 and 13. Typical value of L_7, L_5 and L_6 are 0.9 μH, 0.2 μH, and 0.06 μH respectively. The values of 15 PF and 56 PF for C_{15} and C_{14} determine resonant frequencies of these tuned circuits to be around the centre frequency of the band. Note that while tuning various coils the jumper at pin 4 is removed to disable application of AGC voltage to the controlled amplifiers. A signal of 10 mV (rms) (– 40 db) is applied at the input of IF sub-system for tuning various resonant and trap circuits Potentiometer R_{p1} is adjusted to get an output of 6 V p-p.

Typical performance characteristics of the above video IF system using CA 3068 are given below:

Typical dc supply voltage	11 V
Quiescent circuit current	15 to 45 mA
IF sensitivity for video output of 4 V p-p	250 μV
Max video output	7 V p-p
6 db bandwidth	4.5 MHz
Attenuation at 40.4 MHz	40 db
Attenuation at 33.4 MHz	28 db
Sound IF output	500 μV
Dynamic range of IF AGC	55 db
Tuner AGC voltage variation	2.2 V to 4.5 V

REVIEW QUESTIONS

1. Enumerate design requirements of the IF section of a TV receiver. Draw ideal overall response curve of the IF section of a colour receiver and mark on it the following frequencies: picture carrier, colour sub-carrier, sound carrier and adjacent channel interfering frequencies.

2. Explain fully how by stagger tuning the IF stages, desired gain bandwidth requirements are met without having to increase the number of cascaded stages.

3. What are bifilar-wound coils ? Explain with circuit diagram how such coils are usefully employed for interstage coupling. What is link coupling and why is it used to couple the output from the tuner to the first IF stage ?

4. Explain fully the need for vestigial sideband correction and the manner in which it is achieved. Illustrate your answer by using numerical frequency values corresponding to any television channel in Band-I

5. Why is FM sound signal attenuated by about 26 db with respect to the maximum signal amplitude ? Explain various trap (rejection) circuits that are usually employed to attenuate the sound signal and for correcting effects of vestigial side-band transmission.

6. Explain main features of the three section picture IF amplifier configuration shown in Fig. 23.13. Why are trap circuits provided at the input of first IF stage ?

7. Draw the circuit diagram of a typical three stage IF section employing transistors. Explain how the circuit functions to obtain the desired bandwidth and required wave-shape of the overall IF response curve.

8. Draw block schematic diagram of an IF sub-system employing the IC CA 3068. Enumerate its main features and explain how it performs the functions assigned to it.

24

Receiver Power Supplies

Television receiver power supply provides low voltage (LV) and high voltage (HV) dc sources. In addition, filament power source for vacuum tubes also forms part of the power supply.

Low Voltage Supply

The low voltage dc that is needed for feeding various sections of the receiver circuit is obtained by rectifying and filtering the mains supply. It may be classified as unregulated and regulated dc supply. Most tube-type TV receiver power supplies are unregulated. Small changes in dc output voltage due to normal fluctuations in the mains supply, do not affect the performance of the receiver drastically, because the circuits are designed to operate satisfactorily for reasonable variations in the B_+ voltage. Solid state receivers employ a regulated low voltage dc source because transistors and ICs are very sensitive to over voltage.

In receivers, that employ only tubes, B_+ voltage is generally between 100 and 350 volts with moderate current demands. For transistor receivers, a V_{CC} source between 15 to 100 volts is necessary, but the current drain is quite high. However, receivers that use hybrid circuitry, need a dc supply from 15 to 250 volts.

For the horizontal output stage and other auxiliary circuits a dc supply over 600 V is required. This is met from the 'boosted' B_+ supply, that is economically generated in the horizontal output circuit by augmenting the existing B_+ supply.

EHT Supply

In monochrome receivers an EHT supply of 10 to 16 KV is required at the final anode of the picture tube. The colour picture tube needs a still higher dc voltage, that ranges from 15 to 25 KV. The current drain is, however, very small and seldom exceeds one milliampere. As already explained in chapter 20, EHT is generated by stepping up and rectifying high voltage retrace pulses produced in the horizontal output transformer of the deflection circuit. The design of the horizontal output stage determines regulation of the high voltage supply. In colour receivers, besides providing a regulated low voltage (B_+) supply, a regulator is also used for high voltage supply to prevent changes in focus and convergence in the colour picture tube. Note that if only one power supply of the usual type is decided upon for both low and high voltage requirements it would turn out to be bulky, very expensive and quite out of proportion with other sections of the receiver.

The power drawn from the ac source is about 100 watts in hybrid monochrome receivers but rises to nearly 200 watts in colour receivers of this type. Present day monochrome receivers which employ only transistor and ICs consume as low as 60 watts of mains power.

24.1 LOW VOLTAGE POWER SUPPLIES

Low voltage power supplies either operate from a transformer or are supplied directly from the line. The latter type of receivers are sometimes referred to as 'line connected' while those using a transformer are said to be 'line isolated'.

Transformers have separate high voltage and low voltage secondaries. The low-voltage secondary windings supply power to heat tube filaments. High voltage secondary windings drive the rectifier circuits, which after filtering produce B_+ supply.

The line connected receiver is so termed because one side of the ac supply line is connected to the receiver chassis. This could be dangerous if a user came in contact simultaneously with the receiver chassis and some external ground, unless the mains cord is plugged to connect neutral or earth wire to the chassis. Voltage doublers are used in line connected receivers to obtain higher dc voltages.

Most high quality receivers employ a mains transformer because, though expensive, it gives many advantages. The main merits of using a transformer are as under:

(a) Since the ac source is completely isolated by separate primary and secondary windings, there is no shock hazard in touching the chassis or other exposed metal parts.

(b) The B$_+$ voltage generation is efficient because full-wave rectification can be employed by using a centre tapped high voltage secondary.

(c) A separate filament winding can be provided for parallel connections of heater circuits and so tubes having different filament ratings can be used.

(d) The dc voltage bias that is often needed on the heaters of picture and damper tubes can be easily fed to the isolated filament winding.

(e) With a mains input transformer if tappings are provided on the primary, the ac voltage to the receiver can be changed to counteract persistent low or high ac mains voltage changes. Such a provision if made is referred to as an in-built voltage regulator.

In some receiver designs an autotransformer is employed for stepping up the voltage. Though, this is economical in transformer cost, it does not provide isolation from the ac mains, unless is made sure that the 'hot' side of the line is not connected to the chassis.

Power Supply Input Circuit

A typical circuit arrangement of the input section of a power supply employing a transformer is shown in Fig. 24.1. Besides the features already discussed, a hash filter is used at the mains-supply input points. The power line acts as an antenna and may feed undesirable signals into the receiver, from motors, neon signs or any sparking contacts. Such a disturbance is called hash because it does not have a specific frequency. The inductors block any line transients and other disturbing frequencies, while the capacitors acts as a short to such line disturbances. In addition, the filter helps in confining to itself the signals generated within the receiver. These signals would otherwise interfere with other sets through the mains line. Examples of such signals generated within a receiver are horizontal and vertical sweep voltage and their harmonics. In some power supply designs only two capacitors are used as line filter. The impedance of the filter capacitors is sufficiently high at 50 Hz to effectively isolate the chassis ground from the ac line.

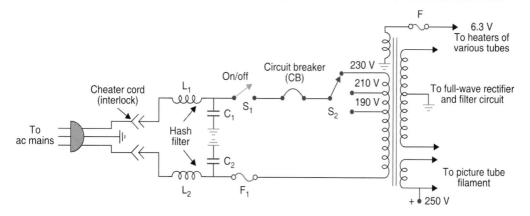

Fig. 24.1. *Input circuit of low voltage power supply of a television receiver.*

Figure 24.1 also shows switches, fuses and a circuit breaker. The switch S_1 is the receiver on-off switch and S_2 is a rear switch which selects a tapping on primary of the transformer. It is set in accordance with the available mains voltage.

Fuses used in TV receivers are of two types—fast acting and time delay. A fast acting (normal below) fuse is used in circuits which do not have surge and transient currents. It is constructed of a single strand of wire that has no heat sink to dissipate momentary overloads. Such fuses are normally rated at 200 to 300% of the full load current.

Time delay fuses are designed to pass transients and surges without opening. Such fuses will, however, blow with sustained overloads or short circuits. These are constructed out a fuse wire that is under spring tension. The spring is imbedded in a solder allow which acts like a heat sink for momentary current surges. When a sustained overload occurs, the solder melts and the fuse opens. In the circuit, F_1 is normally a fast acting fuse while F_2 is of the time delay type.

The circuit breaker (CB) is for overload protection. The contact in the circuit breaker is through a spring loaded bimetallic strip which bends when heated. During normal operation the strip stays in the closed position and current flowing through the switch will not warm it sufficiently to cause it to open. This level of current is called the 'hold current'. If there is an overload, say a partial short in the transformer, the current through the bimetallic strip will be sufficient to bend it by heating and snap open the switch contact. Thus any excessive power drain will trip it, preventing further damage to the receiver and possibly to the house wiring. Once the circuit breaker opens, it will remain so, until it is closed by pushing a button protruding from the circuit breaker. Occasionally the circuit breaker will trip due to temporary surge voltages. It is a good practice to reset the breaker and observe further trippings if any. If the breaker trips again as soon as the receiver is turned on, it is an indication that the power supply section is defective and needs servicing

Interlock plug. When the back cover of the receiver is removed the ac power cord is disconnected by the interlock plug (see Fig. 24.1). This precaution is necessary against any shock hazard on account of the high voltage that exists in the picture tube circuit. In order to operate the receiver when the cover is off, a 'cheater cord' is used which is a substitute line cord with its own plug. The input circuit of line connected power supplies also incorporate most of the above described features.

Rectifiers. Earlier TV receivers employed vacuum tubes as rectifiers in both low and high voltage power supplies. As selenium, germanium and silicon rectifiers became available at cheaper rates, they gradually replaced tubes. Almost all present day receivers use silicon rectifiers. These are manufactured with current ratings ranging from 500 mA to 10 Amps or more. Large units are stud-mounted.

The advantage of a semiconductor diode is its very low internal voltage drop besides the absence of any filament heating power. A silicon diode has a forward voltage drop of about one volt as compared to nearly 20 volts in vacuum tube diodes. However, one precaution must be observed when using silicon rectifiers. Their forward resistance is so low that when the set is turned on a large surge current flows through them to the filter capacitors. Therefore, to limit the diode current below its maximum current rating a resistor of about 10 ohms should be inserted in series with the diode circuits.

24.2 TYPES OF RECTIFIER CIRCUITS

Figure 24.2 shows a full-wave centre-tap and a full-wave bridge rectifier circuit that are commonly used with an isolating power transformer.

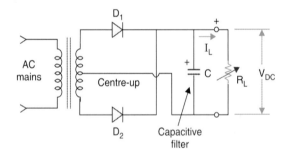

Fig. 24.2 (a). Full wave rectifier with capacitor input filter.

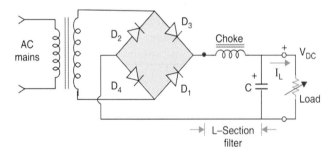

Fig. 24.2 (b). Bridge rectifier with choke input.

The rectifier configurations that can be line-connected and do not employ transformers are illustrated in Fig. 24.3. For the bridge rectifier circuit (Fig. 24.3 (a)), one lead of the dc output cannot be grounded unless both the ac input lines are insulated from the chassis. The half-wave rectifier circuit (Fig. 24.3 (b)) is much less efficient and needs large values of filter components as compared to a full-wave rectifier circuit for the same percentage ripple. However, in this configuration one side of the line and that of the filtered output can be grounded as shown in the circuit.

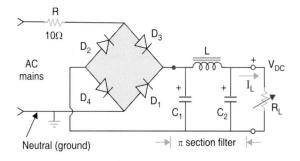

Fig. 24.3 (a). Bridge rectifier (direct on line) with p section filter.

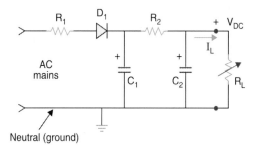

Fig. 24.3 (b). Half-wave rectifier (direct on line) with RC (p) filter.

In all the rectifier circuits discussed so far the positive side of the rectified output has been shown above the ground (chassis) potential. Note, that there is no rule that chassis ground should be most negative point. In any rectifier circuit the polarity of the dc output voltage can be interchanged by simply reversing the direction of all the rectifier diodes employed in the circuit. In fact such a reversal is necessary when *p-n-p* transistors are employed in any section of the receiver. Similarly in some circuits a negative biasing voltage is needed and so a reversal of polarity becomes necessary.

Filter Circuits

A suitable filter circuit is added to the rectifier output for filtering out the ac ripple content. Thus at the output terminals of the filter an almost pure dc source becomes available.

A capacitor filter (Fig. 24.2 (*a*)) tends to deliver a dc output voltage close to the peak value of the ac input voltage wave. However, unless a very large capacitor is used, its regulation is poor and ripple content rise with any increase in load current. Thus it is used only when load current is small and stays constant.

A choke input or inverted 'L' filter (Fig. 24.2 (*b*)) has somewhat better regulation. The ripple content in the output voltage becomes independent, if the load current is maintained larger then a certain minimum value. Therefore, this type of filter is often employed when the load current is more than about 200 mA. In contrast, a 'P_i' section filter (Fig. 24.3(*a*)) provides higher dc voltage from the same ac input and has a very low ripple factor. It is often used for load currents that lie between 50 to 200 mA. Since its regulation is poor, it is mostly useful for fixed loads. A resistor in place of the choke (inductor) can also be used and is equally effective if value of the resistor equals reactance of the choke at the ripple frequency. The resulting R-C filter is shown in Fig. 24.3(b). Its use is limited to very low load currents because of the dc voltage drop across the resistor. However, because of the lower cost of the resistor than that of the choke, it is used in many circuits.

Voltage Doublers

By a series combination of half-wave diodes and their filter capacitors, the amount of dc output voltage can be doubled, tripled or quadrupled. In practical applications the voltage doubler is often used for the B_+ source with a half-wave line connected supply as shown in Fig. 24.4 (a). The circuit operation is quite simple. During the negative half-cycle of the input ac voltage, diode D_1 is forward biased and conducts to charge C_1 with positive polarity towards the anode of D_2. During the positive half-cycle, D_1 gets reverse biased but diode D_2 is forward biased by a voltage equal to the instantaneous addition of the varying line voltage and fixed voltage across the capacitor. The current that flows through D_2 charges the capacitor C_2 with the polarity marked across it. Under steady state conditions and with R_L removed, C_1 charges to a value equal to the peak value of the input ac voltage. The capacitor C_2 charges to twice the peak of the input mains voltage. With load resistor restored across the output terminals, capacitors C_1 and C_2 discharge a little during the interval when the diodes are not conducting and thus the actual output voltage is somewhat less than the peak-to-peak value of the input wave.

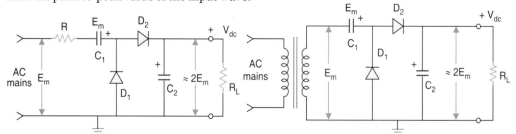

Fig. 24.4 (a). Half wave voltage doubler.   Fig. 24.4 (b). Half-wave voltage doubler with a step-up transformer.

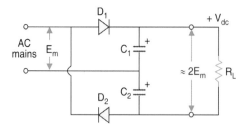

Fig. 24.4 (c). Full-wave voltage doubler.

An important feature of this doubler is that one side of ac input and the dc output are common. For this reason the half-wave doubler is generally used in a line-connected supply without a power transformer. As shown in the figure, the grounded side of the dc output is common to the neutral or ground side of the ac input. The common connection reduces the problem of stray pick-up of 50 Hz hum in the receiver.

Another half-wave voltage doubler circuit is shown in Fig. 24.4 (b). The circuit operation is the same as described above. However, by using an input transformer the line voltage can be stepped up to obtain a still higher dc output voltage.

Full-wave Doubler

Consider the circuit of Fig. 24.4 (c). During the positive half-cycle D_1 conducts to charge C_1 with the polarity marked across it. In the next half-cycle (negative) D_2 conducts to charge C_2 with negative polarity towards the ground terminals. Thus if the load resistor R_L is quite large both C_1 and C_2 charge to the peak value to deliver a dc output voltage equal to twice the peak value of the secondary voltage. This circuit is seldom used in TV receivers since it does not have a common connection for the ac input and the dc output.

The idea of voltage doubling can be extended to obtain still higher dc voltages. However, it is not often used because the voltage regulation becomes worse unless very large capacitors are employed.

24.3 HEATER CIRCUITS

For receivers that use tubes, the heaters can be connected either in parallel on in series. When a low voltage winding exists on the mains transformer or when a separate filament transformer is provided, all the heaters are connected in parallel across that winding. The filament current rating of the tubes may be different but they must have the same voltage rating. One end of the filament winding is tied to chassis ground to save wiring. Each heater returns to the ground end of the winding by an individual ground connection. This is illustrated in Fig. 24.5 (a). A decoupling filter L_1, C_1 is used to prevent feedback of signals between RF and IF amplifier through the common heater line. The inductor or choke consists of a few turns of thick wire and has an inductance around 1 μH. Ferrite beads are also used instead of RF chokes. The reactance of the choke is negligible at 50 Hz and therefore heaters remain effectively in parallel across the filament winding. The picture tube heater has its own filament winding. A dc bias voltage is added to this heater supply to raise the filament potential close to that of the cathode. This minimizes the possibility of arcing between these two electrodes.

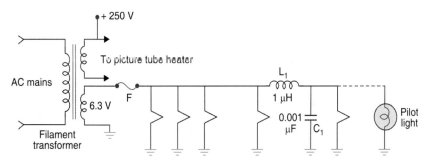

Fig. 24.5 (a). Parallel heater circuit.

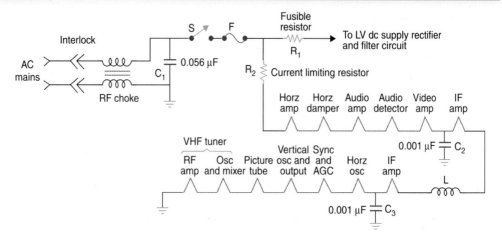

Fig. 24.5 (b). Series heater circuit.

Series Heaters

In Fig. 24.5 (*b*) all the heaters including the picture tube are in series across the supply mains. In this arrangement the current rating of all the tubes must be same though their voltage ratings can be different. The sum of the heater voltage drops should be approximately equal to the line supply voltage. In case it is not so, a series resistor is included in the heater circuit chain. Note that the high voltage rectifier is not in the series string as it receives filament power from the horizontal output transformer.

All the tubes in series must have the same warm-up time otherwise this results in repeated burn-out of some filaments. Some heater strings have a thermistor in series to limit the current until the tubes become hot, since the filament resistance is much less when cold than when it is hot.

Order of series heater connections. Though all the series filaments have the same current, a point closer to the grounded end of the string has lesser potential difference to chassis ground. Therefore, less hum voltage can be coupled into the signal circuits by heater cathode leakage in the tube. This is why the tubes in the VHF tuner are last in the chain while the horizontal deflection amplifier and damper tubes are first. This aspect is illustrated in the series filament circuit shown in Fig. 24.5 (*b*).

24.4   VOLTAGE REGULATORS

Voltage regulators are used for keeping the rectified dc output voltage constant despite variations in the dc load current or in the ac input voltage. The use of a voltage regulator also reduces ripple content in the dc supply and lowers the ac output impedance of the power supply. Low impedance minimizes coupling between amplifier stages through the common B_+ line. Regulation is also important in order to maintain constant raster size. In colour receivers regulation of the EHT supply is equally important to maintain focus and convergence with changes in brightness. Even in monochrome receivers some regulation of the EHT supply is necessary to avoid blooming and breathing of the reproduced picture. Breathing refers to blooming at a regular slow rate.

Types of Voltage Regulators

One method of keeping the output voltage constant is to build a power supply with much more current capacity than actually required. However, this turns out to be very expensive and is normally not favoured. Another approach is to regulate the ac input voltage. The input transformer for the LV supply is designed to have a saturable core to limit any increase in the ac input voltage. Similarly, to prevent any decrease in voltage, an oil-filled capacitor of about 4 μF is connected across the secondary to tune it to resonance. However, this method does not provide a very effective regulation of the dc supply.

The most common method, that is quite cheap and yields excellent regulation is the use of electronic regulators. The various types that find application is good quality receivers are briefly described below:

(a) *Zener diode voltage regulator.* Figure 24.6 (a) is the circuit of a zener regulator designed to feed the sound IC of a TV receiver. The zener (Fig. 24.6 (b)) operates under reverse bias to maintain a fixed voltage (equal to its breakdown voltage) across its terminals.

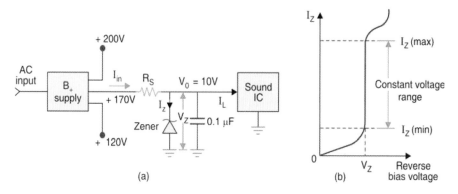

Fig. 24.6. *Zener voltage regulator (a) circuit (b) Zener V-I characteristics.*

The input voltage is kept higher than the required dc output voltage. Usually $I_{Z(\text{max})} = 10$ $I_{Z(\text{min})}$ in the useful range where $V_Z = V_0$ remains constant. Assuming V_{in} to be constant, I_{in} must remain same to keep V_0 constant. Thus any change in I_L should result in a corresponding change in I_Z to keep the voltage drop across R_S constant. However, when V_{in} changes I_{in} adjusts to a new value to keep the drop across R_S equal to the difference between V_{in} and V_0. The essential conditions to be met for keeping the zener in its break down region are: (i) I_L should not exceed a value which results in an I_Z less than the breakdown current, i.e., $I_{Z(\text{min})}$. (ii) when the load circuit opens, i.e., $I_L = 0$, then I_Z should not exceed the maximum rated current of the zener. Too large or too small an input voltage can also result in failure of the regulator to deliver a constant output voltage. Since the zener is connected in shunt across the load, this regulator is known as a *shunt regulator*.

The zener regulator, though simple and inexpensive, suffers from several limitations. The first problem is the power dissipated in the zener which varies with changes in load current. The second disadvantage is its somewhat non-ideal current-voltage characteristics. The voltage across the zener changes by a small percentage of the nominal output voltage with large changes in I_Z.

(b) *Series regulator.* A series emitter follower regulator overcomes the above drawbacks but is not very efficient. Figure 24.7 illustrates the principle of an improved series regulator which employs an error voltage amplifier. Voltage across the zener acts as reference voltage. The value of R_1 is chosen to keep the zener in its breakdown region. Any change in the output voltage changes V_{AB} proportionally. This results in a net change in V_{BE} of transistor Q_2. Because of transistor action, even a small change in the base-emitter voltage results in a large change in collector current, which effectively amounts to amplifying the error signal. Since R_3 is chosen to be large to keep I_{R3} almost constant, any change in I_{C2} is accompanied by a corresponding change in I_{B1}. Thus, any increase in the output voltage V_0, will result in an increase in I_{C2}, which in turn will decrease I_{B1} to reduce I_{E1} ($I_{C1} \approx I_{E1} \approx I_L$). A decrease in I_{E1} will reduce the load current thus restoring the voltage across R_L (load) back to its nominal value V_0. Similarly when V_0 tends to decrease the feed back loop action will increase $I_{E1} (\approx I_L)$ to keep the output voltage constant.

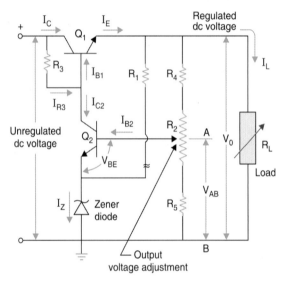

Fig. 24.7. *Basic series voltage regulator.*

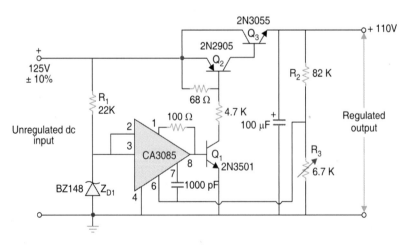

Fig. 24.8. *High voltage regulator employing IC CA 3085.*

(c) IC voltage regulator. Many voltage regulator ICs are available which provide excellent regulation over a wide range of input and output voltages. BEL CA 3085 is one such IC. It has a current capability up to 100 mA at 6 V dc output. By employing CA3085A and CA3085B higher voltage and current rating can be obtained. Figure 24.8 is the circuit of a high voltage regulator employing CA3085. As shown there, additional pass transistor have been used to achieve high voltage operation. Note that the input voltage to the IC proper has been reduced to a convenient value through R_1, ZD_1 so that it operates within its maximum ratings. Similarly R_2, R_3 combination determines the magnitude of the regulated output voltage for feeding the comparator in the IC.

24.5 LOW VOLTAGE POWER SUPPLY CIRCUITS

There is a wide range of low voltage power supply circuits used in TV receivers. Two representative circuits are discussed.

(1) Transformer Power Supply with Series Heater Circuit

In some power supply designs for all-tube receivers a transformer is used for generating B_+ voltage but the tube heaters are connected in series across the ac line. This approach being economical is often used. Figure 24.9 shows the circuit arrangement. Note that a separate filament winding is provided for the picture tube heater. Resistors R_1 and R_2 form a potential divider to provide necessary bias on the picture tube heater.

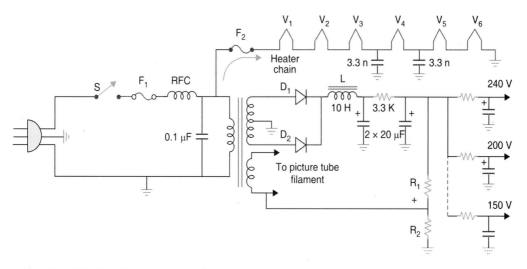

Fig. 24.9. *Transformer connected power supply with heaters in series across the ac supply.*

(2) Line Connected Power Supply

Figure 24.10 is the circuit of a line connected power supply often used in hybrid receivers. It employs a half-wave rectifier (D_1) to feed dc voltage to the vacuum tube circuits. Various loads are isolated through decoupling networks. The diode D_3 isolates the two 240 V $(A$ and $B)$ dc circuits from each other. Both D_1 and D_2 are bypassed by capacitors (3.3 nF) to suppress high frequency surge voltages.

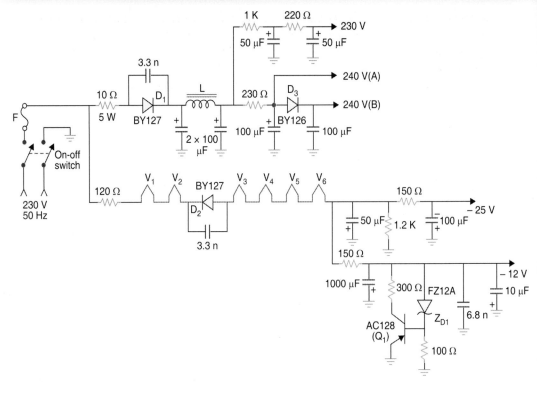

Fig. 24.10. *Line connected power supply for a hybrid television receiver.*

The heater circuit is designed to feed negative low voltage supply to transistor (*p-n-p*) circuits besides heating the filaments. This is achieved by connecting a diode (D_2) in series with the heater string and the low voltage load circuit. The diode conducts in only one direction, thus limiting the heater current to the desired value and also affecting rectification for the low voltage dc supply. The transistor Q_1 with zener diode Z_{D1} operates as a shunt voltage regulator to supply a stabilized dc source at – 12 V.

24.6 HIGH VOLTAGE POWER SUPPLY

Practically all television receivers produce EHT supply by rectifying the sharp retrace pulses that are generated in the high voltage winding of the horizontal output transformer. Details of horizontal output circuit and the manner in which high voltage pulses occur have already been explained in Chapter 20. However, it may be noted that the flyback transformer and rectifier are generally in a separate metal cage. In addition to offering protection against shock, the enclosure is a dust proof cover and is a shield against *X*-rays.

The rectifier tube is often mounted on insulating supports to keep the high voltage leads away from the chassis. At the bottom of the high voltage tube socket, an interconnecting loop of heavy wire, known as corona ring is often used to minimize high voltage arcing around these terminals. Silicon diode rectifiers are now available in sealed stacks and are replacing vacuum diode rectifiers.

Boosted B₊ Supply. As explained earlier the damper diode conducts through the boost capacitor C_B (see Fig. 20.11) to develop a dc voltage across it. This in series with B₊ supply results in boosted B₊₊ supply. It feeds the horizontal output stage that employs a vacuum tube and also to other auxiliary circuits.

24.7 STABILIZED THYRISTOR POWER SUPPLY

An economical power supply for television receivers can also be constructed by using a thyristor as a controlled rectifier. This enables a dc potential in the range of 100 to 240 V to be derived from ac mains without a transformer and with minimal dissipation in the used devices. The basic thyristor power supply circuit is shown in the block diagram of Fig. 24.11. The control circuit triggers the thyristor in the latter half of positive half-cycle of the mains waveform so that mean dc voltage presented to the output filter can take a value less than the peak mains voltage.

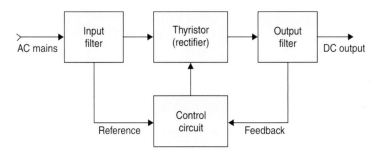

Fig. 24.11. *Block diagram of a thyristor power supply.*

An input filter is used to limit the peak thyristor current to a permitted value to reduce RF switching currents flowing back into the mains. It is usually a section of the control circuit IC. The output low-pass filter takes the form of a single RC combination because of the power levels involved. An active filter using a series power transistor may be used to reduce the size of smoothing capacitors.

Thyristor power supplies have superior performance as compared to their hybrid counterparts in terms of stabilization and heat dissipation. Such power supplies have been successfully used in both monochrome and colour receivers to deliver stabilized 200 V dc for load currents ranging from 300 to 400 mA. Figure 24.12 is the simplified circuit of such a thyristor controlled power supply. The initiation of conduction of the thyristor is controlled by a phase-shifting network. For this purpose mains waveform is delayed by the series network R_5, C_2 and applied to the trigger diac. A diac is a bidirectional thyristor with no gate. Its characteristic has a negative resistance region which starts at a critical potential called the breakdown voltage. When the voltage across C_2 reaches the diac breakdown voltage, the diac conducts and current pulse is applied to the gate of the thyristor causing it to conducts. To stabilize the output the trigger point must be varied as a function of mains input and dc output. The transistor Q_1 is used for this purpose. It controls the charge current of C_2 and hence the trigger point for the thyristor. The base of the transistor is connected to a potential divider across the mains input. If the ac input (mains) voltage tends to increase, the base current increases. The resulting increase in collector current of Q_1 reduces charge current through the capacitor C_2. This delays the trigger point and tends to maintain the output voltage constant. Additional feedback is provided from output through the resistor R_8. Any tendency for the output to rise due to a change in load current causes base current of Q_1 to be increased which reduces the charge current and delay the thyristor trigger point. The reverse would happen if either ac mains voltage or dc output voltage happen to decrease. Hence the effects of mains voltage variations and load changes on the dc output voltage are reduced to a minimum. R_4 is for adjustment of the output voltage to the exact desired value.

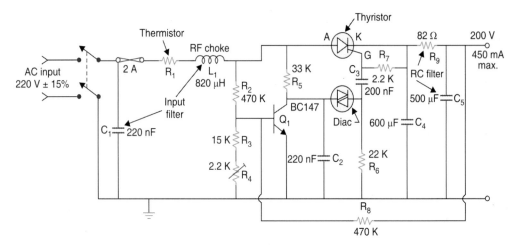

Fig. 24.12. *Stabilized thyristor power supply circuit.*

The L_1, C_1 network, besides suppressing ratio interference, protects the thyristor from possible damage caused by mains transients. The switch-on surge current of some 100 A is reduced by the inclusion of a thermistor R_1 to protect the on-off switch.

24.8 SWITCH MODE POWER SUPPLY (SMPS)

The use of ICs and modular construction is very common in modern television receivers. This has led to the introduction of switching mode power supplies to meet the dc requirements of such receivers. These are smaller, lighter and dissipate less power than equivalent series regulated supplies.

Basic Principle

In a switched mode supply the regulating elements consist of series connected transistors that act as rapidly opening and closing switches. The input ac is first converted (Fig. 24.13) to unregulated dc, which, in turn is chopped by the switching elements operating at a rapid rate, typically 20 KHz. The resultant 20 KHz pulse train is transformer coupled to an output network which provides final rectification and smoothing of the dc output. Regulation is accomplished by control circuits which vary the duty cycle (on-off periods) of the switching elements if the output voltage tends to vary.

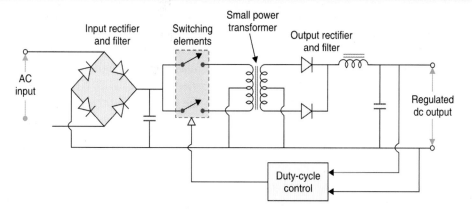

Fig. 24.13. Basic circuit of a switched mode regulated power supply.

Operating Advantages and Disadvantages

The advantages of a SMPS over a conventional regulated supply are:

(*i*) The switching transistors are basically on-off devices and hence dissipate very little power when either on (saturated) or off (non-conducting). Efficiencies ranging from 65 to 85 percent are typical of such supplies as compared to 30 to 45 percent efficiencies for linear supplies.

(*ii*) On account of the higher switching rate (20 KHz) the power transformer, inductor and filter capacitors are much smaller and lighter than those required for operation at power line frequencies. Typically a switching power supply is less than one third in size and weight of a comparable series regulated supply.

(*iii*) A switched-mode supply can operate under low ac input voltage. It has a relatively long hold-up period if input power is lost momentarily. This is so because more energy can be stored in its input filter capacitors.

Disadvantages. Although the advantages are impressive a SMPS has the following inherent disadvantages:

(*i*) Electromagnetic interference (EMI) is a natural by-product of the on-off switching within these supplies. This interference can get coupled to various sections of the receiver and hinder their normal operation. For this reason, switching supplies have built-in shields and filter networks which substantially reduce EMI and also control output ripple and noise. In addition, special shields are provided around those sections of the receiver circuitry which are highly susceptible to electromagnetic interference.

(*ii*) The control circuitry is expensive, quite complex and somewhat less reliable.

Typical Circuit of a SMPS

Figure 24.14 shows simplified circuit and associated waveforms of a typical switching mode power supply. Regulation is achieved by a pair of push-pull switching transistors (Q_1 and Q_2) operating under the control of a feedback network consisting of a pulse-width modulator and a voltage comparison amplifier. The waveforms illustrate the manner in which the duty-cycle is controlled to deliver a constant dc output voltage. The voltage comparison amplifier continuously compares a fraction of the output voltage with a stable reference source V_{r1} and develops a control voltage ($V_{control}$) for the turn-on comparator. The comparator compares $V_{control}$ with a triangular ramp

Chapter 24

waveform occurring at a frequency of 40 KHz. When the ramp voltage is more positive than the control level, a turn-on signal is generated. As shown in the waveforms, any increase or decrease in the control voltage ($V_{control}$) will very width of the turn-on voltage and this in turn will alter the width of drive pulses to both Q_1 and Q_2. The drive pulses pass through steering logic which ensures alternate switching of Q_1 and Q_2. Thus each switch operates at 20 KHz, *i.e.*, one-half of the ramp frequency. When Q_1 is on, current flows in the upper half of the primary winding of transformer T_1 and completes its path through its centre tap. Similarly when Q_2 conducts current flows in opposite direction through the lower half of the same winding to complete its return path thus providing transformer action.

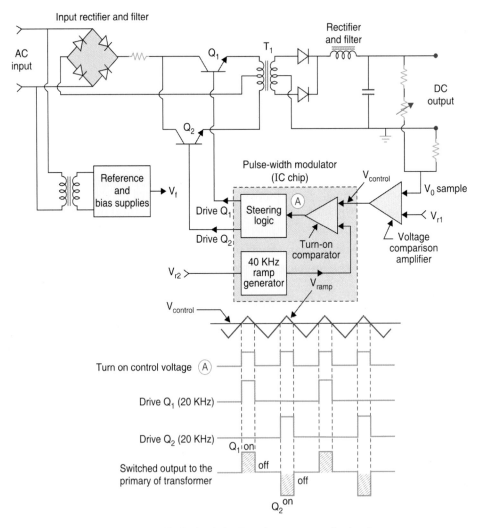

Fig. 24.14. *Typical switched-mode power supply circuit.*

Since the dc output voltage is proportional to the duty-cycle of current through the transformer, increasing the 'on' periods of the switching transistor will increase the output voltage and vice-versa. Thus the control voltage automatically monitors the duty-cycle to maintain a constant dc output voltage despite any input voltage or load current variations. In some such supplies only one transistor is used as the chopping and control element. In such designs a 20 KHz clock pulse is used to time the on-off periods.

The comparator, ramp generator and steering or control logic usually form part of a dedicated IC. Though, not shown, in the modular chip it contains additional circuits for over voltage protection, over-current protection and prevention of any inrush of ac current.

REVIEW QUESTIONS

1. Distinguish between 'line connected' and 'line isolated' power supplies. Describe the merits and demerits of each type.

2. Draw the circuit diagram of the input side of a typical low voltage power supply and explain besides other components the need for (*i*) a circuit breaker, (*ii*) fusible resistors, and (*iii*) a hash filter.

3. Draw circuit configurations of different filtering arrangements and explain special applications of each type. Which filter circuit is often employed in TV receiver power supplies ?

4. Explain with suitable circuit diagrams the action of half-wave and full-wave voltage double circuits. What is their special application in television receiver power supplies ?

5. Compare the series and parallel arrangement of feeding the heater circuits in a TV receiver. Why is a particular order used when connecting tube filaments in sereis ? Why is the filament of the picture tube generally fed from an isolated low voltage ac source ?

6. Why is it necessary to provide regulated dc source to solid-state circuits in a TV receiver ? Draw suitable circuit diagrams of atleast two types of voltage regulators and explain their operation.

7. Explain how a thyristor can be used to provide a regulated dc supply to TV receivers. Draw a typical circuit and explain its operation.

8. What is the basic principle of a switched mode power supply ? What are its merits and demerits ? Explain with a suitable circuit diagram the operating principle of such a regulated power supply.

25

Essentials of Colour Television

We all know how pleasing it is to see a picture in natural colours or watch a colour film in comparison with its black and white version. In fact monochrome reception of natural daylight scenes and pictures taken in black and white are totally unrealistic because they lack colour. However, we accept them because we have been conditioned to do so by constant exposure and lifelong usage to black and white drawings, photographs, films an monochrome television.

It is desirable that a TV system should produce a picture with realistic colours, adequate brightness and good definition that can be easily perceived by our eyes. A monochrome picture does contain the brightness information of the televised scene but lacks in colour detail of the various parts of the picture. The have a colour picture it is thus necessary to add colour to the picture produced on a white raster.

It may be recalled that in a monochrome TV system, the problem of picking up simultaneous information from the entire scene about the brightness levels is solved by scanning. In colour TV it becomes necessary to pick up and reproduce additional information about the colours in the scene. It did not seem to be an easy task and the difficulties in achieving this seemed unsurmountable. However, this challenge was met by the scientists and engineers working on it in U.S.A. All colour TV systems in use at present owe their origin to the NTSC system which was invented and developed in the United States of America. The Radio Corporation of America (RCA) played a prominent role in the development of colour TV. The invention and fabrication of the tricolour shadowmask tube was their major single contribution which made a high quality colour television system possible.

25.1 COMPATIBILITY

Regular colour TV broadcast could not be started till 1954 because of the stringent requirement of making colour TV compatible with the existing monochrome system.

Compatibility implies that (*i*) the colour television signal must produce a normal black and white picture on a monochrome receiver without any modification of the receiver circuitry and (*ii*) a colour receiver must be able to produce a black and white picture from a normal monochrome signal. This is referred to as reverse compatibility.

To achieve this, that is to make the system fully compatible the composite colour signal must meet the following requirements:

(*i*) It should occupy the same bandwidth as the corresponding monochrome signal.

(*ii*) The location and spacing of picture and sound carrier frequencies should remain the same.

(*iii*) The colour signal should have the same luminance (brightness) information as would a monochrome signal, transmitting the same scene.

(*iv*) The composite colour signal should contain colour information together with the ancillary signals needed to allow this to be decoded.

(*v*) The colour information should be carried in such a way that it does not affect the picture reproduced on the screen of a monochrome receiver.

(*vi*) The system must employ the same deflection frequencies and sync signals as used for monochrome transmission and reception.

In order to meet the above requirements it becomes necessary to encode the colour information of the scene in such a way that it can be transmitted within the same channel bandwidth of 7 MHz and without disturbing the brightness signal. Similarly at the receiving end a decoder must be used to recover the colour signal back in its original form for feeding it to the tricolour picture tube.

Before going into details of encoding and decoding the picture signal, it is essential to gain a good understanding of the fundamental properties of light. It is also necessary to understand mixing of colours to produce different hues on the picture screen together with limitations of the human eye to perceive them. Furthermore a knowledge of the techniques employed to determine different colours in a scene and to generate corresponding signal voltages by the colour television camera is equally essential. Therefore this chapter is mainly devoted to these aspects of colour TV, while transmission and reception of colour pictures is explained in the following chapter.

25.2   NATURAL LIGHT

When white light from the sum is examined it is found that the radiation does not consist of a single wavelength but it comprises of a band of frequencies. In fact white light is a very small

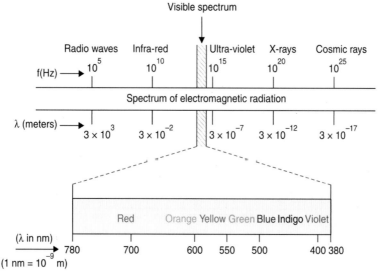

Fig. 25.1. Region of sunlight in the electromagnetic spectrum.

part of the large spectrum of electromagnetic waves which, in total, extend from very low to beyond 10^{25} Hz. The visible spectrum extends over only an octave that centers around a frequency of the order of 5×10^{14} Hz. When radiation from the entire visible spectrum reaches the eye in suitable proportions we see white light. If, however, part of the range is filtered out, and only the remainder of the visible spectrum reaches the eye, we see a colour. The entire electromagnetic spectrum is shown in Fig. 25.1 where the visible spectrum has been expanded and shown separately to demonstrate the range of colours it contains. Note that the various colours merge into one another with no precise boundaries.

25.3 COLOUR PERCEPTION

All objects that we observe are focused sharply by the lens system of the eye on its retina. The retina which is located at the back side of the eye has light sensitive organs which measure the visual sensations. The retina is connected with the optic nerve which conducts the light stimuli as sensed by the organs to the optical centre of the brain.

According to the theory formulated by Helmholtz the light sensitive organs are of two types—rods and cones. The rods provide brightness sensation and thus perceive objects only in various shades of grey from black to white. The cones that are sensitive to colour are broadly in three different groups. One set of cones detects the presence of blue colour in the object focused on the retina, the second set perceives red colour and the third is sensitive to the green range. Each set of cones, may be thought of as being 'tuned' to only a small band of frequencies and so absorb energy from a definite range of electromagnetic radiation to convey the sensation of corresponding colour or range of colour. The combined relative luminosity curve showing relative sensation of brightness produced by individual spectral colours radiated at a constant energy level is shown in Fig. 25.2. It will be seen from the plot that the sensitivity of the human eye is greatest for green light, decreasing towards both the red and blue ends of the spectrum. In fact the maximum is located at about 550 nm, a yellow green, where the spectral energy maximum of sunlight is also located.

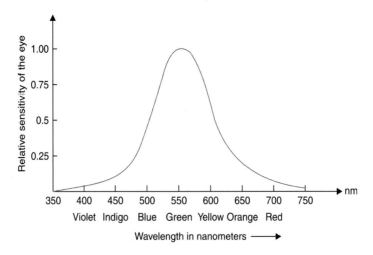

Fig. 25.2. *Approximate relative response of the eye to different colours.*

25.4 THREE COLOUR THEORY

All light sensations to the eye are divided (provided there is an adequate brightness stimulus on the operative cones) into three main groups. The optic nerve system then integrates the different colour impressions in accordance with the curve shown in Fig. 25.2 to perceive the actual colour of the object being seen. *This is known as additive mixing and forms the basis of any colour television system. A yellow colour, for example, can be distinctly seen by the eye when the red and green groups of the cones are excited at the same time with corresponding intensity ratio. Similarly and colour other than red, green and blue will excite different sets of cones to generate the cumulative sensation of that colour. A white colour is then perceived by the additive mixing of the sensations from all the three sets of cones.

Mixing of Colours

Mixing of colours can take place in two ways—subtractive mixing and additive mixing. In subtractive mixing, reflecting properties of pigments are used, which absorb all wavelengths but for their characteristic colour wavelengths. When pigments of two or more colours are mixed, they reflect wavelengths which are common to both. Since the pigments are not quite saturated (pure in colour) they reflect a fairly wide band of wavelengths. This type of mixing takes place in painting and colour printing.

In additive mixing which forms the basis of colour television, light from two or more colours obtained either from independent sources or through filters can create a combined sensation of a different colour. Thus different colours are created by mixing pure colours and not by subtracting parts from white.

The additive mixing of three primary colours—red, green and blue in adjustable intensities can create most of the colours encountered in everyday life. The impression of white light can also be created by choosing suitable intensities of these colours.

Red, green and blue are called primary colours. These are used as basic colours in television. By pairwise additive mixing of the primary colours the following complementary colours are produced:

Red + Green = Yellow

Red + Blue = Magenta (purplish red shade)

Blue + Green = Cyan (greenish blue shade)

Fig. 25.3 (c) depicts the location of primary and complementary colours on the colour circle.

If a complementary is added in appropriate proportion to the primary which it itself does not contain, white is produced. This is illustrated in Fig. 25.3 (a) where each circle corresponds to one primary colour. Fig. 25.3 (b) shows the effect of colour mixing. Similarly Fig. 25.3 (b), illustrates the process of subtractive mixing. Note that as additive mixing of the three primary colours produces white, their subtractive mixing results in black.

---

*Research has shown that the actual neural process of colour perception is substantially different from the tricolour process. However, all colour reproduction processes in television or printing use variations of this process and is found satisfactory.

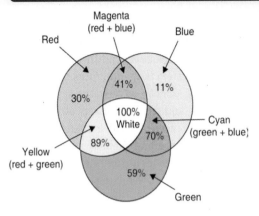

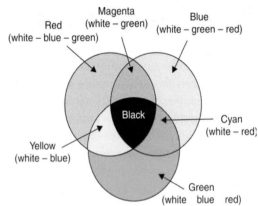

Fig. 25.3 (a). Additive colour mixing. The diagram shows the effect of projecting green, red and blue beams on a white screen in such a way that they overlap.

Fig. 25.3 (b). Subtractive colour mixing. The diagram shows the effect of mixing colour pigments under white light.

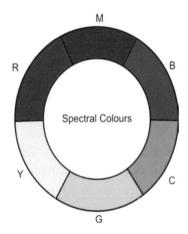

Fig. 25.3 (c). Colour circle with primary and complementary colours.

Grassman's Law

The eye is not able to distinguish each of the colours that mix to form a new colour but instead perceives only the resultant colour. Thus the eye behaves as though the output of the three types of cones are additive. The subjective impression which is gained when green, blue and red lights reach the eye simultaneously, may be matched by a single light source having the same colour. In addition to this, the brightness (luminance) impression created by the combined light source is numerically equal to the sum of the brightnesses (luminances) of the three primaries that constitute the single light. This property of the eye of producing a response which depends on the algebraic sum of the red, green and blue inputs is known as Grassman's Law. White has been seen to be reproduced by adding red, green and blue lights. The intensity of each colour may be varied. This enables simple rules of addition and subtraction.

Tristimulus Values of Spectral Colours

Based on the spectral response curve of Fig. 25.2 and extensive tests with a large number of observers, the primary spectral colours and their intensities required to produce different colours by mixing have been standardized. The red green and blue have been fixed at wavelengths of 700 nm, 546.1 nm and 438.8 nm respectively. The component values (or fluxes) of the three primary colours to produce various other colours have also been standardized and are called the tristimulus values of the different spectral colours. The reference white for colour television has been chosen to be a mixture of 30% red, 59% green and 11% blue. These percentages for the light fluxes are based on the sensitivity of the eye to different colours.

Thus one lumen (lm) of white light = 0.3 lm of red + 0.59 lm of green + 0.11 lm of blue. In accordance with the law of colour additive mixing one lm of white light (see Fig. 25.3 (a)) is also = 0.89 lm of yellow + 0.11 lm of blue or = 0.7 lm of cyan + 0.3 lm of red or = 0.41 lm of magenta + 0.59 lm of green. It may be noted that if the concentration of luminous flux is reduced by a common factor from all the constituent colours, the resultant colour will still be white, though its level of brightness will decrease. The brightness of different spectral colours is associated with that of white. Yellow, for example, appears 89% as bright as the reference white, reflecting the addition of 59% brightness of green and 30% brightness of red. Similarly any other combination of primary colours will produce a different colour with a different relative brightness with reference to white which has been taken as 100 percent.

25.5 LUMINANCE, HUE AND SATURATION

Any colour has three characteristics to specify its visual information. These are (i) luminance, (ii) hue or tint, and (iii) saturation. These are defined as follows:

(i) Luminance or Brightness

This is the amount of light intensity as perceived by the eye regardless of the colour. In black and white pictures, better lighted parts have more luminance than the dark areas. Different colours also have shades of luminance in the sense that though equally illuminated appear more or less bright as indicated by the relative brightness response curve of Fig. 25.2. Thus on a monochrome TV screen, dark red colour will appear as black, yellow as white and a light blue colour as grey.

(ii) Hue

This is the predominant spectral colour of the received light. Thus the colour of any object is distinguished by its hue or tint. The green leaves have green hue and red tomatoes have red hue. Different hues result from different wavelengths of spectral radiation and are perceived as such by the sets of cones in the retina.

(iii) Saturation

This is the spectral purity of the colour light. Since single hue colours occur rarely alone, this indicates the amounts of other colours present. Thus saturation may be taken as an indication of how little the colour is diluted by white. A fully saturated colour has no white. As an example, vivid green is fully saturated and when diluted by white it becomes light green. The hue and saturation of a colour put together is known as chrominance. Note that it does not contain the brightness information. Chrominance is also called chroma.

Chapter 25

Chapter 25

Chromaticity Diagram. Chromaticity diagram is a convenient space coordinate representation of all the spectral colours and their mixtures based on the tristimulus values of the primary colours contained by them. Fig. 25.4 (*a*) is a two dimensional representation of hue and saturation in the *X-Y* plane. If a three dimensional representation is drawn, the '*Z*' axis will show relative brightness of the colour.

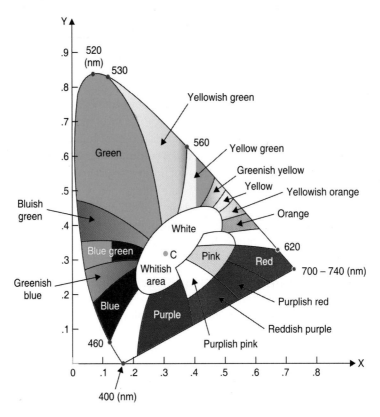

Fig. 25.4 (a). Chromaticity diagram. Note that red, green and blue have been standardized at wavelengths of 700, 646.1 and 438.8 nanometers respectively. X and Y denote colour coordinates. For example white lies at X = 0.31 and Y = 0.32.

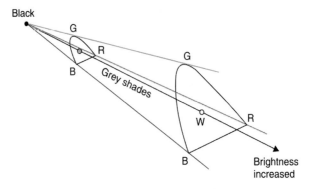

Fig. 25.4 (b). Representation of luminance (brightness).

As seen in the figure the chromaticity diagram is formed by all the rainbow colours arranged along a horseshoe-shaped triangular curve. The various saturated pure spectral colours are represented along the perimeter of the curve, the corners representing the three primary colours—red, green and blue. As the central area of the triangular curve is approached, the colours become desaturated representing mixing of colours or a white light. The white lies on the central point 'C' with coordinates $x = 0.31$ and $y = 0.32$. Actually there is no specific white light—sunlight, skylight, daylight are all forms of white light. The illuminant 'C' marked in Fig. 25.4 (a) represents a particular white light formed by combining hues having wavelength: 700 nm (red) 546.1 nm (green) and 438.8 nm (blue) with proper intensities. This shade of white which has been chosen to represent white in TV transmission and reception also corresponds to the subjective impression formed in the human eye by seeing a mixture of 30 percent of red colour, 59 percent of green colour and 11 percent of the blue colour at wavelengths specified above.

A practical advantage of the chromaticity diagram is that, it is possible to determine the result of additive mixing of any two or more colour lights by simple geometric construction. The colour diagram contains all colours of equal brightness. Since brightness is represented by the 'Z' axis, as brightness increase, the colour diagram becomes larger as shown in Fig. 25.4 (b).

25.6 COLOUR TELEVISION CAMERA

Figure 25.5 shows a simple block schematic of a colour TV camera. It essentially consists of three camera tubes in which each tube receives selectively filtered primary colours. Each camera tube develops a signal voltage proportional to the respective colour intensity received by it. Light from the scene is processed by the objective lens system. The image formed by the lens is split into three images by means of glass prisms. These prisms are designed as diachroic mirrors. A diachroic mirror passes one wavelength and rejects other wavelengths (colours of light). Thus red, green, and blue colour images are formed. The rays from each of the light splitters also pass through colour filters called trimming filters. These filters provide highly precise primary colour images which are converted into video signals by image-orthicon or vidicon camera tubes. Thus the three colour signals are generated. These are called Red (R), Green (G) and Blue (B) signals.

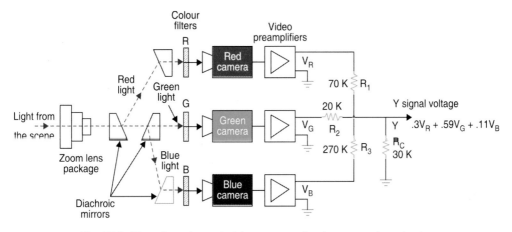

Fig. 25.5. Plan of a colour television camera showing generation of colour signals and Y matrix for obtaining the luminance (brightness) signal.

Simultaneous scanning of the three camera tubes is accomplished by a master deflection oscillator and sync generator which drives all the three tubes. The three video signals produced by the camera represent three primaries of the colour diagram. By selective use of these signals, all colours in the visible spectrum can be reproduced on the screen of a special (colour) picture tube.

Colour Signal Generation

At any instant during the scanning process the transmitted signal must indicate the proportions, of red, green and blue lights which are present in the element being scanned. Besides this, to fulfil the requirements of compatibility, the luminance signal which represents the brightness of the elements being scanned must also be generated and transmitted along with the colour signals. Figure 25.5 illustrates the method of generating these signals. The camera output voltages are labelled as V_R, V_G and V_B but generally the prefix V is omitted and only the symbols R, G, and B are used to represent these voltages. With the specified source of white light the three cameras are adjusted to give equal output voltage.

Gamma Correction

To compensate for the non-linearity of the system including TV camera and picture tubes, a correction is applied to the voltages produced by the three camera tubes. The output voltages are then referred as R', G' and B'. However, in our discussion we will ignore such a distinction and use the same symbols *i.e.*, R, G and B to represent gamma corrected output voltages. Furthermore, for convenience of explanation the camera outputs corresponding to maximum intensity (100%) of standard white light to be handled are assumed adjusted at an arbitrary value of one volt. Then on grey shades, *i.e.*, on white of lesser brightness, R, G and B voltages will remain equal but at amplitude less than one volt.

25.7   THE LUMINANCE SIGNAL

To generate the monochrome or brightness signal that represents the luminance of the scene, the three camera outputs are added through a resistance matrix (see Fig. 25.5) in the proportion of 0.3, 0.59 and 0.11 of R, G and B respectively. This is because with white light which contains the three primary colours in the above ratio, the camera outputs were adjusted to give equal voltages. The signal voltage that develops across the common resistance R_C represents the brightness of the scene and is referred to as 'Y' signal. Therefore,

$$Y = 0.3\ R + 0.59\ G + 0.11\ B$$

Colour Voltage Amplitudes

Figure 25.6 (*a*) illustrates the nature of output from the three cameras when a horizontal line across a picture having vertical bars of red, green and blue colours is scanned. Note that at any one instant only one camera delivers output voltage corresponding to the colour being scanned.

In Fig. 25.6 (*b*) different values of red colour voltage are illustrated. Here the red pink and pale pink which are different shades of red have decreasing values of colour intensity. Therefore the corresponding output voltages have decreasing amplitudes. Thus we can say that R, G or B voltage indicates information of the specific colour while their relative amplitudes depend on the level of saturation of that colour.

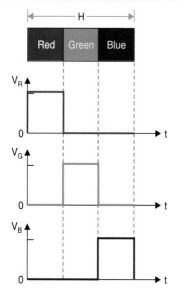

Fig. 25.6 (a). *Camera video output voltages for red green and blue colour bars. 'H' indicates one horizontal scanning line.*

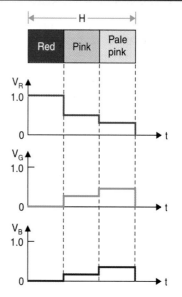

Fig. 25.6 (b). *Decreasing amplitudes of the red camera output indicates effect of colour desaturation.*

Next consider the scanning across a picture that has yellow and white bars besides the three pure colour bars. The voltages of the three camera outputs are drawn below the colour bar pattern in Fig. 25.7. Notice the values shown for yellow, as an example of a complementary colour. Since yellow includes red and green, video voltage is produced for both these primary colours. Since there is no blue in yellow, the blue camera output voltage is at zero for the yellow bar. The white bar at the right includes all the three primary colours and so all the three cameras develop output voltage when this bar is scanned.

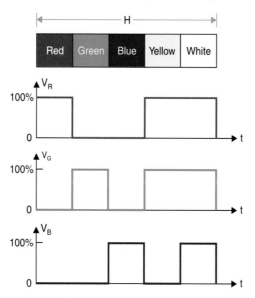

Fig. 25.7. *Red, green and blue camera video output voltages for the bar pattern shown above.*

'Y' Signal Amplitude

As already stated the 'Y' signal contains brightness variations of the picture information, and is formed by adding the three camera outputs in the ratio, $Y = 0.3\,R + 0.59\,G + 0.11\,B$.

These percentages correspond to the relative brightness of the three primary colours. Therefore a scene reproduced in black and white by the 'Y' signal looks the same as when it is televised in monochrome. Figure 25.8 (a) illustrates how the 'Y' signal voltage is formed from the specified proportions of R, G, and B voltages for the colour bar pattern. The addition, as already explained is carried out (see Fig. 25.5) by the resistance matrix. Note that the 'Y' signal for white has the maximum amplitude (1.0 or 100%) because it includes R, G and B.

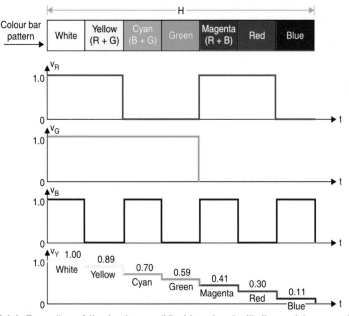

Fig. 25.8 (a). *Formation of the luminance (Y) video signal with the resistance matrix from colour camera outputs. Note that yellow appears 89% (.3R + .59G) bright as compared to 100% for white. Similarly cyan, green, magenta, red and blue appear 70%, 59%, 41%, 30% and 11% bright respectively as compared to assumed 100% brightness for white.*

For the other bars the magnitude of Y decreases or changes in accordance with the colour or colours that form the bars (see Fig. 25.8 (b)). All the voltage values can be calculated as illustrated by the writeup below Fig. 25.8 (a). If only this 'Y' signal is used to reproduce the pattern, it will appear as monochrome bars shading-off from white at the left to grey in the centre and black at the right. These values correspond to the staircase pattern of Y voltage shown in the figure. Note the progressive decrease in voltage for the relative brightness of the various colour bars.

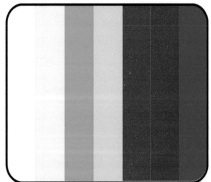

Fig. 25.8 (b). *Colour bar pattern with colours in order of decreasing brightness : white, yellow, cyan, green agenta, red and blue*

Production of Colour Difference Voltages

The 'Y' signal is modulated and transmitted as is done in a monochrome television system. However, instead of outputs are combined with the Y signal to obtain what is known as colour difference signals. Colour difference voltages are derived by subtracting the luminance voltage from the colour voltages. Only $(R - Y)$ and $(B - Y)$ are produced. It is only necessary to transmit two of the three colour difference signals since the third may be derived from the other two. The reason for not choosing $(G - Y)$ for transmission and how the green signal is recovered are explained in a later section of this chapter. The circuit of Fig. 25.5 is reproduced in Fig. 25.9 to explain the generation of $(B - Y)$ and $(R - Y)$ voltages. The voltage V_Y as obtained from the resistance matrix is low because R_C is chosen to be small to avoid crosstalk. Hence it is amplified before it leaves the camera subchassis. Also the amplified Y signal is inverted to obtain $-Y$ as the output. This is passed on to the two adder circuits. One adder circuit adds the red camera output to $-Y$ to obtain the $(R - Y)$ signal. Similarly the second adder combines the blue camera output to $-Y$ and delivers $(B - Y)$ as its output. This is illustrated in Fig. 25.9. The difference signals thus obtained bear information both about the hue and saturation of different colours.

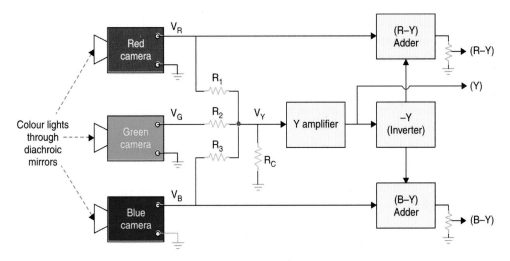

Fig. 25.9. Production of luminance and colour-difference signals.
Note that $Y = 0.3R + 0.59G + 0.11B$
$(R - Y) = 0.7R - 0.59G - 0.11B$
and $(B - Y) = 0.89B - 0.59G - 0.3R$.

Compatibility Considerations

The colour difference signals equal zero when white or grey shades are being transmitted. This is illustrated by two examples.

(a) On peak whites let $R = G = B = 1$ volt

Then $\quad\quad\quad\quad\quad Y = 0.59G + 0.3R + 0.11B = 0.59 + 0.3 + 0.11 = 1$ (volt)

$\therefore \quad (R - Y) = 1 - 1 = 0$, volt and $(B - Y) = 1 - 1 = 0$ volt

(b) On any grey shade let $R = G = B = v$ volts $(v < 1)$

Then $\quad\quad\quad\quad\quad Y = 0.59v + 0.3v + 0.11v = v$

$\therefore \quad (R - Y) = v - v = 0$ volt and $(B - Y) = v - v = 0$ volt

Thus it is seen that colour difference signals during the white or grey content of a colour scene of during the monochrome transmission completely disappear and this is an aid to compatibility in colour TV systems.

25.8 VALUES OF LUMINANCE (Y) AND COLOUR DIFFERENCE SIGNALS ON COLOURS

When televising colour scenes even when voltages R, G and B are not equal, the 'Y' signal still represents monochrome equivalent of the colour because the proportions 0.3, 0.59 and 0.11 taken of R, G and B respectively still represent the contribution which red, green and blue lights make to the luminance. This aspect can be illustrated by considering some specific colours.

Desaturated Purple

Consider a desaturated purple colour, which is a shade of magenta. Since the hue is magenta (purple) it implies that it is a mixture of red and blue. Two word desaturated indicates that some white light is also there. The white light content will develop all the three *i.e.*, R, G and B voltages, the magnitudes of which will depend on the intensity of desaturation of the colour. Thus R and B voltages will dominate and both must be of greater amplitude than G. As an illustration let $R = 0.7$, $G = 0.2$ and $B = 0.6$ volts. The white content is represented by equal quantities of the three primaries and the actual amount must be indicated by the smallest voltage of the three, that is, by the magnitude of G.

Thus white is due to 0.2 R, 0.2 G and 0.2 B. The remaining, 0.5 R and 0.4 B together represent the magenta hue.

(*i*) The luminance signal $Y = 0.3 R + 0.59 G + 0.11 B$. Substituting the values of R, G, and B we get $Y = 0.3 (0.7) + 0.59 (0.2) + 0.11(0.6) = 0.394$ (volts).

(*ii*) The colour difference signals are:

$$(R - Y) = 0.7 - 0.394 = + 0.306 \text{ (volts)}$$
$$(B - Y) = 0.6 - 0.394 = + 0.206 \text{ (volts)}$$

(*iii*) *Reception at the colour receiver*—At the receiver after demodulation, the signals, Y, $(B - Y)$ and $(R - Y)$, become available. Then by a process of matrixing the voltages B and R are obtained as:

$$R = (R - Y) + Y = 0.306 + 0.394 = 0.7 \text{ V}$$
$$B = (B - Y) + Y = 0.206 + 0.394 = 0.6 \text{ V}$$

(*iv*) $(G-Y)$ *matrix*—The missing signal $(G-Y)$ that is not transmitted can be recovered by using a suitable matrix based on the explanation given below:

$$Y = 0.3 R + 0.59G + 0.11B$$

also $(0.3 + 0.59 + 0.11)Y = 0.3R + 0.59G + 0.11B$

Rearranging the above expression we get:

$$0.59(G - Y) = - 0.3 (R - Y) - 0.11 (B - Y)$$

∴ $$(G - Y) = \frac{-0.3}{0.59} (R - Y) - \frac{0.11}{0.59} (B - Y) = - 0.51(R - Y) - 0.186 (B - Y)$$

Substituting the values of $(R - Y)$ and $(B - Y)$

$$(G - Y) = -(0.51 \times 0.306) - 0.186(0.206) = -0.15606 - 0.038216 = -0.194$$

$\therefore$ $G = (G - Y) + Y = -0.194 + 0.394 = 0.2$, and this checks with the given value.

(*v*) *Reception on a monochrome receiver*—Since the value of luminance signal $Y = 0.394$ V, and the peak white corresponds to 1 volt (100%) the magenta will show up as a fairly dull grey in a black and white picture. This is as would be expected for this colour.

Desaturated Orange

A desaturated orange having the same degree of desaturation as in the previous example is considered now. Taking $R = 0.7$, $G = 0.6$, and $B = 0.2$, it is obvious that the output voltages due to white are $R = 0.2$, $G = 0.2$ and $B = 0.2$. Then red and green colours which dominate and represent the actual colour content with, $R = 0.5$ and $G = 0.4$ give the orange hue. Proceeding as in the previous example we get:

(*i*) Luminance signal $Y = 0.3\,R + 0.59G + 0.11\,B$. Substituting the values of R, G and B we get $Y = 0.586$ volt.

(*ii*) Similarly, the colour difference signal magnitudes are:

$$(R - Y) = (0.7 - 0.586) = +0.114$$
$$(B - Y) = (0.2 - 0.586) = -0.386$$

and
$$(G - Y) = -0.51(R - Y) - 0.186(B - Y) = 0.014$$

(*iii*) At the receiver by matrixing we get

$$R = (R - Y) + Y = 0.7$$
$$G = (G - Y) + Y = 0.6$$
$$B = (B - Y) + Y = 0.2$$

This checks with the voltages developed by the three camera tubes at the transmitting end.

(*iv*) *Reception on a monochrome receiver*—Only the luminance signal is received and, as expected, with $Y = 0.586$, the orange hue will appear as bright grey.

25.9   POLARITY OF THE COLOUR DIFFERENCE SIGNALS

As has been demonstrated by the above two examples, both $(R - Y)$ and $(B - Y)$ can be either positive or negative depending on the hue they represent. The reason is that for any primary, its complement contains the other two primaries. Thus a primary and its complement can be considered as opposite to each other and hence the colour difference signals turn out to be of opposite polarities. This is illustrated by the colour phasor diagram of Fig. 25.10. Observe that a purplish-red hue is represented by $+ (R - Y)$ while its complement, a bluish-green hue corresponds to $- (R - Y)$. Similarly $+ (B - Y)$ and $- (B - Y)$ represent purplish-blue and greenish-yellow hues respectively (See Fig. 25.10). Note that green colour is obtained by a combination of $- (R - Y)$ and $- (B - Y)$ while cyan is obtained by a combination of $- (R - Y)$ and $+ (B - Y)$ signals. Furthermore, any one of the three primaries or their complementaries can be obtained by a combination of two of the above four signals. It may also be noted that the colour difference video signals have no brightness component and represent only the different hues.

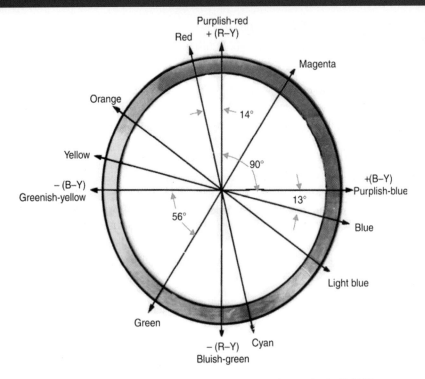

Fig. 25.10. *Colour circle showing location and magnitude (100%) of primary and complementary colours.*

Unsuitability of (G – Y) Signal for Transmission

As shown earlier, $(G-Y) = -0.51(R-Y) - 0.186(B-Y)$. Since the required amplitudes of both $(R-Y)$ and $(B-Y)$ are less than unity, they may be derived using simple resistor attenuators across the respective signal paths. However, if $(G-Y)$ is to be one of the two transmitted signals then

(*i*) if $(R-Y)$ s the missing signal, its matrix would have to be based on the expression:

$$(R-Y) = \frac{-0.59}{0.3}(G-Y) - \frac{0.11}{0.3}(B-Y)$$

The factor 0.59/0.3 (= 1.97) implies gain in the matrix and thus would need an extra amplifier.

(*ii*) Similarly if $(B-Y)$ is not ransmitted, the matrix formula would be:

$$(B-Y) = \frac{-0.59}{0.11}(G-Y) - \frac{0.3}{0.11}(R-Y)$$

The factor 0.59/0.11 = 5.4 and 0.3/0.11 = 2.7, both imply gain and two extra amplifiers would be necessary in the matrices. This shows that it would be technically less convenient and uneconomical to use $(G-Y)$ as one of the colour difference signals for transmission.

In addition, since the proportion of G in Y is relatively large in most cases, the amplitude of $(G-Y)$ is small. It is either the smallest of the three colour difference signals, or is atmost equal to the smaller of the other two. The smaller amplitude together with the need for gain in the matrix would make *S/N* ratio problems more difficult then when $(R-Y)$ and $(B-Y)$ are chosen for transmission.

25.10 COLOUR TELEVISION DISPLAY TUBES

The colour television picture tube screen is coated with three different phosphors, one for each of the chosen red, green and blue primaries. The three phosphors are physically separate from one another and each is energized by an electron beam of intensity that is proportional to the respective colour voltage reproduced in the television receiver. The object is to produce three coincident rasters with produce the red, green and blue contents of the transmitted picture. While seeing from a normal viewing distance the eye integrates the three colour information to convey the sensation of the hue at each part of the picture. Based on the gun configuration and the manner in which phosphors are arranged on the screen, three different types of colour picture tubes have been developed. These are:

1. Delta-gun colour picture tube
2. Guns-in-line or Precision-in-line (*P-I-L*) colour picture tube.
3. Single sun or Trintron Colour picture tube.

25.11 DELTA-GUN COLOUR PICTURE TUBE

This tube was first developed by the Radio Corporation of America (R.C.A.). It employs three separate guns (see Fig. 25.11 (*a*)), one for each phosphor. The guns are equally spaced at 120° interval with respect to each other and tilted inwards in relation to the axis of the tube. They form an equilateral triangular configuration.

As shown in Fig. 25.11 (*b*) the tube employs a screen where three colour phosphor dots are arranged in groups known as triads. Each phosphor dot corresponds to one of the three primary colours. The triads are repeated and depending on the size of the picture tube, approximately 1,000,000 such dots forming nearly 333,000 triads are deposited on the glass face plate. About one cm behind the tube screen (see Figs. 25.11 (*b*) and (*c*)) is located a thin perforated metal sheet known as the shadow mask. The mask has one hole for every phosphor dot triad on the screen. The various holes are so oriented that electrons of the three beams on passing through any one hole will hit only the corresponding colour phosphor dots on the screen. The ratio of electrons passing through the holes to those reaching the shadow mask is only about 20 percent. The remaining 80 percent of the total beam current energy is dissipated as a heat loss in the shadow mask. While the electron transparency in other types of colour picture tubes is more, still, relatively large beam currents have to be maintained in all colour tubes compared to monochrome tubes. This explains why higher anode voltages are needed in colour picture tubes than are necessary in monochrome tubes.

Generation of Colour Rasters

The overall colour seen is determined both by the intensity of each beam and the phosphors which are being bombarded. If only one beam is 'on' and the remaining two are cut-off, dots of only one colour phosphor get excited. Thus the raster will be seen to have only one of the primary colours. Similarly, if one beam is cut-off and the remaining two are kept on, the rasters produced by excitation of the phosphors of two colours will combine to create the impression of a complementary colour. The exact hue will be determined by the relative strengths of the two beams. When all the three guns are active simultaneously, lighter shades are produced on the screen. The is so because red, green and blue combine in some measure to form white, and this combines with whatever colours are present to desaturate them. Naturally, intensity of the colour produced depends on the intensity of beam currents. Black in a picture is just the absence of excitation when all the three beams are cut-off. If the amplitude of colour difference signals drops to zero, the only signal left to control the three guns would be the Y signal and thus a black and white (monochrome) picture will be produced on the screen.

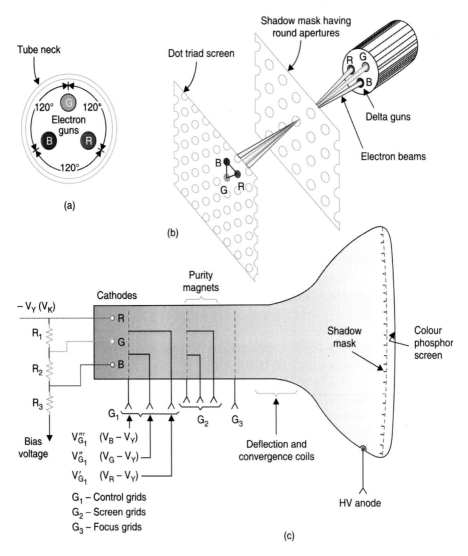

Fig. 25.11. Delta-gun colour picture tube (a) guns viewed from the base (b) electron beams, shadow mask and dot-triad phosphor screen (c) schematic diagram showing application of 'Y' and colour difference signals between the cathodes and control grids.

Primary Colour Signals

The demodulators in the receiver recover $(B - Y)$ and $(R - Y)$ video signals. The $(G - Y)$ colour video signal is obtained from these two through a suitable matrix. All the three colour difference signals are then fed to the three grids of colour picture tube (see Fig. 24.11 (c)). The inverted luminance signal $(-Y)$ is applied at the junction of the three cathodes. The signal voltages subtract from each other to develop control voltages for the three guns, *i.e.*,

$$V'_{G1} - V_k = (V_R - V_Y) - (-V_Y) = V_R$$
$$V''_{G1} - V_k = (V_G - V_Y) - (-V_Y) = V_G$$

and $\qquad\qquad V'''_{G1} - V_k = (V_B - V_Y) - (-V_Y) = V_B$

In some receiver designs the Y signal is subtracted in the matrix and the resulting colour voltages are directly applied to the corresponding control grids. The cathode is then returned to a fixed negative voltage.

25.12   PURITY AND CONVERGENCE

While deflecting the three beams by vertical and horizontal deflecting coils it is necessary to ensure that each beam produces a pure colour and all the three colour rasters fully overlap each other. For obtaining colour purity each beam should land at the centre of the corresponding phosphor dot irrespective of the location of the beams on the raster. This needs precise alignment

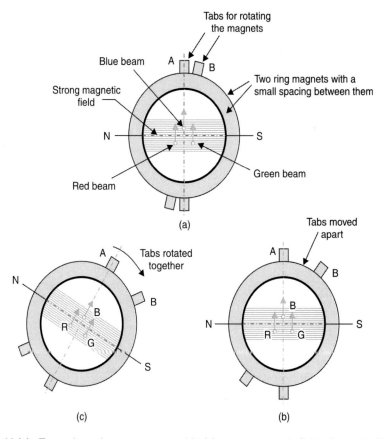

Fig. 25.12 (a). Two pole purity magnet assembly (a) strong magnetic field when tabs (A and B) are nearly together (b) spreading the tabs reduces magnetic field (c) rotating the magnets together rotates the magnetic field to cause change in the direction of beam deflection.

of the colour beam and is carried out by a circular magnet assembly known as the purity magnet. It is mounted externally on the neck of the tube and close to the deflection yoke. The purity magnet assembly consists of several flat washer like magnets held together by spring clamps in such a way that these can be rotated freely. The tabs on the magnets can be moved apart to reduce resultant field strength. This is illustrated for a two pole magnet in Fig. 25.12 (*a*). As shown in the same figure, the tabs when moved together change the direction of magnetic field. Two, four and six pole magnet units are employed to achieve individual and collective beam deflections. Thus to affect purity and static convergence the beams can be

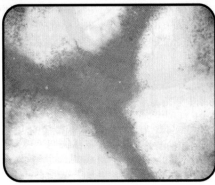

Fig. 25.12 (b). Incorrect adjustment of the purity magnet

deflected up or down, right or left and diagonally by suitably orienting the purity magnets.

Yoke Position

The position of the yoke on the tube neck determines the location of the deflection centre of the electron beams. A wrong setting will result in poor purity due to improper entry angles of the beams into the mask openings. Since deflection due to yoke fields affects the landing of the beams on the screen more towards the edges of the tube, the yoke is moved along the neck of the tube to improve purity in those regions.

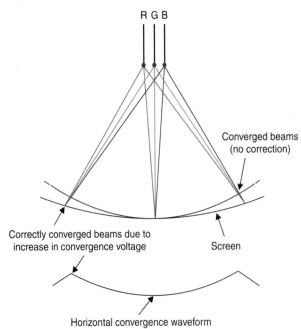

Convergence

The technique of bringing the beams together so that they hit the same part of the screen at the same time to produce three coincident rasters is referred to as convergence. Convergence errors are caused by (*i*) non-coincident convergence planes, (*ii*) nonuniformity of the deflection field and (*iii*) flat surface of the picture tube screen. Figure 25.13 illustrates correct and incorrect convergence of beams. Proper convergence is achieved by postional adjustment of the individual beams. It falls into two parts referred to as (*i*) static and (*ii*) dynamic convergence. Static convergence involves movement of the beams by permanent magnetic fields which, once correctly set, bring the beams into convergence in the central area of the screen.

Fig. 25.13. Over-convergence of electron beams at the screen edges unless a corrective deflection field is utilized.

Convergence over the rest of the screen is achieved by continuously varying (dynamic) magnetic fields, the instantaneous strengths of which depend upon the positions of the spots on the screen. These fields are set up by electromagnets which carry currents at horizontal (line) and vertical (field) frequencies. In practice convergence coils are often connected in series with respective yoke windings.

Pincushion Correction

The use of permanent magnets for the elimination of pincushion distortion is not feasible for colour receivers because the magnets would tend to introduce purity problems. Therefore, dynamic pincushion correction is used with colour picture tubes. Such a correction automatically increases horizontal width and vertical size in those regions of the raster that are shrunken because of pincushion distortion.

Degaussing

The main cause of poor purity is the susceptibility of the mask and its mounting frame to become magnetized by the earth's magnetic field and/or by any other strong magnetic fields. The effect of these localized magnetic fields is the deviation of electron beams from their normal path. Thus the beams strike wrong phosphors causing poor purity especially at the edges of the screen. To prevent such effects the picture tubes are magnetically shielded. It is done by placing a thin silicon steel (mu metal) housing around the bell of the tube. Since the mask structure and shield material have non-zero magnetic retentivity they get weakly magnetized by the earth's magnetic field and other extraneous fields such as from magnetic toys, domestic electrical apparatus etc. Thus over a time despite initial adjustments or whenever the colour receiver is moved from the one location to another the stray field changes to affect purity. To overcome this drawback, some form of automatic degaussing is incorporated in all colour receivers. Degaussing means demagnetizing iron and steel parts of the picture tube mountings. A magnetic object can be demagnetized by placing it in an alternating magnetic field which becomes weaker over a period of time. This way the magnetized object is forced to assume the strength of the external degaussing field and becomes weaker and weaker as the degaussing field diminishes. A degaussing coil is used for this purpose. It is wrapped round the tube bowl close to the rim-band of the screen. The circuit is so designed that when the receiver is first switched on, a strong mains current passes through the coil and then dies away to an insignificant level after a few moments. This way the effects of localized manetic fields are removed each time the receiver is used.

Drawbacks of the Delta-gun Tube

While the delta-gun colour picture tube has been in use for nearly two decades it suffers from the following drawbacks:

(*i*) Convergence is difficult and involves considerable circuit complexity and service adjustments. In most delta-gun tubes, four static convergence magnets and a dynamic convergence assembly are employed. In all about 12 pre-set controls become necessary to achieve proper vertical and horizontal convergence over the entire screen.

(*ii*) The focus cannot be sharp over the entire screen because the focus and convergence planes cannot remain coincident for the three beams which emanate from guns positioned at 120° with respect to each other around the tube axis.

Chapter 25

(*iii*) The electron transparency of the mask is very low since it intercepts over 80 percent of the beam currents.

Therefore the delta-gun tube has been superseded by the P.I.L. and Trintron colour picture tubes. It is now manufactured mostly for replacement purposes.

25.13   PRECISION-IN-LINE (P.I.L.) COLOUR PICTURE TUBE

This tube as the name suggests has three guns which are aligned precisely in a horizontal line. The gun and mask structure of the P.I.L. tube together with yoke mounting details are illustrated in Fig. 25.14. The in-line gun configuration helps in simplifying convergence adjustments. As shown in the figure colour phosphors are deposited on the screen in the form of vertical strips

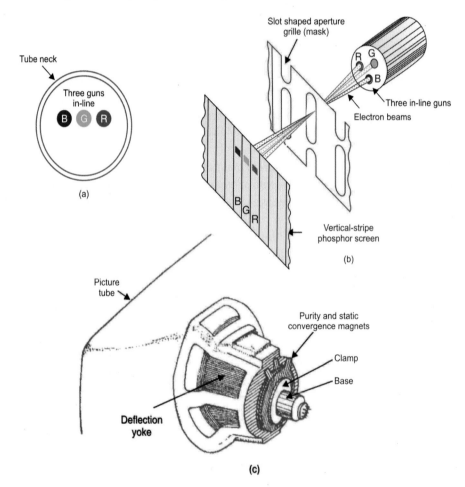

Fig. 25.14. *Precision in-line (P-I-L) or cathodes-in-line colour picture tube*
(a) in-line guns (b) electron beams, aperture grille and striped three colour phosphor screen
(c) mountings on neck and bowl of the tube.

in triads. (*R*, *G*, *B*) which are repeated along the breadth of the tube. To obtain the same colour fineness as in a delta-gun tube the horizontal spacing between the strips of the same colour in adjacent triads is made equal to that between the dots of the same colour in the delta-gun tube. As shown in Fig. 25.14 (*b*), the aperture mask has vertical slots corresponding to colour phosphor stripes. One vertical line of slots is for one group of fine strips of red green and blue phosphors. Since all the three electron beams are on the same plane, the beam in the centre (green) moves along the axis of the tube. However, because of inward tilt of the right and left guns the blue and red beams travel at an angle and meet the central beam at the aperture grille mask. The slots in the mask are so designed that each beam strikes its own phosphor and is prevented from landing on other colour phosphors.

The P.I.L. tube is more efficient, *i.e.*, has higher electron transparency and needs fewer convergence adjustments on account of the in-line gun structure. It is manufactured with minor variations under different trade names in several countries and is the most used tube in present day colour receivers.

Purity and Static Convergence

Modern P.I.L. tubes are manufactured with such precision that hardly any purity and static convergence adjustments are necessary. However, to correct for small errors due to mounting tolerances and stray magnetic fields multipole permanent magnet units are provided. Such a unit is mounted on the neck of the tube next to the deflection yoke. The various magnets are suitably oriented to achieve colour purity, static convergence and straightness of horizontal raster lines.

Convergence Errors

The need for the three rasters to be accurately superposed on each other with no east to west (lateral) or north to south (vertical) displacement (*i.e.*, in proper colour register) puts stringent constraints on distribution of the deflection field. Non-optimum distribution of the field along with the fact that the screen is a nearly flat surface (and not spherical) can produce two kinds of lack of colour registration, commonly known as convergence errors. These are (*a*) astigmatism and (*b*) coma effects.

(*a*) *Astigmatism.* As shown in Fig. 25.15 (*a*), with a uniform field the rays from a vertical row of points (*P, G, Q*) converge short of the screen producing a vertical focal line on the screen. However, rays (*R, G, B*) from a row of horizontal points converge beyond the screen producing a horizontal line on the screen. Such an effect is known as astigmatism and causes convergence errors. In the case of a P.I.L. tube, beams *R, G* and *B* emerge only from a horizontal line and so any vertical astigmatism will have no effect. However, if horizontal astigmatism is to be avoided, the horizontal focus must be a point on the screen and not a line for any deflection. A given change in the shape of the deflection field produces opposite changes in vertical and horizontal astigmatic effects. Since vertical astigmatism is of no consequence in a *P-I-L* tube, suitable field adjustments can be made to ensure that the beams coming from in-line guns in a horizontal line converge at the same point on the screen irrespective of the deflection angle. Such a correction will however, be at the cost of a much larger vertical astigmatism but as stated above it does not interfere with the correct registration of various colours.

(*b*) *Coma Effect.* Due to nonuniformity of the deflection field all the beams are not deflected by the same amount. As shown in Fig. 25.15 (*b*) the central beam (green) deflects by a smaller amount as compared to the other two beams. For a different nonuniformity of the deflection field, the effect could be just opposite producing too large a displacement of the central beam. Such a distortion is known as coma and results in misconvergence of the beams.

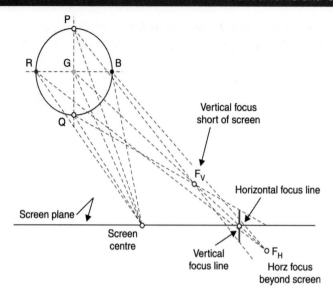

Fig. 25.15 (a). Astigmatism in a uniform deflection field.

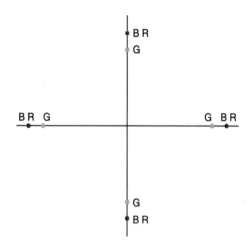

Fig. 25.15 (b). The coma effect—central beam deflection
less (or more) than the side beams.

Field Distribution for Optimum Convergence

In order to correct astigmatic and coma effects different field configurations are necessary. To help understand this, it is useful to visualize the deflection field in roughly two parts, the half of it closer to the screen and the other half closer to the guns. Astigmatism is caused by only that part of the field which is closer to the screen whereas coma effects occur due to nonuniformities of the field all over the deflection area.

To correct for misconvergence due to astigmatism in a P.I.L. tube the horizontal deflection field must be pincushion shaped and the vertical deflection field barrel shaped (Fig. 26.16 (a)) in the half near the screen. But such a field configuration produces undue amounts of coma error. To circumvent this, the gun end of the field can be modified in such a way that the coma produced by the screen end field is just neutralized, thus giving a scan that is free from all convergence errors. This needs that the field distribution at the gun end be barrel shaped horizontally and pincushion vertically. The above two requirements are illustrated in Fig. 26.16 (*b*) which shows necessary details of horizontal and vertical field distributions relative to each other along the axis of the tube. Though complicated, the above constraints of field distribution are met by proper deflection coil design and thus both astigmatic and coma effects are eliminated.

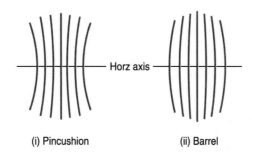

(i) Pincushion   (ii) Barrel

Fig. 25.16 (a). Field configurations (i) pincushion horizontal deflection field (ii) barrel shaped vertical deflection field.

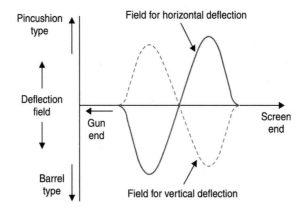

Fig. 25.16 (b). Quantitative visualization of deflection field distribution for astigmatism and coma corrections in a P-I-L picture tube.

25.14 THE DEFLECTION UNIT

Based on the field convergence techniques described above several picture tube manufacturers introduced deflection units for providing automatic self-convergence. However, such designs were initially limited to relatively small screen picture tubes. Around 1973 Philips introduced a self converging picture tube and deflection coil combination system for 110° deflection and screen size up to 66 cm (26″). It makes use of the in-line gun array in conjunction with a specially designed saddle shaped deflection coil assembly. The system is known as 20 *AX* and needs the least number of corrections for obtaining purity and true convergence. The yoke assembly is provided with a plastic ring for axial alignment over a distance of about 5 mm. Similarly there is a provision to rotate the yoke assembly by a small angle ($\approx 7°$) to correct raster orientation.

In this and other similar systems marketed by other manufacturers, the three electron guns, mask and deflection yoke assembly are constructed with such precision that hardly any purity and convergence adjustments are necessary. However, small errors do occur on account of mounting tolerances, stray magnetic field effects and nonuniform field distribution. Small corrections are therefore inevitable in such assemblies.

25.15   PURITY AND STATIC CONVERGENCE ADJUSTMENTS

As stated earlier a multipole magnetic assembly is provided along with the yoke. The assembly incorporates four ring shaped permanent magnet units. The magnetic rings comprise of (*i*) two pairs of 2-pole magnets, (*ii*) one pair of 4-pole magnets and (*iii*) one pair of 6-pole magnets. Each pair consists of an inner and outer ring of identical magnetic configuration. As in any purity magnet assembly, rotating one of the two rings varies the resultant magnetic field strength and rotating them together varies direction of the resultant field. For mutual rotation in opposite direction the rings are coupled by small pinion gears. Details of various adjustments for a typical multipole unit are as under:

(*a*) *Horizontal colour purity.* As illustrated in Fig. 25.17 horizontal colour purity is obtained by varying field strength of the 2-pole magnet situated between the 4-pole and 6-pole magnets.

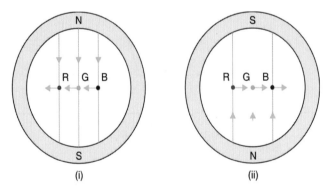

(i)   (ii)

Fig. 25.17. Horizontal colour purity. All the three beams move
equally in the horizontal direction.

(*b*) *Static convergence.* It is obtained by varying the field strength and direction of the 4-pole and 6-pole magnet pairs. The 4-pole field moves the outer electron beams (red and blue) equally in oppsite directions. The 6-pole field moves the outer electron beams equally in the same direction. The central beam (green) is unaffected. Magnetic field direction and consequent deflections are illustrated in Fig. 25.18.

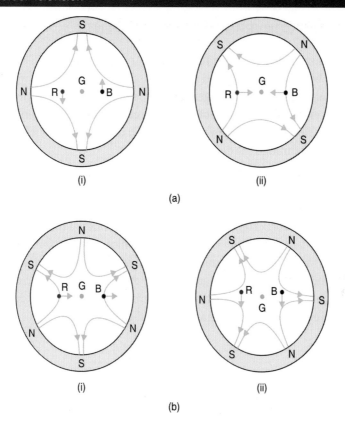

Fig. 25.18. Static convergence (a) the outer electron beams (R and B) move equally in opposite directions (b) the outer electron beams move equally in the same direction.

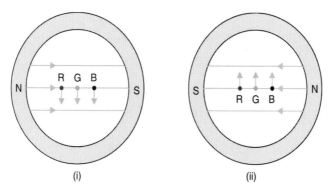

Fig. 25.19. Raster (horizontal axis) symmetry. All the three beams move equally in the vertical direction.

(c) *Raster symmetry.* Horizontal axis or raster symmetry is adjusted by varying the field strength of the 2-pole magnet located at the rear of the multi-pole unit. All the three beams (Fig. 25.19) are equally moved in a vertical direction.

25.16   DYNAMIC CONVERGENCE ADJUSTMENTS

The display unit consisting of picture tube and deflection unit is inherently self-converging. However, small adjustments become necessary and are provided. For this purpose two types of four-pole dynamic magnetic fields are used. One is generated by additional windings on the yoke ring of the deflection unit. It is energized by adjustable sawtooth currents synchronized with scanning. The other type of dynamic field is generated by sawtooth and parabolic currents which are synchronized with scanning and flow through the deflection coils.

(*i*) *Line symmetry.* Figure 25.20 (*a*) shows a situation in which the plane where the beams are converged automatically is slightly tilted with respect to the screen plane due to some small left-right asymmetry in the distribution of the horizontal deflection field. As a result, horizontal convergence errors of opposite signs occur at the sides of the screen. The same type of error can be caused by a horizontal deviation of the undeflected beams from the screen centre. As shown in Fig. 25.20 (*b*) such an error can be corrected by a 4-pole field aligned diagonally with respect to the deflection fields. This field is generated by driving a sawtooth current at line frequency through an additional four-pole winding provided around the core of the deflection yoke. The sawtooth current is obtained directly from the line deflection circuit.

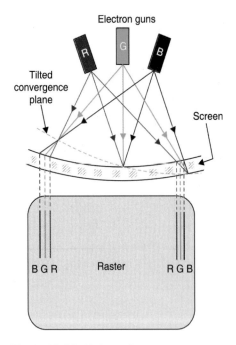

Fig. 25.20 (a). Horizontal convergence errors.

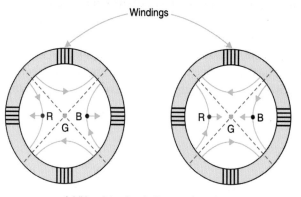

Additional 4-pole windings on the yoke core

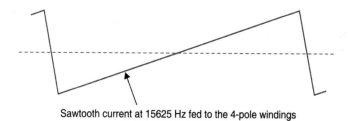

Sawtooth current at 15625 Hz fed to the 4-pole windings

Fig. 25.20 (b). *Line symmetry correction (horizontal red-to-blue distance)
at the ends of horizontal axis.*

(ii) Field symmetry. As illustrated in Fig. 25.21 (*a*) vertical displacement of the plane of the beams with respect to the centre of the vertical deflection causes horizontal convergence errors during vertical deflection. These errors can be corrected by feeding a rectified sawtooth current at field frequency through the additional 4-pole winding on the deflection unit.

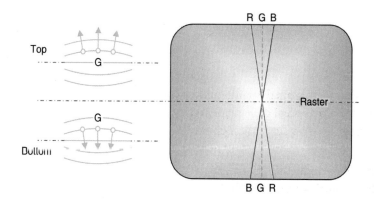

Fig. 25.21 (a). *Field symmetry. Red-blue cross-over of the vertical lines cause
horizontal convergence errors at the ends of vertical axis.*

(iii) Line balance. Vertical displacement of the plane of beams with respect to centre of the horizontal deflection field causes cross-over of the horizontal red and blue lines. This is illustrated in Fig. 25.21 (*b*). The same type of error can also be caused by top-bottom asymmetry of the horizontal deflection field. It can be corrected by a four-pole field which is aligned orthogonally with respect to the deflection fields. Such a field is generated by unbalancing (see corresponding figures) the line deflection current through the two halves of the horizontal deflection coils.

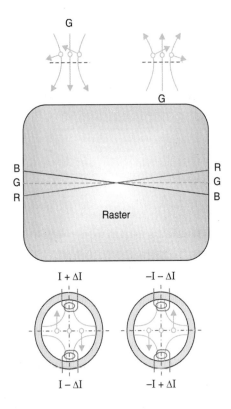

Fig. 25.21 (b). *Red-blue cross-over of horizontal lines causes vertical convergence errors at the end of horizontal axis.*

(iv) Line balance parabola. If the plane of the beams (electron guns) happens to be rotated with respect to its normal orientation, a parabolic vertical convergence error occurs during both horizontal and vertical deflection (Fig. 25.22). This error can be corrected by feeding a parabolic current at line frequency through the line deflection coils.

(v) Field balance at top and bottom. Left-right asymmetry of the vertical deflection field or horizontal deviation of the undeflected beams from the screen centre causes vertical convergence errors during vertical deflection. This is illustrated (Fig. 25.23) separately for top and bottom of the screen. The correction at the top is made by unbalancing the field defelction coils during the first-half of the field scan. Similarly correction at the bottom is made by unbalancing the field deflection coils during the second-half of the field scan.

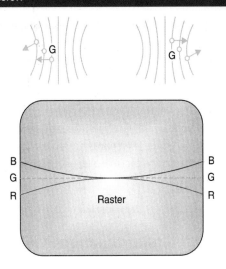

Fig. 25.22. *Rotation of electron beams causes vertical convergence errors of a parabolic nature.*

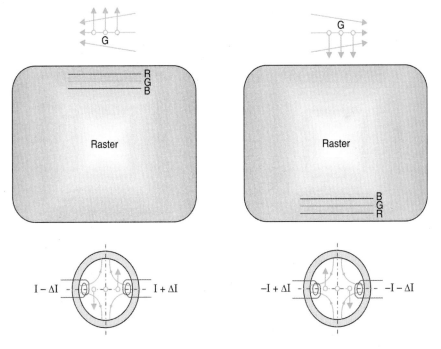

Fig. 25.23 (a). *Vertical misconvergence at the top.*

Fig. 25.23 (b). *Vertical misconvergence at the bottom.*

25.17 TRINTRON COLOUR PICTURE TUBE

The Trintron or three in-line cathodes colour picture tube was developed by 'SONY' Corporation of Japan around 1970. It employs a single gun having three in-line cathodes. This simplifies constructional problems since only one electron gun assembly is to be accommodated. The three phosphor triads are arranged in vertical strips as in the P.I.L. tube. Each strip is only a few thousandth of a centimetre wide. A metal aperture grille like mask is provided very close to the screen. It has one vertical slot for each phosphor triad. The grille is easy to manufacture and has greater electron transparency as compared to both delta-gun and P.I.L. tubes. The beam and mask structure, together with constructional and focusing details of the Trintron are shown in Fig. 25.24. The three beams are bent by an electrostatic lens system and appear to emerge from the same point in the lens assembly. Since the beams have a common focus plane a sharper image is obtained with good focus over the entire picture area. All this simplifies convergence problems and fewer adjustments are necessary.

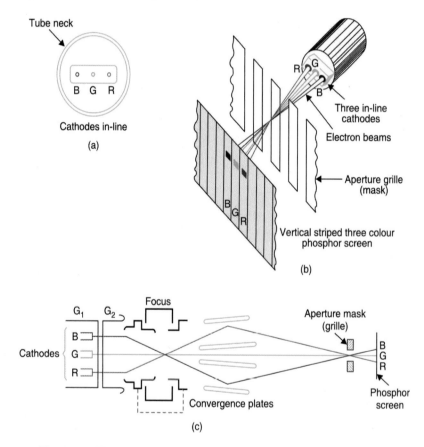

Fig. 25.24. *Trintron (cathodes in-line) colour picture tube (a) gun structure*
(b) electron beams, vertical-striped three colour phosphor screen
(c) constructional, focus and convergence details.

The latest version of Trintron was perfected in 1978. It incorporates a low magnification electron gun assembly, long focusing electrodes and a large aperture lens system. The new high precision deflection yoke with minimum convergence adjustments provides a high quality picture with very good resolution over large screen display tubes.

25.18 | PINCUSHION CORRECTION TECHNIQUES

As mentioned earlier, dynamic pincushion corrections are necessary in colour picture tubes. Figure 25.25 (a) is the sketch of a raster with a much exaggerated pincushion distortion. The necessary correction is achieved by introducing some cross modulation between the two deflection fields.

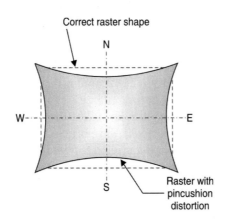

Fig. 25.25 (a). *Pincushion distortion.*

E-W Correction

To correct *E-W* (horizontal) pincushioning, the horizontal deflection sawtooth current must be amplitude modulated at a vertical rate so that when the electron beam is at the top or bottom

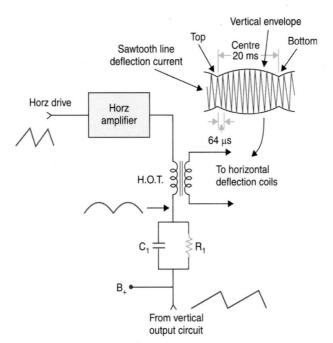

Fig. 25.25 (b). *Horizontal (E – W) pincushion correction circuit and waveforms.*

of the raster, the horizontal amplitude is minimum and when it is at the centre of the vertical deflection interval the horizontal sawtooth amplitude is maximum. To achieve this a parabolic voltage obtained by integrating the vertical sawtooth voltage (network R_1, C_1 in Fig. 25.25 (b)) is inserted in series with the dc supply to the horizontal deflection circuit. As a result, amplitude of individual cycles of the 15625 Hz horizontal output varies in step with the series connected 50 Hz parabolic voltage. As shown in Fig. 25.25 (b) the modified horizontal sawtooth waveshape over a period of the vertical cycle (20 ms) has the effect of pulling out the raster at the centre to correct E-W pincushioning.

N-S Correction

The top and bottom or N-S pincushion correction is provided by forcing the vertical sawtooth current to pulsate in amplitude at the horizontal scanning rate. During top and bottom scanning of the raster a parabolic waveform at the horizontal rate is superimposed on the vertical deflection sawtooth. In fact this increases vertical size during the time the beam is moving through the midpoint of its horizontal scan. The parabolic waveform at the top of the raster is of opposite polarity to that at the bottom since the raster stretch required at the top is opposite to the needed at the bottom. The amplitude of the parabolic waveform required for top and bottom pincushion correction decreases to zero as vertical deflection passes through the centre of the raster. The basic principle of obtaining necessary deflection waveshapes is the same as for E-W correction. Figure 25.25 (c) shows the basic circuit and associated waveforms.

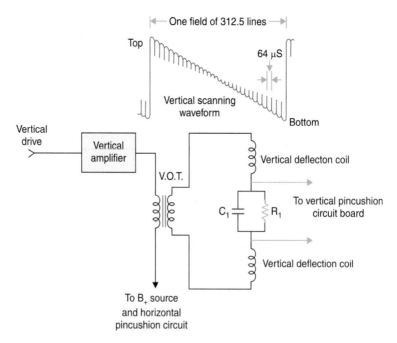

Fig. 25.25 (c). *Vertical (N – S) pincushion correction circuit and waveforms.*

25.19   AUTOMATIC DEGAUSSING (ADG) CIRCUIT

There are many degaussing circuits in use. Figure 25.26 (a) shows details of a popular automatic degaussing circuit. It uses a thermistor and a varistor for controlling the flow of alternating current through the degaussing coil. When the receiver is turned on the ac voltage drop across the thermistor is quite high (about 60 volts) and this causes a large current to flow through the degaussing coil. Because of this heavy current, the thermistor heats up, its resistance falls and voltage drop across it decreases. As a result, voltage across the varistor decreases thereby increasing its resistance. This in turn reduces ac current through the coil to a very low value. The circuit components are so chosen that initial surge of current through the degaussing coil is close to 4 amperes and drops to about 25 mA in less than a second. This is illustrated in Fig. 25.26 (b). Once the thermistor heats up degaussing ends and normal ac voltage is restored to the B_+ rectifier circuit.

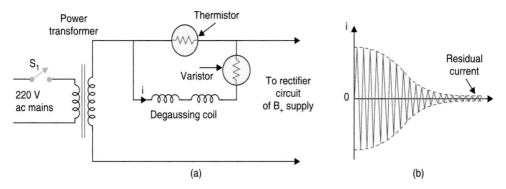

Fig. 25.26. Automatic degaussing (ADG) (a) typical circuit (b) variation of current in the degaussing coil when receiver is just switched on.

25.20   GREY SCALE TRACKING

It may be recalled that red, green and blue lights must combine in definite proportions to produce white light. The three phosphors have different efficiencies. Also the three guns may not have identical I_p/V_{gk} characteristics and cut-off points. Therefore, it becomes necessary to incorporate suitable adjustments such that monochrome information is reproduced correctly (with no colour tint) for all settings of the contrast control. In practice, this amounts to two distinct steps:

(a) Adjustment of Low Lights

Non-coincident I_p/V_{gk} characteristics and consequent difference in cut-off points of the three guns, results in appearance of coloured tint instead of pure dark grey shades in areas of low brightness. To correct this it is necessary to bring the cut-off points in coincidence. This is achieved by making the screen grid (i.e., 1st anode) voltage different from each other. Potentiometers are normally provided for this purpose in the dc voltage supply to the three screen grids.

(b) Adjustment of High Lights

It is equally necessary to ensure that all other levels of white are also correctly reproduced. This amounts to compensating for the slightly different slopes and also for the substantially different phosphor efficiencies. This is achieved by varying the video (luminance) signal drive to the three guns. Since the red phosphor has the lowest efficiency, maximum video signal is fed to the red cathode and then by the use of potentiometers video signal amplitudes to the green and blue guns are varied (see Fig. 25.11) to obtain optimum reproduction of high lights.

References

1. J. Kasshock, Deflection in the 20 AX systems, *IEEE Trans.* BTR, Vol. 20, No. 4, Nov. 1974.
2. P.G.J. Barton, The 20 AX system and picture tube, *IEEE Trans.* BTR, Vol. 20, No. 4 Nov. 1974.

REVIEW QUESTIONS

1. What do you understand by compatibility in TV transmission ? Enumerate essential requirements that must be met to make a colour system fully compatible.
2. Explain briefly how the human eye perceives brightness and colour sensations. Comment on the spectral response of the human eye.
3. Explain the terms—(i) primary colours, (ii) complementary colours, (iii) additive colour mixing, (iv) hue, (v) saturation, (iv) luminance, and (vii) chrominance.
4. Describe with a diagram the construction of a colour TV camera and its optical system. Why are the outputs of all the three camera tubes set equal when standard white light is made incident ?
5. Explain how the 'Y' and colour difference signals are developed from camera outputs. Why is the 'Y' signal set = $0.3R + 0.59G - 0.11B$?
6. Why is the $(G - Y)$ difference signal not chosen for transmission ? Explain how it is obtained in the receiver for modulating the corresponding beam of the picture tube.
7. Explain how colour difference signals disappear at the output of the signal combining matrix on white and grey shades. What is the significance of 'Y' signal in colour transmission and reception ?
8. While televising a static desaturated colour scene, the camera outputs were found to have the following amplitudes:
$$V_R = 0.7 \text{ V}, V_G = 0.6 \text{ V} \quad \text{and} \quad V_B = 0.3 \text{ V}$$
What is the basic hue of the scene ? Compute values of $(G–Y)$ and Y signals and establish that true hue and brightness will be reproduced in the receiver. What shade will such a signal produce in a monochrome receiver ?
9. Describe with suitable diagrams the gun arrangement and constructional details of a delta-gun colour picture tube. Why is it necessary to connect a very high voltage at the final anode of a colour picture tube ?
10. Define purity and convergence and explain why elaborate static and dynamic corrections become necessary to obtain colour purity and coincident rasters.
11. What are the drawbacks of a delta-gun tube ? Describe constructional details of a P.I.L. tube and explain how it is different from a delta gun colour tube. What are its distinguishing features ?
12. Define astigmatism and coma errors with reference to a P.I.L. tube and describe the field distribution which must be obtained to overcome such distortions in the reproduced picture.
13. Describe how a multipole ring magnets assembly can be used to deflect the electron beam in different directions. Describe purity and static convergence adjustments which are normally provided with a P.I.L. tube-deflection assembly.
14. Describe briefly the dynamic convergence errors in a P.I.L. tube. Explain the corrections that are normally made to achieve almost distortionless reproduction of colours on the screen of a P.I.L. tube. Explain why vertical $(N-S)$ raster shap correction is not necessary in such a picture tube.

15. Describe essential features of a Trintron colour picture tube. Explain why is it considered superior both the delta-gun and P.I.L. picture tubes.

16. What is pincushion correction and explain how both $E - W$ and $N - S$ pincushions corrections are carried out.

17. What do you understand by degaussing and why it is necessary ? Describe with a typical circuit how degaussing is affected each time the receiver is switched on.

18. What do you understand by grey scale tracking ? Explain how it is carried out both for low and high light levels of the video signal.

26

Colour Signal Transmission and Reception

Three different systems of colour television (CTV) emerged after prolonged research and experimentation. These are:

(*i*) The American NTSC (National Television Systems Committee) system.

(*ii*) The German PAL (Phase Alteration by Line) system.

(*iii*) The French SECAM (Sequential Couleures a memoire) system.

When quality of the reproduced picture and cost of equipment are both taken into account, it becomes difficult to establish the superiority of one system over the other. Therefore, all the three CTV systems have found acceptance in different countries and the choice has been mostly influenced by the monochrome system already in use in the country.

Since India adopted the 625 line CCIR (B standards) monochrome system it has chosen to introduce the PAL system* (B & G standards) because of compatibility between the two, and also due to its somewhat superior performance over the other two systems.

In many respects transmission and reception techniques employed in the NTSC and PAL systems are similar. These are, therefore, treated together before going into encoding and decoding details of each system. The SECAM system, being much different from the other two, is described separately in the later part of this chapter.

26.1 COLOUR SIGNAL TRANSMISSION

The colour video signal contains two independent informations, that of hue and saturation. It is a difficult matter to modulate them to one and the same carrier in such a way that these can be easily recovered at the receiver without affecting each other. The problem is accentuated by the need to fit this colour signal into a standard TV channel which is almost fully occupied by the 'Y' signal. However, to satisfy compatibility requirements the problem has been ingeniously solved by combining the colour information into a single variable and by employing what is known as frequency interleaving.

---

*In the CCIR PAL I standards the picture and sound carrier are 6 MHz apart and the channel bandwidth is 8 MHz. The only difference between PAL-B and PAL-G is in the channel bandwidth. PAL-B with channel bandwidth of 7 MHz does not provide any inter-channel gap, whereas PAL-G with channel bandwidth of 8 MHz provides a band gap of 1 MHz in-between successive channels.

Frequency Interleaving

Frequency interleaving in television transmission is possible because of the relationship of the video signal to the scanning frequencies which are used to develop it. It has been determined that the energy content of the video signal is contained in individual energy 'bundles' which occur at harmonics of the line frequency (15.625, 31.250 ... KHz) the components of each bundle being separated by a multiplier of the field frequency (50, 100, ... Hz). The shape of each energy bundle shows a peak at the exact harmonics of the horizontal scanning frequency. This is illustrated in Fig. 26.1. As shown there, the lower amplitude excursions that occur on either side of the peaks are spaced at 50 Hz intervals and represent harmonics of the vertical scanning rate. The vertical sidebands contain less energy than the horizontal because of the lower rate of vertical scanning. Note that the energy content progressively decreases with increase in the order of hormonics and is very small beyond 3.5 MHz from the picture carrier.

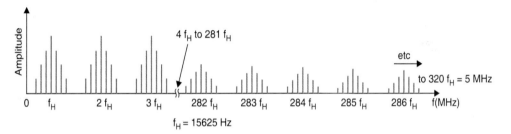

Fig. 26.1. Composition of video information at multiples of line frequency.

It can also be shown that when the actual video signal is introduced between the line sync pedestals, the overall spectra still remains 'bundled' around the harmonics of the line frequency and the spectrum of individual bundles become a mixture of continuous portion due to the video signal are discrete frequencies due to the field sync as explained earlier. Therefore, a part of the bandwidth in the monochrome television signal goes unused because of spacing between the bundles. This suggests that the available space could be occupied by another signal. It is here where the colour information is located by modulating the colour difference signals with a carrier frequency called 'colour subcarrier'. The carrier frequency is so chosen that its sideband frequencies fall exactly mid-way between the harmonics of the line frequency. This requires that the frequency of the subcarrier must be an odd multiple of half the line frequency. The resultant energy clusters that contain colour information are shown in Fig. 26.2 by dotted chain lines along with the Y signal energy bands. In order to avoid crosstalk with the picture signal, the frequency of the subcarrier is chosen rather on the high side of the channel bandwidth. It is 567 times one-half the line frequency in the PAL system. This comes to: $(2 \times 283 + 1)\, 15625/2 - 4.43$ MHz. Note that in the American 525 line system, owing to smaller bandwidth of the channel, the subcarrier employed is 455 times one-half the line frequency $i.e.$, $(2 \times 227 + 1)\, 15750/2$ and is approximately equal to 3.58 MHz.

Chapter 26

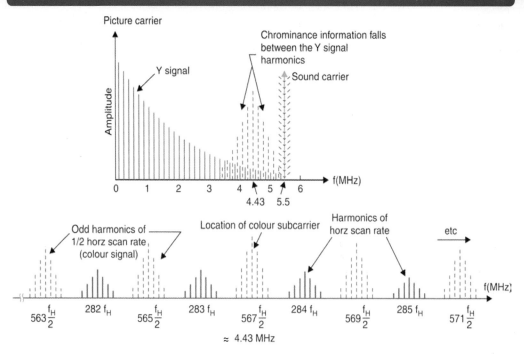

Fig. 26.2. *Interleaving of the colour signal.*

26.2   BANDWIDTH FOR COLOUR SIGNAL TRANSMISSION

The Y signal is transmitted with full frequency bandwidth of 5 MHz for maximum horizontal details in monochrome. However, such a large frequency spectrum is not necessary for colour video signals. The reason being, that for very small details, the eye can perceive only the brightness but not the colour. Detailed studies have shown that perception of colours by the human eye, which are produced by combinations of the three primary colours is limited to objects which have relatively large coloured areas ($\approx$ 1/25th of the screen width or more). On scanning they generate video frequencies which do not exceed 0.5 MHz. Further, for medium size objects or areas which produce a video frequency spectrum between 0.5 and 1.5 MHz, only two primary colours are needed. This is so, because for finer details the eye fails to distinguish purple (magenta) and green-yellow hues from greys. As the coloured areas become very small in size (width), the red and cyan hues also become indistinguishable from greys. Thus for very fine colour details produced by frequencies from 1.5 MHz to 5 MHz, all persons with normal vision are colour blind and see only changes in brightness even for coloured areas. Therefore, maximum bandwidth necessary for colour signal transmission is around 3 MHz ($\pm$ 1.5 MHz).

26.3   MODULATION OF COLOUR DIFFERENCE SIGNALS

The problem of transmitting $(B\text{-}Y)$ and $(R\text{-}Y)$ video signals simultaneously with one carrier frequency is solved by creating two carrier frequencies from the same colour subcarrier without any change in its numerical value. Two separate modulators are used, one for the $(B\text{-}Y)$ and the other for the $(R\text{-}Y)$ signal. However, the carrier frequency fed to one modulator is given a relative phase shift of 90° with respect to the other before applying it to the modulator. Thus, the two equal subcarrier frequencies which are obtained from a common generator are said to

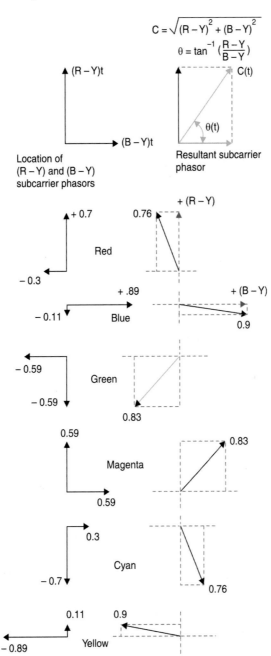

$$C = \sqrt{(R-Y)^2 + (B-Y)^2}$$

$$\theta = \tan^{-1}\left(\frac{R-Y}{B-Y}\right)$$

Location of $(R-Y)$ and $(B-Y)$ subcarrier phasors

Resultant subcarrier phasor

Fig. 26.3. Quadrature amplitude modulated colour difference signals and the position of resultant subcarrier phasor for the primary and complementary colours. Note that the magnitudes shown correspond to unweighted values of colour difference signals.

be in quadrature and the method of modulation is known as quadrature modulation. After modulation the two outputs are combined to yield C, the resultant subcarrier phasor. Since the amplitude of C, the chrominance signal, corresponds to the magnitudes of colour difference signals, its instantaneous value represents colour saturation at that instant. Maximum amplitude corresponds to greatest saturation and zero amplitude to no saturation *i.e.*, white. Similarly, the instantaneous value of the C phasor angle (θ) which may vary from 0° to 360° represents hue of the colour at that moment. Thus the chrominance signal contains full information about saturation and hue of various colours. This being a crucial point in colour signal transmission, is illustrated by a few examples. However, it would be necessary to first express $(R\text{-}Y)$ and $(B\text{-}Y)$ in terms of the three camera output voltages. This is done by substituting $Y = 0.59G + 0.3R + 0.11B$ in these expressions. Thus $(R\text{-}Y)$ becomes $R - 0.59G - 0.3R - 0.11B = 0.7R - 0.59G - 0.11B$. Similarly, $(B\text{-}Y)$ becomes $B - 0.59G - 0.3R - 0.11B = 0.89B - 0.59G - 0.3R$.

Now suppose that only pure red colour is being scanned by the colour camera. This would result in an output from the red camera only, while the green and blue outputs will be zero. Therefore, $(R\text{-}Y)$ signal will become simply $+ 0.7R$ and $(B\text{-}Y)$ signal will be reduced to $- 0.3R$. The resultant location of the subcarrier phasor after modulation is illustrated in Fig. 26.3. Note that the resultant phasor is counter clockwise to the position of $+ (R\text{-}Y)$ phasor.

Next consider that the colour camera scans a pure blue colour scene. This yields $(R\text{-}Y) = - 0.11B$ and $(B\text{-}Y) = 0.89\,B$. The resultant phasor for this colour lags $+ (B\text{-}Y)$ vector by a small angle. Similarly the location and magnitude for any colour can be found out. This is illustrated in Fig. 26.3 for the primary and complementary colours.

Another point that needs attention is the effect of desaturation on the colour phasors. Since desaturation results in reduction of the amplitudes of both $(B\text{-}Y)$ and $(R\text{-}Y)$ phasors, the resultant chrominance phasor accordingly changes its magnitude depending on the degree of desaturation. Thus any change in the purity of a colour is indicated by a change in the magnitude of the resultant subcarrier phasor.

Colour Burst Signal

Suppressed carrier double sideband working is the normal practice for modulating colour-difference signals with the colour subcarrier frequency. This is achieved by employing balanced modulators. The carrier is suppressed to minimize interference produced by the chrominance signals both on monochrome receivers when they are receiving colour transmissions and in the luminance channel of colour receivers themselves. As explained in an earlier chapter the ratio of the sideband power to carrier power increases with the depth of modulation. However, even at 100% modulation two-thirds of the total power is in the carrier and only one-third is the useful sideband power. Thus suppressing the carrier clearly eliminates the main potential source of interference. In addition of this, the colour-difference signals which constitute the modulating information are zero when the picture detail is non-coloured (*i.e.*, grey, black or white shades) and so at such times the sidebands also disappear leaving no chrominance component in the video signal.

As explained above the transmitted does not contain the subcarrier frequency but it is necessary to generate it in the receiver with correct frequency and phase relationship for proper detection of the colour sidebands. To ensure this, a short sample of the subcarrier oscillator, (8 to 11 cycles) called the 'colour burst' is sent to the receiver along with sync signals. This is located in the back porch of the horizontal blanking pedestal. The colour burst does not interfere with the horizontal sync because it is lower in amplitude and follows the sync pulses. Its exact location is shown in Fig. 26.4. The colour burst is gated out at the receiver and is used in conjuction with a phase comparator circuit to lock the local subcarrier oscillator frequency and phase with that at the transmitter. As the burst signal must maintain a constant phase relationship with the scanning signals to ensure proper frequency interleaving, the horizontal and vertical sync pulses are also derived from the subcarrier through frequency divider circuits.

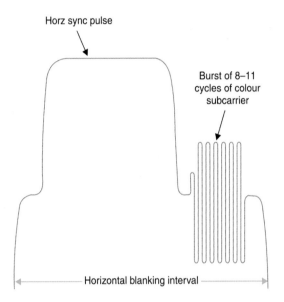

Fig. 26.4. Location of colour burst on the back porch of each horizontal sync pulse.

26.4   WEIGHTING FACTORS

The resultant chrominance signal phasor (C) is added to the luminance signal (Y) before modulating it with the channel carrier for transmission. The amplitude, *i.e.*, level line of Y signal becomes the zero line for this purpose. Such an addition is illustrated in Fig. 26.5 for a theoretical 100 percent saturated, 100 percent amplitude colour bar signal. The peak-to-peak amplitude of green signal (± 0.83) gets added to the corresponding luminance amplitude of 0.59. For the red signal the chrominance amplitude of ± 0.76 adds to its brightness of 0.3. Similarly other colours add to their corresponding luminance values to form the chroma signal. However, observe that it is not practicable to transmit this chroma waveform because the signal peaks would exceed the limits of 100 percent modulation. This means that on modulation with the picture carrier some of the colour signal amplitudes would exceed the limits of maximum sync tips on one side and white level on the other. For example, in the case of magenta signal, the chrominance value of + 0.83 when added to its luminance amplitude of 0.41 exceeds the limits of 100 percent modulation of both white and black levels. Similarly blue signal amplitude greatly exceeds the black level and will cause a high degree of overmodulation.

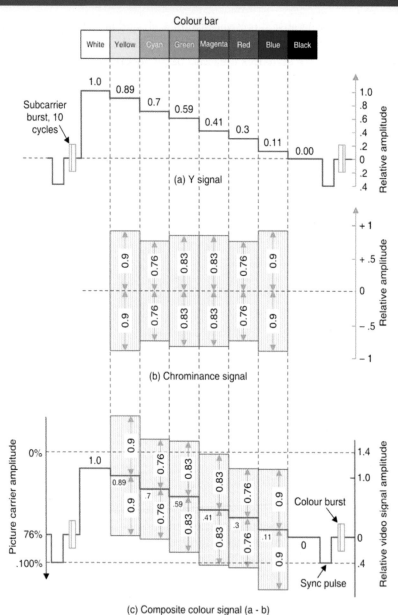

Fig. 26.5. *Generation of composite colour signal for a theoretical 100% saturated, 100% amplitude colour bar signal. As seen from the relative amplitude scales, the composite video signal would cause gross over-modulation. Therefore in practice the colour difference signals are reduced in amplitude to avoid any excessive over-modulation.*

If overmodulation is permitted the reproduced colours will get objectionably distorted. Therefore, to avoid overmodulation on 100 percent saturation colour values, it is necessary to reduce the amplitude of colour difference video signal before modulating them with the colour subcarrier. Accordingly, both $(R–Y)$ and $(B–Y)$ components of the colour video signal are scaled down by multiplying them with what are known as 'weighting factors'. Those used are 0.877 for the $(R–Y)$

component and 0.493 for the (B–Y) component. The compensated values are obtained by using potentiometers at the outputs of (R–Y) and (B–Y) adders (see Fig. 25.9). Note that no reduction is made in the amplitude of Y signal. It may also be noted that since the transmitter radiates weighted chrominance signal values, these must be increased to the uncompensated values at the colour TV receiver for proper reproduction of different hues. This is carried out by adjusting gains of the colour difference signal amplifiers. The unweighted and weighted values of colour difference signals are given below in Table 26.1.

Table 26.1

Colour	Luminance signal (λ)	$B-Y$	$R-Y$	$G-Y$	C_{SC}	$B-Y$	$R-Y$	C_{SC}
			Unweighted				Weighted	
White	1	0	0	0	0	0	0	0
Yellow	0.89	−.89	+.11	+.11	.9	−.4385	+.096	0.44
Cyan	0.7	+.3	−.7	+.3	.76	+.148	−.614	0.63
Green	0.59	−.59	−.59	+.41	.83	−.29	−.517	0.59
Magenta	0.41	+.59	+.59	−.41	.83	+.29	+.517	0.59
Red	0.3	−.3	+.7	−.3	.76	−.148	+.614	0.63
Blue	0.11	+.89	−.11	−.11	.9	+.4388	−.096	0.44
Black	0	0	0	0	0	0	0	0

$$C_{SC} = \sqrt{(B-Y)^2 + (R-Y)^2}$$

$(B-Y)$ weighted = 0.493 $(B-Y)$ unweighted

$(R-Y)$ weighted = 0.877 $(R-Y)$ unweighted

26.5 FORMATION OF THE CHROMINANCE SIGNAL

Using the information of Table 26.1, Fig. 26.6 illustrates the formation of the chroma signal for a colour bar pattern after the colour difference signals have been scaled down in accordance with corresponding weighting factors. Note that new amplitudes of the chrominance subcarrier signals are 0.63 for red and cyan, 0.59 for green and magenta and 0.44 for blue and yellow. These amplitudes will still cause overmodulation to about 33%. This is permitted, because in practice, the saturation of hues in natural and staged scenes seldom exceeds 75 percent. Since the amplitude of chroma signal is proportional to the saturation of hue, maximum chroma signal amplitudes are seldom encountered in practice. Therefore, the weighted chroma values result in a complete colour signal that will rarely, if ever, overmodulate the picture carrier of a CTV transmitter. Hence it is not necessary to further decrease the signal amplitudes by employing higher weighting factors.

Chapter 26

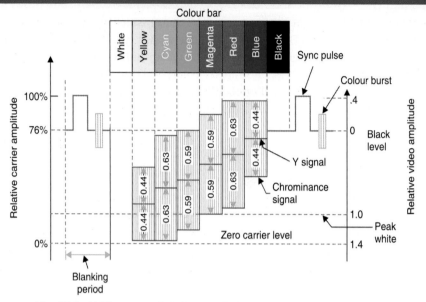

Fig. 26.6. *100% saturated, 100% amplitude colour-bar signal in which the colour difference signals are reduced by weighting factors to restrict the chrominance signal excursions to 33% beyond black and peak white levels.*

Chroma Signal Phasor Diagram

The compensation (readjustment) of chroma signal values results in a change of chroma phase angles. In the NTSC system it is a common practice to measure phase angles relative to the

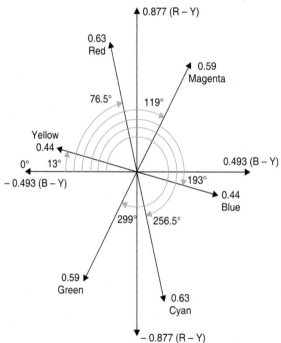

Fig. 26.7. *Magnitude and phase relationships of compensated chrominance signals for the primary and complementary colours.*

$-(B–Y)$ phasor. This location has been designated 0° or the reference phase position on the phasor diagram (see Fig. 26.7) because this is also the phase of the colour burst that is transmitted on the back porch of each horizontal sync pulse. Referring to Fig. 26.7 the compensated colour magenta is represented by a phasor at an angle of 119°. In the same manner the diagram indicates phase angles and amplitudes of other colour signals. Note that primary colours are 120° apart and complementary colours differ in phase by 180° from their corresponding primary colours.

26.6 NTSC COLOUR TV SYSTEM

The NTSC colour system is compatible with the American 525 line monochrome system. In order to save bandwidth, advantage is taken of the fact that eye's resolution of colours along the reddish blue-yellowish green axis on the colour circle is much less than those colours which lie around the yellowish red-greenish blue axis. Therefore two new colour video signals, which correspond to these colour regions, are generated. These are designated as I and Q signals. The I signal lies in a region 33° counter clockwise to $+ (R – Y)$ where the eye has maximum colour resolution. It* is derived from the $(R – Y)$ and $(B – Y)$ signals and is equal to $0.60R – 0.28G – 0.32B$. As shown in Fig. 26.8 it is located at an angle of 57° with respect to the colour burst in the balanced modulator circuits. Similarly the Q** signal is derived from colour difference signals by suitable matrix and equals $0.21R –0.52G + 0.31B$. It is located 33° counter

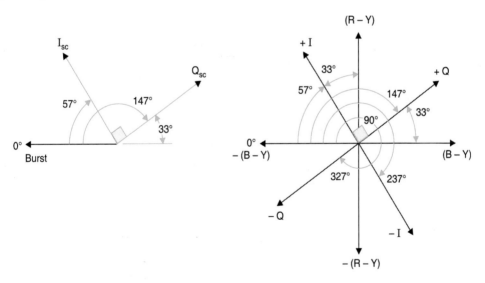

Fig. 26.8. Phasor diagrams of the I and Q signals in the NTSC system.

clockwise to the $(B – Y)$ signal and is thus in quadrature with the I signal. As illustrated in Fig. 26.8 the Q signal covers the regions around magenta (reddish-blue) and yellow-green shades. Similarly orange hues correspond to phase angles centred around $+ I$ and the complementary blue-green (cyan) hues are located around the diametrically opposite $– I$ signal. Since the eye is capable of resolving fine details in these regions, I signal is allowed to possess frequencies up to 1.5

---

*$I = 0.74(R – Y) – 0.27(B – Y)$.

**$Q = 0.48(R – Y) + 0.41(B – Y)$.

MHz. However, the eye is least sensitive to colours that lie around the ±Q signals, and therefore it is allowed a bandwidth of only ±0.5 MHz with respect to the colour subcarrier.

It may be noted that both I and Q signals are active up to 0.5 MHz and being at right angles to each other, combine to produce all the colours contained in the chrominance signal. However, the Q signal drops out after 0.5 MHz and only I signal remains between 0.5 and 1.5 MHz to produce colours, the finer details of which the eyes can easily perceive.

To help understand this fact it may be recalled that only one colour difference signal is needed for producing colours which are a mixture of only two colours. Thus the Q signal is not necessary for producing colours lying in the region of orange (red + green) and cyan (green + blue) hues. Hence at any instant when $Q = 0$ and only I signal is active the colours produced on the screen will run the gamut from reddish orange to bluish green.

Bandwidth Reduction

Double sideband transmission is allowed for the Q signal and it occupies a channel bandwidth of 1 MHz (± 0.5 MHz). However, for the I signal the upper sideband is restricted to a maximum of 0.5 MHz while the lower sideband is allowed to extent up to 1.5 MHz. As such it is a form of vestigial sideband transmission. Thus in all, a bandwidth of 2 MHz is necessary for colour signal transmission. This is a saving of 1 MHz as compared to a bandpass requirement of 3 MHz if ($B -Y$) and ($R - Y$) are directly transmitted.

It is now obvious that in the NTSC system, advantage is taken of the limitations of the human eye to restrict the colour signal bandwidth, which in turn results in reduced interference with the sound and picture signal sidebands. The reduction in colour signal sidebands is also dictated by the relatively narrow channel bandwidth of 6 MHz in the American TV system.

Exact Colour Subcarrier Frequency

The colour subcarrier frequency in the NTSC system has been chosen to have an exact value equal to 3.579545 MHz. The reason for fixing it with such a precision is to maintain compatibility between monochrome and colour systems. Any interference between the chrominance signal and higher video frequencies is minimized by employing suppressed carrier (colour subcarrier) transmission and by using a notch filter in the path of the luminance signal. However, when a colour transmission is received on a monochrome receiver a dot pattern structure appears along each raster line on the receiver screen. This is caused by the colour signal frequencies that lie within the pass-band of the video section of the receiver. As illustrated below such an interference can be eliminated if the subcarrier frequency is maintained at the exact value mentioned above.

Assume that the interfering colour signal has a sinusoidal variation which rides on the average brightness level of the monochrome signal. This produces white and black dots on the screen. If the colour subcarrier happens to be a multiple of the line frequency ($n \times f_h$) the phase position of the disturbing colour frequency will be same on successive even or odd fields. Thus black and white dots will be produced at the same spots on the screen and will be seen as a persistent dot pattern interference. However, if a half-line offset is provided by fixing the sub-carrier frequency to be an odd multiple of the half-line frequency, the disturbing colour signal frequency will have opposite polarity on successive odd and even fields. Thus as the same spot on the display screen a bright dot image will follow a dark one alternately. The cumulative effect of this on the eye would get averaged out and the dot pattern will be suppressed.

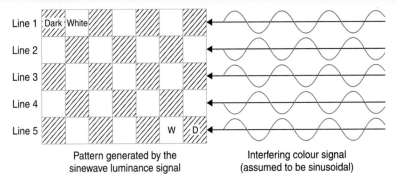

Line 1 Dark White

Line 2

Line 3

Line 4

Line 5   W D

Pattern generated by the   Interfering colour signal
sinewave luminance signal   (assumed to be sinusoidal)

Fig. 26.9. *Illustration of the technique used to reduce dot-pattern*
interference due to the luminance signal.

As an illustration of this phenomenon assume that a simple five line scanning system is being used. Figure 26.9 shows the effect of sinewave luminance signal that is an odd harmonic of one-half of the scanning frequency. Each negative excursion of the signal at the cathode of the picture tube will produce a unit area of brightness on the screen while the positive going excursions of the signal will cause unit dark areas on the picture. In the illustration under consideration where the sinewave completes 3.5 cycles during one active horizontal line, four dark areas and three areas of brightness will be produced during the first line scan. Because of the extra half-cycle the next horizontal scan begins with an area of brightness and the entire line contains only three dark areas. the same off-set occurs on each succeeding line, producing a checkerboard pattern on the screen. Since the scanning rate in the example utilizes an odd number of horizontal scans for each complete presentation, the luminance signal will be 180° out of phase with the previous signal as line number one is again scanned. Thus the pattern obtained on the screen will be the reverse of that which was generated originally. The total effect of the above process on the human eye is one of cancellation. Since in actual practice scanning takes place at a very fast rate, the presistency of the eye blends the patterns together with the effect that visibility of the dot structure is considerably reduced and goes unnoticed on a monochrome receiver.

The compatibility considerations thus dictate that the colour sub-carrier frequency (f_{sc}) should be maintained at 3.583125, $i.e.$, $(2 \times 227 + 1) \times 15750/2$ MHz.

However, the problem does not end here, because the sound carrier and the colour sub-carrier beat with each other in the detector and an objectionable beat note of 0.92 MHz is produced (4.5 – 3.58 = 0.92). This interferes with the reproduced picture. In order to cancel its cumulative effect it is necessary that the sound carrier frequency must be an exact multiple of an even harmonic of the line frequency. The location of the sound carrier at 4.5 MHz away from the picture carrier cannot be disturbed for compatibility reasons and so 4.5 MHz is made to be the 286th harmonic of the horizontal deflection frequency. Therefore, $f_h = 4.5$ MHz/286 = 15734.26 Hz is chosen. Note that this is closest to the value of 15750 Hz used for horizontal scanning for monochrome transmission. A change in the line frequency necessitates a change in the field frequency since 262.5 lines must be scanned per field. Therefore the field frequency (f_v) is changed to be 15734.26/262.5 = 59.94 Hz in place of 60 Hz.

The slight difference of 15.74 Hz in the line frequency (15750 – 15734.26 = 15.74 Hz) and of 0.06 Hz in the field frequency (60 – 59.94 = 0.06 Hz) has practically no effect on the deflection oscillators because an oscillator that can be triggered by 60 Hz pulses can also be synchronized to produce 59.94 Hz output. Similarly the AFC circuit can easily adjust the line frequency to a slightly different value while receiving colour transmission.

As explained earlier the colour sub-carrier frequency must be an odd multiple of half the line frequency to suppress dot pattern interference. Therefore f_{sc} is fixed at $(2n + 1)\, f_h/2$, $i.e.$, $455 \times 15734.26/2 = 3.579545$ MHz. To obtain this exact colour sub-carrier frequency, a crystal controlled oscillator is provided.

Encoding of Colour Picture Information

Figure 26.10 illustrates the enconding process of colour signals at the NTSC transmitter. A suitable matrix is used to get both I and Q signals directly from the three camera outputs. Since $I = 0.60R - 0.28G - 0.32B$, the green and blue camera outputs are inverted before feeding them to the appropriate matrix. Similarly for $Q = 0.21R - 0.52G + 0.31B$, in invertor is placed at the output of green camera before mixing it with the other two camera outputs.

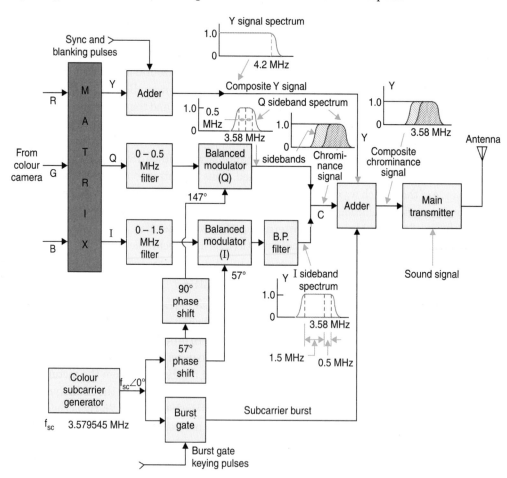

Fig. 26.10. Functional diagram of a NTSC coder.

The bandwidths of both I and Q are restricted before feeding them to the corresponding balanced modulators. The subcarrier to the I modulator is phase shifted 57° clockwise with respect to the colour burst. The carrier is shifted by another 90° before applying it to the Q modulator. Thus relative phase shift of 90° between the two subcarriers is maintained for quadrature amplitude modulation.

It is the characteristic of a balanced modulator that while it suppresses the carrier and provides frequency translation, both the amplitude and phase of its output are directly related to the instantaneous amplitude and phase of the modulating signal. Thus, with the subcarrier phase angles shifted to the locations of I and Q, the outputs from both the modulators retain full identity of the modulating colour difference signals. The sideband restricted output from the I modulator combines with the output of Q modulator to form the chrominance signal. It is then combined with the composite Y signal and colour burst in an adder to form composite chrominance signal. The output from the adder feeds into the main transmitter and modulates the channel picture carrier frequency. Note that colour subcarrier has the same frequency (3.579545 MHz) for all the stations whereas the assigned picture carrier frequency is different for each channel.

26.7   NTSC COLOUR RECEIVER

A simplified block diagram of the NTSC colour receiver is shown in Fig. 26.11. The signal from the selected channel is processed in the usual way by the tuner, IF and video detector stages. The sound signal is separately detected, demodulated and amplified before feeding it to the loudspeaker. Similarly AGC, sync separator and deflection circuits have the same form as in monochrome receivers except for the inclusion of purity, convergence and pincushion correction circuits.

At the output of video detector the composite video and chrominance signals reappear in their original premodulated form. The Y signal is processed as in a monochrome receiver except that the video amplifier needs a delay line. The delay line introduces a delay of about 500 ns which is necessary to ensure time coincidence of the luminance and chroma signals because of the restricted bandwidth of the latter.

Decoding of the Chroma (C) Signal

The block diagram of Fig. 26.11 shows more details of the colour section of the receiver. The chroma signal is available along with other components of the composite signal at the output of the video preamplifier. It should be noted that the chrominance signal has colour information during active trace time of the picture and the burst occurs during blanking time when there is no picture. Thus, although the 'C' signal and burst are both at 3.58 MHz, they are not present at the same time.

Chrominance Bandpass Amplifier

The purpose of the bandpass amplifier is to separate the chrominance signal from the composite video signal, amplify it and then pass it on to the synchronous demodulators. The amplifier has fixed tuning with a bandpass wide enough ($\approx$ 2 MHz) to pass the chroma signal. The colour burst is prevented from appearing at its output by horizontal blanking pulses which disable the bandpass amplifier during the horizontal blanking intervals. The blanking pulses are generally applied to the colour killer circuit which is turn biases-off the chrominance amplifier during these periods.

Chapter 26

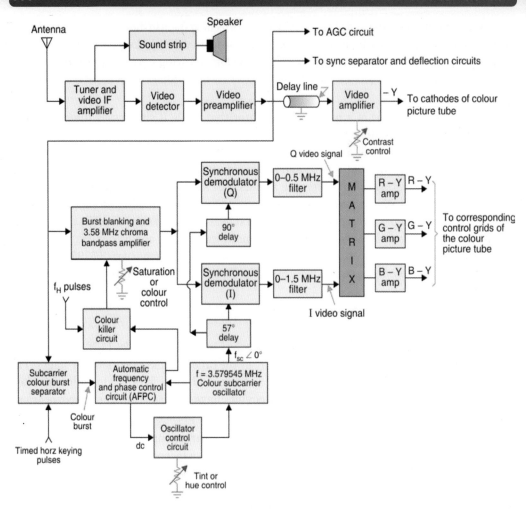

Fig. 26.11. *Simplified block diagram of a NTSC receiver.*

Colour Demodulators

Synchronous demodulators are used to detect the modulating signal. Such a demodulator may be thought of as a combination of phase and amplitude detectors because the output is dependent on both phase and amplitude of the chroma signal. As shown in the block diagram (Fig. 26.11) each demodulator has two input signals, the chroma which is to be demodulated and a constant amplitude output from the local subcarrier oscillator. The oscillator output is coupled to the demodulators by phase-shifting networks. The I demodulator oscillator voltage has a phase of 57° with respect to the burst phase ($f_{sc} \angle 0°$) and so has the correct delay to detect the I colour difference signal. Similarly the oscillator voltage to the Q demodulator is delayed by 147° (57° + 90°) for detecting the Q colour-difference signal. Thus the I and Q synchronous demodulators convert the chroma signal (a vector quantity) into its right-angle components (polar to rectangular conversion).

The Colour Matrix

This matrix is designed to produce $(R-Y)$, $(G-Y)$ and $(B-Y)$ signals from the I and Q video signals. Colour difference signal amplifiers are required to perform two functions. While amplifying the signals they also compensate for the chroma signal compression (weighting factors) that was introduced at the transmitter as a means of preventing overmodulation. The $(R-Y)$ amplifier provides a relative boost of $1.14 = 1/87.7\%$ while the $(B-Y)$ amplifier does so by a factor of $2.03 = 1/49\%$. Similarly the $(G-Y)$ amplifier reduces its output level to become $0.7(70\%)$ in a relative sense.

The grids and cathode of the picture tube constitute another matrix. The grids are fed positive colour difference signals and the cathode receives $-Y$ signal. The resultant voltages between the three grids and cathode become: $(R-Y)-(-Y)=R$, $(G-Y)-(-Y)=G$ and $(B-Y)-(-Y)=B$ and so correspond to the original red, green and blue signals generated by the colour camera at the transmitting end.

Burst Separator

The burst separator circuit has the function of extracting 8 to 11 cycles of reference colour burst which are transmitted on the back porch of every horizontal sync pulse. The circuit is tuned to the subcarrier frequency and is keyed 'on' during the flyback time by pulses derived from the horizontal output stage. The burst output is fed to the colour phase discriminator circuit also known as automatic frequency and phase control (AFPC) circuit.

Colour Subcarrier Oscillator

Its function is to generate a carrier wave output at 3.579545 MHz and feed it to the demodulators. The subcarrier frequency is maintained at its correct value and phase by the AFPC circuit. Thus, in a way the AFPC circuit holds the hue of reproduced colours at their correct values.

Colour Killer Circuit

As the name suggests this circuit becomes 'on' and disables the chroma bandpass amplifier during monochrome reception. Thus it prevents any spurious signals which happen to fall within the bandpass of the chroma amplifier from getting through the demodulators and causing coloured interference on the screen. This colour noise is called 'confetti' and looks like snow but with large spots in colour. The receiver thus automatically recognizes a colour or monochrome signal by the presence or absence of the colour sync burst. This voltage is processed in the AFPC circuit to provide a dc bias that cuts off the colour killer circuit. Thus when the colour killer circuit is off the chroma bandpass amplifier is 'on' for colour information. In some receiver designs the colour demodulators are disabled instead of the chroma bandpass amplifier during monochrome reception.

Manual Colour Controls

The two additional operating controls necessary in the NTSC colour receivers are colour (saturation) level control and tint (hue) control. These are provided on the front panel of the colour receiver. The colour control changes the gain of the chrominance bandpass amplifier and thus controls the intensity or amount of colour in the picture. The tint control varies phase of the 3.58 MHz oscillator with respect to the colour sync burst. This circuit can be either in the oscillator control or AFPC circuit.

Chapter 26

26.8 LIMITATIONS OF THE NTSC SYSTEM

The NTSC system is sensitive to transmission path differences which introduce phase errors that results in colour changes in the picture. At the transmitter, phase changes in the chroma signal take place when changeover between programmes of local and television network systems takes place and when video tape recorders are switched on. The chroma phase angle is also effected by the level of the signal while passing through various circuits. In addition crosstalk between demodulator outputs at the receiver causes colour distortion. All this requires the use of an automatic tint control (ATC) circuit with provision of a manually operated tint control.

26.9 PAL COLOUR TELEVISION SYSTEM

The PAL system which is a variant of the NTSC system, was developed at the Telefunken Laboratories in the Federal Republic of Germany. In this system, the phase error susceptibility of the NTSC system has been largely eliminated. The main features of the PAL system are:

(*i*) The weighted $(B - Y)$ and $(R - Y)$ signals are modulated without being given a phase shift of 33° as is done in the NTSC system.

(*ii*) On modulation both the colour difference quadrature signals are allowed the same bandwidth of about 1.3 MHz. This results in better colour reproduction. However, the chroma signal is of vestigial sideband type. The upper sideband attenuation slope starts at 0.57 MHz, *i.e.*, $(5 - 4.43 = 0.57$ MHz) but the lower sideband extends to 1.3 MHz before attenuation begins.

(*iii*) The colour subcarrier frequency is chosen to be 4.43361875 MHz. It is an odd multiple of one-quarter of the line frequency instead of the half-line offset as used in the NTSC system. This results in somewhat better cancellation of the dot pattern interference.

(*iv*) The weighted $(B - Y)$ and $(R - Y)$ signals are modulated with the subcarrier in the same way as in the NTSC system (QAM) but with the difference, that phase of the subcarrier to one of the modulators (V modulator) is reversed from $+ 90°$ to $- 90°$ at the line frequency. In fact the system derives its name, phase alteration by line (*i.e.*, PAL), from this mode of modulation. This technique of modulation cancels hue errors which result from unequal phase shifts in the transmitted signal.

As explained earlier the $(B - Y)$ and $(R - Y)$ subcarrier components in the chrominance signal are scaled down by multiplying them with the 'weighting' factors. For brevity the weighted signals are then referred to as U and V components of the chrominance signal where $U = 0.493\ (B - Y)$ and $V = 0.877\ (R - Y)$. Thus as illustrated in Fig. 26.12 (*a*)

$$C_{PAL} = U \sin \omega_s t \pm V \cos \omega_s t$$

$$= \sqrt{U^2 + V^2}\ \sin (\omega_s t \pm \theta) \text{ where } \tan \theta = V/U$$

The switching action naturally occurs during the line blanking interval to avoid any visible disturbance.

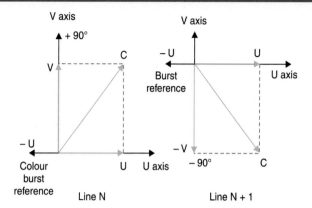

Fig. 26.12 (a). Sequence of modulation i.e., phase change of 'V' signal
on alternate lines in the PAL colour system.

The PAL Burst

If the PAL signal were applied to an NTSC type decoder, the $(B - Y)$ output would be U as required but the $(R - Y)$ output would alternate as $+ V$ and $- V$ from line to line. Therefore, the V demodulator must be switched at half the horizontal (line) frequency rate to give '$+ V$' only on all successive lines. Clearly the PAL receiver must be told how to achieve the correct switching mode. A colour burst (10 cycles at 4.43 MHz) is sent out at the start of each line. Its function is to synchronize the receiver colour oscillator for reinsertion of the correct carrier into the U and V demodulators.

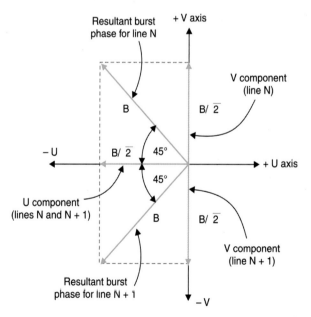

Fig. 26.12 (b). Illustration of PAL colour burst swing.

While in NTSC the burst has the phase of $- (B - Y)$ and a peak-to-peak amplitude equal to that of the sync, in PAL the burst is made up to two components (see Fig. 26.12 (b)), a $- (B - Y)$ component as in NTSC but with only $1/\sqrt{2}$ of the NTSC amplitude and an $(R - Y)$ component which like all the $(R - Y)$ information is reversed in phase from line to line. This $\pm (R$

$-Y$) burst signal has an amplitude equal to that of the $-(B-Y)$ burst signal, so that the resultant burst amplitude is the same as in NTSC. Note that the burst phase actually swings $\mp 45°$ (see Fig. 26.12 (b)) about the $-(B-Y)$ axis from line to line. However the sign of $(R-Y)$ burst component indicates the same sign as that of the $(R-Y)$ picture signal. Thus the necessary switching mode information is always available. Since the colour burst shifts on alternate lines by $\pm 45°$ about the zero reference phase it is often called the swinging burst.

As already pointed out the chroma signal is susceptible to phase shift errors both at the transmitter and in the transmission path. This effect is sometimes called 'differential phase error' and its presence results in changes of hue in the reproduced picture. This actually results from a phase shift of the colour sideband frequencies with respect to colour burst phase. The PAL system has a built-in protection against such errors provided the picture content remains almost the same from line to line. This is illustrated by phasor diagrams. Figure 26.13 (a) shows phasors representing particular U and V chroma amplitudes for two consecutive lines of a field. Since there is no phase error the resultant phasor (R) has the same amplitude on both the lines. Detection along the U axis in one synchronous detector and along the V axis in another, accompanied by sign switching in the latter case yields the required U and V colour signals. Thus correct hues are produced in the picture.

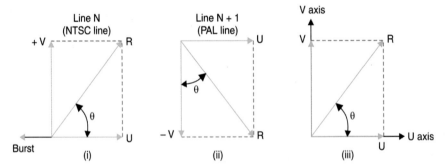

Fig. 26.13 (a). No phase error—(i) and (ii) show subcarrier phasors on two consecutive lines while (iii) depicts location of resultant phasor at the demodulator. Note that the phasor $(U \pm jV)$ chosen for this illustration represents a near magenta shade.

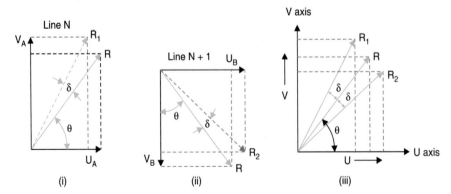

Fig. 26.13 (b). Cancellation of phase error.(d). (i) and (ii) show subcarrier phasors on two consecutive lines when a phase shift occurs. (iii) location of resultant phasor at the demodulator.

Now suppose that during transmission the phasor R suffers a phase shift by an angle δ. As shown in Fig. 26.13 (b) (i), the corresponding changes in the magnitude of U and V would mean a permanent hue error in the NTSC system. However, in the PAL system (Fig. 26.13 (b) (ii)) the resultant phasor at the demodulator will swing between R_1 and R_2 as illustrated in Fig. 26.13 (b) (iii). It is now obvious that the phase error would cancel out if the two lines are displayed at the same time. In actual practice however, the lines are scanned in sequence and not simultaneously. The colours produced by two successive lines, therefore, will be slightly on either side of the actual hue.

Since the lines are scanned at a very fast rate the eye due to persistence of vision will perceive a colour that lies between the two produced by R_1 and R_2 respectively. Thus the colour seen would more or less be the actual colour. It is here, where the PAL system claims superiority over the NTSC system.

26.11   PAL-D COLOUR SYSTEM

The use of eye as the averaging mechanism for the correct hue is the basis of 'simple PAL' colour system. However, beyond a certain limit the eye does see the effect of colour changes on alternate lines, and so the system needs modification. Remarkable improvement occurs in the system if a delay line is employed to do the averaging first and then present the colour to the eye. This is known as PAL-D or Delay Line PAL method and is most commonly used in PAL colour receivers. As an illustration of the PAL-D averaging technique, Fig. 26.14 shows the basic circuit used for separating individual U and V products from the chrominance signal. For convenience, both U and V have been assumed to be positive, that is, they correspond to some shade of magenta (purple). Thus for the first line when the V modulator product is at 90° to the $+U$ axis, the phasor can be expressed as $(U + jV)$. This is called the NTSC line. But on the alternate (next) line when the V phase is switched to $-90°$, the phasor becomes $(U - jV)$ and the corresponding line is then called the PAL line.

As shown in Fig. 26.14 (a), a delay line and, adding and subtracting circuits are interposed between the chrominance amplifier and demodulators. The object of the delay line is to delay the chrominance signal by almost exactly one line period of 64 μs. The chrominance amplifier feeds the chrominance signal to the adder, the subtractor and the delay line. The delay line in turn feeds its output to both the adder and subtractor circuits. The adder and subtractor circuits, therefore, receive the two signals simultaneously. These may be referred to at any given time as the direct line and delay line signals. For the chosen hue, if the present incoming line is an NTSC line, the signal entering the delay-line and also reaching the adder and subtractor is $(U + jV)$. But then the previous line must have been the PAL line i.e., $(U - jV)$ and this signal is simultaneously available from the delay line. The result is (see Fig. 26.14 (a)) that the signal information of two picture lines, though transmitted in sequence, are presented to the adder and subtractor circuits simultaneously. The adder yields a signal consisting of U information only but with twice the amplitude $(2U)$. Similarly, the subtraction circuit produces a signal consisting only of V information, with an amplitude twice that of the 'V' modulation product.

To permit precise addition and subtraction of direct and delayed line signals, the delay line must introduce a delay which is equivalent to the duration of an exact number of half-cycles of the chrominance signal. This requirement would not be met if the line introduced a delay of exactly 64 μs. At a frequency of 4.43361875 MHz the number of cycles which take place in 64 μs are $4.43361875 \times 10^6 \times 64 \times 10^{-6} = 283.7485998$ Hz, ≈ 283.75 Hz. A delay line which introduces a delay equal to the duration of 283.5 subcarrier cycles is therefore suitable. This is equal to a time delay of 63.943 μs $(1/f_{sc} \times 283.5)$.

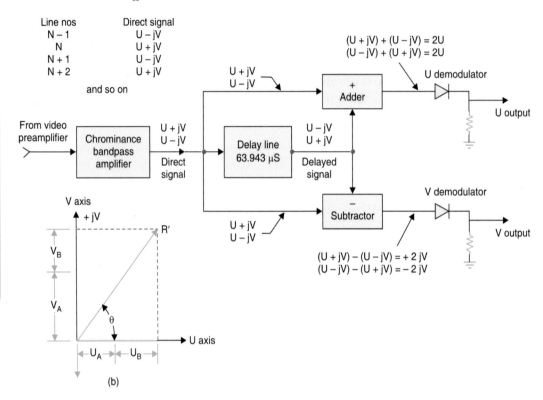

Fig. 26.14. (a) Basic principle of Pal-D demodulation. (b) Summation of consecutive line phasors in a PAL-D receiver.

The addition and subtraction of consecutive line phasors can also be illustrated vectorially by phasor diagrams. Figure 26.14 (b) is such a phasor diagram pertaining to the phase error illustration of Fig. 26.13 (b). The U and V signals thus obtained are fed to their respective synchronous demodulators for recovery of the colour difference signals.

Choice of Colour Subcarrier Frequency

If the sub-carrier frequency is chosen on the half-line offset basis as is done in the NTSC system, an annoying vertical line-up of dots occurs on certain hues. This is due to phase reversal of the sub-carrier at line frequency. To overcome this difficulty, a quarter-line offset is given instead and f_{sc} is made an odd multiple of one quarter of the line frequency. For optimum results this is slightly modified by adding 25 Hz to it, to provide a phase reversal on each successive field. Thus the actual relationship between f_{sc}, f_h and f_v can be expressed as

$$f_{sc} = \frac{f_h}{4}(2 \times 567 + 1) + \frac{f_v}{2} = \frac{f_h}{4}(1135) + \frac{f_v}{2} \quad \text{or} \quad = f_h(284 - 1/4) + \frac{f_v}{2}$$

This on substituting the values of f_h and f_v gives

$$f_{sc} = 4.43361875 \text{ MHz}.$$

It may be mentioned that though formation of any dot pattern on the screen must be suppressed by appropriate choice of sub-carrier frequency, it is of low visibility and appears only when colour transmission is received on a monochrome receiver. Since the chrominance signal disappears on grey or white details in a picture, the interference appears only in coloured areas of the scene. All colours appear as shades of grey on a monochrome receiver and thus the dot effect is visible only in those coloured parts of the picture which are reproduced in lighter shades of grey.

Colour Subcarrier Generation

The colour subcarrier frequency of 4.43361875 MHz is generated with a crystal controlled oscillator. In order to accomplish minimum raster disturbance through the colour subcarrier it is necessary to maintain correct frequency relationships between the scanning freqeuncies and the subcarrier frequency. It is therefore usual to count down from the subcarrier frequency to twice the line frequency $(2f_h)$ pulses which are normally fed to monochrome sync pulse generators. There are several ways in which frequency division can be accomplished. In the early days of colour television it was necessary to choose frequencies which had low-order factors so that division could take place in easy stages, but such constraints are no longer necessary. The PAL subcarrier frequency is first generated directly. Then the 25 Hz output obtained from halving the field frequency is subtracted from the subcarrier frequency. The frequency thus obtained is first divided by 5 and then by 227. Such a large division is practicable by using a chain of binary counters and feeding back certain of the counted down pulses to earlier points of the chain that the count down is changed from a division of 2^8 to a division of 227.

26.12 THE PAL CODER

Figure 26.15 is the functional diagram of a PAL coder. The gamma corrected R, G and B signals are matrixed to form the Y and the weighted colour difference signals. The bandwidths of both ($B - Y$) and ($R - Y$) video signals are restricted to about 1.3 MHz by appropriate low-pass filters. In this process these signals suffer a small delay relative to the Y signal. In order to compensate for this delay, a delay line is inserted in the path of Y signal.

The weighted colour difference video signals from the filters are fed to corresponding balanced modulators. The sinusoidal sub-carrier is fed directly to the U modulator but passes through a ± 90° phase switching circuit on alternate lines before entering the V modulator. Since one switching cycle takes two lines, the squarewave switching signal from the multivibrator to the electronic phase switch is of half-line frequency *i.e.*, approximately 7.8 KHz. The double sideband suppressed carrier signals from the modulators are added to yield the quadrature amplitude modulated (Q.A.M.) chrominance (C) signal. This passes through a filter which removes harmonics of the subcarrier frequency and restricts the upper and lower sidebands to appropriate values. The output of the filter feeds into an adder circuit where it is combined with the luminance and sync signals to form a composite colour video signal. The bandwidth and location of the composite colour signals (U and V) is shown along with the Y signal in Fig. 26.15.

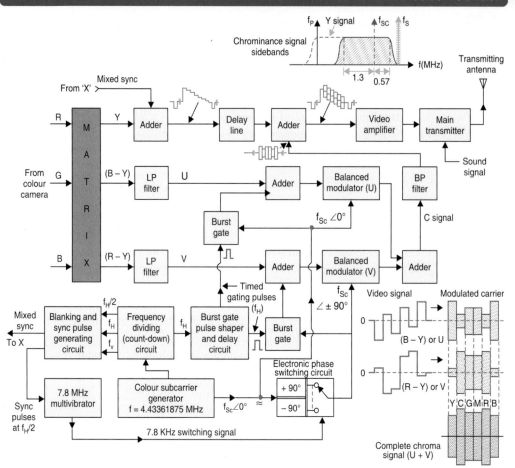

Fig. 26.15. Basic organization of the PAL coder.

Notice that the colour burst signal is also fed to the modulators along with the U and V signals through the adders. The burst signals are obtained from the circuits that feed the colour subcarrier signal to the two modulators. However, before feeding the burst signals to the U and V adders these are passed through separate burst gates. Each burst gate is controlled by delayed pulses at f_H rate obtained from the frequency dividing circuit. The gating pulses appear during the back porch period. Thus, during these intervals the $(B-Y)$ i.e., U modulator yields a subcarrier burst along $-U$ while the $(R-Y)$ i.e., V modulator gives a burst of the same amplitude but having a phase of $\pm 90°$ on alternate lines relative to the $-U$ phasor. At the outputs of the two modulators, the two burst components combine in the adder to yield an output which is the vector sum of the two burst inputs. This is a subcarrier sinewave (≈ 10 cycles) at $+45°$ on one line and $-45°$ on the next line with reference to $-U$ phasor.

The colourplexed composite signal thus formed is fed to the main transmitter to modulate the station channel picture carrier in the normal way. The sound signal after being frequency modulated with the channel sound carrier frequency also forms part of the RF signal that is finally radiated through the transmitter antenna system.

26.13 PAL-D COLOUR RECEIVER

Various designs of PAL decoder have been developed. The one shown in the colour receiver block diagram of Fig. 26.16 is a commonly used arrangement. It will be noticed that the general pattern of signal flow is very close to that of the NTSC receiver. Necessary details of various sections of the receiver are discussed.

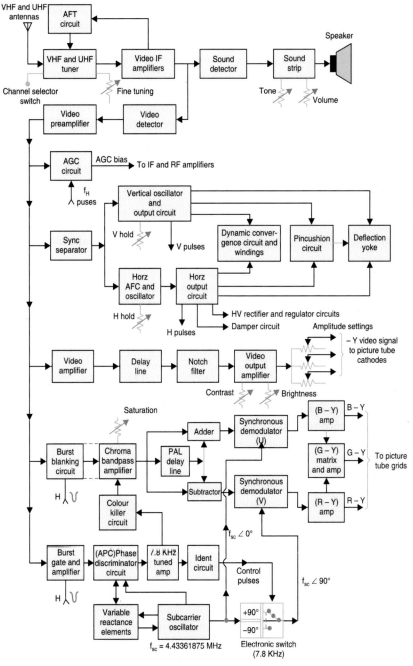

Fig. 26.16. Block diagram of a PAL-D colour receiver.

1. Tuner

It is necessary to maintain local oscillator frequency at the correct value to obtain exact colour burst frequency for proper reproduction of different colours in the picture. Therefore, colour receiver tuners employ an additional circuit known as automatic frequency tuning (AFT). This circuit actually controls the local oscillator frequency to obtain a picture IF of exactly 38.9 MHz at the converter output. The discriminator in the AFT circuit measure the intermediate frequency and develops a dc control voltage proportional to the frequency deviations if any. This error voltage is applied to the reactance section of the local oscillator to maintain its frequency at the correct value. More details of AFT are given in the next chapter along with other special circuits.

2. Sound Strip

The frequency modulated sound IF signal is processed in the usual way to obtain audio output. The volume and tone controls are associated with the audio amplifier, the output of which feeds into the loudspeaker. Thus the sound strip of a colour receiver is exactly the same as in a black and white receiver.

3. AGC, Sync-separator and Deflection Circuits

The AGC and sync-separator circuits function in the same way as in a monochrome receiver. However, the deflection circuits, besides developing normal horizontal and field scanning currents also provide necessary wave-forms for dynamic convergence and pincushion correction. In addition, pulses from the horizontal output transformer are fed to several circuits in the colour section of the receiver.

4. Luminance Channel

The video amplifier in the luminance channel is dc coupled and has the same bandwidth as in the monochrome receiver. It is followed by a delay line to compensate for the additional delay the colour signal suffers because of limited bandpass of the chrominance. This ensures time coincidence of the luminance and chrominance signals. The channel also includes a notch filter which attenuates the subcarrier by about 10 db. This helps to suppress the appearance of any dot structure on the screen along with the colour picture. The inverted composite video signal available at the output of luminance channel is fed to the junction of three cathodes of the picture tube. This part of the circuit also includes drive adjustment necessary for setting of the black level and obtaining correct reproduction of colours.

5. Colour Signal Processing

The signal available at the output of video detector is given some amplification (video preamplifier) before feeding it to the various sections. All modern receivers use ICs for processing the colour signal. However, for a better understanding, the operation of each stage is described with the help of discrete component circuitry.

(a) *Chrominance bandpass amplifier.* As noted previously the chroma bandpass amplifier selects the chrominance signal and rejects other unwanted components of the composite signal. The burst blanking, colour level control and colour killer switch also form part of this multistage amplifier.

(*i*) **Burst Blanking.** The output from the video preamplifier (Fig. 26.17) is fed to the first stage of chroma bandpass amplifier through an emitter follower stage (Q_1). Negative going horizontal blanking pulses are coupled to the base of Q_1 through diode D_1. The pulses drive Q_1 into cut-off during colour burst intervals and thus prevent it from reaching the demodulators.

(*ii*) **Bandpass Stage.** The emitter follower output is fed to the bandpass stage through a tuned circuit consisting of L_1 and C_3. The necessary bandwidth centered around 4.43 MHz is adjusted by R_5 and R_6. The tuning of L_1 also incorporates necessary correction on account of vestigial sideband transmission of the chrominance signal.

(*iii*) **Automatic Colour Control (ACC).** The biasing of amplifier (Q_2) in Fig. 26.17 is determined by the dc control voltage fed to it by the ACC circuit. The ACC circuit is similar to the AGC circuit used for automatic gain control of RF and IF stages of the receiver. It develops a dc control voltage that is proportional to the amplitude of colour burst. This voltage when fed at the input of Q_2 shifts its operating point to change the stage gain. Thus net overall chroma signal output from the bandpass amplifier tends to remain constant.

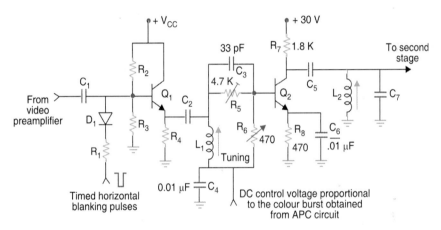

Fig. 26.17. Colour burst blanking circuit and 1st stage of the chroma bandpass amplifier.

(*iv*) **Manual Colour (Saturation) Control.** As shown in Fig. 26.18 the chroma signal from the first stage is applied through R_1 and R_2 to the emitter of Q_3 and cathode of D_2, the colour control diode. The diode is forward biased by a voltage divider formed by R_3, R_4 and R_5 the colour control potentiometer. When the diode is excessively forward biased (R_5 at + 30 V position) it behaves like a short circuit. Under this condition the chroma signal gets shorted via C_1 to ground and there is no input to the demodulators. As a result, a black and white picture is produced on the screen. If R_5 is so adjusted that forward bias on D_1 is almost zero, the diode would present a very high impedance to the signal. Under this condition all the available signal feeds into the amplifier and a large signal voltage appears at the output of the chroma bandpass amplifier. This is turn produces a picture with maximum saturation. At other settings of R_5, conductance of D_1 would cause the signal current to divide between the emitter of Q_3 and C_1 resulting in intermediate levels of picture colour saturation. No tint or hue control is provided in PAL receivers because of the inbuilt provision for phase shift cancellation. In same receiver designs the saturation control is combined with the contrast control.

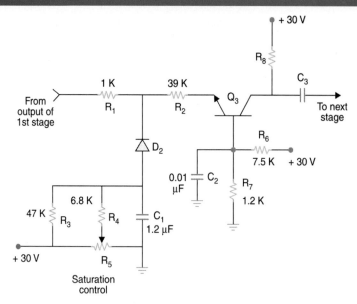

Fig. 26.18. Saturation control in the chroma bandpass amplifier.

(v) **Colour Killer Circuit.** The colour killer and associated circuits are shown in Fig. 26.19. The forward bias of Q_5, the last stage of bandpass amplifier depends on the state of the colour killer circuit. When a colour signal is being received, the 7.8 KHz (switching rate of the $(R-Y)$ signal) component is available at the APC (automatic phase control) circuit of the reference subcarrier oscillator. It is applied via C_1 to the base of tuned amplifier Q_6. The amplified 7.8 KHz signal is ac coupled to Q_7. Diode D_3 conducts on negative half cycles charges the capacitor C_2 with the polarity marked across it. The discharge current from this capacitor provides forward bias to Q_7, the emitter follower. Such an action results in a square wave signal at the output of Q_7. It is coupled back via a 680 ohm resistor to the tuned circuit in the collector of Q_6. This provides positive feedback and thus improves the quality factor of the tuned circuit.

The colour killer diode D_4 rectifies the square-wave output from the emitter of Q_7. The associated RC filter circuit provides a positive dc voltage at point 'A' and this serves a source of forward bias to the chrominance amplifier Q_5. Diode D_5 is switched on by this bias and so clamps the voltage produced at 'A' by the potential divider (3.3 K and 680 ohm) across the + 15 V line.

When a monochrome transmission is received there is no 7.8 KHz input to the colour killer diode D_4 and no positive voltage is developed at its cathode (point A). Both D_5 and the base emitter junction of Q_5 are now back biased by the – 20 V potential returned at 'A' via the 220 K resistor. The chrominance signal channel, therefore, remains interrupted.

(b) *Separation of U and V modulation products.* As explained in section 26.11 the addition of two picture lines radiated in sequence but presented to the adder circuit simultaneously yields a signal consisting of only U information but with an amplitude equal to twice the amplitude of the chrominance signal's U modulation product. Similarly the subtraction of the two lines produces a signal consisting of V information with an amplitude equal to twice that of the V modulation product.

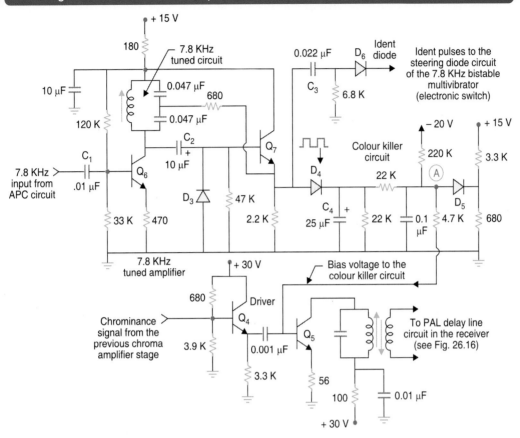

Fig. 26.19. Colour killer and allied circuits.

Synchronous Demodulators. The output from the adder and substractor (Fig. 26.16) consists of two independent double sideband, suppressed carrier RF signals. These are the U modulation product and the line-by-line phase inverted V modulation product. The two individual RF signals are fed to their respective demodulators. Each of the demodulators also receives a controlled subcarrier of correct phase to allow recovery of the colour difference signal waveforms. It may be noted that the modulators do not have to handle Q.A.M. ($u \pm jv$) RF signals as is the case in the NTSC system. Therefore, it is not absolutely necessary to employ synchronous demodulators. However, synchronous demodulators are preferred in practice because they yield an accurate and constant no-colour zero voltage level above and below which (sometimes positive and sometimes negative) the colour different signal voltage varies.

(c) Colour difference amplifiers and matrixing. There are two approaches to driving the colour picture tube. In one scheme, the three colour difference signals are amplified and fed to the appropriate grids of the picture tube. The $- Y$ signal is fed to the junction of three cathodes for matrixing. In another approach $R, G,$ and B video signals are obtained directly by a suitable matrix from the modulator outputs. Each colour signal is then separately amplified and applied with negative polarity ($- R, - G, - B$) to the respective cathodes. The grids are then returned to suitable negative dc potentials. This, *i.e.*, R, G, B method is preferred in transistor amplifiers because less drive voltages are necessary when $R, G,$ and B are directly fed to the picture tube.

The use of dc amplifiers would involve difficulties, especially in maintaining a constant 'no colour' voltage level at the picture tube grids. It is usual to employ ac amplifiers and then establish a constant no-colour level by employing dc clamps. The clamping is affected during line blanking periods by pulses derived from the line time base circuitry. Thus the dc level is set at the beginning of each active line period. The discharge time constants of the ac coupling networks are chosen to be quite large so that the dc level does not change significantly between the beginning and end of one active line period.

6. Subcarrier Generation and Control

The primary purpose of this section is to produce a subcarrier of correct frequency to replace the subcarrier suppressed in the two chrominance signal balanced modulators at the transmitter end encoder. Not only must the generated subcarrier be of exactly the right frequency but it must also be of the same phase reference as the original subcarrier. A crystal oscillator is used in the receiver and this is forced to work at the correct frequency and phase by the action of an automatic frequency and phase control circuit. This is usually called the APC circuit. The APC circuit compares the burst and locally generated reference subcarrier to develop a control voltage. The burst signal is obtained through the burst gate amplifier circuit. The identification circuit and electronic switch for line switching the subcarrier to the V modulator also form part of the subcarrier generation circuitry.

   (a) *Burst gate amplifier.* The burst gate or burst amplifier as it is often called separates the colour burst from the chrominance signal. It is essentially a gated class B or C operated tuned amplifier with a centre frequency equal to 4.43 MHz. The bandpass of the amplifier is approximately 0.6 MHz. The burst gate shown in Fig. 26.20 uses a *p-n-p* transistor in the common emitter configuration. The biasing circuit is so arranged that transistor Q_{11} is normally reverse biased. The circuit has two inputs. One of the inputs is a large amplitude delayed horizontal retrace pulse obtained from the flyback transformer. It momentarily drives the amplifier out of cut-off and allows the amplifier to function normally. The other input is the chrominance signal obtained from the video preamplifier.

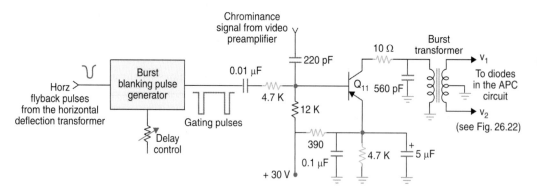

Fig. 26.20. Burst gate amplifier.

For proper operation the timing of the horizontal pulse and the colour burst must be such that the amplifier is turned on and kept operative during the time when the colour burst is at the input of the burst gate. Under this condition the output of the burst gate contains only the amplified colour burst. The horizontal retrace pulse obtained at the flyback transformer is generated during the horizontal sync pulse interval. Since this occurs before the colour burst, a pulse-shaping and time-delay network is placed between the flyback transformer and the burst gate. With correct delay time, the pulse will occur at the same time as the burst and the amplifier output will have maximum burst amplitude and duration. The sinusoidal burst signal available at the centre-tapped secondary of the output transformer provides anti-phase inputs to the phase detector diodes in the APC circuit.

(b) *Reference subcarrier oscillator.* A typical circuit of a reference subcarrier oscillator is shown in Fig. 26.21. It is a crystal controlled oscillator of the inverted Colpitt's type. The tuned circuit, *i.e.*, the crystal is connected between the base and emitter circuit of transistor Q_{10}. Normally the total external capacitance required across the crystal to make the oscillator work at 4.43361875 MHz is around 20 pF. The capacitor C_5 and the variable capacitance (varactor) diode D_9 provide this. The oscillator has a resonant circuit in the collector lead of Q_{10} which is tuned 4.43 MHz. This removes the harmonic content in the oscillator output. From the collector the subcarrier signal at 4.43361875 MHz is ac coupled to the input of an amplifier which amplifies the oscillator output and passes it on to the U and V demodulator circuits. The amplifier output is also fed to an emitter follower (Q_9) for providing input to the APC circuit.

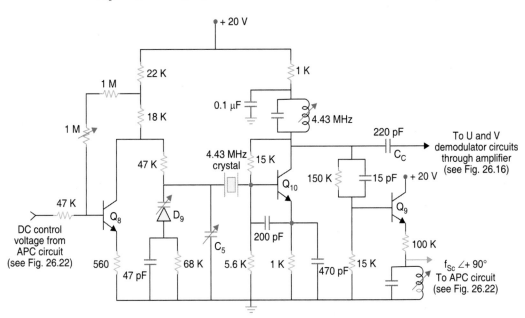

Fig. 26.21. *Reference subcarrier oscillator and allied circuits.*

(c) *Automatic phase control (APC) circuit.* The simplified schematic diagram of a typical phase discriminator circuit is shown in Fig. 26.22. It receives two inputs, the locally generated reference subcarrier and the transmitted burst. The burst output is available at the centre-tapped transformer (see Fig. 26.20) of the burst gate amplifier. In the absence of any subcarrier input, the two diodes, D_7 and D_8 conduct equally to charge C_7 and C_8 with equal voltages but of opposite polarity. If the circuit is perfectly balanced and the slider of the balancing control is in the centre, no control voltage is developed.

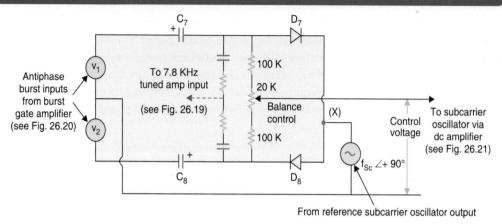

Fig. 26.22. Simplified circuit of the phase detector in the APC circuit.

The effect of the reference oscillator input at the common connection of the APC circuit (marked X) may now be considered.

With the subcarrier oscillator frequency exactly correct, its phase is arranged to be shifted by 90° with reference to the incoming burst. Thus the two inputs to the APC circuit are in quadrature. The charges on the two capacitors will still remain equal since the subcarrier signal passes through zero potential when the diodes are pulsed into conduction by the peaks of the burst signal. The line-by-line phase alternation of ± 45° about the 180° axis of the burst signal does not affect the circuit balance because mean phase is still 180°. In fact the burst shift of ± 45° favours one diode on one line but to the other during the next line. Thus the mean output voltage remains zero.

There is, however, always a 7.8 KHz ac component superimposed on the mean level of the output voltage. This is taken off and fed to the 7.8 KHz tuned amplifier, the output of which feeds the identification and colour killer circuits. It may be noted that the 7.8 KHz signal is not passed on by the dc amplifier (Q_8) (Fig. 26.21) to the reference oscillator control element because the frequency response of the circuit is kept quite low.

If the oscillator frequency tends to increase the reference signal to the common connection now passes through zero ahead of the centre of the time interval when the burst pulses the two diodes. The net input to D_8 becomes greater than D_7 and charge on C_8 exceeds that on C_7. This results in a net positive control voltage. This positive output is inverted by the dc amplifier Q_8. The negative going output from Q_8 reduces reverse bias across the varactor diode D_9 with the result that the capacitance of D_9 increases. This in turn reduces the oscillator frequency.

If the oscillator frequency tends to decrease, analogous arguments will establish that reverse bias across D_9 will increase thereby decreasing its capacitance. This will lead to increase in the oscillator frequency. Thus the APC circuit continuously senses the subcarrier frequency and applies necessary correction. The maximum oscillator frequency error an APC circuit can correct when first switched on to an incoming signal is known as its 'pull in' range. This is typically ± 400 KHz in this circuit.

(*d*) *Identification (Ident) circuit.* In Fig. 26.16 a single pole two-way switch has been shown for alternately reversing phase of the subcarrier output before applying it to the V demodulator. The switch is actually a bistable multivibrator controlled by line pulses from the flyback transformer. It thus switches at the required rate of 7.8 KHz. However, it is necessary that the instant and sequence of switching is synchronized with that of the swinging burst at the transmitter. An identification signal is developed for this purpose. In Fig. 26.19 the 7.8 KHz square wave from the emitter of Q_7 is coupled via C_3 (0.022 µF) to the anode of diode D_6, the 'Indent' diode. The diode D_6 conducts on positive half cycles and develops a positive dc voltage at its cathode. This identification voltage is applied to the coupling network of the 7.8 KHz bistable multivibrator (electronic switch) through the steering diodes. It overrides one is every two of the negative going line trigger pulses to maintain correct sequence of switching.

26.14   MERITS AND DEMERITS OF THE PAL SYSTEM

The problem of differential phase errors has been successfully overcome in the PAL system. This is its main merit. In addition the use of PAL-D technique in receivers for electrically accumulating adjacent line colour signals considerably circumvents hue errors. Thus a manual hue control becomes unnecessary. However, the delay line technique of reception also involves a reduction in the vertical resolution of the chrominance signal but the effect is less pronounced because the two chrominance signals are radiated continuously and the receiver interpolates between the signals of two consecutive lines.

The use of phase alternation by line technique and associated control circuitry together with the need of a delay line in the receiver makes the PAL system more complicated and expensive. The receiver cost is higher for the PAL colour system. In addition, the PAL system presents problems in magnetic recording since a complete colour coding sequence requires eight fields instead of four necessary in the NTSC system.

26.15   SECAM SYSTEM

The SECAM system was developed in France. The fundamental difference between the SECAM system on the one hand and the NTSC and PAL systems on the other is that the latter transmit and receive two chrominance signals simultaneously while the SECAM system is "sequential a memoire", *i.e.*, only one of the two colour difference signals is transmitted at a time. The subcarrier is frequency modulated by the colour difference signals before transmission. The magnitude of frequency deviation represents saturation of the colour and rate of deviation its fineness.

If the red difference signal is transmitted on one line then the blue difference signal is transmitted on the following line. This sequence is repeated for the remaining lines of the raster. Because of the odd number of lines per picture, if nth line carriers $(R - Y)$ signal during one picture, it will carry $(B - Y)$ signal during scanning of the following picture. At the receiver an ultrasonic delay line of 64 µs is used as a one line memory device to produce decoded output of both the colour difference signals simultaneously. The modulated signals are routed to their correct demodulators by an electronic switch operating at the rate of line frequency. The switch is driven by a bistable multivibrator triggered from the receiver's horizontal deflection circuitry. The determination of proper sequence of colour lines in each field is accomplished by identification (Ident) pulses which are generated and transmitted during vertical blanking intervals.

SECAM III

During the course of development the SECAM system has passed through several stages and the commonly used system is known as SECAM III. It is a 625 line 50 field system with a channel bandwidth of 8 MHz. The sound carrier is + 5.5 MHz relative to the picture carrier. The nominal colour subcarrier frequency is 4.4375 MHz. As explained later, actually two subcarrier frequencies are used. The Y signal is obtained from the camera outputs in the same way as in the NTSC and PAL systems. However, different weighting factors are used and the weighted colour difference signals are termed D_R and D_B where $|D_R| = 1.9 (R - Y)$ and $|D_B| = 1.5 (B - Y)$.

Modulation of the Subcarrier

The use of FM for the subcarrier means that phase distortion in the transmission path will not change the hue of picture areas. Limiters are used in the receiver to remove amplitude variations in the subcarrier. The location of subcarrier, 4.4375 MHz away from the picture carrier reduces interference and improves resolution. In order to keep the most common large deviations away from the upper end of the video band, a positive frequency deviation of the subcarrier is allowed for a negative value of $(R - Y)$. Similarly for the blue difference signals a positive deviation of the subcarrier frequency indicates a positive $(B - Y)$ value. Therefore, the weighted colour signals are: $D_R = - 1.9 (R - Y)$ and $D_B = 1.5 (B - Y)$. The minus sign for D_R indicates that negative values of $(R - Y)$ are required to give rise to positive frequency deviations when the subcarrier is modulated.

In order to suppress the visibility of a dot pattern on monochrome reception, two different subcarriers are used. For the red difference signal it is $282 f_H = 4.40625$ MHz and for the blue difference signal it is $272 f_H = 4.250$ MHz.

Pre-emphasis

The colour difference signals are bandwidth limited to 1.5 MHz. As is usual with frequency modulated signals the SECAM chrominance signals are pre-emphasised before they are transmitted. On modulation the subcarrier is allowed a linear deviation $= 280 D_R$ KHz for the red difference signals and $230 D_B$ KHz for the blue difference signals. The maximum deviation allowed is 500 KHz in one direction and 350 KHz in the other direction for each signal although the limits are in opposite directions for the two chrome signals.

After modulating the carrier with the pre-emphasised and weighted colour difference signals $(D_R$ and $D_B)$, another form of pre-emphasis is carried out on the signals. This takes the form of increasing the amplitude of the subcarrier as its deviation increases. Such a pre-emphasis is called high-frequency pre-emphasis. It further improves signal to noise ratio and interference is very much reduced.

Line Identification Signal

The switching of D_R and D_B signals line-by-line takes place during the line sync pulse period. The sequence of switching continues without interruption from one field to the next and is maintained through the field blanking interval. However, it is necessary for the receiver to be able to deduce as to which line is being transmitted. Such an identification of the proper sequence of colour lines in each field in accomplished by identification pulses that are generated during vertical blanking periods. The signal consists of a sawtooth modulated subcarrier (see Fig. 26.23) which is positive going for a red colour-difference signal and negative going for the blue colour-difference signal. At the receiver the Ident pulses generate positive and negative control signals for regulating the instant and sequence of switching.

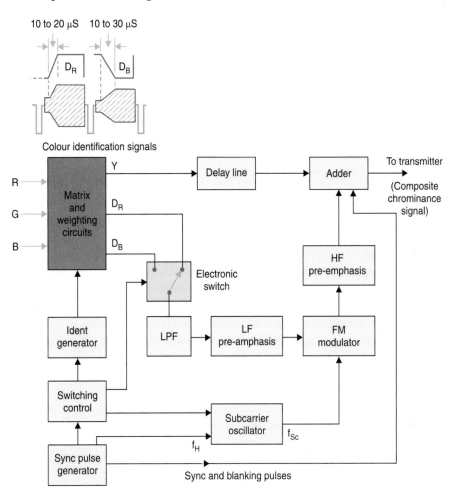

Fig. 26.23. Functional diagram of a SECAM III Coder.

SECAM Coder

Figure 26.23 is a simplified functional diagram of a SECAM III coder. The colour camera signals are fed into a matrix where they are combined to form the luminance ($Y = 0.3R + 0.59G + 0.11B$) and colour-difference signals. The SECAM weighting and sign factors are applied to the colour-difference signals so that the same subcarrier modulator can be used for both the chrominance (D_R and D_B) signals. The Ident signal is also added in the same matrix.

An electronic switch which changes its mode during every line blanking interval directs D_R and D_B signals to the frequency modulator in a sequential manner, *i.e.*, when D_R is being transmitted on the line, then D_B is not used and vice versa.

Sync Pulse Generation and Control

The line frequency pulses from the sync pulse generator are passed through selective filters which pick out the 272nd and 282nd harmonics of f_H. These harmonics are amplified and used as the two subcarrier references. The sync pulse generator also synchronizes the switching control unit which in turn supplies operating pulses to the electronic switch for choosing between D_R and D_B signals. The switching control also operates the circuit which produces modulated waveforms of the Ident signal. These are added to the chrominance signals during field blanking period and before they are processed for modulation.

The output from the electronic switch passes through a low-pass filter which limits the bandwidth to 1.5 MHz. The bandwidth limited signals are pre-emphasized and then used to frequency modulate the subcarrier. The modulator output passes through a high frequency pre-emphasis filter having a bell-shaped response before being added to the Y signal. The sync and blanking pulses are also fed to the same adder. The adder output yields composite chrominance signal which is passed on to the main transmitter.

SECAM Decoder

SECAM receivers are similar in most respects to the NTSC and PAL colour receivers and employ the same type of colour picture tubes. The functional diagram of a SECAM III decoder is shown in Fig. 26.24. The chroma signal is first filtered from the composite colour signal. The bandpass filter, besides rejecting unwanted low frequency luminance components, has inverse characteristics to that of the bell-shaped high frequency pre-emphasis filter used in the coder. The output from the bandpass filter is amplified and fed to the electronic line-by-line switch via two parallel paths. The 64 µs delay lines ensures that each transmitted signal is used twice, one on the line on which it is transmitted and a second time on the succeeding line of that field. The electronic switch ensures that D_R signals, whether coming by the direct path or the delayed path, always go to the D_R demodulator. Similarly D_B signals are routed only to the D_B demodulator.

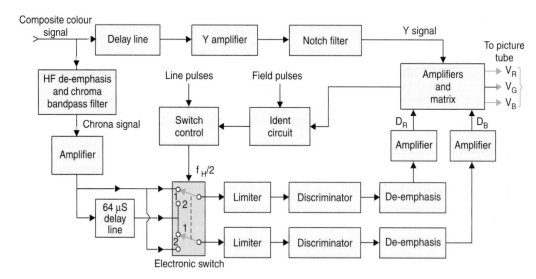

Fig. 26.24. Functional diagram of a SECAM III decoder.

The switch is operated by line frequency pulses. In case phasing of the switch turns out to be wrong, *i.e.*, it is directing D_R and D_B signals to the wrong demodulators, the output of each demodulator during the Indent signal period becomes positive instead of negative going. A sensing circuit in the Ident module then changes the switching phase.

The electronic switch directs the frequency modulated signals to limiters and frequency discriminators. The discriminators have a wider bandwidth than that employed for detecting commercial FM sound broadcasts. After demodulation the colour difference signals are de-emphasized with the same time constant as employed while pre-emphasing.

As in other receivers the matrix networks combine the colour difference signals with the Y signal to give primary colour signals R, G and B which control the three electronic beams of the picture tube.

It may be noted that a SECAM receiver requires only two controls—brightness and constrast, both for monochrome and colour reception. The saturation and hue controls are not needed because the system is immune to these distortions. This is so because the colour signals are constant amplitude, frequency modulated signals and the frequency deviations which carry colour information are not affected during transmission.

NIR SECAM System

The advent of the commercially available delay lines primarily developed for the SECAM and PAL system led to the development of other SECAM systems. In the later versions known as SECAM IV and SECAM V, quadrature amplitude modulation (as in the NTSC system) and synchronous detectors are used instead of the frequency modulators and discriminators. These were developed at the Russian National Institute for Research (NIR) and are sometimes referred to as NIR-SECAM systems. Here the transmitted chrominance signals amplitude is not proportional to the vector sum of the two chrominance components but is proportional to the square root of the vector sum. The signal are, however, transmitted on a sequential basis which is the distinguishing feature of all SECAM systems. At the receiver, one line time delay is used in the decoder so that both types of transmitted signals are available during all the lines of the picture. A change-over switch is required to route the constant one phase signal over one path and the varying or other phase signal over another path. The rest of the receiver circuit is similar and produces R, B and G signal voltages at the electrodes of the picture tube to reproduce all the colours on the screen.

Chapter 26

26.16 MERITS AND DEMERITS OF SECAM SYSTEMS

Several advantages accrue because of frequency modulation of the subcarrier and transmission of one line signal at a time. Because of FM, SECAM receivers are immune to phase distortion. Since both the chrominance signals are not present at the same time, there is no possibility of cross-talk between the colour difference signals. There is no need for the use of Q.A.M. at the transmitter and synchronous detectors at the receiver. The subcarrier enjoys all the advantages of FM. The receiver does not need ATC and ACC circuits. A separate manual saturation control and a hue control are not necessary. The contrast control also serves as the saturation control. All this makes the SECAM receiver simple and cheaper as compared to NTSC and PAL receivers.

It may be argued that the vertical resolution of the SECAM system is inferior since one line signal combines with that of the previous to produce colours. However, subjective tests do not bring out this deficiency since our visual perception for colours is rather poor.

In addition, while SECAM is a relatively easy signal to record there is one serious drawback in this system. Here luminance is represented by the amplitude of a voltage but hue and saturation are represented by the deviation of the sub-carrier. When a composite signal involving luminance and chrominance is faded out in studio operation it is the luminance signal that is readily attenuated and not the chrominance. This makes the colour more saturated during fade to black. Thus a pink colour will change to red during fade-out. This is not the case in NTSC or PAL systems. Mixing and lap dissolve presents similar problems.

In conclusion it may be said that all television systems are compromises since changing one parameter may improve one aspect of performance but degrade another, for example increasing bandwidth improves resolution of the picture but also increases noise. In fact when all factors are taken into account it is difficult to justify the absolute superiority of one system over the other. In many cases political and economic factors have been the apparent considerations in adopting a particular monochrome and the compatible colour system. It can therefore be safely concluded that the three colour systems will co-exist. Possibly some consensus on international exchange will be reached in due course of time.

REVIEW QUESTIONS

1. Explain how by frequency interleaving the colour information is accommodated within the same channel bandwidth of 7 MHz.

2. Explain with a block diagram how both $(B-Y)$ and $(R-Y)$ signals are combined around the same sub-carrier frequency by quadrature modulation. Why is the colour signal bandwidth requirement much less than that of the Y signal ?

3. Discuss the factors which influence the choice of sub-carrier frequency in a colour TV system. Justify the choice of 3.579545 MHz as the subcarrier frequency in the NTSC system. How does it affect the line and field frequencies ?

4. Why are the modulated sub-carrier vectors shifted by 33° to constitute Q and I signals in the NTSC system ? Why different bandwidths are assigned to Q and I signals ?

5. Explain how the differential phase-error is continuously corrected in the PAL system while affecting Q.A.M. of the colour difference signals. Establish the value of f_{sc} as used in the PAL system. How does it help to suppress the appearance of a dot pattern structure on the screen ?

6. Explain with a suitable block diagram the encoding process in the PAL colour system. Why is the colour burst signal transmitted after each scanning line ?

7. Explain the delay line method of separating U and V signals in a PAL receiver. What is the function of a colour killer circuit in the path of chrominance signal in the receiver ?

8. Describe the manner in which the APC circuit functions to keep the reference subcarrier frequency constant in the PAL system. What is IDENT signal and how is it separated in the receiver ? Explain how it is used to control the electronic switch and colour killer circuit.

9. What is the basic difference between the SECAM and other colour TV system. Describe briefly the encoding and decoding processes of the SECAM III system.

10. Discuss the relative merits and demerits of the three television systems. Explain the factors which influence the choice of any one of the three systems.

27

Remote Control and Special Circuits

In some television receivers special circuits are provided to accomplish certain tasks not ordinarily performed by the standard receiver circuitry. For example, remote control is one such facility which enables the television viewer to operate from a distance (without leaving his or her seat) most of the controls situated on the front panel of the receiver. Such a unit adds much to the cost of the receiver and is generally made optional even with high quality receivers. Similarly, in some receiver designs, additional circuits are used, as a refinement to the functions already performed by the existing system. Automatic fine tuning (AFT) is one such provision to obtain automatic control of fine tuning of the receiver. As explained in the previous chapter, it is a must in colour receivers. Other such circuits include touch tuning, automatic brightness control, instant-on circuit etc. A number of such circuits, besides remote control, are discussed to acquaint the reader with the basic principles of such techniques.

27.1 REMOTE CONTROL

Based on the design of the remote control unit and type and make of the receiver, it is possible to control as many as five different functions. These include volume-up and on-off, volume-down, channel selection, colour-up, colour-down, etc. Though a variety of remote control television systems are in use, basically all are composed of the same three primary sections as illustrated in Fig. 27.1. A transmitter box, commonly known as a 'Bonger' or 'Clicker', used at a distance from the receiver, sends out a signal that is intercepted by a transducer provided in the receiver. The nature of the signal received is interpreted by the sensing system for, say, channel selection or for on-off and volume control purposes. The received signals are processed to control mechanical or electronic units. A mechanical unit consists of relays and ratchets besides a reversible motor and gear trains. In electronic systems dc voltages are used to vary the controlling functions. For example, volume and colour controls are obtained by varying the operating bias of the amplifiers that form part of these circuits. DC control of channel selection makes use of varactor tuning. In later designs of remote control, digital ICs have been used to process the signals received from the transmitting box. In addition, recent developments in high speed IC counters have made the frequency synthesizer approach applicable to television tuning and this has considerably simplified design of remote control units.

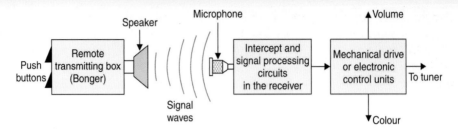

Fig. 27.1. Basic remote control system.

Transmitting Unit

The transmitting unit may be mechanically actuated or electronic in nature. It generates ultrasonic sound waves at frequencies around 40 KHz. In the air these waves are directed towards the receiver where they operate the pick-up transducer which is normally a piezo-electric microphone. The use of frequencies around 40 KHz is advantageous because these are too high for the human ear and can be confined by the walls to the rooms, with very little possibility of interference to receivers in other rooms or beyond the receiver at which they are aimed. Furthermore, it is quite easy to generate acoustic signals by purely mechanical means without the need of a battery or any other powering source.

Control Frequencies

There are no standard values but the frequency range normally employed for remote control functions lies between 37 KHz to 44 KHz. Some typical values for the main functions are:

41.25 KHz–Channel selection

37.75 KHz–Volume up

43.25 KHz–Volume down

The contrast and brightness controls are usually not included for remote control because AGC circuit automatically changes the receiver gain to maintain the desired contrast and brightness in the reproduced picture.

Generation of Frequencies

Since each function needs a distinct signal to actuate it, several frequencies are to be generated to perform different functions. A mechanical source or an electronic oscillator may be used to generate ultrasonic sound waves.

The Bonger

The most common method employed to generate different frequencies by mechanical means is the use of vibrating rods. As shown in Fig. 27.2 the transmitting box (Bonger) contains several cylindrical aluminium rods of slightly different lengths. For each rod a separate control button is provided on the front panel of the hand-held remote control box. When any control button is pushed a spring loaded hammer strikes the corresponding rod and sets it into longitudinal mode vibrations at a definite frequency. The sound energy thus generated is radiated through an open grille provided at the top of the box. For example, an aluminium rod of 7.5 cm length has a fundamental resonant frequency of about 40 KHz and would thus radiate energy at this frequency.

Therefore, to control three or four functions within the television receiver, three or four rods of somewhat different lengths are employed, each actuated by a separate hammer and push button. Once a rod has been struck and the energy transmitted, it is then desirable to dampen the vibrations of the rod as quickly as possible. This is achieved by a small spring which comes into play soon after the rod is set into motion by the hammer on the depression of the push button.

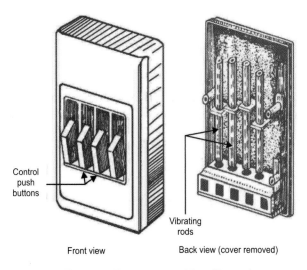

Control push buttons

Vibrating rods

Front view   Back view (cover removed)

Fig. 27.2. Remote control box (Bonger).

Electronic Transmitter

As already stated ultrasonic waves can also be generated electronically. Such an arrangement comprises an electronic oscillator and a loudspeaker. This is shown in Fig. 27.3 where the transistor Q_1, along with the output transformer and feedback network, forms a Hartley oscillator. Note that the transistor base and collector circuits are open when all the switches are open and hence there is no output from the oscillator. When any switch is depressed a different capacitor gets connected across the secondary of the output transformer, which in turn changes the resonant frequency of the oscillator tank circuit. The oscillator output is delivered to a piezo-electric loudspeaker that radiates the ultrasonic sound energy.

Chapter 27

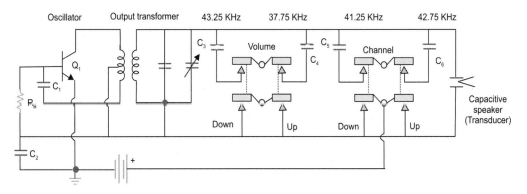

Oscillator   Output transformer   43.25 KHz   37.75 KHz   41.25 KHz   42.75 KHz

Fig. 27.3. Schematic diagram of a four-tone electronic transmitter.
Depression of any button connects a different capacitor across the oscillator tank circuit.

Pick-up Transducer

At the receiver the ultrasonic energy is picked up by a crystal type microphone, which converts sound energy into corresponding electrical signals. The microphone employs a barium-titanate crystal element that has piezoelectric properties. A small bar of this material is placed between two conducting electrodes. When it is mechanically strained by the striking sound waves along one axis a proportionate voltage is generated along the other axis at the frequency of transmitted sound waves. A small 'U' shaped piece of aluminium is added to the assembly. Beyond this assembly and at a distance of approximately one-quarter wavelength corresponding to a frequency of about 40 KHz, a small rectangular horn is attached. Both the one-quarter wavelength space and the horn serve to match the impedance of the barium titanate assembly to the air. The combination of this microphone and the amplifiers in the system provide sufficient sensitivity for operating various controls from a distance of about 7 to 10 metres.

Remote Control Receivers

The remote control receiving systems may be broadly classified as electromechanical and electronic.

27.2 ELECTROMECHANICAL CONTROL SYSTEM

The block diagram of Fig. 27.4 shows three different functions each controlled by an electromechanical circuit. The transmitter has provision to generate three ultrasonic signals having frequencies 41.25 KHz, 37.75 KHz, and 43.25 KHz. The microphone picks up the radiated tone and converts it into an electrical signal. The broad-band (30 to 50 KHz) high gain ($\approx 10^6$) amplifier amplifies this signal and delivers it to an array of tuned circuits. Each tuned circuit is resonant to one of the three frequencies that can be radiated by the transmitter. The output of each circuit is coupled to the armature of a functional relay through a driver. Thus, depending on the function to be performed, the corresponding series tuned circuit develops a large voltage across its inductor. As shown in Fig. 27.5 the tuned circuit L_1, C_1 on receipt of a signal at 37.75 KHz will cause driver transistor Q_1 to pass operating current through the armature of relay K_1. The operation of K_1 connects 220 V ac across the upper winding of the driver motor. At the same time the phase shifting network R_2, C_3, comes in series with the lower winding. The resulting rotating magnetic field develops torque to rotate the motor armature in such a direction that the volume control shaft moves clockwise to raise the sound output. The motor speed is normally reduced to about 10 rmp before coupling it to the associated control potentiometer. When signal at 43.25 KHz is received, relay K_2 will close thereby reversing the direction of rotating field. This will result in the rotation of volume control shaft in the opposite direction and audio output will be lowered. Similarly identical circuits can be added and actuated by different frequency tones to control colour saturation and other functions.

In the case of channel selection by remote control the tuning motor turns the station selector shaft in only one direction. Unused channels can be skipped by indexing or programming an index wheel, that turns with the tuner shaft in such a way, that the motor stops only for the tuner shaft positions of the channels used, but passes through the tuning position of channels not in use. This is easily possible by providing an earth return circuit to the motor at channel locations that are to be skipped.

Chapter 27

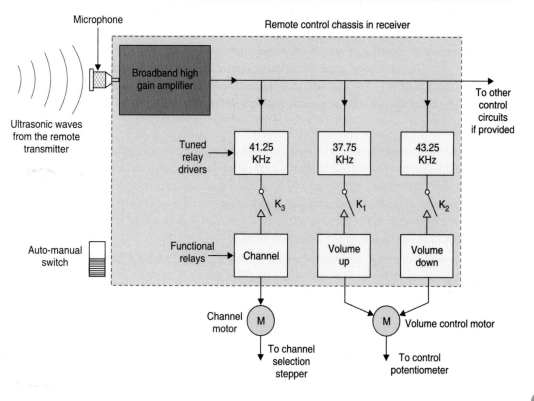

Fig. 27.4 *Block diagram of a remote control system for three receiver functions.*

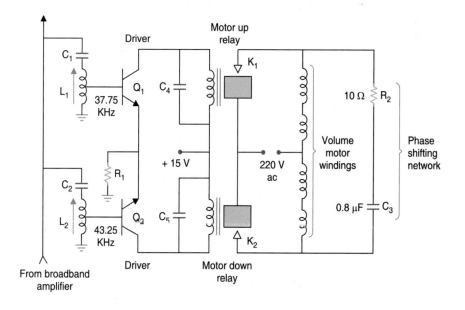

Fig. 27.5 *Volume control motor and its associated circuits.*

27.3   ELECTRONIC CONTROL SYSTEMS

The later versions of remote control either use only one motor or are fully electronic and employ dc voltages to vary the controlled positions. For example, a varactor tuner that does not have a rotating switch needs only a dc potential for channel selection. Similarly other functions can be controlled by changing dc conditions in the controlled circuits.

Memory Circuit

In motor operated remote control systems the levels of volume, colour and channel selected are 'remembered' from one viewing session to the next by the mechanical positions of various controls. In all electronic (motorless) remote control systems a memory circuit is used to 'remember' the control positions established during the last viewing session. If memory is not provided, turning-off the receiver returns all the control voltages to zero and it becomes necessary to readjust all controls each time the receiver is switched-on for use.

The schematic diagram of a memory circuit used to control sound volume is shown in Fig. 27.6. The 'memory module' consists of a neon bulb (NE), a low leakage capacitor C_6 and a MOSFET Q_1. The neon lamp acts as an open circuit unless the voltage across it reaches the ionizing potential. Similarly, the input impedance of the MOSFET, being of the order of 10^8 ohms, behaves almost like an open circuit. Thus any charge once stored in the memory capacitor C_6 will stay on for a long time.

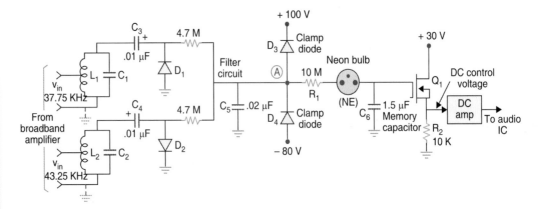

Fig. 27.6. *Memory circuit for volume control in an all electronic remote system.*

A dc control voltage is used to activate the memory circuit. For example, when sound volume is to be increased, the tuned circuit L_1, C_1 couples a large voltage at 37.75 KHz to the diode D_1. The rectified positive voltage is filtered by C_5 and applied to the junction of D_3 and D_4. As soon as this voltage exceeds 100 V, D_3 turns on and clamps the junction point (A) to the external 100 V supply. Thus a 100 V dc source gets applied across the neon lamp and associated circuit. About 80 volts are dropped across the neon diode when it conducts and the remaining 20 volts become available to charge C_6 via R_1. The magnitude of the voltage that builds up across the memory capacitor depends on how long the 'volume up' key of the transmitter is depressed. However, this voltage does not exceed 5 to 6 volts because of the large time constant of the charging circuit. The charge across C_6 can stay on for several months after the receiver is switched off.

The voltage across C_6 is the bias for Q_1, the MOSFET memory follower. The voltage across R_2, the source resistance of Q_1, is proportional to the voltage across C_6. Thus the voltage change caused by the depression of 'volume up' button gets applied to the dc amplifier which in turn changes the bias on the audio IC to increase power delivered to the loudspeaker.

In order to decrease the volume, the transmitter is activated to produce a tone at 43.25 KHz. This is selected by the filter L_2, C_2 and rectified by diode D_2. Since D_2 is connected in opposite direction from that of D_1, a negative control voltage is developed across the memory capacitor. Note that the negative clamp voltage (-80 V) is lower than the positive clamp so that the FET gate voltage does not exceed its cut-of voltage.

The decreased output voltage across R_2 results in a change of bias voltage to the audio IC in the opposite direction thereby reducing sound output from the loudspeaker. Similarly the output across R_2 may be utilized for operating a one shot-flip-flop, that energizes a relay to accomplish the on-off function.

For controlling other functions identical circuitry is provided and its output is fed to corresponding circuit points in the receiver. Note that for electronic tuning several input circuits would be necessary to charge the memory capacitor to different voltages for each channel.

Remote Control Circuits Employing Digital ICs

Now, when a large variety of digital ICs are available at a reasonable cost, many manufacturers of remote control equipment have introduced new control circuits, where almost all electromechanical units have been replaced by electronic circuitry employing ICs. As an illustration, receiver's 'VOLUME-UP' and 'VOLUME-DOWN' functions when controlled electronically operate in the following manner:

When the receiver is off and a remote 'on-volume-up' signal is received, a digital keyer starts counting pulses from a slow rate digital clock provided in the same IC. Fifteen discrete counts complete the sequence of switching on the receiver and raising the sound volume to its maximum level. The digital numbers, as counted by the keyer, are converted to discrete voltage levels by a D/A (digital-to-analog) converter. The voltage level that becomes available after count number one on a particular pin of the IC, is used for switching-on the receiver. In effect, this control voltage (dc) turns on a transistor, that is initially off, to feed current to a lamp that is connected in its emitter circuit. The lamp illuminates a light dependent resistor (LDR) that forms part of the gate circuit of a thyristor. When the lamp turns on, the resistance of the LDR decreases to allow enough gate current and the thyristor is turned on. The thyristor is in series with the mains supply circuit to the receiver and as soon as it turns on, power is switched on to the receiver.

As long as the 'on-volume-up' control is kept depressed the sound volume increases in steps. At each successive count (1 to 14) the control voltage increases to initiate action that results in increase of volume. Normally, volume changes proceed at the rate of three levels per second. To turn the receiver off, the 'volume-down-off' switch is depressed. The keyer counts up till count 15 is reached. At this stage the keyer is timed to be gated out, which in turn results in cutting off supply to the receiver.

All Electronic System

As explained earlier, the command signals at different frequencies are separated by tuned circuits and then routed to the appropriate circuitry for performing corresponding functions. However, in an all electronic system, this function is performed by a specially designed IC called a 'digital decoder'. The internal circuitry of this IC contains a counter that is turned on, on receipt of any command signal. It is designed to count any pulses present, for a predetermined count period normally 1/50th of a second, and compare this with a previous count. In fact the incoming signal is counted seven times. If all values agree within prescribed tolerance, the counter considers the input valid and a confidence counter is incremented by one. When confidence counter reaches seven, the voltage on the IC pin, corresponding to the frequency count, is activated and performs the required function. Thus different voltage will appear at corresponding pins to perform different functions. All the command frequencies are exact harmonics of 50 Hz. This harmonic relationship provides noise immunity by preventing erroneous triggering.

27.4   ELECTRONIC TOUCH TUNING

Recently several new approaches to the design of TV channel tuning have emerged. The ideal tuning system should incorporate (*i*) individual channel selection, (*ii*) no channel alignment, (*iii*) digital channel display, (*iv*) remote control and (*v*) touch type control. The only feasible approach which can meet the above requirements is an all electronic tuning system. It employs a varactor tuner and a totally electronic system for channel selection, tuning and identification.

The requirement of parity tuning for both VHF and UHF channels has led to the use of digital and dedicated ICs in such control circuits. Some designs employ micro-processors for initiation, control, selection and interface purposes, However, it is not necessary to provide tuning for all the 80 and odd UHF channels because more than six such channels are seldom available in any area or location. The more popular designs centre around approximately 18 channels–the twelve standard VHF channels and about six variable UHF channels which are selected on the basis of local area requirements. In one typical design the VHF potentiometers are set to correspond to a given channel readout. The UHF channels, however, are tuned in by the potentiometers and the corresponding channel indicator number is set to agree with the channel tuned with a single patchboard switching matrix.

Channel Selection

Channel entry is by means of a touch channel selection keyboard called *touch tuning*. It uses very low pressure sensitive switches arranged much the same way as in a pocket calculator keyboard. The basic organization of the keyboard circuit is shown in Fig. 27.7. With an inactive keyboard the input to Q_1 is low and the oscillator is disabled. When a key is depressed, the base of Q_1 is pulled high (forward biased) via varactor potentiometers connected to + 30 V. This actuates the gate in the oscillator circuit and allows it to oscillate. The oscillator output pulses cycle the counter until it addresses the decoder output (active low) of the depressed key. As this key decodes the output becomes low and the oscillator is stopped via Q_1. The associated circuits are programmed for individual channel selection.

The counter in this circuit can be sequentially advanced for external sources such as a remote control tuner by gating a single pulse to the counter.

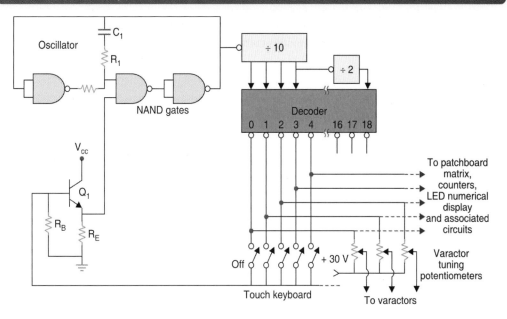

Fig. 27.7. Basic keyboard organization of an 18-channel electronic 'Touch Tuning' channel selection system.

Channel Indication

The decoder (details not shown) simultaneously triggers a counter which counts in the binary system. The output of the binary counter is fed to a decimal decoder that selects the proper drive according to the equivalent decimal number. The channel number is then displayed by a readout unit on the channel indicator. Tuning systems which employ microprocessors have the necessary interface circuitry to generate corresponding video signals. The channel number is thus displayed on the picture tube screen after channel selection. Such a provision is known as 'direct address package' (DAP). In some designs, the time of the day is also displayed along with the channel number in one corner of the picture tube.

27.5 FREQUENCY SYNTHESIZER TV TUNER

Recent developments in high speed IC counters have made the frequency synthesizer approach practicable for television tuning. A frequency synthesizer TV tuner is a closed loop system, self compensating for varactor tolerances and component drifts. It provides easy interface with digital displays and simplifies remote control design. It is highly reliable and has no alignment problems.

Tho block diagram of Fig. 27.8 illustrates the frequency synthesizer system for electronic channel selection. The dotted chain block represents a standard VHF or UHF varactor tuner. The output of the local voltage controlled oscillator (VCO) is amplified to an acceptable level for driving a digital emitter coupled logic (ECL) prescaler (divider). The purpose of the prescaler is to divide down the local oscillator high frequencies to a range that can be processed and counted by standard TTL logic. The output of the prescaler is fed to the programmable divider which is the heart of the system. Its output is passed on to a frequency and phase comparator where it is

Chapter 27

compared with a crystal controlled reference frequency. Then the comparator output is fed to an integrator-amplifier, which in turn provides voltage control to the varactor tuner input.

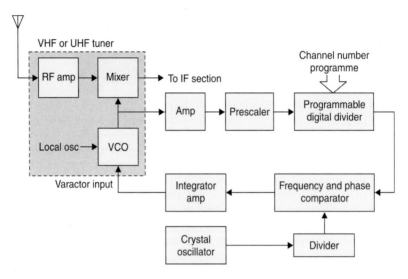

Fig. 27.8. *Block diagram of a frequency synthesizer TV tuner.*

The divide ratio of the programmable counter determines the desired channel. At the comparator, there are two significant signals–the counted output and the reference signal. When these two signals differ in frequency or phase, a dc error voltage is applied to the tuner VCO with a suitable polarity in order to reduce the frequency and phase error to zero. At this time the VCO is at the selected channel frequency. A touch tuning type entry system is used for channel selection and a modern seven segment display is employed for channel indication.

27.6 AUTOMATIC FINE TUNING (AFT)

The local oscillator frequency in the RF tuner is set to provide exact IF frequencies. However, despite many remedial measures to improve the stability of the oscillator circuit, some drift does occur on account of ambient temperature changes, component aging, power supply voltage fluctuations and so on. For a monochrome receiver a moderate amount of change in the local oscillator frequency can be tolerated without much effect on the reproduced picture and sound output. The fine tuning control is adjusted occasionally to get a sharp picture. The sound output is automatically corrected because of the use of intercarrier sound system. However, requirements of frequency stability of the local oscillator in a colour TV receiver are much more stringent. This is so, because proper fine tuning of colour receivers is judged both by the sharpness of the picture and quality of colour.

If the local oscillator frequency becomes higher than the correct value, the picture IF, subcarrier IF and sound IF frequencies will also become higher by the same amount and fall on incorrect positions on the IF response curve. This will result in poor picture quality because the amplitudes of low-frequency video sidebands clustered around the picture IF will decrease. At the same time chrominance signal sidebands clustered around the location of subcarrier will receive more gain and hence become stronger than normal. Similarly, if the local oscillator frequency changes to become less than the desired value opposite effects would result and colour reproduction will become weak. In case the decrease in frequency is more than 1 MHz, the colour burst may be lost and only a black and white picture will be seen on the screen. Similar troubles can also result in receivers that employ remote control tuning because of non-availability of fine tuning control and imperfections of the mechanical/electronic system employed for channel selection.

Automatic Frequency Control (AFC)

In order to simplify the operation from the point of view of setting the fine tuning control correctly, and to overcome the problem of local oscillator drift, all colour receivers and those employing remote control use an AFT circuit.

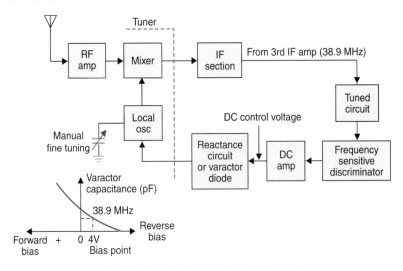

Fig. 27.9 Block diagram of an AFT circuit.

The AFT circuit is actually automatic frequency control (AFC) on the local oscillator in the tuner. This control aims at obtaining a picture IF frequency of exactly 38.9 MHz at the converter output. This is possible only if the local oscillator frequency in the tuner is maintained at a value which is exactly 38.9 MHz higher than the incoming channel carrier frequency. To achieve this, the IF frequency, as obtained from the IF amplifier section, is measured by a discriminator circuit that forms part of the AFT control. The block diagram of such an AFT circuit is illustrated in Fig. 27.9. Here the discriminator has basic function similar to that described in Chapter 21 for FM sound detection. Its output is a dc correction voltage that indicates the deviation of the IF frequency from its exact value of 38.9 MHz. Since this frequency depends on the oscillator input to the mixer, the AFT voltage indicates the error in local oscillator frequency. The 'S' shaped response curve of the discriminator is very steep and frequency shifts as little as 50 KHz are clearly indicated. The AFT control voltage is zero at balance i.e. when the IF frequency is exactly 38.9 MHz. For frequency deviations on either side of 38.9 MHz the net dc correcting voltage is either positive or negative depending on whether the frequency, is above or below the correct value. The dc control voltage thus developed is applied to a reactance circuit or to a varactor diode that forms part of the local oscillator tank circuit. The junction capacitance of the varactor diode varies with the applied dc control voltage and thus changes the resonant frequency of the oscillator tuned circuit to shift the frequency to the correct value.

Chapter 27

AFT or AFC Circuit

A typical automatic frequency control circuit together with the discriminator characteristics is shown in Fig. 27.10 (a). Signal from the IF amplifier section of the receiver is coupled by C_1 and L_1 to the base of transistor Q_1. This stage is an IF amplifier that couples its output to the 'Foster-Seeley' or phase shift discriminator. The transformer T_1 is tuned to 38.9 MHz. When the input signal is at exact IF frequency the control voltage developed across the output terminals A and B is zero. However, if the local oscillator frequency is not at its nominal value for the particular selected channel, the picture IF frequency will be either above or below 38.9 MHz. In this event the discriminator output voltage will be either a positive or a negative dc voltage. In some circuit designs this is amplified by a dc amplifier before being applied to the varactor in the tuner assembly. A typical value for the AFT error voltage is ±6 volts and can be measured across the output points A and B.

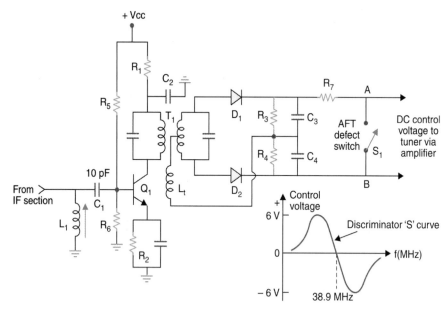

Fig. 27.10 (a). A typical automatic frequency control (AFC) circuit.

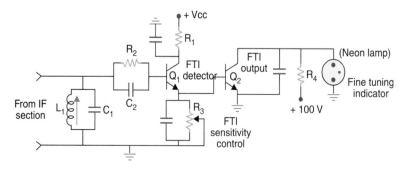

Fig. 27.10 (b). A typical fine tuning indicator (FTI) circuit.

AFT Defeat Switch

The switch S_1 shown across the output terminals is known as the 'AFT defeat switch' and can be closed to turn off the AFT circuit for manual fine tuning adjustments.

FTI Circuit

In some AFT circuits a fine tuning indicator (FTI) is provided to give a visual indication of any incorrect fine tuning. A neon lamp fixed at a suitable location on the front panel lights up to indicate that tuning is 'off' and must be corrected. A simplified circuit of such a control is shown in Fig. 27.10(b). A sample of the IF signal, after one stage of amplification if necessary, is applied across the parallel resonant circuit L_1, C_1 that is tuned to the IF frequency. The voltage developed across the tuned circuit is fed to the base of transistor Q_1 through a self-bias network R_2, C_2. The voltage developed across the emitter resistor R_3 drives the output transistor Q_2. The transistor Q_1 is connected as an emitter follower and provides both current gain and isolation between the tuned circuit and low input resistance of the output transistor. In the absence of any input signal the transistor Q_2 stays at cut-off and full 100 V dc gets applied across the neon lamp which is enough to ionize the gas in it. When the IF frequency is at its correct value of 38.9 MHz, maximum voltage is developed across the tuned circuit. This results in enough drive voltage across R_3 and transistor Q_2 goes into full conduction. The resulting collector current reduces the voltage across the neon lamp below its ionizing potential and it goes off. If the tuning is off the input to the base circuit of Q_1 decreases because, when the parallel resonant circuit does not receive a 38.9 MHz signal its impedance drops considerably thereby shunting the base input of Q_1. This results in reduced input to Q_2 and it goes to cut-off. Now there is no shunting effect across the indicating lamp and it lights up. Thus when the indicator light is on it is a sign that the fine tuning is off and needs adjustment.

27.7 BOOSTER AMPLIFIERS

As radio communication systems developed the receivers became more and more sensitive. Early TV receivers needed 200 to 300 μV of signal for a picture of reasonable quality. The present day sets give a similar picture with an antenna signal strength of about 50 μV in the lower VHF range.

In areas where the signal strength is somewhat less, raising the antenna or using an antenna that is directional and has higher gain results in an acceptable picture. However, in deep fringe areas where the signal from the desired station is very weak and fails to produce any worthwhile picture, an additional RF amplifier external to the receiver becomes necessary. Such amplifiers are known as booster amplifiers and are normally mounted on the antenna mast close to the antenna terminals. A booster amplifiers is a broad-band transistor RF amplifier designed to have a reasonable gain but a very high internal signal-to-noise ratio. It may be emphasized that a booster capable of providing a high gain but incapable of providing a good signal-to-noise ratio will give a picture with lot of snow. Similarly, a booster amplifier having minimum internal noise but low gain will fail to provide a satisfactory picture. Thus a booster amplifier must have both the attributes, *i.e.* reasonable gain and high signal-to-noise ratio.

In booster amplifiers that employ only one transistor, either a grounded-emitter or a grounded-base configuration is used. Though the grounded-emitter arrangement gives a somewhat higher gain the grounded-base configuration is more stable and is therefore preferred. Such amplifiers are untuned and are designed to give a small boost to signals from several nearby stations. However, when the interest is in a single station a tuned amplifier can be used to yield higher gain.

The circuit diagram shown in Fig. 27.11 is that of a booster amplifier specially designed to receive channel 4 (61 to 68 MHz) station in deep fringe areas. It gives a gain of about 20 db, with a noise figure between 2.00 to 2.50 db. The first stage is designed with a minimum noise figure (NF) criterion. The second stage is designed to give necessary gain and the last stage is for matching its output impedance to the lead-in cable. Between the first and second stages a tuned LC circuit has been provided for proper tuning of the frequency band of 61 to 68 MHz. A brass core is used for tuning this circuit. The small amount of power required by the booster amplifier is fed through the transmission line that couples the antenna to the receiver and is switched on along with the receiver.

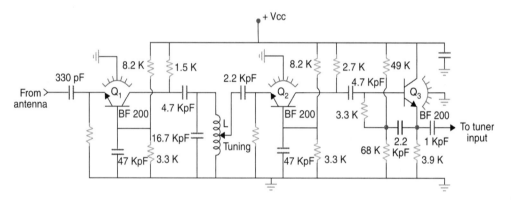

Fig. 27.11. Booster amplifier circuit for channel 4 (61-68 MHz).

27.8   AUTOMATIC BRIGHTNESS CONTROL

The amount of light in the room where the TV set is in use determines the setting of the brightness control for a satisfactory picture. The contrast control is also linked, to some extent, with the brightness setting and often needs readjustment when the brightness control is varied. The brighter the room the greater is the need to advance the settings of both brightness and contrast controls for a bright picture and effective reproduction of the various shades of grey in the scene.

A circuit which automatically accomplishes this is known as 'automatic brightness control' (A.B.C.). It forms a part of the video amplifier circuitry. A typical circuit arrangement is illustrated in Fig. 27.12. The controlling element is an LDR (light dependent resistance) which is actually photo-sensitive material whose resistance varies with changes in light intensity falling on it. The LDR is mounted on the front panel of the cabinet to sense changes of illumination in the room. It forms a part of the video amplifier circuit that controls the screen-grid voltage of the video amplifier tube and the bias voltage (grid to cathode potential) of the picture tube. As shown in Fig. 27.12 resistor R_1 (which is in parallel with the LDR) and resistor R_2 form a potential divider across the B+ supply to feed necessary dc voltages both at G_2 of V_1 and control grid of the picture tube. When illumination in the room increases, say, when lights are switched on or during daylight, the resistance of the LDR decreases. Because of the potentiometer arrangement this increases the screen grid voltage to tube V_1. The increased screen grid voltage provides greater contrast because of enhanced amplification of the composite video signal. At the same time the increased potential also appears at the grid of the picture tube. This decreases the net dc negative bias, between the grid and cathode of the picture tube, which in turn results in increased screen brightness. For a decrease in light intensity the LDR resistance increases and the resultant fall in the video amplifier screen grid and picture tube grid voltages decrease both contrast and brightness in proportion to changes in the general illumination of the room.

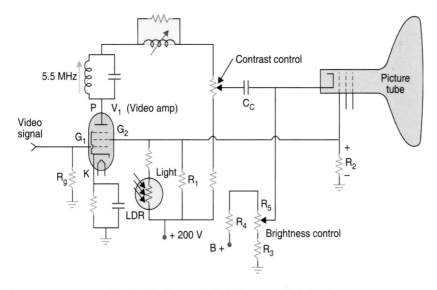

Fig. 27.12. Automatic brightness control circuit.

In video amplifiers that employ transistors, the LDR is placed in parallel with the emitter resistance. Any change in the effective emitter resistance, due to changes in room illumination, alters both the gain and quiescent dc potential at the collector, which is directly coupled to the picture tube circuitry. Thus any variation in brightness automatically affects both the contrast and brightness of the reproduced picture. In some designs a switch is provided which disables the 'ABC' circuit if necessary.

27.9 INSTANT-ON CIRCUITRY

In television receivers that employ tubes there is some delay in reception after the switch is turned on because of the warm-up period required by the picture and other tubes before they assume normal operation. Though transistors virtually require no warm-up time, in receivers which employ all transistors there is also a small delay in reception because of the time required by the picture tube cathode to attain full emission. To overcome this delay and permit instantaneous reception when the receiver switch is turned on, the filament circuitry is modified. The circuit of Fig. 27.13 (a) shows power supply arrangement of a all transistor television receiver. The diodes D_1 and D_2 connected to the secondary winding L_2 together with the filter network meet dc voltage requirements of all the stages. However, a separate low voltage winding L_3 is provided for the filament (heater) of the picture tube. To provide instant-on feature the master switch S_1 on the primary side of the mains transformer is kept closed all the time unless the set is not to be used for several days or still longer periods. When the double pole single-throw (DPST) switch S_2 is in the open position, the centre tap of winding L_1 remains disconnected and thus no dc voltage is produced by the full-wave rectifier circuit. However, in the filament winding the switch is shunted by diode D_3 and about one-third of the normal filament current continues to flow through the picture tube heater, even when S_2 is off. This is enough to permit the picture to come on almost instantly when S_2 is closed, which in its 'on' position shunts D_3 and permits application of full filament voltage to the picture tube.

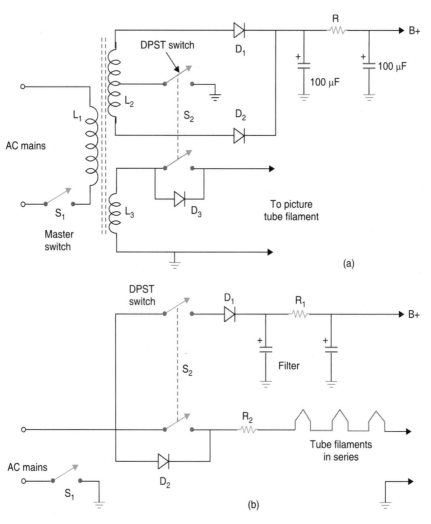

Fig. 27.13. Instant-on circuits (a) power supply employing mains transformer (b) direct on line power supply.

For tube-type television receivers that do not employ mains transformer the circuit arrangement employed to provide instant-on feature is shown in Fig. 27.13 (b). As shown there, the upper switch of the DPST assembly, when off, disconnects power from the dc voltage rectifier circuit while the lower switch permits partial supply to the heater string through D_2 that shunts it. When switch S_2 is thrown on, both dc supply and full filament current are immediately restored and the picture appears on the screen without any warm-up delay. Though the receiver, when put under instant-on facility (master switch closed), consumes about 30 watts of power there is no sudden surge of current through the filaments when the receiver is turned on. This results in extended life of the tubes. Besides this, the heat that is generated within the cabinet when the set is not in use (S_2 off) helps reduce undesired effects of dampness on various windings and components in areas of high humidity.

27.10 PICTURE-TUBE BOOSTERS

After a prolonged use the emission from the cathode of the picture tube declines and the reproduced picture exhibits symptoms of low contrast and reduced brilliancy. To extend the life of such picture tubes their filament voltage is increased which in turn raises the heat applied to the cathode, so that the latter then assumes normal emission. In some receiver designs a built-in provision is made for raising the filament voltage from its normal value of 6 volts to about 8 volts. This is shown in Fig. 27.14 where the 'brightness' screw provided with the receiver can be fixed in the socket to bypass the 3 ohm resistor normally used to initially keep the voltage at 6 volts at the heater terminals. In the absence of such a provision a separate booster transformer can be used instead of the 6 volts winding on the mains transformer. A suitable arrangement can also be provided in receivers that have all the filaments (heaters) in a string across the mains supply. It has been observed that the reactivated life of the picture tube gets extended by several years with the use of such boosters.

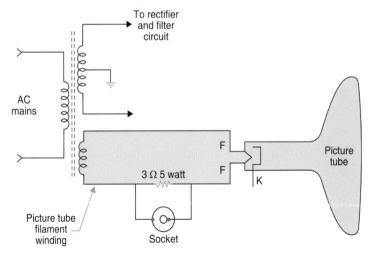

Fig 27.14. Picture tube booster arrangement.

Chapter 27

REVIEW QUESTIONS

1. What are the main components of a television remote control system ? Describe by a block diagram how such a system functions.

2. Why are the frequencies of command signals chosen in the 40 KHz range ? Explain how are these generated both by a mechanical 'bonger' and an electronic oscillator.

3. Explain briefly the transducers used for transmitting and receiving command signals. How are the command signals separated at the receiver ?

4. Describe with suitable diagrams how (i) remote volume control and (ii) remote channel selection are carried out on receipt of command signals. Explain how a memory circuit is used for channel selection in a varactor controlled tuner.

5. Enumerate the needs an ideal electronic channel tuning system should meet. Explain briefly the "Touch Tuning" method of channel selection and identification.

6. Describe with a suitable block diagram the basic principle of a frequency synthesizer TV tuner. Explain how the selected channel number is displayed on the picture tube screen.

7. Explain with a circuit diagram the functioning of an automatic fine tuning control. Describe how the fine tuning indicator circuit operates to give indication of incorrect fine tuning.

8. Write descriptive notes on the following :

   (i) booster amplifiers

   (ii) automatic brightness control

   (iii) instant-on circuits

   (iv) picture tube boosters.

Chapter 27

28

Alignment and Servicing Equipment

The alignment procedure and servicing techniques necessary for television receivers are far more exacting than the methods commonly employed for tuning and repairing broadcast receivers. Thus, while a cathode ray oscilloscope (CRO) and a vacuum tube voltmeter (VTVM) may be considered a luxury for radio receiver work, it is not so for television receiver manufacture and servicing. In order to meet particular needs of TV receivers, special equipment and versions of basic electronic instruments have been developed. These include a sweep-cum-marker generator, a video pattern generator, a wobbuloscope and oscilloscopes having special facilities. In addition to this, various probes are available for high frequency work and high voltage measurements.

Most of these instruments are beyond the reach of a common service engineer because of their prohibitive cost. He may have to manage with just a pattern generator, a multimeter and a few other assorted tools. In any case no instrument can be fully effective unless its principle of operation, use and limitations are fully understood. It is all the more necessary for a technician who cannot afford to possess costly equipment to fully acquaint himself with the operation and use of the instruments he possesses. Therefore, this chapter is devoted to basic principle and applications of all important testing instruments. The alignment and servicing of television receivers and discussed in the next two chapters.

28.1 MULTIMETER

This is a commonly used instrument for measuring voltages, currents and resistances in electronic circuits. It employs a sensitive permanent magnet moving-coil movement for indicating the quantity under measurement. Various switches are provided on the front panel of the instrument for converting it into a multirange voltmeter, milliammeter or ohmmeter. On ac ranges, semiconductor diodes convert alternating voltages and currents to corresponding dc quantities before measuring them.

Voltage Measurements

Almost all electronic circuits are high impedance low power configurations. Therefore, it is essential that the voltmeter when connected across the circuit should not load it. This is only possible if its impedance (resistance) is much higher than the circuit impedance. To achieve this, movements which draw only 50 to 100 μA for full scale deflection are employed in multimeters. The input resistance of a voltmeter is defined as the resistance the meter offers per volt of the value indicated by the range of the meter. Thus a voltmeter having an input resistance of 10 K ohms per volt will have an input resistance of (250 × 10 K) 2500 K, when set on 250 V range. A sensitivity of the order of 20 K ohms/volt is essential for reasonably accurate measurements in television receivers. For example in AGC voltage measurements, a meter having low input resistance will not only yield erroneous readings but also upset the prevailing circuit conditions.

The meter has less input resistance on corresponding ac ranges because more ac input is necessary to provide equivalent dc current for meter deflection. A multimeter can measure ac voltages upto about 50 KHz only. Beyond this frequency the lead and circuit capacitance together with limitations of the meter bridge rectifier affect the accuracy considerably. All meters are normally calibrated to read r.m.s. values of the applied ac voltage.

Current Measurement

The same movement is used for voltage and current measurements. Since the movement can carry only a very small current, shunts are provided to extend its range. The values of shunt resistances are so chosen that for the current indicated on the range switch, the meter draws full-scale deflection current. For ac current measurements, a bridge rectifier converts ac to dc before applying it to the same circuit configuration.

Resistance Measurement

A commonly used ohmmeter circuit employing the same movement is shown in Fig. 28.1. The terminals marked A and B are initially shorted and R_2 (zero adjustment) varied till the meter records full scale deflection. This corresponds to zero external resistance. The unknown resistance is connected across the same two terminals. The current indicated by the movement is inversely proportional to the value of resistance under measurement. Thus the current magnitudes can be calibrated to indicate resistance values. With terminals A and B open, no current flows through the meter and zero current corresponds to infinite resistance or open circuit. Since the circuit current is proportional to $1/R_x$, the ohms (resistance) scale is highly non linear. On account of excessive scale crowding near the zero current value, higher value resistances cannot be accurately measured unless a separate range is provided. This requires higher battery voltage in the meter.

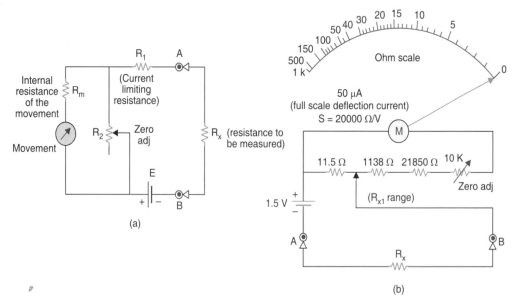

Fig. 28.1. Ohmmeter section of a multimeter (a) basic circuit (b) actual circuit configuration and scale.

28.2 VACUUM TUBE VOLTMETER (VTVM)

A VTVM has a very high input impedance so that its loading effect is much less. It is very useful for measurement of dc voltages across high resistance circuits and ac voltages at higher frequencies. VTVMs are generally of two types:

(i) Rectifier-Amplifier Type

The schematic diagram of Fig. 28.2 illustrates the principle of this type of voltmeter. At its input a rectifier converts the ac voltage under measurement to a proportionate dc voltage. The RC time constant of the load across the diode is made very large as compared to the period of the input ac signal. The dc output voltage developed across the combination is nearly equal to the peak value of the positive half-cycle of the input wave. The dc voltage thus obtained is amplified by a direct coupled amplifier before feeding it to the movement. To limit the magnitude of input voltage to the dc amplifier a multi-step attenuator is used between the rectifier circuit and amplifier. This attenuator serves as a range selector.

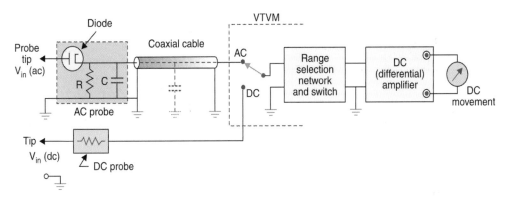

Fig. 28.2. Block diagram of a rectifier amplifier (peak-reading) type of VTVM.

The load resistance R across the rectifier can be made very high to obtain an input impedance of the order of tens of mega-ohms. The meter draws very little current because the voltage across the RC combination stays almost equal to the positive peak value of the input signal. The power required by the movement to cause deflection is provided by the dc power supply that feeds the amplifier. Though the meter measures peak value of the input voltage it is calibrated to indicate r.m.s. value of the sinusoidal input voltage. For inputs other than sinusoidal a correction factor would be necessary to know the correct amplitude of the signal under measurement. Since a dc amplifier cannot provide large gain the sensitivity of this voltmeters is limited to about one volt full scale deflection. It can measure as voltages at frequencies up to 50 MHz. Its range can be extended to more than 100 MHz with the help of a special probe where the rectifier block is shifted closer to the point of measurement to ward off cable capacitance effects. This is illustrated in Fig. 28.2.

DC voltage measurement. The section comprising of range selector switch, dc amplifier and movement of the rectifier amplifier VTVM is essentially a dc voltmeter. A separate probe with an isolating resistance is provided for dc voltage measurements. An ac/dc switch is used for change over from ac to dc voltage measurements. The input resistance for dc voltage measurements is made large by using a high impedance circuit at the input terminals of the direct couple amplifier. An input resistance of the order of several mega-ohms can easily be obtained by this arrangement. Thus a rectifier-amplifier VTVM can be used both for ac and dc voltage measurements.

(ii) Amplifier-Rectifier Type

In the amplifier-rectifier type of VTVM, the ac voltage to be measured is first amplified by a wide-band ac amplifier stabilized by negative feedback. The output from the amplifier is rectified and the resulting direct current is used for deflection on the dc movement. Figure 28.3 shows the basic circuit arrangement of this VTVM. Because of rectification this type of VTVM gives an indication proportional to the average amplitude of the input wave. However, the scale is calibrated to read r.m.s. values.

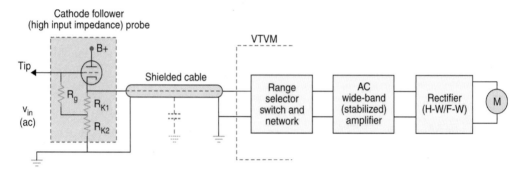

Fig. 28.3. *Block diagram of an amplifier-rectifier (average reading) type of VTVM.*

Since reactive coupling is used in the amplifier the frequency range is limited to about 10 MHz. A probe containing a cathode follower or a source follower when used along with this meter raises the input impedance of the voltmeter to several megaohms. Since an ac coupled amplifier can easily provide a large gain, a sensitivity of the order of 1 mV full scale deflection can be easily obtained. Thus, a very high sensitivity, high input impedance and fairly broad frequency range make this VTVM very useful for ac voltage measurements in audio and video circuits. This meter is also know as *AC* milli-voltmeter. Note that there is no provision for dc voltage measurements in this type of electronic voltmeter.

Resistance measurement with a VTVM. All rectifier-amplifier type VTVMs have a provision for measurement of any unknown resistance. This is done by placing the unknown resistance in series with an internal battery and a resistive network which serves as the range selector for resistance measurements. A separate switch connects this network to the input of the dc amplifier. The voltage developed across the resistance to be measured is amplified and interpreted as resistance on the corresponding scale. Note that the resistance scale in a VTVM reads increasing resistance from left to right which is opposite to the scale markings on a multimeter. This is so because a VTVM reads a higher resistance as a higher voltage whereas a multimeter indicates a higher resistance as a smaller current.

Some VTVMs incorporate a facility to measure dc current besides voltage and resistance measurements. This is achieved by converting current to a corresponding voltage drop and then measuring it by the dc VTVM. The scale, however, is calibrated to indicate input current magnitudes.

(iii) Electronic Multimeters

Electronic multimeters are new type of VTVMs which employ transistors, FETs and other semiconductor devices instead of vacuum tubes. Such electronic meters are easily portable like the conventional multimeters because of low voltage (battery) dc supply requirements. They have the same building blocks, *i.e.* measuring circuits, amplifier circuits and an analog meter movement. The input impedance is usually very high (≈ 11 M) and is same for all the voltage ranges.

28.3   DIGITAL MULTIMETERS

A digital multimeter (DMM) displays voltage, current or resistance measurements, as discrete numerals instead of a pointer deflection on a continuous scale as in an analog meter. Numerical readout is advantageous because it reduces parallax, human reading and interpolation errors. Features such as autoranging and automatic polarity further reduce measurement errors and any possible instrument damage through overload or reversed polarity.

While there are numerous techniques used in DMMs, the basic principle of operation is the same. The electrical quantity to be measured must be converted to a number of pulses proportional to the unknown voltage. These pulses are counted and then displayed on a seven-segment display.

The basic circuit for dc voltage measurements is illustrated in Fig. 28.4. The input voltage V_x generates a pulse whose width is proportional to this voltage. The leading edge of the pulse allows the clock pulses to go to the counter, and the trailing edge stops them. The number of pulses counted will be proportional to the voltage under measurement. Referring to Fig. 28.4, $T = KV_x$ where K is a constant of proportionality. If $V_x = 7.388$ V and $K = 0.001$, then $T = 7.388 \times .001 = 7.388$ ms. If the clock or oscillator has a frequency $f = 1000$ KHz, then the number of pulses 'n' that pass to the counter are : $n = T \times$ number of pulses per ms. Therefore, $n = 7.388 \times 10^3 = 7388$. On a four-digit meter, the reading will be 7.388. The decimal point is determined by the range selector unless there is an autoranging in which case internal circuitry makes the determination.

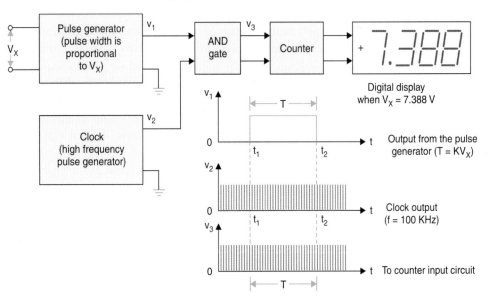

Fig. 28.4. *Functional diagram of a digital voltmeter.*

DMMs are becoming increasingly popular as their price has become more competitive.

28.4 CATHODE RAY OSCILLOSCOPE (CRO)

A CRO provides a visual display of any time varying electrical signal and so it may be regarded as a fast *X-Y* recorder where the electron beam of the cathode-ray tube is the pencil and the phosphor coating on its screen is the sheet on which the plot gets drawn. The electron gun in the C.R. tube has external provision both for focusing the beam to produce a well-defined trace and controlling its intensity for varying the brilliancy of the trace which appears on the screen. These controls are provided on the front panel of the scope.

Two pairs of deflecting plates mounted at right angles to each other, ahead of the electron gun, provide deflection of the electron beam both along the horizontal (*X*) axis and vertical (*Y*) axis of the screen. Electrostatic deflection is employed. By varying dc potential across the deflecting plates the spot can be shifted anywhere on the screen. To achieve this, two additional controls known as 'horizontal shift' and 'vertical shift' are located on the front panel of the oscilloscope.

The deflection sensitivity (volt/cm of deflection of the electron beam) is very low and so amplifiers are always used to achieve the desired deflection. Gain controls for both horizontal and vertical deflection amplifiers are located on the front panel to facilitate adjustment of breadth and height of the trace obtained on the screen.

Time-Base

A large proportion of scope applications require a time-base that produces linear motion of the beam along the horizontal axis. This is achieved by applying a sawtooth sweep voltage across the horizontal pair of deflecting plates. This is generated within the scope by a special circuit. The ramp of the sweep voltage produces a bright solid line on the screen while its sudden fall after the trace period results in a quick retrace or flyback. Auxiliary circuits are used to blank the retrace.

All general purpose scopes employ a cathode coupled multivibrator circuit to generate sawtooth voltage. The 'OFF' and 'ON' periods of a multivibrator are used to control the charge and discharge of a capacitor to develop sawtooth voltage. Since the voltage developed across the capacitor is very low, it is amplified by the horizontal amplifier before applying it to the horizontal deflecting plates. Thus the gain control of the horizontal amplifier serves to change the horizontal scale of the displayed pattern. The sweep circuit (oscillator) has two controls for setting the sweep frequency —one for coarse adjustment in discrete steps and the other for fine adjustments between any two coarse steps. This is achieved by changing the time constant of the sweep circuit. These two controls (coarse timebase and fine frequency) are also available on the front panel as shown in Fig. 28.5.

Visual Display of Signals

Simultaneous application of a time varying voltage to the vertical deflecting plates (direct or through the amplifier) and a sweep voltage to the horizontal deflecting plates results in a trace on the screen of the signal applied to the vertical input terminals. If the sweep frequency is a sub-multiple of the frequency of the signal applied to the vertical deflecting plates, the spot will follow identical paths on successive sweeps and the pattern will appear as a stationary plot of the unknown waveform. The number of cycles of the applied signal which appear on the screen depend on the ratio of the vertical signal frequency and the sweep frequency. If the unknown signal frequency is not an exact multiple of the sweep frequency the pattern moves back and forth on the screen. However, the pattern can be made stationary by manipulating the fine frequency control.

Chapter 28

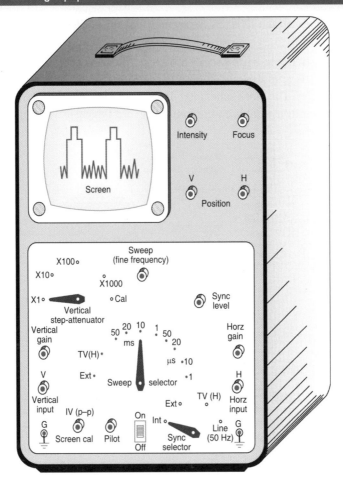

Fig. 28.5. Front panel view and operating controls of a general purpose oscilloscope.

Synchronization

To keep the pattern locked at the set rate, a part of the vertical input signal is processed and applied to the input of the sweep generator to control its frequency, so that the two remain in step. This is termed as 'internal synchronization'. A switch on the front panel permits the application of either internal or external synchronization. Such a provision is helpful while aligning television receivers with a sweep generator.

In some oscilloscopes a provision for 'trigger mode' sweep is also available. This enables precise locking of the time base with any portion of the input signal. Such a facility is helpful for studying one-shot outputs or transients.

Special Time-Base Circuits

Oscilloscopes suitable for TV manufacture and servicing include additional special sweep circuits which generate horizontal sweep (sawtooth) voltage at 25 Hz and 7812.5 Hz. These when triggered from the video input signal applied at the vertical input provide a stationary trace of two successive vertical or horizontal cycles of the video signal on the screen. These special time base modes are

Chapter 28

termed as TVV (TV Vertical) and TVH (TV Horizontal) and are labelled as such on the corresponding sweep switch positions.

In most scopes, attenuators are provided both at the vertical (Y) and horizontal (X) input terminals to reduce the amplitude of the input signal, if necessary, before applying it to the corresponding amplifier. The block schematic diagram of Fig. 28.6 shows the location of all the controls that are normally available on the front panel of a typical oscilloscope.

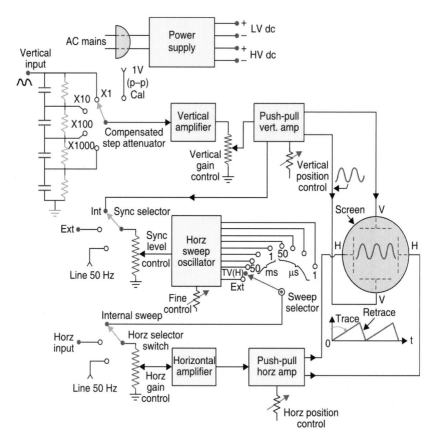

Fig. 28.6. Simplified block diagram of a cathode ray oscilloscope.

CRO Applications

(a) As a voltmeter. In certain circuits where a relatively low impedance of the voltmeter will cause loading or detuning of the circuit a CRO can be effectively used as a voltmeter. As shown in Fig. 28.7 (*a*), the unknown voltage is applied across the Y-input terminals and the time-base circuit is switched off. This results in a vertical line on the screen. The length of this line is measured on the graticule inscribed on a clear plastic plate which is placed over the outside face of the CRT screen. The graticule is initially calibrated by applying an ac voltage of known amplitude. Note that the vertical line on the screen denotes peak-to-peak value of the applied signal.

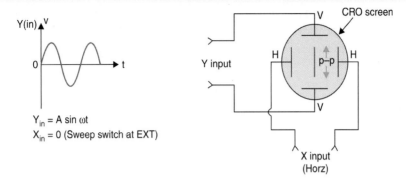

Fig. 28.7 (a). Voltage measurement with a CRO.

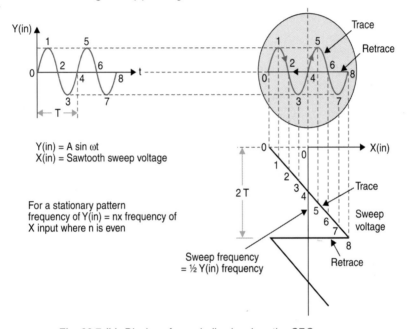

$Y(in) = A \sin \omega t$
$X(in) = $ Sawtooth sweep voltage

For a stationary pattern
frequency of Y(in) = nx frequency of
X input where n is even

Sweep frequency
= ½ Y(in) frequency

Fig. 28.7 (b). Display of a periodic signal on the CRO screen.

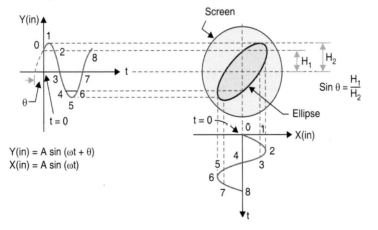

$Y(in) = A \sin (\omega t + \theta)$
$X(in) = A \sin (\omega t)$

$\sin \theta = \dfrac{H_1}{H_2}$

Fig. 28.7 (c). Phase-shift measurement with a CRO.

Chapter 28

In TV receivers this method can be usefully employed for measuring peak-to-peak voltages in the horizontal and vertical sweep circuits where a VTVM will not give any correct indication of such nonsinusoidal waveforms.

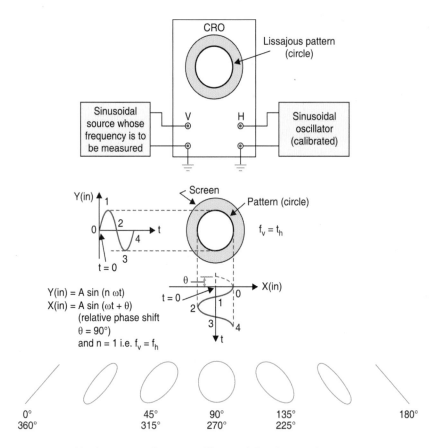

Lissajous patterns for n = 1 at different relative phase angles.

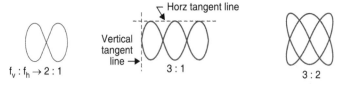

Lissajous patterns when $f_v > f_h$.

$$f_v = \frac{f_h \cdot t_h}{t_v}$$

f_v = frequency into vertical input
f_h = frequency into horizontal input
t_h = number of loops which touch the horz tangent line
t_v = number of loops which touch the vertical tangent line

Fig. 28.8. Frequency measurement with a CRO.

(*b*) *Time function display* (sec Fig. 28.7 (*b*)). As already explained any complex signal can be displayed on the CRO screen by suitably manipulating the time base and synchronizing controls. However, the limitations are (*i*) bandwidth of the vertical amplifier and (*ii*) maximum sweep frequency of the available scope. For an accurate display of the video signal, the vertical amplifier must have a bandwidth that extends at least up to 5 MHz and has high vertical sensitivity. The means to provide these attributes makes the oscilloscope quite expensive. Somewhat costlier scopes have dc coupled vertical amplifiers and trigger-mode coupled timebase circuits. The amplifier bandwidth usually extends from dc to 10 MHz with a sensitivity of 5 to 20 mV/cm. The time-base has calibrated ranges from about 1 μ*s*/cm to 0.1 *s*/cm. The horizontal amplifier is also dc coupled.

(*c*) *Phase-shift measurement.* The circuit arrangement for measurement of phase-shift between the input and output of any circuit having reactive elements is illustrated in Fig. 28.7 (*c*). Such a technique can be usefully employed for checking phase-shift characteristics of a video amplifier. Note that the ellipse produced on the scope screen is independent of the magnitudes of input and output signals. When the major axis of the ellipse lies in the first or third quadrant, the phase angle is between 0° to 90° or 270° to 360°. However, when the major axis lies in the second or fourth quadrant, the phase angle range is 90° to 180° or 180° to 270°. For a phase angle $\theta < \pi$, the spot moves clockwise and when $\theta > \pi$, the spot moves anticlockwise.

(*d*) *Frequency measurement.* The unknown frequency of any signal can be measured with a CRO if an oscillator with accurate calibration is available. The method of connecting the unknown and known frequency sources is illustrated in Fig. 28.8. The resulting 'Lissajous patterns' indicate the ratio between the two frequencies. The patterns shown in the figure are for sinusoidal sources. The measurement of horizontal and vertical sweep frequencies in a TV receiver presents a problem because of their trapezoidal waveshape. Though with a little practice, the somewhat complex pattern that appears on the screen can be correctly interpreted, it is easier to employ the comparison method. Two complete cycles of the receiver sweep signal whose frequency is to be measured are first obtained on the scope screen. The sweep signal is then removed and instead the output of a calibrated sinusoidal oscillator is connected. Its frequency is varied to again obtain two complete cycles without disturbing the timebase controls. The frequency indicated on the oscillator scale is the frequency of the signal under measurement. If the scope has a calibrated timebase, the unknown frequency (period) can be read directly.

TV Picture on Oscilloscope Screen

While discussing the applications of a CRO it is interesting to learn how the cathode ray tube in the oscilloscope can be used to produce a picture on its screen in green colour. The circuit connections are shown in Fig. 28.9. The procedure to obtain the picture is as under:

(*i*) Connect the video signal from a suitable point in the video amplifier of the given TV receiver to the intensity modulation or '*Z*' axis input terminal of the scope. This terminal couples the signal to the control grid of the cathode ray tube. It is normally located at the back panel of most scopes.

(*ii*) Extend the 50 Hz deflection voltage from the TV receiver to the '*Y*' input terminals of the scope. Similarly connect horizontal sweep voltage of the receiver to the '*X*' input terminals of the scope with timebase switch set on 'external'. This should be done through an isolating resistance to prevent detuning of the horizontal oscillator. The sweep output of the scope timebase circuit, when set at 15625 Hz can also be used instead of the output form horizontal section of the receiver.

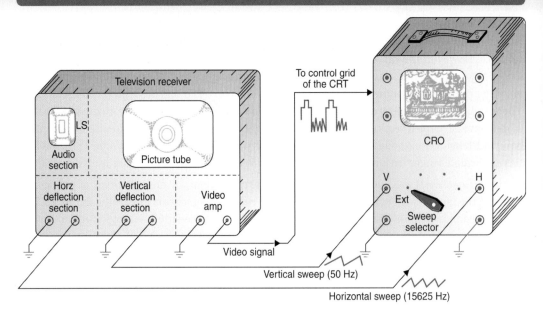

Fig. 28.9. Display of television pictures on the screen of an oscilloscope.

(*iii*) Vary the horizontal and vertical gain controls of the scope to obtain a raster with an aspect ratio of 4 : 3. A video signal of 25 V P-P amplitude will produce a reasonable picture. In case a negative picture is obtained connect the video signal from another point in the video amplifier where the polarity is opposite to that obtained earlier.

CRO Probes

It is essential to use a pair of conductors to connect the signal to be observed to the vertical input terminals of the scope. Normally a shielded cable is used to prevent pick up of any interfering signals or hum due to hand capacitance. The capacitance of such a cable together with the input capacitance of the scope is of the order of 60 to 70 pF. Though the shunting effect of this capacitance would not alter the wave shape of any audio signal but will affect the complex nature of the video signal when connected for viewing it on the scope screen. Thus while observing high frequency signals a simple 'direct probe' as it is called, will not serve the purpose.

Low Capacitance Probe

To avoid undesirable loading by the leads and scope capacitance, a high impedance or low capacitance probe is used. Figure 28.10 shows the probe connections and its equivalent circuit. It can be shown that if the time constants R_1C_1 and R_2C_2 are made equal, all the frequencies in any complex input signal would be equally effected during transmission from the signal source point to the scope input terminals. In the figure R_2 (1 MΩ) and C''_2 ($\approx$ 20 pF) represent input impedance of a typical oscilloscope's vertical amplifier circuit. The resistor R_1 is normally chosen to be 9 MΩ to increase the scope impedance to 10 MΩ. The capacitor C_1 is varied to make $R_1C_1 = R_2C_2$. As is obvious from the circuit configuration of the probe, the input signal gets attenuated by a factor of 10 : 1. Thus the probe can be used only at those points in the receiver where the signal amplitude in quite large. Video, sync, and deflection voltages are examples of signals where this probe can be used effectively. A switch is sometimes provided across the probe, the closure of which, converts it into a direct probe.

Chapter 28

Source Follower High Impedance Probe

In high frequency circuits where the amplitude of the signal is low, a source follower probe (see Fig. 28.11) is successfully used. This probe presents high impedance at its input terminals and does not cause any signal attenuation because of almost unity gain of the configuration employed. The probe is also known as an active probe because it is essentially a small signal wideband amplifier configuration for faithful transmission of the signal to the scope input terminals.

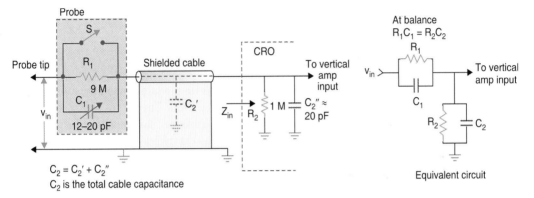

Fig. 28.10. Low capacitance or high input impedance (compensated) probe.

It may be noted that the above described probes can also be used with VTVM and FET voltmeters in high frequency circuits.

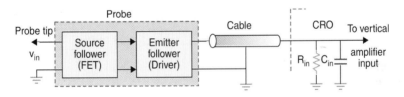

Fig. 28.11. Active or source follower probe.

Detector or Demodulator Probe

The circuit of such a probe is shown in Fig. 28.12. Basically it is a diode rectifier and RF filter which detects the low frequency envelope of any amplitude modulated carrier. The resulting output is a low frequency signal which can be applied to any CRO or meter for alignment and analysis of high frequency tuned circuits. It is very useful for the tuning of RF and IF stages in radio and TV receivers.

The above description of various applications of a CRO and its probes clearly establishes its versatility as a very useful instrument for signal tracing, alignment and servicing of TV receivers. It may be noted that though a scope which can handle signals in the video frequency range is desirable but one with a bandwidth of about 1 MHz or ever less is good enough for alignment and servicing. A less expensive CRO is better than none and a technician who is fully familiar with its use, capability and limitations soon learns to make the maximum use of the instrument he possesses.

Chapter 28

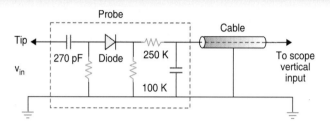

Fig. 28.12. Demodulator probe.

28.5 VIDEO PATTERN GENERATOR

A pattern generator provides video signals, direct and with RF modulation, on the standard TV channels for alignment, testing and servicing of television receivers. The output signal is designed to produce simple geometric patterns like vertical and horizontal bars, chequer-board, cross-hatch, dots etc. These patterns are used for linearity and video amplifier adjustments. In addition to this, FM sound signal is also provided in pattern generators for aligning sound section of the receiver.

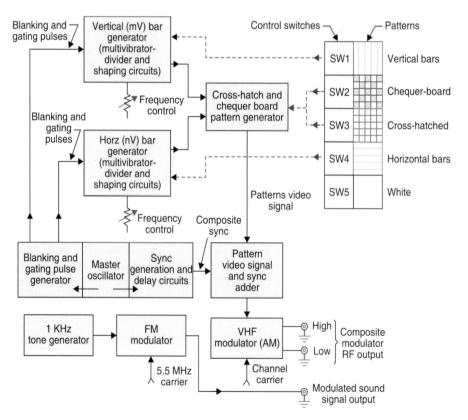

Fig. 28.13. Simplified functional block diagram of a pattern-cum-sound signal generator.

A simplified functional block diagram of the pattern cum FM sound signal generator is shown in Fig. 28.13. The generator employs two stable chains of multivibrators, dividers and pulse shaping circuits, one below the line frequency to produce a series of horizontal bars and the other above 15625 Hz to produce vertical bars. The signals are modified into short duration pulses which when fed to the video section of the receiver along with the sync pulse train produce fine lines on the screen.

Multivibrators produce square wave video signal at m times the horizontal frequency to provide m vertical black and white bars. After every m cycles, the horizontal blanking pulse triggers the multivibrator for synchronizing the bar signal on every line. A control on the front panel of the pattern generator enables variation of multivibrator frequency to change the number of bars.

Similarly square wave pulses derived either from 50 Hz mains or from the master oscillator are used to trigger another set of multivibrators to generate squarewave video signal that is n times the vertical frequency. These on feeding the video amplifier produce horizontal black and white bars. The number of horizontal bars can also be varied by a potentiometer that controls the switching rate of the corresponding multivibrator. The bar pattern signal is combined with the sync and blanking pulses in the video adder to produce composite video signals before feeding it to the modulator.

Provision of switches in the signal paths of the two multivibrators enables generation of various patterns. If both mH and nV switches are off, there will be a blank white raster. With mH switch on, vertical bars will be produced and when only nV switch is on, horizontal bars are generated. With both switches on, a cross hatch pattern will be produced.

The horizontal bar pattern is used for checking vertical linearity. These bars should be equally spaced throughout the screen for linearity. If necessary frame linearity preset controls in the receiver can be varied to get best possible results. Similarly the vertical bar pattern can be used for checking and setting horizontal linearity. With the cross-hatch pattern formed by vertical and horizontal lines, the linearity may be adjusted more precisely because any unequal spacing of the lines is easily discernible. If the pattern rolls up or down or loses horizontal synchronization, the frequency preset controls in the receiver oscillator circuits can be adjusted till the pattern becomes stationary. Picture centering and aspect ratio can also be checked with the cross-hatch pattern by counting the number of squares on the vertical and horizontal sides of the screen. In practice the picture tube screen is usually so placed that it faces a mirror fixed opposite to it on the work bench. The technician can thus observe the patterns in the mirror while making adjustments at the back side of the receiver.

Besides linearity and sweep frequency checks the pattern generator can also be used for detecting any spurious oscillations in sweep generation circuits, interaction between the two oscillators, poor interlacing, barrel and pincushion effects etc. As already mentioned almost all pattern generators provide additional facilities for aligning sound section and checking RF and IF stages of the receiver. Modulated picture signal on limited number of channels is also available for injecting into the RF section of the receiver. Similarly FM sound signal with a carrier frequency of 5.5 MHz ± 100 KHz modulated by 1 KHz tone is provided for aligning sound IF and discriminator circuits. A 75/300 ohms VHF balun is usually available as a standard accessory to the pattern generator.

Chapter 28

28.6 SWEEP GENERATOR

The familiar RF generator when used for alignment and testing of RF and IF stages of a TV receiver permits recording of circuit performance at one frequency at a time. Therefore, plotting of the total response curve point by point over the entire channel bandwidth becomes a laborious process and takes a long time. To overcome this difficulty, a special RF generator known as sweep generator is used. It delivers RF output voltage at a constant amplitude which sweeps across a range of frequencies and continuously repeats this at a predetermined rate.

The sweep generator is designed to cover the entire VHF and UHF spectrum. Any frequency in these bands can be selected as the centre frequency by a dial located on the instrument panel.

Sweep Width

Frequency sweep is obtained by connecting a varactor diode across the high frequency oscillator circuit. A modified triangular voltage at 50 Hz* is used to drive the varactor diode. Thus the frequency sweeps on either side of the oscillator's centre frequency at the rate of driving voltage frequency. The amplitude of the driving voltage applied across the varactor diodes can be varied to control maximum frequency deviation on either side of the carrier frequency. This is known as sweep-width and can be adjusted to the desired value up to a maximum of about ± 15 MHz. A width control (potentiometer) is provided for this purpose.

Alignment Procedure

The output of the sweep generator is connected to the input terminals of the tuned circuit under test. The frequency and 'sweep-width' dials are adjusted to sweep the range which lies in the passband of the circuit. With an input signal of constant amplitude, the output voltage varies in accordance with the frequency-gain characteristics of the circuit. The magnitude of the output voltage varies with time as the oscillator frequency sweeps back and forth through the centre frequency. The RF output is detected either by video detector in the receiver or by a demodulator probe. The detected output varies at the sweep rate of 50 Hz and its instantaneous amplitude changes in accordance with the circuit characteristics. Thus the output signal is a low frequency signal at 50 Hz and can be displayed on an ordinary scope provided the low frequency response of its vertical amplifier is flat down to at least 50 Hz.

Time-base for Visual Display

Since the HF oscillator frequency sweeps at the rate of 50 Hz, the detected output voltage also varies at the same rate and completes its back and forth cycle in 20 ms (1/50th sec.). In order to obtain a linear display of the detected signal the scope sweep voltage must vary accordingly and have a frequency of 50 Hz. Therefore, to obtain exact correspondence a triangular sweep voltage is developed and made available on the sweep generator along with other outputs.

The basic circuit configuration used for alignment of any tuned circuit with a sweep generator and scope is shown in Fig. 28.14. The time-base switch of the scope is set on 'external' and its horizontal input terminals are connected to the 'sweep output' terminals on the sweep generator. Thus the application of a 50 Hz triangular voltage to the horizontal deflecting plates results in a linear display on the scope screen, both during trace and retrace periods. The resulting patterns are shown in Fig. 28.15.

*In U.S.A. where the supply frequency is 60 Hz a sweep rate of 60 or 120 is used.

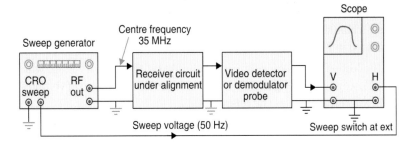

Fig. 28.14. Test equipment connections for alignment of RF, IF sections of the receiver.

Effect of Phase Shift

If there is a slight phase difference between the 50 Hz voltage used for modulating the RF signal and that available for horizontal deflection of the scope beam, the retrace portion of the curve will not follow precisely the trace path. This is illustrated in Fig. 28.15 (c). To correct this deviation, a phase control is made available on the front panel of the sweep generator.

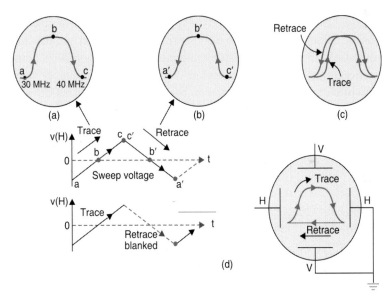

Fig. 28.15. Development of the response curve on the scope screen (a) trace (b) retrace (c) effect of phase shift (d) effect of retrace blanking or use of sawtooth sweep.

Phase Reversal

A phase reversal switch is also provided on the sweep generator. This is to ensure that the low frequency side of the response curve lies on the left side of the oscilloscope screen. If displayed otherwise it would cause unnecessary confusion. In some sweep generators a polarity switch is also provided to reverse the polarity of the trace if necessary.

Retrace Blanking

Modern sweep generators employ a sawtooth sweep with negligible retrace time instead of a triangular sweep. Besides eliminating any phase shift problems the ramp sweep produces a zero reference line on the CRO screen during retrace periods. In sweep generators that employ a triangular sweep, a blanking switch is used to blank the retrace part of the pattern. The resulting pattern is illustrated in Fig. 28.15 (*d*).

The vertical deflection on the scope screen is calibrated with a known voltage which is available either on the oscilloscope or sweep generator. This enables measurement of voltages at all critical frequencies with the retrace line as the reference.

28.7 MARKER GENERATOR

Though the sweep generator provides a visual display of the characteristics of the circuit or amplifier, this alone is inadequate because it does not give any precise information of the frequency on the traced curve. For this, a separate RF generator known as *marker generator* is used. This generator, though essentially an RF signal generator in the VHF and UHF bands, is of much higher accuracy than the familiar signal generators. The output of the marker generator is set at the desired frequency within the pass-band of the circuit under test. As indicated in Fig. 28.16, the outputs from the two generators are mixed together before applying to the input terminals of the circuit under test. The two signals heterodyne to produce outputs at the sum and difference of the two frequencies. These appear at the detector output and represent amplitude characteristics of the circuit. Because of the limited frequency response of the scope's vertical amplifier, the sum and most difference frequency signals generated, when the sweep frequency varies over a wide range, fail to produce any vertical deflection on the screen. However, as the sweep frequency approaches and just crosses the marker frequency, the difference frequency signals which are produced, lie within the pass band of the vertical amplifier. Thus they produce a pip on the screen along with the trace generated by the low frequency output of the detector. In order to get a sharp pip, a suitable capacitor is shunted across the vertical input terminals of the scope.

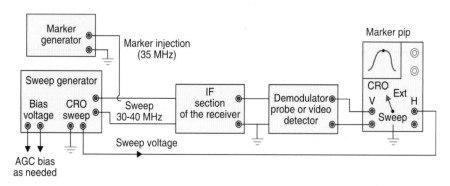

Fig. 28.16. Pre-injection of marker generator output in the sweep generator output.

Since the pip is produced at the marker frequency, it can be shifted to any point on the response curve by varying the marker generator frequency. This enables exact location of the trap and IF frequencies while tuning the RF and IF stages of the receiver.

Marker generator, besides providing dial controlled frequency often have a provision to generate crystal controlled fixed frequency outputs at several important frequencies. These additional frequencies in a generator designed for CCIR 625 line system include : 31.4 MHz (band edge),

31.9 MHz (trap frequency), 33.4 MHz (sound IF), 34.47 MHz (colour IF), 36.15 MHz (band centre), 38.9 MHz (picture IF), 40.4 MHz (trap frequency) and 41.4 MHz (band edge). These birdy type markers can be switched in, either individually or simultaneously by toggle or push button switches provided for this purpose. Such a provision is very useful for exact alignment of IF section of the receiver.

Marker Injection

There are two methods of adding marker signals to the oscilloscope display set up. The pre-injection method, where the sweep generator and marker generator outputs are mixed before they are applied to the circuit under test, suffers from the following drawbacks :

(*i*) The marker signal has to pass through the tuned circuit being aligned and so the amplitude of the marker pip on the response curve will depend or the gain of the tuned circuit or amplifier at that particular frequency. Thus with a constant marker input a large pip will be produced on top of the response curve, but will slowly decrease in amplitude as it is moved down the skirts (slopes) of the response curve and will eventually disappear at the base line.

(*ii*) To counteract this, if the amplitude of the marker output is kept high, it will overload the circuit under test in its pass band region, resulting in distortion of shape of the curve.

Therefore the pre-injection method has been replaced with what is known as post-injection technique. In this method a fraction of the sweep generator output is separately mixed with the marker generator output. The available output is then mixed with the detected output before applying it to the scope. The circuit connections are illustrated in Fig. 28.17.

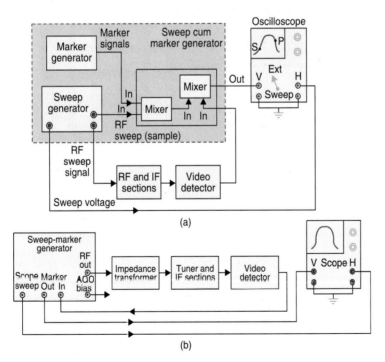

Fig. 28.17. *Post-injection of marker output (a) basic circuit arrangement (b) connections as in a sweep-marker generator.*

Sweep-Marker Generator

Earlier sweep and marker generators were manufactured as separate units. However, now the two instruments are combined into a single instrument known as 'sweep-marker generator'. It includes a built-in marker adder for post-injection of the marker signal. Such units also provide a variable dc bias source for feeding a fixed bias to the RF and IF sections of the receiver. This is necessary during alignment when the internal AGC bias must be disabled for correct tuning.

The frequency spectrum covered by a quality sweep generator includes VHF and UHF ranges in both manual and auto modes. The RF output is about 0.5 V rms across 75 ohms with a continuously variable attenuator up to 50 db or 0 to 60 db in 10 db steps. The marker section provides crystal controlled output at all important frequencies. The output can be internally modulated with a 1 KHz tone when necessary. These accessories include a demodulator probe, a balun and special cables with matched terminations.

Telonic (USA) 1011 B VHF sweep generator (see Fig. 28.18) needs special mention because of its versatility and reliability. It covers a frequency range of 5-300 MHz in manual, auto and program modes. Its auto-track feature (option 402 B) eliminates manual adjustment of the sweep generator. This option automatically tunes the centre frequency of sweep generator to track the IF of the tuner. Similarly the autotrace option 409 B accepts detected output of the tuner and yields a constant display to the scope. Simultaneously actual input amplitude is indicated by the height of a pulse on the display.

Fig. 28.18. Telonic 1011 B VHF sweep generator.

The generator can provide different marker systems. Option 206B produces a 'birdy' type marker whenever the frequency of the sweep generator is equal to the frequency of an externally applied CW signal. Another option 297B produces a birdy maker at the user's desired frequency. One to seven markers are available. Similarly option 337 B provides IF centred pulse pair markers at any spacing from 1 to 12 MHz for local oscillator adjustments. Yet another choice, option 342 B, produces a pulse marker pair at sound and video of the selected channel.

The main specifications of the Telonic 1011 B sweep generator are as under :

Frequency range	5-300 MHz
Sweep width	5-30 MHz in manual, auto and program modes
Sweep rate	Sawtooth, electronically line synchronized to 50/60 Hz line with capability to sweep at half the line rate.
RF output	0.5 V rms, 0.25 V with continuously variable attenuator (option 103 B)
Flatness	± 0.5 dB over full range.
Spurious and harmonics	– 30 db max.
Line voltage	105/125 V or 210/250 V

Television Wobbuloscope

This instrument combines a sweep generator, a marker generator and an oscilloscope all in one. It is a very useful single unit for alignment of RF, IF and video sections of a TV receiver. It may not have all the features of a high quality sweep generator but is an economical and compact piece of equipment specially designed for television servicing. The oscilloscope usually has provision for TV-V and TV-H sweep modes. A RF output down to 1 MHz is also available for video amplifier testing.

Analyst Generator

An analyst generator not only provides all the outputs which are available on a pattern generator but also duplicates all essential signals of the television receiver. Besides providing RF, IF and video signals in the form of patterns, composite video signal including sync, intercarrier sound, separate sync, audio, flyback and yoke test signals are also available. These outputs are normally used for trouble shooting and not for alignment purposes.

28.8 THE COLOUR BAR GENERATOR

The composite colour video signal at the output of video detector consists of luminance (Y) signal, the chrominance signal, the colour burst, sync pulses and blanking pulses. In practice, the amplitude of video signal is continuously varying due to the changing picture content. Therefore, such a waveform is not useful for adjustments and trouble shooting purposes. The colour bar generator acts as a substitute transmitter and supplies the receiver a known non-varying colour pattern signal for alignment and servicing purposes.

The Gated Rainbow Colour Bar Generator

The gated colour bar generator develops such a composite video signal that it produces a rainbow colour bar pattern on the receiver screen. The pattern consists of 10 colour bars ranging in colour from shades of red on the left side, through blue in the centre, to green in the far right side. The colour bar pattern, the associated video waveform and corresponding phase relationships of the gated pattern are illustrated in Fig. 28.19. Each colour in the bar pattern has been identified and lined up with the associated modulating waveform. Note that the composite video signal for the pattern consists of a horizontal sync pulse and 11 equal amplitude bursts of the colour subcarrier frequency. The burst to the right of the horizontal sync pulse is the colour burst. The other bursts (1 to 10) differ in phase from one another and correspond to different colours in the bar pattern.

Chapter 28

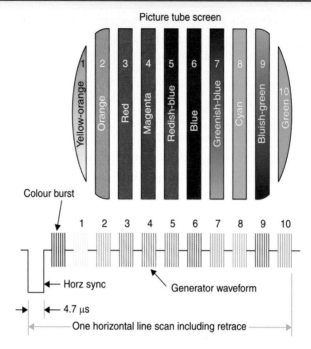

Fig. 28.19 (a). *Colour bar pattern display on the screen of a colour picture tube and corresponding output waveform of the rainbow generator.*

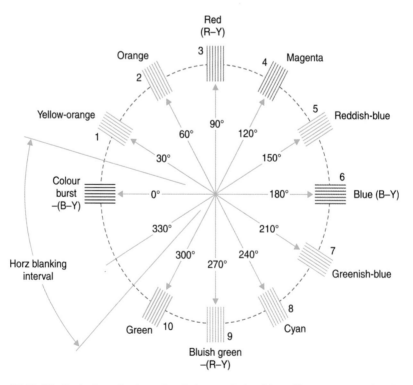

Fig. 28.19 (b). *Illustration of colour signal phase relationships with respect to colour burst.*

The basic principle of the colour bar generator is quite simple. Any two signals of different frequencies have a phase difference that is continuously changing. As shown in Fig. 28.20 a crystal oscillator is provided to generate a frequency = 4.41799375 MHz. This is 15625 Hz lower than the colour subcarrier frequency (4.43361875 – 4.41799375 = 15625 Hz). Since the difference in frequency is equal to the horizontal scanning rate, the relative phase between the two carrier frequencies will change by 360° per horizontal line. Thus the effective carrier signal at 4.41799375 MHz will appear as a signal that is constantly changing in phase (360° during each horizontal line time) when compared to the 4.43361875 MHz reference oscillator in the television receiver. It is the phase of the chrominance signal which determines the colour seen and therefore such a frequency relationship provides the colour bar signal. Since there is a complete change of phase of 360° for each horizontal sweep, a complete range of colours is produced during each horizontal line. Each line displays all the colours simultaneously since the phase between the frequency of crystal oscillator and that of the horizontal scanning rate frequency is zero at the beginning of each such line and advances to become 360° at the end of each horizontal sweep stroke.

As shown in the block diagram of Fig. 28.20 the colour bar pattern is produced by gating 'on' and 'off' the 4.41799375 MHz oscillator at a rate 12 times higher than the horizontal sweep frequency (15625 × 12 = 187.5 KHz). The gating at a frequency of 187.5 KHz produces colour bars with blanks between each colour bar. The colour bars have a duration corresponding to 15° and are 30° apart all around the colour spectrum. When viewed on the picture tube screen of a normally operated colour receiver, these bars appear as shown in Fig. 28.19 (a). (The bars are also shown in colour plate No. 5).

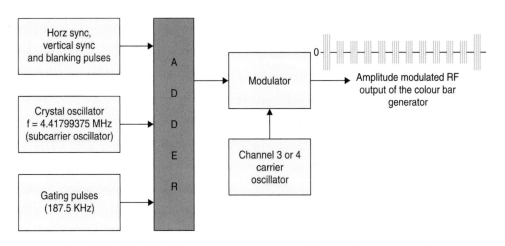

Fig. 28.20. *Simplified block diagram of a gated rainbow colour bar generator.*

It may be noted that out of the twelve gated bursts, only 10 show on the picture tube as colour bars. This is so, because one of the bursts occurs at the same time as the horizontal sync pulse and is thus eliminated. The other burst occurs immediately after the horizontal sync pulse and becomes the colour sync burst which is used to control the subcarrier reference oscillator in the television receiver. Thus phase angles from 0° to 30°, that correspond to the colour burst and colours such as yellow (13°) are blanked out during horizontal retrace and do not appear on the picture tube screen.

The adder while gating the crystal oscillator output also combines horizontal sync, vertical sync and blanking pulses to it. The composite colour video signal available at the output of the adder can be fed directly to the chrominance bandpass amplifier in the TV receiver. However, it is usually amplitude modulated with the carrier of either channel 3 or 4. The option is given so that the service engineer can select a channel that does not have a local live transmission that might cause interference.

Receiver Operation with Colour Bar Signal

The modulated RF output of the colour bar generator is fed to the antenna terminals of the receiver. This signal is processed by the various RF and IF sections of the receiver and detected at the video detector. As shown in Fig. 28.21, it is then applied to the colour demodulators via the chroma bandpass amplifier. If the alignment of all the stages is proper the waveform feeding the colour demodulators will be essentially the same as that which modulates the RF carrier in the colour bar generator. The main difference is that the horizontal sync which occur at the line rate are removed by the 4.43 MHz tuned circuits in the bandpass amplifier.

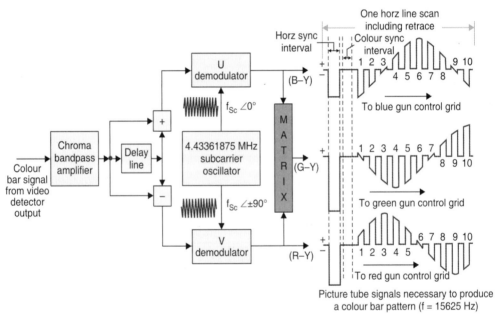

Fig. 28.21. *Simplified block diagram of colour section of the receiver showing passage of colour bar signal.*

The input to each demodulator consists of gated colour bar signal and a locally generated constant amplitude subcarrier oscillator signal. The demodulator output after matrixing yield $(R - Y)$, $(G - Y)$, and $(B - Y)$ colour difference signals. As shown by the waveforms drawn along the matrix in Fig. 28.21, the three signals are out of phase with respect to each other and are serrated because of interruptions by the gating signal. These signals are amplified before feeding to the corresponding control grids of the colour picture tube. Not that at the instant the voltage on the blue gun control grid is maximum positive (bar 6), it is zero and negative at the red and green guns respectively. Such a combination will increase the blue gun beam current, reduce considerably the red gun beam current and almost cut-off the green beam. The net result will be, that a blue bar will be produced on the screen. Similarly, it can be argued that the 3rd bar will attain a red colour and the 10th bar will be green. The other bars will vary in colour, depending on the relative amplitude of the pulse to each control grid of the picture tube.

NTSC Colour Bar Generator

Another type of colour bar generator is the NTSC colour generator. It is compatible with the 525 line American TV system and its output signal also includes the Y signal component. The Y signal corresponds to the shades of grey that would be produced in a black and white receiver by the seven bars of colour that constitute of the pattern. Each of the seven colour bars corresponds to a burst of 3.58 MHz. The amplitude of the burst represents saturation of the colour bar and its phase corresponds to the colour. The NTSC colour bar pattern consists of three primary colours, their complementary colours at 100 percent saturation and a white bar.

A colour bar generator is very useful for colour signal tracking in various sections of the receiver. It is also used for initial adjustments of colour sync, range of tint, colour controls, etc.

Colour-Bar Pattern Generator

This instrument is a combination of a pattern generator and a colour bar generator. It is useful for black and white and colour receiver alignment and servicing. The Philips PM 5501 (see Fig. 28.22) is one such generator specially designed for use at customer's house. It provides the five essential test patterns, necessary to make installations, fast checks and repairs on both monochrome and colour receivers. The patterns are selected by push buttons. The instrument is extremely light and portable.

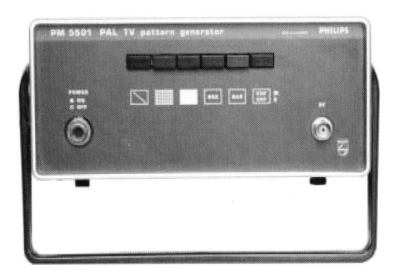

Fig. 28.22. Philips PM 5501 colour-bar pattern generator.

The PM5501 operates according to the CCIR 625 line system G, TV standards. To ensure fast operation, two fixed RF signals can be selected, one on VHF channel 7 and the other for UHF channel 30. When required, channel 7 may be altered to any frequency in Band III (170 to 230

MHz). The same applies to UHF channel 30, which can be changed internally to any channel in the frequency range, 470 to 600 MHz.

For accurately tuning and checking the sound performance of a colour television receiver (CTV), the generator has a 1 KHz tone available. This tone also makes checks on interference between luminance-sound and chroma-sound.

The main technical specifications of this colour bar pattern generator are given below.

Test signals

(a) 8 bars, linearized, grey scale

(b) cross-hatch pattern

(c) 100% white pattern (with burst)

(d) red pattern (50% saturated)

(e) standard colour bar with white reference 75% contrast (internally changeable to full bars)

Video carrier

(a) VHF band III : 170–230 MHz

(b) UHF band IV : 470–600 MHz

RF output

> 100 mV p-p (75 ohm impedance)

Video modulation—Am, negative

Sound carrier

Frequency—5.5 MHz (or 6 MHz by internal adjustment)

Modulation—FM

Internal signal—1 KHz sine wave

FM sweep 40 KHz on 5.5 MHz.

Chroma : PAL—G and I standards

Power : 115–230 V, 50–60 Hz, 6 watt,

Dimensions ($w \times h \times d$) : 23 × 11 × 21 cm,

Weight : 1.25 Kg.

28.9 VECTROSCOPE

This test instrument combines a keyed colour-bar generator with an oscilloscope and is used for alignment and testing the colour section of a TV receiver. The amplitude and phase of the chrominance signal represent colour saturation and hue of the scene. The information can also be displayed on the oscilloscope screen in the form of a Lissajous pattern. The resultant display is called a vectrogram. A separate colour-bar generator can also be used with a conventional CRO to produce a vectrogram. The necessary connections and the resulting pattern are shown in

Fig. 28.23. With the gated colour bar generator connected at the input terminals of the receiver, serrated $(R-Y)$, and $(B-Y)$ video signals become available at the corresponding control grids of the colour picture tube. These two outputs are connected to the vertical and horizontal inputs of the oscilloscope. Since both $(R-Y)$ and $(B-Y)$ inputs are interrupted sinewaves and have a phase difference of 90° with respect to each other, the resultant Lissajous pattern will be in the basic form of a circle which collapses towards the centre during serrations in the signals. Assuming ideal input signal waveforms the formation of vectrogram is illustrated in Fig. 28.24. Since there are ten colour bursts, the pattern displays ten petals. The horizontal sync and colour burst do not appear in the display because these are blanked out during retrace intervals. The position of each petal represents phase angle of each colour in the colour bar pattern. For example petal 1, petal 3, $(R-Y)$, petal 6 $(B-Y)$ and petal 10 $(G-Y)$ correspond to angles 30°, 90°, 180° and 300° respectively. The vectroscopes usually have an overlay sheet on the scope screen and is marked with segment numbers and corresponding phase angles. This enables the user to identify different colours and interrupt the size and shape of each petal.

In actual practice, the tops of $(R-Y)$ and $(B-Y)$ bar signals do not have sharp corners and the resultant pattern is somewhat feathered and rounded at the periphery. This is depicted in the actual pattern shown on the oscilloscope screen in Fig. 28.23. Corner rounding and feathering occurs due to limited high frequency response and non-zero rise time of the amplifiers in the colour bar generator and oscilloscope.

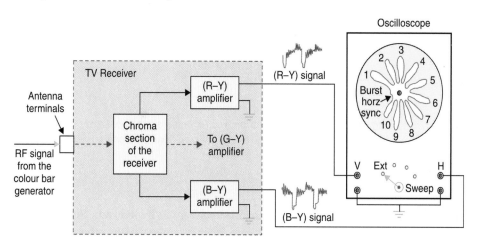

Fig. 28.23. Circuit connections for producing a vectrogram on the scope screen.

With the vectrogram display, chroma troubles can be ascertained and servicing expedited. For example, loss of $(R-Y)$ signal will cause the vectrogram to be one horizontal line only. Similarly absence of $(B-Y)$ will result in a single vertical line on the screen. Any change of the receiver colour control will alter the amplitudes of both $(R-Y)$ and $(B-Y)$ signals and this will cause the diameter of the pattern to change. The receiver's fine tuning will affect the size and shape of the reproduced pattern. Proper fine tuning will produce the largest and best shaped vectrogram. If some of the petals are longer that others, non-linear distortion is indicated. If the petal tops are flattened, some circuit overloading is occuring in the receiver. Numerous other checks can be made with the vectroscope to localize defective colour stages, mistuned bandpass amplifiers maladjusted circuitry in the subcarrier oscillator section and inoperative colour stages.

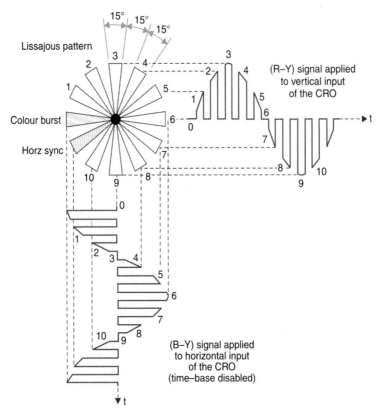

Fig. 28.24. *Ideal keyed rainbow vectrogram (lissajous pattern) produced by (R—Y) and (B—Y) colour difference signals.*

28.10 HIGH VOLTAGE PROBE

A high voltage probe is necessary for measuring EHT voltages in TV receivers. Since the maximum voltage range on multimeters seldom exceeds 1 KV, it becomes necessary to attenuate high voltages before connecting them to a multimeter.

Most voltage probes use a compensated string of resistive network to form a voltage divider. The attenuation ratio is usually 100 : 1. However, for measuring voltages of the order of 30 KV or so, resistance type of probe is not suitable because of stray conduction paths set up by the resistors. Therefore capacitor type probes are used to provide an attenuation of 1000 : 1. Such probes contain two or more capacitors with values selected to provide desired voltage reduction and to match the input impedance of the meter. Special high voltage probes have been developed for TV service work which have built-in meters to measure voltages up to 30 KV. For safety such probes have a separate earth terminal which should be connected to ground when carrying out EHT measurement.

Chapter 28

In addition to all the instruments and accessories described above a regulated power supply, a tube tester and a transistor tester are equally necessary pieces of equipment. With the increased use of ICs it is desirable to posses an IC tester and additional low voltage regulated power supplies.

REVIEW QUESTIONS

1. Describe with suitable circuit diagrams how voltage, current and resistance measurements are made in a multimeter. Why a movement having very low full scale deflection current is employed in multimeters ?

2. Why are VTVMs preferred for both dc and ac voltage measurements in electronic circuits ? Why is the input resistance of a VTVM much higher than a multimeter when set to measure volts ?

3. Discuss relative merits of the two main types of electronic voltmeters. Why is an amplifier-rectifier type of VTVM preferred over a rectifier-amplifier type for measurements in video circuits ? What is the function of probes in dc and ac voltage measurements ?

4. Explain with suitable circuit diagrams how a CRO can be employed for measurement of phase-shift and frequency. Why is a probe necessary for faithful display of a video signal ?

5. Draw a circuit diagram to illustrate the use of CRO for displaying a television picture on its screen instead of on receiver's own picture tube. Explain all the necessary steps and precautions that must be observed for obtaining a well defined picture with a correct aspect ratio.

6. Describe how various patterns are generated in a video pattern generator. Explain typical applications of this instrument for testing and aligning a TV receiver.

7. Explain the need and method of sweeping the carrier frequency around its chosen central value. Describe briefly the use of all controls which are normally provided on the front panel of a sweep-cum-marker generator.

8. Explain with a diagram of connections how a sweep generator can be employed to obtain the response curve of a wide band tuned circuit on a CRO.

9. Why is it necessary to inject marker signals into the sweep generator output ? Why is post-injection of marker generator output preferred over the pre-injection method ? What is the significance of crystal controlled spot frequencies in a marker generator ?

10. Describe the basic principle of a colour bar generator and explain how it is employed to check the chroma section of a colour receiver.

11. Explain how Lissajous patterns obtained on a vectroscope can be interpreted to localize colour receiver faults.

12. Write short notes on the following : (*i*) TV wobbuloscope, (*ii*) Analyst generator (*iii*) Demodulator probe, (*iv*) EHT probe.

Chapter 28

Receiver Circuits and Alignment

High component density and complexity of circuits in television receivers justify the use of integrated circuits to replace complex circuit blocks. The use of ICs reduces component, assembly, and production line costs. On account of these merits and better performance, there is an increasing trend to use more and more ICs in television receivers. All present day TV receivers incorporate two or more ICs to replace discrete circuitry in low power and complex sections of their circuits.

Comprehensive details of various sections of the receiver employing tubes, transistors and ICs are discussed in the previous chapters. In order to show how various blocks fit together, complete circuit of a monochrome receiver is explained. In addition, circuit organization of both monochrome and colour receivers employing ICs in most sections of the receiver are also discussed. Tuning and alignment of different sub-assemblies of the receiver is explained along with circuit details.

29.1   MONOCHROME TV RECEIVER CIRCUIT

Out of the numerous receiver circuits that are in use, one developed by Bharat Electronics Ltd. (BEL) has been chosen for detailed discussion. It is a multichannel fully solid-state receiver conforming to CCIR 625-B system. It employs three ICs. The design of the receiver is so simple that it can be assembled* with a minimum of tooling and test facilities. Such an exercise can be very instructive for fully grasping alignment and servicing techniques of a television receiver. Fig. 29.1 on the foldout page shows the circuit diagram of this receiver. A brief description of the circuit follows:

(a) Tuner

The receiver employs a turret type tuner and provides all channels between 3 to 10 in the VHF range. A high pass filter with a cut-off frequency of 40 MHz is used at the input to reduce interference due to IF signals. The tuner operates from a + 12 V supply and has an effective AGC range of 50 db.

---

*The technical manual of the receiver can be obtained on request from M/s Bharat Electronics Ltd. Jalahalli, P.O., Bangalore 560013 India. The manual gives all necessary circuit details, coil data, list of components, assembly and alignment details.

(b) Video IF Section

The video IF sub-system consists of IC, BEL CA*3068 and other associated components housed in a modular box to avoid any possible RF interference. All interconnections to this module are either through feed through capacitors or insulated lead throughs. The IF sub-assembly provides (*i*) a gain around 75 db to the incoming signal from the tuner, (*ii*) required selectivity and bandwidth, (*iii*) attenuation to adjacent channel interfering frequencies, (*iv*) attenuation of 26 db to sound IF for intercarrier sound, (*v*) sound IF and video outputs and (*vi*) AGC voltage to the IF section and tuner.

(c) Sound IF Sub-system and Audio Section

The important functions performed by this sub-system are:

(*i*) to amplify intercarrier IF signal available from the picture IF amplifier,

(*ii*) to recover sound signal,

(*iii*) to amplify the sound signal and deliver at least 2 watts of audio power to the loud-speaker.

The circuit of BEL CA**3065 IC consists of (*i*) a regulated power supply, (*ii*) a sound IF amplifier-limiter, (*iii*) an FM limiter, (*iv*) an electronic attenuator and a buffer amplifier, and (*v*) an audio driver.

The sound section operates from a + 12 V supply. It employs transistor BC 148B as a boot-strapped driver and matched transistor pair, 2N5296 and 2N6110, at the output stage. The bandwidth of the audio amplifier is from 40 Hz to 15 KHz and can deliver 2 watts of useful audio power.

(d) Video Amplifier and Picture Tube Biasing

This section of the receiver uses transistors BF 195 C as driver (buffer), BD115 as video amplifier and BC147B in the blanking circuit. The video amplifier delivers 90 V p-p signal to the cathode of picture tube $500\text{-}C_1P_4$. The current limiting is provided by diode 0A79 and associated circuitry. Transistor Q 503 switches amplifier transistor Q502 only during the time when video signal is present and turns it off during horizontal and vertical sync periods. The horizontal blanking pulses of 60 V p-p and vertical blanking pulses of 40 V p-p are applied to the base of transistor Q503. High voltage output of 1.1 KV from the horizontal output circuit is rectified and fed to focusing grids of the picture tube through a potential divider. Brightness control operates from a + 200 V supply. The contrast control functions by varying input signal amplified to the video amplifier transistor Q 502.

(e) Horizontal Oscillator Sub-system

This section employs transistor Q 401 (BC 148 A), IC, CA 920 and associated passive components. Composite video signal from the IF section is applied to pin 8 of IC 401 (CA 920). The functions of the horizontal subsystem are:

(*i*) to generate a line frequency signal, the frequency of which can be current controlled,

(*ii*) to separate sync information from the composite signal,

(*iii*) to compare the phase of sync pulses with that of the oscillator output and generate a control voltage for automatic tuning of the oscillator,

---

*For more details on CA 3068 refer section 23.8.

**For more details on CA3065 refer section 21.7.

Chapter 29

Chapter 29

(*iv*) phase comparison between the oscillator waveform and middle of the line flyback pulse, to generate a control voltage for correction of the switching delay time in the horizontal driving and output stages, and

(*v*) shaping and amplification of the oscillator output to obtain pulses capable of driving the horizontal deflection driver circuit.

(f) Horizontal Output Circuit

Output from the horizontal oscillator is applied to the base of horizontal driver transistor Q802 through a coupling capacitor (C 803). The transistor switches from cut-off to saturation when a pulse is applied to its base and provides the necessary drive for Q803 (BU 205). The output transistor (Q803) drives the line output transformer to provide deflection current to the yoke coils. In addition, the output circuit (*i*) generates flyback pulses for blanking, AFC and AGC keying, (*ii*) provides auxiliary power supplies and generates high voltage (+ 18 KV) for anode of the picture tube and (*iii*) produces 1.1 KV dc for the focusing grids. The heater supply of 6.3 V for the picture tube is taken across winding 10-11 of the line output transformer.

(g) Vertical Deflection Circuit

As shown in Fig. 29.1, the circuit operates from a 40 V dc supply obtained through the line output transformer. This sub-system is a self oscillatory synchronized oscillator with a matched pair of output transistors. Clipped vertical sync pulses are fed at the base of Q 70 (BC 148 C) to provide it with a stable drive. Resistor R 724 senses yoke current and feeds a voltage proportional to this current back to the base of Q 702 (BC148 B) for adjustment of the picture height. Coupling network between the collector of Q 701 and the base of Q 702 incorporates the necessary 'S' correction and provides linearity of the deflection current. Hold control forms part of the input circuit of transistor Q 701.

(h) Power Supply Circuit

The power supply circuit is a conventional transformerless half-wave rectifier and filter circuit. It provides 200 V for feeding various sections of the receiver.

Front panel control. Four controls *i.e.* on/off volume control, channel selector and fine tuning, contrast control and brightness control are brought out at the front panel of the receiver. In addition, vertical hold and horizontal hold controls are available on the side panel for occasional adjustments.

29.2   MONOCHROME RECEIVER ALIGNMENT

The exact procedure for aligning the tuner and IF stages is based on their design and is perfected after a great deal of experimentation by the manufacturers of television receivers. Therefore, the alignment should be carried out strictly in accordance with the procedure recommended in the literature of the receiver.

It is wrong for a service engineer to tamper with the RF and IF sections of the receiver unless he has full equipment, expertise and service manual of the receiver. In most cases these stages require a 'touch up' alignment in which relatively few adjustments are necessary and an experienced technician can accomplish this task by looking at the picture and hearing the sound in the loudspeaker.

However, if alignment is necessary or a new set is to be aligned it is necessary to carry out the following checks before proceeding with the alignment of RF and IF sections of the receiver.

(i) The vertical and horizontal sweep circuits should be checked for a steady and disturbance free raster.

(ii) Video amplifier and picture tube circuitry should be checked for proper bandwidth and low frequency phase shift characteristics,

(iii) Using a pattern generator or otherwise, necessary adjustments should be made to ensure linearity, correct aspect ratio and proper interlacing.

Alignment Sequence

It is a good practice to carry out receiver alignment in the following sequence:

(i) Setting up of trap circuits at proper frequencies,

(ii) Alignment of IF stages,

(iii) Sound section alignment,

(iv) Tracking of RF and mixer stages,

(v) Tracking of oscillator circuits to obtain best picture and sound output.

Alignment Precautions

While the detailed alignment procedure may differ from receiver to receiver, it is worthwhile to observe the following precautions before commencing alignment of any section of the receiver:

(i) Shielded wire should be used for interconnecting sweep generator, receiver under alignment and scope to avoid stray field pick up effects.

(ii) All connecting leads and cables must be kept as small as possible to minimize high frequency signal losses.

(iii) All the equipment and receiver should have a common ground because a floating ground would result in incorrect alignment. For grounding receivers having a direct on line type of supply, an isolating transformer should be used between the supply mains and receiver.

(iv) It should be ensured that the load resistance across which the scope is connected has one end at ground potential and is also strapped to the ground terminal of the scope.

(v) A low pass filter formed by a cable (see Fig. 29.2) or a suitable probe should be used for connecting signal output from the detector to the scope terminals. The 100 K-ohm resistance in series and the capacitance of the cable form a low-pass filter thus bypassing high frequency pick-ups.

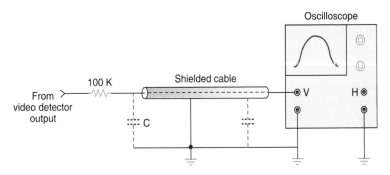

Fig. 29.2. *A low-pass filter circuit in place of a proper probe.*

(*vi*) A balun (Fig. 29.3 (*a*)) should be used for connecting sweep generator output to the receiver input terminals to provide a match. In case a balun is not available a network shown in Fig. 29.3 (*b*) can be used. This not only provides the desired match between the 75-ohm sweep generator output and 300-ohm receiver but also satisfies the balanced input requirement.

Improper termination of the sweep generator cable or faulty grounding of the equipment can be checked by grasping the cable and watching the response curve. A change in the shape or size of the pattern when the cable is held indicates presence of standing waves on the cable. This must be corrected by proper grounding before proceeding with alignment.

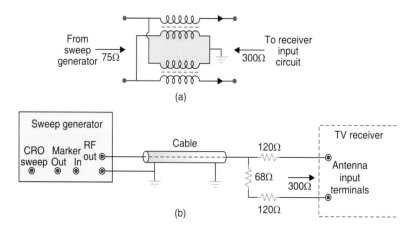

Fig. 29.3. *Impedance match between the sweep generator and receiver input terminals (a) by a balun and (b) by a resistive network.*

(*vii*) Any overloading of the circuit should be avoided. The output from the sweep generator must be kept as low as possible to avoid distortion of the reproduced pattern.

(*viii*) Even when there is no possibility of adjacent channel interference the trap circuits should be correctly adjusted. This is necessary because any incorrect tuning of trap circuits may result in attenuation of the frequencies which are a part of the desired response.

(*ix*) To avoid stray capacitance effects, insulated hexagonal wrench and screw driver should be used for trimmer settings and a narrow shanked screw driver for slug adjustments.

(*x*) No force should be applied on adjustments which have been sealed. Acetone or any other suitable thinner should be applied to soften the cement before making adjustments.

Alignment of Trap Circuits

It is desirable that the trap circuits are tuned before IF alignment because it helps in quicker and more precise alignment of the IF section. The trap circuits can be aligned by setting the sweep generator to deliver amplitude modulated RF output or by employing an RF signal generator (marker generator) modulated with a 1 KHz tone as the signal source. A VTVM or a CRO connected across the detector load resistor serves as the detector indicator. The local oscillator should be disabled to avoid any interference while setting trap circuits. As shown in Fig. 29.4 the signal generator is connected at different input points (shown by arrows) to tune different rejection circuits. Each time the frequency and signal level should be adjusted depending on the frequency to be rejected and the location of signal source. The cores of the resonant (trap) circuits are varied till minimum response is obtained on the scope or VTVM shows least voltage. The sound IF frequency rejection circuit should also be tuned along with other trap circuits to give a small output on the VTVM. This helps in obtaining desired response while tuning IF stages with a sweep generator.

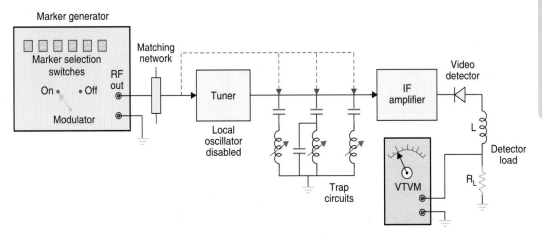

Fig. 29.4. *Alignment of trap circuits.*

IF Section Alignment

The primary purpose of IF amplifier alignment is to obtain a response curve of proper shape, frequency coverage and gain. Correct alignment of the combined picture and sound IF stages is necessary for a good picture and distortion free sound. There are two methods of aligning IF stages:

(*a*) *Marker generator and electronic voltmeter method.* For staggered-tuned IF stages of the receiver, the individual stages are tuned at predetermined frequencies using a CW marker generator and VTVM. The electronic voltmeter is connected at the output of video detector. After adjusting all the trap circuits for minimum output and tuned band-pass circuits for maximum output, a sweep generator, a marker generator and a scope is used to check the overall response curve. Necessary touch-up of tuned circuits is made while comparing the response curve with the one issued by the manufacturer of the receiver.

(*b*) *Sweep cum marker generator and scope method.* One such method is to inject the output of the sweep cum marker generator into the input circuit of the last IF amplifier and observe the response on a scope connected to the video detector output. The tuned circuit of the IF stage is adjusted for proper shape and amplitude following the curves given in the service manual of the receiver. The process is repeated by moving the sweep and marker generator closer to the tuner up to and including the mixer circuit. The additional circuits are aligned each time, following the manufacturer's reference curves.

In another method each IF stage is aligned separately. A demodulator probe is used for detection and a scope for observing the response curve. When the individual stages have been 'coarse' aligned, an overall response is taken and fine adjustments made if required.

Overall IF Response

The overall IF response includes all the IF tuned circuits from the mixer output in the tuner to the video detector. Figure 29.5 shows necessary connections of the sweep-marker generator and CRO to the receiver for obtaining an overall visual response curve of the IF section.

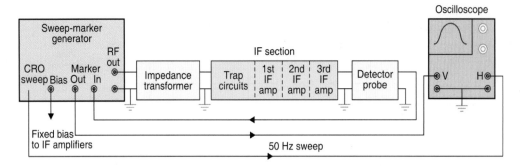

Fig. 29.5. *Set-up for the alignment of video IF section of the receiver.*

Initial adjustments. Before proceeding to adjust double-tuned circuits of the IF section make the following connections and adjustments:

(*i*) Disable the horizontal scanning oscillator to prevent the appearance of any spikes on the response curve. In the absence of any specific service instructions connect a dummy load across the horizontal sweep amplifier to prevent B_+ voltage from going too high.

(*ii*) Disable AGC bias and instead connect the recommended fixed bias from a bias box or from the sweep generator if available.

(*iii*) Connect sweep cum marker generator output to the mixer input (see Fig. 29.5) through an impedance matching transformer. If such a transformer is not available make connection through a 0.002 μF coupling capacitor.

(*iv*) In case the sensitivity of the scope's vertical amplifier is low, take output after one stage of video amplifier through an isolating resistor. Note that the voltage fed to the CRO is not the IF signal but only a dc voltage varying at 50 Hz. It is thus desirable that the scope amplifiers are either dc coupled or have a very good low frequency response.

(*v*) Calibrate the graticule of the CRO screen by feeding a low frequency ac voltage of known amplitude. This is necessary for meaning relative amplitudes of the picture and sound carriers with reference to the maximum amplitude of the response curve. Note that just seeing the response curve does not serve any purpose unless relative amplitudes at different frequencies are correctly set.

(*vi*) Set the sweep-mode switch of the scope on 'external' and connect 50 Hz sweep voltage from the generator to the horizontal input terminals of the scope.

(*vii*) Set the dial of the sweep generator at about 36 MHz and adjust sweep width control to get an overall variation from 28 to 43 MHz.

(*viii*) Disable tuner oscillator to prevent any interference with sweep and marker signals.

(*ix*) Set the marker generator to deliver simultaneous outputs at all important frequencies along the response curve. Make connections for post-injection of these signals through the built-in adder.

(x) Switch on all equipment and receiver. Allow a small warm up time before proceeding to adjust various tuned circuits.

IF Alignment Procedure

(i) Observe the pattern which appears on the scope screen and make horizontal gain control adjustments if necessary to obtain a suitable spread of the pattern.

(ii) To make sure that the IF stages are not being overloaded, vary step attenuator of the sweep generator to see that the height of the curve varies with changes in the RF output voltage. If the shape of the curve changes while doing so, the output of the generator should be further reduced till the amplitude of the curve varies without changing its shape.

(iii) Adjust phase control and polarity switch if necessary to obtain a single trace of desired polarity.

(iv) Use the blanking control to get a reference line during retrace interval of the sweep.

(v) Adjust slugs of the various tuned circuits to obtain an overall response as given in the service manual. In general, the response should be almost flat between 34 to 38 MHz. Use variable markers to locate these frequencies. Retune trap circuits to get almost 50% amplitude at the picture carrier marker of 38.9 MHz with reference to the maximum on the response curve. Similarly adjust the amplitude at the sound carrier marker to about 5 to 10% of the maximum amplitude. Vary the cores of adjacent channel trap circuits to obtain proper slopes at the two ends of the response, taking care that the amplitude at these frequencies is practically zero.

While adjusting tuned transformers, their slugs may produce two resonant points. The adjustment which is obtained with the core farther away from centre of the coil is normally the correct position for optimum band-width. In case the IF section is very much misaligned, each stage starting from the last IF amplifier should be separately tuned before making final adjustments as outlined above.

Sound IF Alignment

The circuit connections for aligning the sound IF and discriminator circuit are shown in Fig. 29.6. For sound IF alignment put the tuner switch on an 'OFF' channel position and proceed as under:

(i) Connect the sweep generator through a 0.001 μF capacitor to the sound IF take-off point.

(ii) Set the sweep generator at the intercarrier sound IF frequency of 5.5 MHz and the sweep width to about 1 MHz. (± .5 MHz).

(iii) Set the marker output at 5.5 MHz. In case a separate marker generator is used, loosely couple it to the same point as the sweep generator.

(iv) Adjust tuned circuit to obtain a proper response curve on the scope screen. The marker should lie at the centre of the curve.

(v) If the marker appears at some other point on the curve or the two slopes of the trace are not identical, again vary the IF transformer cores to get correct response.

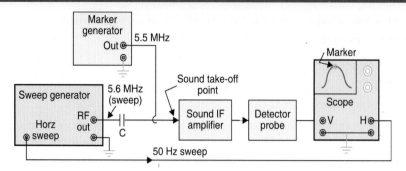

Fig. 29.6. Set up for the alignment of sound IF section of the receiver.

FM Detector Alignment

Without changing the input points and settings of sweep and marker generators shift the CRO at the output of detector circuit (Fig. 29.7 (*a*)). A convenient point is the centre point of the volume control. Use a 47 K-ohm isolating resistor between the scope and volume control take-off-point. Adjust discriminator transformer cores to obtain an 'S' shaped curve as shown in Fig. 29.7 (*b*). In case the two halves of the response curve above and below the base line are not identical, readjust the transformer cores to get the desired response. Note that the disappearance of the marker will indicate correct setting of the detector at the centre (5.5 MHz) frequency.

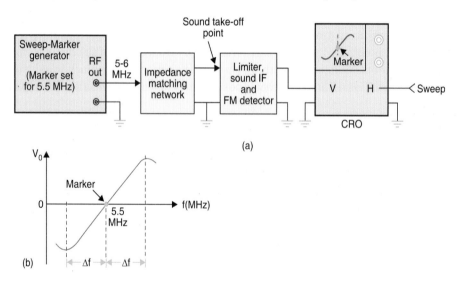

Fig. 29.7. Sound FM detector alignment (a) set-up (b) response curve.

RF Tuner Alignment

The main requirements of tuner alignment are:

    (*i*) Adjustment of tuned circuits at the antenna input terminals, RF amplifier output and mixer input, to obtain a bandwidth of 7 MHz.

    (*ii*) Setting of the local oscillator to correct frequency for each channel.

    The initial precautions are the same as detailed for IF section alignment. In addition the receiver, sweep and marker generators should be placed on a metal plate acting as a ground plane. The equipment and receiver should be properly bonded and grounded to the metal plate.

The circuit connections are shown in Fig. 29.8. The tuner should be left to have its normal AGC bias and the local oscillator should not be disabled. However, the horizontal sweep oscillator may be cut-off to avoid any undesired pick up. The shield cover of the tuner must be left 'on' because its capacitance has an effect an all tuner adjustments. The alignment procedure is as under:

(i) Connect the sweep and marker generator combination to the receiver input terminals through a balun or any other matching termination.

(ii) Set the sweep and marker generator frequencies in accordance with the channel setting on the tuner.

(iii) Adjust width control to obtain a sweep of about 10 MHz.

(iv) Adjust slugs of the tuned circuits (antenna input, RF amplifier output, mixer input) to obtain maximum gain and a symmetrical response curve. The markers on either side will indicate the channel bandwidth.

(v) Proceed as above for the remaining channels and ensure that practically same bandwidth and response curve is obtained on all the channels. Note that for each channel it would be necessary to reset the sweep generator and marker frequencies to new values.

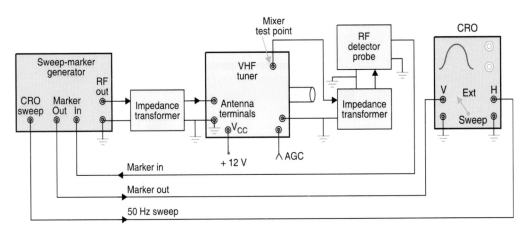

Fig. 29.8. A typical set-up for the alignment of a VHF tuner.

Local Oscillator Adjustments

It is often recommended that the tuner oscillator be adjusted on each active channel before attempting RF tuner alignment. This does not require any equipment. The fine tuning control should be set in the centre of the range and the slug of each channel oscillator varied for best possible picture and sound. The individual oscillator coil slugs are generally accessible, one at a time, through a hole in front of the tuner. It may be noted that in case a tuner coil or wafer is changed it would be necessary to use an RF oscillator to tune local oscillator frequencies to their correct values. Similarly the use of the RF generator would be necessary when the receiver is tuned and aligned for the time after assembly.

29.3 | TELEVISION TEST CHARTS

A wide range of optical test charts have been designed to provide precise functional information both for checking camera channel performance and stringent quality control at all the stages of television equipment and receiver manufacture. These include resolution test chart for evaluating picture sharpness, grey scale or gamma chart for testing contrast capability of the system and linearity test charts to test geometrical distortion and overall linearity. Universal test charts are also available which have a number of patterns combined together to rapidly evaluate linearity, resolution etc. of the television system.

A test pattern generator which is a miniature television station capable of producing non-moving pictures is used for transmitting these charts. All TV stations transmit a particular universal test chart for some time before the commencement of their programmes. This enables the viewers to have a quick check about the performance of their receivers. Though the short duration for which the test chart is transmitted is not enough for fault finding or alignment of the receiver but it does provide enough information about the nature of the fault if any. Many universal test charts have been developed but the one designed by the Radio, Electronics, and Television Manufacturers Association of America (RETMA) is commonly used. This chart is shown in Fig. 29.9. The letters noted on the various parts of the chart are used as reference while explaining the functions of corresponding sections of the chart.

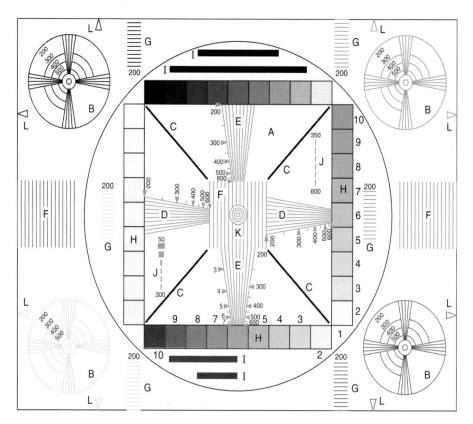

Fig. 29.9. A typical TV test chart.

The chart has a large white circular area at the centre against a grey background and four small circles at the four corners. These circles have different geometrical patterns and stripes within them, which help in the evaluation of picture quality at a quick glance. The purpose of the various geometrical patterns in the chart is explained below:

(*i*) The large white circle (*A*) at the centre indicates by its shape any distortion in horizontal and vertical linearity and correctness of aspect ratio. The small circles (*B*) evaluate geometry and resolution of the picture at the corners. These circles are also used for setting height, width, centering and linearity of the picture. All the circles should appear round and the ones at the corners should be equally visible. The centre of the raster can be set by looking at the location of the four small circles.

(*ii*) The diagonal lines (*C*) in the white circle are for checking interlacing. The lines should be thin and uniform. If interlacing is poor the lines would appear jagged.

The tapered vertical and horizontal wedges of black and white stripes inside the white circle evaluate horizontal and vertical resolution respectively.

(*iii*) The horizontal wedges (*D*) are for checking vertical resolution. The points at which the horizontal stripes appear to be merging with each other is an indication of vertical resolution. This is shown by numbers written along the wedges. With a well adjusted camera lens system and good focus setting in the picture tube, a vertical resolution of more than 400 lines can be obtained in practice.

(*iv*) The vertical wedges (*E*) are for determining horizontal resolution and video bandwidth. The number written close to the point where the vertical stripes just do not appear distinct indicates horizontal resolution. One part of the wedge is marked to read video bandwidth directly. For example, resolution lines of 200 correspond to a video bandwidth of 2.5 MHz and 400 lines correspond to 5 MHz. Resolution wedges are also provided in the four corner circles to give an indication of resolution in the respective areas. The aspect ratio of 4 : 3 is correct if the outer edges of the four central wedges appear to make a perfect square.

(*v*) The broad vertical striped boxes (*F*) in the centre and on the left and right hand sides of the chart have a resolution of 200 lines. These enable adjustments for horizontal linearity of the picture. The spacing between the vertical lines should be same for proper linearity.

(*vi*) The narrow groups of horizontal stripes (*G*) inside and outside the white circle enable evaluation of vertical linearity. For correct linearity the width and spacing of the stripes should be uniformly equal.

(*vii*) The four gradation bars (*H*) are for contract and brightness checks. Each shading step should stand out separate and distinct from others for wide contrast and proper brightness settings. With too little a contrast the shading steps will appear washed out or faded. However, with excessive contrast the shading steps will lost their form and the black shadings will appear very black. The brightness and contrast controls can be properly set by observing these gradation bars.

(*viii*) Black horizontal bars (*I*) located inside the white circle above and below the horizontal gradation bars are for checking low frequency response of the video amplifier. These lines should appear sharply defined with no leading or trailing edges. A smear effect or streaking indicates poor low frequency response.

(*ix*) The single resolution lines (*J*) in the white circle indicate ringing if any in the video amplifier response.

(*x*) The concentric circles (*K*) at the centre are for checking focusing and sharpness of the picture tube beam. The focus adjustment should be such that the central spot is smallest with surrounding circles clearly visible.

(*xi*) The corners of the eight white small triangles (L) along the edges of the chart serve as boundaries of the raster on the picture tube.

(*xii*) The uniform grey background of the white circle is useful for observing any noise interference effects on the picture.

29.4 ALL IC TELEVISION RECEIVERS

Integrated circuit technology has advanced at a very rapid pace during the past decade or so. Dedicated or special purpose ICs are now available for most sections of both monochrome and colour receivers. The only sections for which ICs have not yet been made are the tuner, video output for large screen receivers and horizontal deflection output circuits. No IC has been developed for the tuner because its associated external circuitry has lot of coils and associated switches and this defeats the very purpose of integrating the circuit. Video amplifier and horizontal sweep sections have been fully integrated for small screen receivers, but, it has not been possible sofar to do so for large screen receivers. This is mostly due to fabrication difficulties because of high voltages involved in these circuits. However, with a view to obtain more compactness and lower costs, new composite ICs are being developed which can perform all the functions now assigned to two or more integrated chips. For example, an IC has been perfected which incorporates video preamplifier, noise cancellation circuit, sync separator, horz AFC, vertical driver and vertical output stage. The chip contains 107 transistors, 25 diodes, 157 resistors and 4 capacitors. It is sealed in a 24 pin dual-in-line plastic package. This IC can be used in a 30 cm (12″) black and white television receiver circuit without any additional heat sink. Similarly another chip has been developed which not only houses the entire sound section of the receiver but also contains additional circuit blocks for use in video tape recording and playback circuits.

Monochrome TV Receiver Employing ICs

The schematic diagram of a modern monochrome receiver employing nine ICs is shown in Fig. 29.10. The functions performed by the various ICs are indicated in respective sections. The commonly used IC in each stage together with the exact or near alternatives are listed in each block.

The performance data of all integrated circuit chips is given in the manuals issued by their manufacturers. The functional circuits of all the commonly used ICs in various sections of the receiver have already been discussed in corresponding previous chapters and hence are not repeated here.

It may be noted that VHF-UHF tuner, touch control tuning and remote control circuits have not yet found their way in the black and white television receivers manufactured in India. The obvious reasons are the limited number of channels on which TV programmes are transmitted and high cost of such sophisticated innovations. In fact, with the introduction of colour television, receiver manufactures in the country may further simplify the circuitry of their black and white receivers to cut down costs in order to maintain sales.

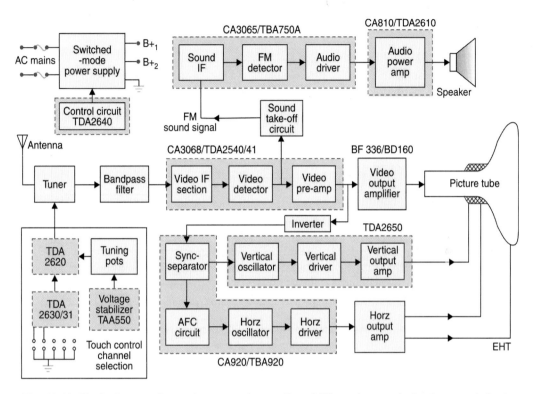

Fig. 29.10. Block diagram of a modern monochrome (B and W) receiver employing integrated circuits in most sections of its circuitry. The various ICs are shown in separate dotted chain boxes.

Modern Colour Television Receivers

In countries where colour transmission started soon after the introduction of black and white television, colour receiver manufacture also passed through 'vacuum tube' hybrid, and solid-state versions. However, at present integrated circuits are used in almost all sections of colour receivers. As mentioned earlier, the ICs meant for the picture (vision) IF, sound and several other sections of the TV receivers have been so designed that they are equally suitable for use in monochrome and colour receiver circuits. This is so, because, except for the chroma decoder section and picture tube circuitry, both *B & W* and colour receivers perform identical functions.

In India, where colour transmission begun late it was natural that colour receiver manufacturers took advantage of the advances in semiconductor technology and opted for circuits which employ ICs in almost all sections of the receiver. The schematic block diagram of a colour receiver employing 15 ICs is shown in Fig. 29.11. The receiver has been broadly divided into several blocks. Each sections is indicated by a dotted chain line boundary. The various sub-sections have been suitably labelled to indicate the function(s) performed by them. The designation and numbers of commonly used ICs are also given in each block. It may be noted that in the IF section surface acoustic wave filters (SAW) have been used instead of the conventional IF bandpass (tuned) circuits. This results in improved performance and better selectivity.

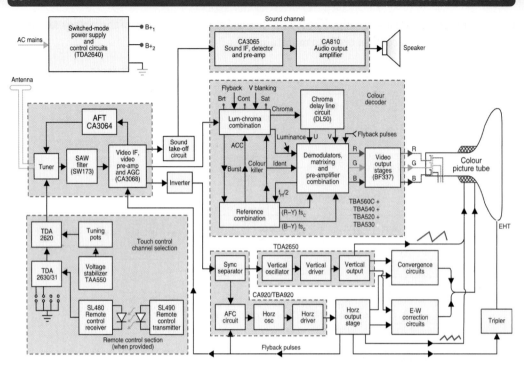

Fig. 29.11. *Schematic block diagram of a PAL-D colour receiver employing touch control channel selection.*

Dedicated ICs are marketed by several manufacturers for various sections of the receiver. However, the operating conditions, V_{CC} supply, pin numbers and in some cases the functions performed are somewhat different from each other. Therefore, their choice for any receiver circuit has to be carefully made to ensure compatibility, correct interfacing and sequence of signal processing. The chart drawn on page 631 gives a list of ICs for various sections of the colour receiver. It may be noted that some ICs are exactly identical to the ones shown in the block diagram, while others are near alternatives. Therefore, the selection for each subsection will depend on the required performance and the choice made for other sections of the receiver.

29.5   ALIGNMENT OF COLOUR RECEIVERS

The precautions and sequence for aligning colour receivers is the same as detailed for monochrome receivers in the earlier sections of this chapter. It is recommended that the manufacturer's service instructions should be strictly followed. The height, width, centring, linearity etc. are adjusted before proceeding with the alignment of various tuned circuits. The RF and IF stages are provided with fixed bias voltages as recommended in the service manual.

Picture Tube Adjustments

As explained in the previous chapter, the in-line picture tube yoke assembly and convergence circuits are so designed that very few adjustments are necessary for optimum purity, and convergence. The various techniques used to achieve this were also explained along with the picture tube constructional details. The procedure outlined below is for 'touch up' adjustments to ensure optimum purity and convergence.

Integrated circuits (ICs) for various sections of a colour receiver

No.	Function	ICs
1.	Automatic frequency tuning (AFT)	*CA 3064 MC 1364
2.	Picture (vision) IF	*CA 3068 TDA 2540/41 TCA 270
3.	IF interstage filters (surface acoustic wave filter) (SAW)	*SW 173 SW 172 SW 174
4.	Sound section (SIF and preamp.)	*CA 3065 TBA 120 TBA 750A
5.	Audio (sound) output	*CA 810 TDA 2610
6.	Sync processing (inclusive of horz osc. and AFC)	*CA 920 TBA 920 TDA 2590
7.	Vertical deflection	*TDA 2650 (Class B) TDA 2600 (switched mode)
8.	Colour decoder (various sections)	⎡*TBA 560C TBA 540 TBA 520 TBA 530
	Other IC combinations for the colour decoder	(*i*) TBA 540 + TBA 560C + TCA 800 (R.G.B. outputs) (*ii*) TDA 2560 + TDA 2522 + TDA 2530 (*y* and colour difference signals)
9.	Switched mode power supply drive (SMP)	*TDA 2640 TDA 2581
10.	Channel selection and control (a) Touch control (b) Remote control (c) μP based control	 *TDA 2620 + TDA 2630/31 + TAA 550 SL 480 + SL 490 and ML 922 CT 2012 + CT 1650 + CT 2030

*ICs labelled in the block diagram of the colour receiver (Fig. 29.11).

Chapter 29

(*a*) *Purity adjustment.* A blank snow-free raster is first obtained. Many receivers have service switches at the rear of the receiver chassis for such purposes. The blue and green guns are switched off leaving a red raster only. The deflection yoke is loosened and moved until a uniform red raster is obtained. If small areas of impurity remain, purity magnets are adjusted as instructed in the alignment literature, for best results. Blue and green rasters are next checked for uniformity by switching-in the other two guns, one at a time. All the guns are then switched on to provide a white raster. The yoke is now tightened in position after ensuring that there is no raster tilt. Colour plate No. 6 shows the effect of incorrect purity adjustments.

(*b*) *Static convergence.* After restoring the receiver to normal functioning a blank white raster is first obtained. A colour pattern generator set for dot pattern is connected to the input of the receiver. The channel selector switch is moved to a channel recommended for such adjustments. The brightness control is moved for a dark background and small white dots. The fine tuning control is adjusted for best picture. Residual colour if any is removed by varying the saturation control. The static convergence adjustments are then made (see section 25.15) so that the colour dots if any superimpose (converge) to form single white dots at the centre. An example of proper static convergence is shown in Fig. 29.12.

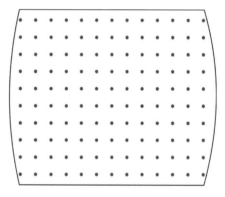

Fig. 29.12. Static convergence-correctly converged dot pattern.

(*c*) *Dynamic convergence.* For dynamic convergence adjustments, the colour pattern generator is now switched for a cross-hatched pattern. The various receiver controls are set for a sharp stable cross-hatched pattern on the receiver screen. In case the convergence is poor, dynamic convergence controls (see section 25.16) are manipulated till a white cross-hatch pattern

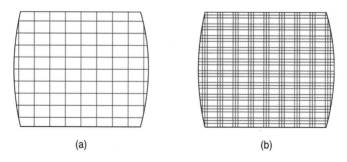

(a)　　　　　　　　　　　　　　　　　　(b)

Fig. 29.13. Dynamic convergence adjustments (a) properly converged cross-hatch pattern (b) misconverged cross-hatch pattern.

is obtained. Usually the blue gun is switched off to converge red and green first. After these two colours converge to yield yellow, the blue gun is moved in and adjustments made for a white cross-hatched pattern. The patterns show in Fig. 29.13 illustrate properly converged and misconverged cross-hatched lines. A mirror mounted on the work bench in front of the receiver screen is often used for observing patterns while making adjustments at the rear of the receiver.

Tuner and IF Section

The procedure for aligning the tuner and IF sections of the receiver is the same as for a monochrome receiver except that an additional marker at 34.47 MHz is switched in to locate the colour IF frequency. While tuning the RF section the channel response curve is made a little more flat topped to fully preserve the colour signal sidebands. Similarly the IF section is tuned to obtain nearly 50 percent amplitude (– 5 db) at the colour IF frequency. The lower slope (skirt) of the IF curve is adjusted for correct reproduction of colour video signals. In the absence of specific instructions, the RF and IF sections should be tuned to obtain response curve as shown in Fig. 22.13 and 23.2.

Sound Section

The sound section of the colour receiver is exactly same as in monochrome receiver. Therefore, its tuning procedure is the same as explained earlier for a B and W receiver.

Chroma Bandpass Amplifier Alignment

The chrominance signal from the antenna on its way to the colour demodulators passes through the tuner, video IF section, video preamplifier and the chroma bandpass amplifier (BPA). Each of these circuits has its own frequency response that affects the operation of other circuits.

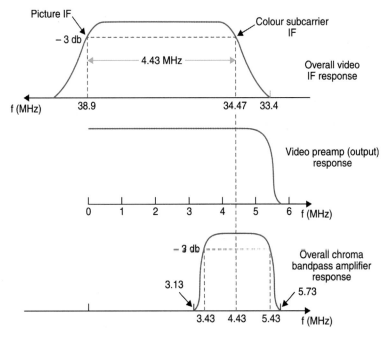

Fig. 29.14. Relative frequency response curves of the video IF amplifier, video pre-amplifier and chroma bandpass amplifier.

The necessary requirements for the RF, IF and video amplifier sections have already been explained. As shown in Fig. 29.14, the upper slope of the overall IF response and the roll-off characteristics of the video preamplifier affect the amplifier and phase of the chrominance signal feeding the bandpass amplifier. The chrominance sidebands which are close to 3.5 MHz have higher amplitudes than those near 5.3 MHz. If these inequalities are not compensated for, severe phase distortion will occur and seriously affect the colour reproduction in the picture.

The method of compensation used to achieve a flat overall response in the chrominance signal frequency range is to adjust the input tuned circuit of the chroma BPA in such a way that its frequency response curve is opposite to the slopes of the IF section and video preamplifier characteristic curves. This needs critical alignment of the bandpass amplifier. This is usually carried out in two steps. First the BPA is sweep aligned in the same way as the video IF section is aligned. This is done to ensure that the interstage and output tuned circuits of the BPA are correctly adjusted for a flat response of ± 1.0 MHz around 4.43 MHz. Then the video sweep modulation technique is used to adjust the input circuit of the BPA for a flat topped overall receiver response. The receiver RF and IF stages are aligned before setting the bandpass characteristics of the chroma amplifier.

The alignment procedure is as under:

A video sweep generator is adjusted for an output with a sweep width of 4 MHz centered around 4.43 MHz and varying at 50 Hz. It is connected to the input of chroma bandpass amplifier through a suitable network. The various tuned circuits ahead of the input network are adjusted with the help of a detector probe and CRO for a flat topped response of ± 1 MHz around 4.43 MHz.

The overall alignment of the receiver is next carried out to ensure correct passage of the chroma signal from receiver input terminals to the demodulators. The circuit hook-up for this and corresponding waveshapes at various points are shown in Fig. 29.15. A video sweep generator is adjusted for an output with a sweep width of 0 to 6 MHz varying at a rate of 50 Hz. Markers at 3.43 MHz, 4.43 MHz, and 5.43 MHz are added to it through an absorption marker. The signal is then fed to an encoder where blanking, sync and colour burst signals are added to it. Thus the signal becomes a standard TV signal.

The composite signal obtained from the encoder is amplitude modulated in an IF modulator with a carrier frequency of 38.9 MHz. The IF modulator output feeds into a converter where the video IF signal is further translated to the desired channel frequency band. The converter is fed with the desired CW channel carrier signal from an RF oscillator. The bandwidth restricted RF signal available at the output of the converter is fed through an attenuator at the input terminals of the receiver under alignment. Its channel selector switch is thrown at the same channel, the carrier frequency of which was employed at the converter.

With tuner and video IF sections properly aligned, the output of the video detector will be a modulated signal that sweeps from 0-6 MHz. This signal passes through the video preamplifier before reaching the input of the BPA. The combined roll-off characteristics of the video IF and video preamplifier sections cause the higher frequencies of the signal spectrum to become attenuated. As already explained, the BPA is tuned to a centre frequency of 4.43 MHz, with a bandwidth of about ± 1 MHz. Thus frequency components below 3 MHz in the input signal are eliminated. The input and output tuned circuits of the BPA are now adjusted to compensate for the roll-off introduced by the video IF and video preamplifier circuits. The adjustments are so made that the BPA output response is flat topped with a bandwidth of ±1.1 MHz around 4.43 MHz.

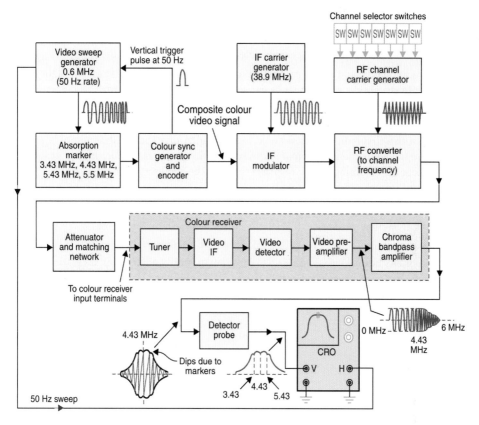

Fig. 29.15. Video sweep modulation method of aligning the chroma bandpass amplifier of a PAL colour receiver.

If the scope used has a vertical bandwidth up to 10 MHz, the sweep modulated envelope can be observed directly on its screen. The other approach is to use a demodulator probe before feeding the signal to the vertical input of the CRO. The resultant response curve will have blanking pulses and a number of marker dips that are due to the absorption markers. Such a display will clearly indicate the overall output response curve of the bandpass amplifier and the location of all important frequencies.

AFT Circuit

The AFT circuit alignment can be carried out on completion of the video IF alignment with only two changes in the connections. The generator output cable is connected direct to the AFT test point and the detector probe is bypassed.

The peak-to peak spacing of the AFT discriminator 'S' curve is usually about 1 MHz, so that at full IF sweep, the discriminator curve will appear compressed. However, it can be expanded by advancing the horizontal gain control of the oscilloscope. With a marker injection at 38.9 MHz the waveshape seen on the scope screen is used for adjusting the AFT tuned circuits.

REVIEW QUESTIONS

1. Explain the merits of employing ICs in TV receivers. Draw the block schematic of a receiver which employs ICs and enumerate the functions performed by each integrated circuit.

2. A universal test chart is shown in Fig. 29.4. Explain briefly various parts of the chart when displayed on the receiver screen enable detection of specific faults in the receiver.

3. What precautions should be observed before commencing alignment of any section of the receiver ? Enumerate the checks which must be carried out before proceeding with alignment of the RF and IF sections of the receiver.

4. Describe the initial adjustments which must be made before aligning IF section of the receiver. Explain with a circuit diagram how would you proceed to align the tuned circuits of IF amplifiers. What are the adjustments which are necessary on the sweep generator, marker generator and oscilloscope for obtaining proper response curve on the CRO screen ?

5. Describe briefly the alignment procedure and precautions for aligning the following sections of the receiver.

   (*i*) sound IF section   (*ii*) FM discriminator circuit

   (*iii*) RF tuner   (*iv*) local oscillator.

6. Explain briefly how you would proceed to carry out (*a*) purity adjustment, (*b*) static convergence and (*c*) dynamic convergence adjustments in a colour receiver.

7. Describe with a suitable circuit diagram the method of aligning the chroma bandpass amplifier of a colour receiver.

30

Receiver Servicing

Trouble shooting and repairing any equipment needs a step-by-step logical approach to locate and correct any fault in its operation. For efficient servicing of television receivers it is not enough to have adequate knowledge of functions of the various blocks of the receiver but it is also essential to have full familiarity with voltages and their waveshapes that are obtained at various points in the circuit when the set is operating normally. It is also important to be fully conversant with the functions and the effect of manipulation of all controls and adjustments provided in a receiver.

A set of proper tools and equipment helps very much in tracing any fault and repairing it. A technician who is fully conversant with the use and limitations of the instruments he possesses, and can intelligently analyze the visual and analog indications he gets with them, can do the job much faster and with good results.

Trouble shooting is a skill in itself and does not come automatically by reading books and remembering circuits alone. Even the handbooks which are exclusively devoted to servicing give only a guideline to locate faults because it is impossible to list all possible varieties of faults and their remedies. Servicing needs long experience and there are no short cuts to it. A determined effort and hard work coupled with a thorough understanding of the operating principles of the receiver is necessary to become a good servicing engineer. An ability to correctly interpret service literature and carry out alignment and adjustments accordingly is equally important to obtain best results.

30.1 TROUBLE SHOOTING PROCEDURE

General procedure for trouble shooting a black and white receiver is given in the chart drawn on the next two pages. The first step towards finding the cause of failure of a receiver is visual inspection of the components, tubes, transistors and ICs mounted on various printed circuit boards, modules and chassis of the receiver. Such an inspection sometimes leads quickly to the defective device or part in the receiver. For example, burnt or charred resistors can often be spotted by visual observation. Similarly overheated transformers, oil or wax filled capacitors and coils can be located by a peculiar small caused by overheating and shorting. In addition a disoldered wire of a coil or component, broken load of a transistor or other similar faults can be found out by a preliminary visual inspection.

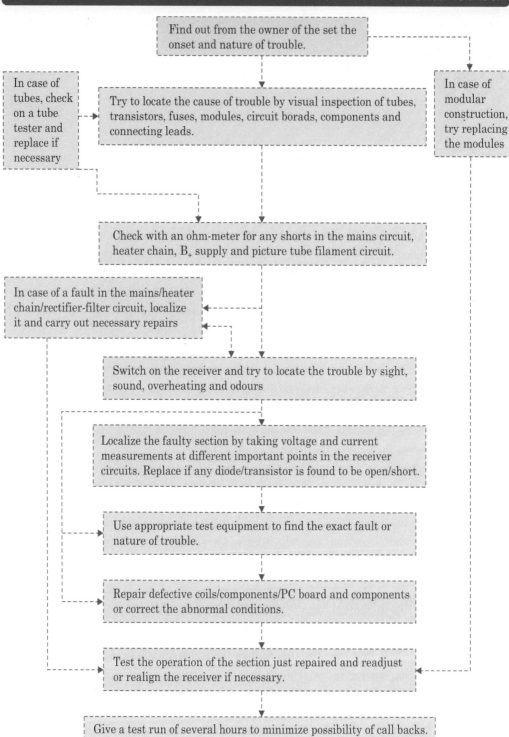

When any such fault is located, it is a good practice to immediately ensure with an ohmmeter that there is no partial or full short in any section of power supply of the receiver. If a faulty capacitor is found to be the cause of excessive current drain, it should be replaced by a new one having the same capacity and voltage ratings. If a component is found burnt or charred it should be replaced by an identical piece. Similarly broken leads, if any, should be carefully soldered at their correct locations. To ensure that the short has been removed, another quick check of the low voltage (LV) power supply should be carried out before connecting the receiver to the mains. Absence of any short on the dc supply line is indicated by a momentary deflection of the ohmmeter pointer (caused by surge charging current of filter capacitors) when the meter is connected across any B_+ point and chassis ground. Even when no visual fault is located, it is advisable to carry out the ohmmeter check before connecting the receiver cord to the mains plug.

As soon as the power is turned on, the tube filaments including that of the picture tube should be checked for proper glow and warm up. Any break in the filament circuit, caused by a broken lead or fused tube filament should be located and repaired/replaced. A burnt tube should be replaced after checking the associated circuit to determine if the tube failed naturally from long use or due to some trouble in the circuit. Similarly, if it is found that a burnt fuse or circuit breaker (contactor) was the cause of discontinuity across the mains lead, the same should be replaced after making sure that there is no short in the mains transformer.

In case any arcing is observed or heard in the high voltage section of the receiver when the set is connected to the mains, it should be switched off immediately. With the receiver off, the high voltage capacitor should be discharged, fault localized and rectified before turning on the receiver again.

In many cases, repairs carried out on the basis of visual inspection and preliminary tests restore normal functioning of the receiver. However, it is often necessary to investigate further to localize the fault. This is carried out by watching the raster or picture on the screen and listening to the sound output from the loudspeaker. In fact the competence of a serviceman is his ability to recognize and isolate trouble by the visual information he gets from the picture tube screen and the aural output from the speaker.

Fault Localization

A television receiver is an economical and technically feasible combination of two receivers in one. The picture and sound signals have a common path from antenna to the video detector after which the two separate out to their respective channels. Another section of the receiver provides necessary signals for producing the raster and maintaining synchronism between the televised scene and reproduced picture. Thus the known path of the video, audio, and sync signals together with the symptoms observed on the screen and noted from the sound output can form a basis for localizing any trouble in one or more sections of the receiver. For example, if a receiver exhibits a distorted picture accompanied by a distorted audio output, it becomes obvious that the defect is in a circuit which is common to both the signals. Thus the fault is localized to the RF and IF sections of the receiver. On the other hand, if only the sound output is missing or appears to be distorted, the fault lies in the sound section following the point of separation of the two signals. Similarly if nothing but a white horizontal line appears on the screen, it becomes evident, that the trouble lies somewhere in the vertical circuit. However, if the picture holds vertically but tears diagonally, the trouble probably lies either in the horizontal sync or horizontal sweep circuit. Thus the method of observing symptoms from picture and sound is very efficient for fault location and saves much servicing time. In many cases it becomes possible to localize the fault to a particular section soon after turning on the receiver.

Chapter 30

Signal Source for Observing Faults

Though any transmission from a local TV station can be used for observing possible faults but continuous change of picture details on the screen makes it difficult to observe minor irregularities in the reproduced picture. A test pattern obtained either from a test chart generator or an analyst generator shows much more than a picture and so is very useful for detecting all types of faults in the picture. However, if these instruments are not available, a pattern generator set to produce a cross-hatched pattern on the screen can be used as input signal source. In any case, an initial observation by tuning in any one of the local channels is helpful for localizing trouble to one of the major functional areas of the receiver. Once this has been done, the normal trouble shooting tools can be used to isolate defective component or components of that section.

30.2 TROUBLE SHOOTING MONOCHROME RECEIVERS

Though it is not practicable to list all symptoms of various faults but most troubles in a black and white receiver can be grouped into functional areas or circuits in the receiver. This approach is often followed and the chart that follows gives all necessary details. The probable cause of any trouble as listed in the chart is based on the assumption that the receiver had been working satisfactorily and suddenly developed the observed symptoms. It may be noted that if alignment and tracking has been tampered with, sound and/or picture could be lost because of off-resonance conditions in the tuned circuits. It is further assumed that all tubes inclusive of picture tube are glowing and there is no fault in the filament circuit(s).

Trouble Shooting Chart Based on Symptoms in a Monochrome Receiver

No.	Symptoms	Defective Stage(s)	Probable Fault(s)
1.	No sound, no raster	L.V. Power supply	B_+ shorted, defective rectifier(s), (failure of low voltage systems).
2.	No sound, no picture, but raster normal	RF tuner, IF section, video section	A bad tube, transistor, IC or any part in the common sections of the receiver, loose antenna or transmission line connections.
3.	No sound, but picture normal	Sound IF, FM detector, audio	Defective sound IF, discriminator, audio amplifier, audio output transformer and loudspeaker.
4.	Sound distorted, but picture normal	Sound IF, FM detector, audio	Misaligned sound IF or discriminator, gassy tubes, leaky transistors, bad IC, short or open capacitors in audio amplifier, shorted output transformer, defective speaker.

(Contd.)...

5.	Sound normal, but no raster	High voltage power supply, picture tube circuit	Failure of high voltage power supply, defective picture tube, high negative bias on picture tube control grid, defective video amplifier.
6.	Sound normal, raster normal but no picture	Video output, video driver, picture tube circuit	Defective tubes, transistors, IC or components in stages after sound take-off.
7.	Sound normal, but picture distorted	IF stages, AGC, vertical and horizontal sweep circuits	Misaligned IF stages, defective AGC tube/transistor, wrong AGC bias setting, nonlinearity in vertical and horizontal sweep circuits.
8.	No vertical deflection	Vertical sweep circuit	Vertical oscillator not working, vertical output transformer open, defective output tube/transistor, low B_+ supply, open vertical deflection coil.
9.	Insufficient vertical height	Vertical sweep circuit	Weak oscillator tube/transistor, low boosted B_+ supply, faulty output transformer, shorted turns in deflection coil, faulty VDR if used, leaky coupling or bypass capacitors, open resistors.
10.	No vertical sync. (Picture rolls)	Sync separator, vertical sweep circuit	Defective in sync separator to vertical input circuit, vertical oscillator frequency too high or too low, misadjusted vertical hold control.
11.	Non-linear vertical sweep (distorted picture)	Vertical sweep circuit	Faulty tubes/transistors, fault in feedback loop, misadjusted vertical linearity control.
12.	Poor interlacing or line pairing	Vertical deflection section	Open capacitor in vertical sync integrator, open decoupling capacitor(s) in oscillator and output stages.
13.	Insufficient width of raster	Horizontal deflection circuit	Weak oscillator tube/transistor, defective (shorted) horizontal output transformer, weak horizontal output tube/transistor.

Chapter 30

(Contd.)...

Chapter 30

14.	Foldover in picture but sound normal	Horizontal deflection circuit	Incorrect horizontal linearity, defective damper, incorrect horizontal sweep.
15.	Non-linearity in the horizontal direction	Horizontal deflection circuit	Incorrect horizontal sweep output, wrong linearity control setting.
16.	Keystone effect in the raster	Horizontal and vertical output circuits	Partial short in vertical/horizontal coils.
17.	Picture not centered but sound normal	Picture tube circuit	Incorrect setting of centering controls, defective centering controls.
18.	Negative picture	AGC circuit, video detector, picture tube circuit	Overloaded video signal, wrong polarity of detector diode.
19.	No horizontal sync (heavy slanting streaks across the screen) but sound normal	Sync separator, AFC, horizontal oscillator	Horizontal sweep out of synchronism, defective sync separator, defective diode in the AFC circuit.
20.	Incorrect picture size, but sound normal	Horizontal and vertical sweep circuits, yoke coils	Weak tubes/transistors in vertical and horizontal sweep output circuits, low B_+ supply, defective yoke.
21.	Poor resolution	IF section, video amplifier	Misaligned IF amplifiers, insufficient video amplifier bandwidth.
22.	Weak picture and sound, but both of good quality	Tuner and IF sections	Defect in antenna system, improperly oriented antenna, defective tubes/transistors in RF and IF sections, defective IC in the IF section.
23.	Picture and sound weak and also of poor quality	RF tuner, IF section	Shift in local oscillator frequency, change in fine tuning control, improper IF alignment.
24.	Hum or motor boating from loudspeaker	Audio section	Defective filter capacitors, defective decoupling capacitor.
25.	Sound bars on picture screen	Video IF section	Misaligned sound IF traps (too much gain at the sound IF frequency).

30.3 SERVICING OF VARIOUS FUNCTIONAL BLOCKS

After the trouble is localized to a single functional area, the next step is to isolate the trouble to a circuit and then to the faulty component or components. It is not advisable to charge into the receiver with a soldering iron and screw driver without a logical search for the cause of trouble. The circuit diagram, if available, should first be studied and various measurements taken before deciding to change any component or removing any device for checking it. This saves time and needs least use of the soldering iron.

The major faults that occur in a monochrome TV receiver and their servicing procedure, based on the available indications and observations, is explained below. For convenience and ease of understanding various faults have been divided according to main sections of the receiver.

1. Low Voltage Power Supply (B_+)

Low voltage power supply feeds dc voltage to various sections of the receiver and filament voltage to tubes. All solid state receivers normally have a regulated dc source and a separate filament transformer for heater of the picture tube. It is important to realize that any fault in the B_+ supply affects performance of all sections of the receiver. Therefore it is necessary to ensure that this part of the receiver is performing satisfactorily before attempting to localize faults in other sections of the receiver.

Low voltage power supplies either operate from a transformer or are derived from the line. Whenever servicing a line connected receiver, an isolation transformer with 1 : 1 voltage ratio must be used to avoid accidental shocks and ensure protection to test instruments.

(a) An inoperative (dead) set. This implies that there is no light on the screen and no sound from the loudspeaker. The tube filaments may or may not be glowing or some of the tubes may only be glowing.

Transformer type power supply—Check for:

    (*i*) defective thermal cut-out (if provided), fuse or on-off switch,

    (*ii*) defective power cord, cable or interlock,

    (*iii*) burnt filament of the rectifier tube (if used) or weak rectifier,

    (*iv*) open filament winding(s),

    (*v*) defective transformer secondary or open centre tap on secondary,

    (*vi*) open filter choke,

    (*vii*) shorted filter capacitor(s),

    (*viii*) open filament winding of picture tube.

Line connected power supply

    (*i*) defective thermal cut-out, fuse or on-off switch,

    (*ii*) defective power cord plug, cable, or interlock,

    (*iii*) open or short in the filament circuit,

    (*iv*) shorted filament bypass capacitor,

    (*v*) open silicon rectifiers,

    (*vi*) open filter choke or resistor in B_+ line,

    (*vii*) open filament winding of picture tube.

Note. A short in some section of the B_+ load will cause all voltages to read low and often cause the fuse to blow out or the circuit breaker (if provided) to trip. Excessive current drain due to shorts in the load can be traced by opening connection at different branching points of B_+ line and monitoring dc voltage with a voltmeter. When there is a partial short, the trouble is most likely at the point of lowest voltage. A multimeter can be used to locate the defective component in the power supply line.

(b) Insufficient width and height of raster (Fig. 30.1)—This occurs due to low B_+ supply. The vertical and horizontal deflection circuits are normally stabilized for ± 10% mains voltage variations. However, if the mains voltage drops more than 10% the deflection circuits are affected. The dc supply can also drop due to a fault in its rectifier-filter circuits. Check for:

  (i) open filter capacitor,

  (ii) open diode in voltage doubler or bridge rectifier circuit,

  (iii) excessive current drain at some point in B_+ line.

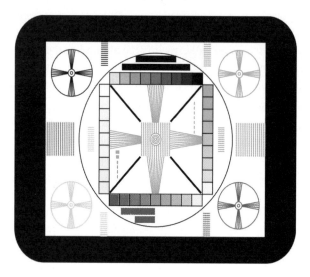

Fig. 30.1. Shrunken raster (insufficient width and height) due to low B_+ supply.

(c) Hum bar(s) or bend in the raster and/or hum in sound output from the speaker (Fig. 30.2)—Check for:

  (i) open output filter capacitors,

  (ii) filament to cathode leakage in tube receivers.

(d) Poor sync, folded picture, non linearity in vertical scanning, motorboating in sound output—Check for:

  (i) open filter capacitor(s).

  (ii) faulty decoupling capacitor(s).

  (iii) poor regulation of B_+ supply.

2. High Voltage Power Supply and Picture Tube Circuitry

Before proceeding to locate faults in EHT and picture tube circuits, it should be ensured that the horizontal sweep amplifier and its damper circuit are functioning normally. Note that without horizontal scanning there is no high voltage and no brightness.

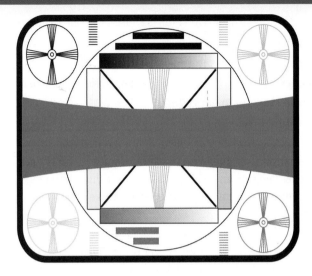

Fig. 30.2. Hum bar in picture due to 50 Hz hum voltage in the video signal.

E.H.T. Supply

(a) *Not raster*—Check for:

(i) open or shorted picture tube filament,

(ii) defective (open or short) HV rectifier,

(iii) defective damper diode,

(iv) shorted or open EHT winding,

(v) defective picture tube.

Note. A dim picture may be produced due to a weak EHT rectifier or a partial short in EHT winding on the horizontal output transformer (H.O.T.).

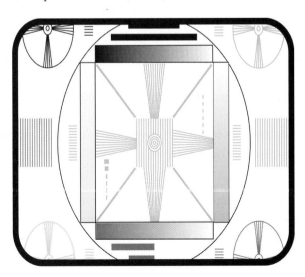

Fig. 30.3. Blooming of the reproduced picture.

(b) Blooming of picture (Fig. 30.3)—This is caused by poor EHT regulation. Check for:

    (*i*) leaky HV capacitor,

    (*ii*) weak HV rectifier,

    (*iii*) high resistance in the EHT lead due to a defective series resistor,

    (*iv*) insufficient rectifier filament voltage,

    (*v*) defective picture tube.

(c) Break-up of raster, jumping of picture, horizontal tearing, series of light streaks on the screen—These faults are due to arcing and corona in the EHT section. An arcing occurs when potential difference between any two points is greater than the voltage, the dielectric between them can withstand. Corona is normally associated with an electric discharge from a sharp point in a high voltage circuit.

Arcing may be seen on the picture tube screen or directly from the source of arc. Similarly corona may be heard as a sizzling or frying noise. It may also be recognized by a purplish-blue discharge at the source or by the presence of ozone gas smell. Although corona may not have an immediate effect on the picture, it will cause gradual deterioration of components if left unchecked. For any visible arcing check for:

    (*i*) high voltage leads lying too close to chassis ground,

    (*ii*) improper grounding of external aquadag coating of the picture tube,

    (*iii*) partial short at the mountings of high voltage rectifier unit.

Picture Tube Circuitry

No raster, sound normal—Check for:

    (*i*) defective picture tube.

    (*ii*) incorrect voltages at various electrodes of the picture tube (rare),

    (*iii*) incorrect setting of picture tube controls because of defect(s) in external circuitry,

Manufacturers usually supply data for rated voltages at various electrodes. In the absence of such information check the following with a voltmeter.

    (*i*) with cathode as reference the control grid voltage should be between – 30 V and – 120 V, and

    (*ii*) the second grid voltage should be about 400 volts while that of the focus grid (G_4) between 0-400 V.

    (*iii*) loose/broken connections in picture tube socket, yoke plugs, and high voltage leads,

    (*iv*) defective video amplifier coupling circuit.

3. Horizontal Sweep Circuits

The horizontal sweep circuit operates at 15625 Hz and is synchronized by an AFC circuit. This section is also the source for EHT supply. In solid state receivers, other dc voltages are also obtained from the horizontal sweep system.

(a) Single vertical line in the centre of the screen, sound normal (Fig. 30.4)—Check for:

    (*i*) open horizontal deflection coil.

    (*ii*) open linearity coil,

    (*iii*) open dc blocking capacitor.

Fig. 30.4. Single vertical line in the centre of the raster due to failure of horizontal sweep circuit.

(b) *Poor horizontal linearity*—Check for:

   (*i*) incorrect drive voltage amplitude,

   (*ii*) incorrect drive voltage waveform,

   (*iii*) leaky coupling capacitor between horizontal oscillator and output stage,

   (*iv*) defective output tube/transistor,

   (*v*) shorted linearity control.

(c) *Horizontal foldover* (Fig. 30.5)—Foldover is a severe form of nonlinearity. Check for:

   (*i*) defective output tube/transistor,

   (*ii*) faulty biasing circuit of horizontal output stage,

   (*iii*) defective oscillator or output circuit.

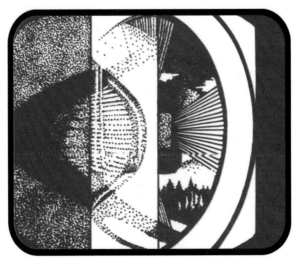

Fig. 30.5. Horizontal foldover at left side of the raster.

Chapter 30

(d) *Horizontal keystone effect* (Fig. 30.6)—This is indicated by a dim and narrow raster with its bottom wider than the top or vice versa. Check for:

(i) defective horizontal yoke (shorted turns),

(ii) shorted balancing capacitor across one section of the horizontal deflection coils.

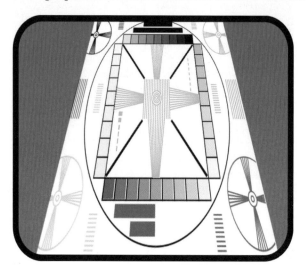

Fig. 30.6. Horizontal keystoning effect. The pattern shown is the result of a shorted horizontal coil in the deflection yoke.

(e) *Ringing* (Fig. 30.7)

(i) defective tube/transistor,

(ii) linearity coil damping resistance open,

(iii) defective L.O.T.,

(iv) incorrect yoke balancing capacitor,

(v) open capacitor in the cathode of the damper tube.

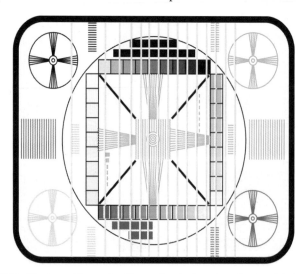

Fig. 30.7. Ringing-white vertical bars due to improper horizontal damping. The pattern shown is due to open capacitor in cathode circuit of the damper diode (tube).

(f) Loss of horizontal sync (horizontal pulling) (Fig. 30.8)

   (*i*) defective sync separator circuit,

   (*ii*) low sync amplitude,

   (*iii*) defective AFC circuit,

   (*iv*) defect in anti-hunt circuit,

   (*v*) incorrect horz oscillator frequency beyond control of the AFC circuit.

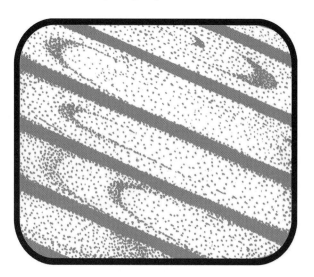

Fig. 30.8. *Loss of horizontal synchronization.*

4. Vertical Sweep Circuit

This circuit feeds vertical deflection coils and operates at 50 Hz. It is synchronized by sync pulses developed at the output of a low-pass filter.

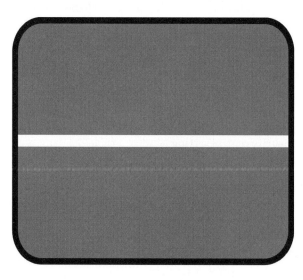

Fig. 30.9. *Loss of vertical sweep.*

(*a*) *Bright horizontal line on the screen, sound normal* (Fig. 30.9)—Check for:

    (*i*) open vertical deflection coil connections,

    (*ii*) faulty tubes/transistors in the circuit,

    (*iii*) faulty IC if employed,

    (*iv*) vertical oscillator not functioning.

(*b*) *Insufficient picture height* (Fig. 30.10)

    (*i*) weak oscillator and/or output tubes/transistors,

    (*ii*) defective IC if employed.

    (*iii*) faulty output transformer,

    (*iv*) misadjusted height control,

    (*v*) shorted turns in deflection coil.

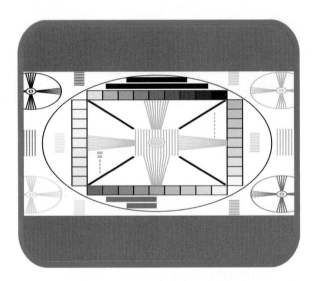

Fig. 30.10. *Insufficient height of the picture.*

(*c*) *Poor vertical linearity (height satisfactory)*—Check for:

    (*i*) low B_+ voltage,

    (*ii*) defective tube/transistors/IC,

    (*iii*) leaky coupling capacitor,

    (*iv*) incorrect linearity control setting,

    (*v*) wrong drive voltage due to fault in negative feedback loop of the oscillator,

    (*vi*) partial short in output transformer.

(*d*) *Picture rolls vertically* (Fig. 30.11)—Check for:

    (*i*) oscillator frequency too high or too low,

    (*ii*) incorrect setting of hold control,

    (*iii*) fault in vertical sync processing circuit.

Chapter 30

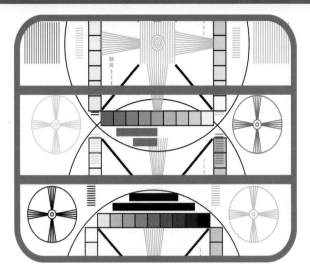

Fig. 30.11. Picture rolls vertically at a fast rate due to complete loss of vertical synchronization.

(e) *Poor interlacing* (Fig. 30.12)—Check for:

  (i) open capacitor in vertical sync integrator,

  (ii) open decoupling capacitor in vertical or horizontal sweep circuits.

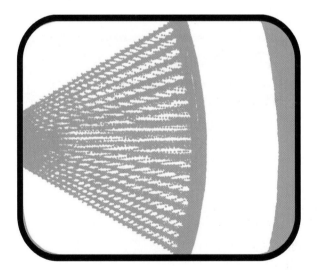

Fig. 30.12. Horizontal wedge showing poor interlacing.

(f) *Vertical keystoning* (Fig. 30.13)—Check for:

  (i) shorted turns in vertical deflection coils,

  (ii) defective vertical output transformer,

  (iii) defective thermistor across deflection coils.

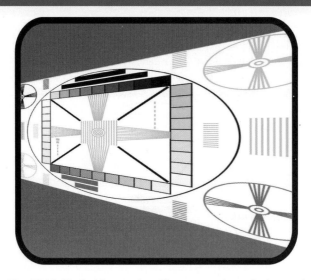

Fig. 30.13. Vertical keystoning. The pattern shown is the result
of defective vertical coil in the deflection yoke.

(*g*) *Cramping or foldover at the bottom of raster* (Fig. 30.14)—Check for :

  (*i*) incorrect bias on output tube/transistor,

  (*ii*) distorted sweep voltage waveshape,

  (*iii*) low B_+ supply.

Fig. 30.14. Foldover at bottom of the raster.

(*h*) *Excessive vertical size of the picture*—Check for :

  (*i*) low picture tube anode voltage,

  (*ii*) low value of sweep generator saw-tooth charging capacitor.

5. Sync Separator and Processing Circuit

The sync pulses are separated from the composite video signal by a sync separator. Filter circuits separate vertical and horizontal sync pulses from each other. The vertical sync pulses are integrated to form vertical oscillator control voltage. However, horizontal sync pulses are fed to the AFC circuit which develops a dc control voltage for the horizontal oscillator. The noise pulse effect is thus reduced.

(*a*) *Poor or complete loss of both vertical and horizontal sync, picture breaks into strips and rolls vertically* (Fig. 30.15)—Check for:

      (*i*) defective sync separator,

      (*ii*) defective coupling capacitor to sync circuit,

      (*iii*) faulty tube/transistor/IC used in the system.

Fig. 30.15. *Complete loss of both vertical and horizontal synchronization. The picture breaks into strips and rolls vertically.*

(*b*) *Loss of vertical sync (picture rolls vertically)* (Fig. 30.16)—Check for:

      (*i*) defective integrating circuit (leaky capacitor or open resistance),

      (*ii*) short, open or leaky coupling capacitor.

(*c*) *Loss of horizontal sync (picture breaks into strips and rolls in horizontal direction*—Check for:

      (*i*) defective diodes in AFC circuit,

      (*ii*) defective coupling capacitor,

      (*iii*) open diode load resistance,

      (*iv*) absence of flyback pulses from L.O.T.

(*d*) *Vertical jitter (picture jumps up or down)*—Check for:

      (*i*) open capcitor in low pass filter (integrating circuit),

      (*ii*) defective sync separator tube/transistor/IC,

Chapter 30

(*iii*) incorrect bias on sync tube/transistor,

(*iv*) incorrect B_+ to the circuit,

(*v*) open by-pass capacitor.

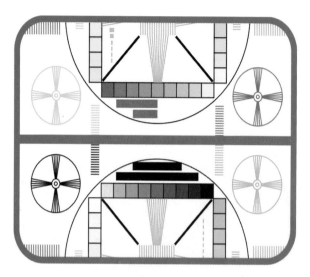

Fig. 30.16. *Vertical scanning not synchronized–picture slowly slips frames vertically.*

(*e*) *Picture pulling (bend in the picture with edges of raster straight)* (Fig. 30.17)— Check for:

(*i*) incorrect bias to sync separator,

(*ii*) excessive hum in vertical sync.

Fig. 30.17. *Picture pulling-bend in the picture.*

Note. With excessive hum in vertical sync, picture tends to show vertical blanking bar across the middle of the picture.

(*f*) *Gear-tooth effects (cogwheel distortion)*—This is usually caused by hunting in the horizontal AFC circuit. Check for:

      (*i*) incorrect component values in AFC circuit,

      (*ii*) leaky anti-hunt capacitor.

    (*g*) *Horizontal pulling* (Fig. 30.18)—Check for:

      (*i*) defective AFC diodes,

      (*ii*) leaky or open anti-hunt capacitor,

      (*iii*) leaky sync coupling capacitor,

      (*iv*) defective sync tube/transistor/IC.

Fig. 30.18. Horizontal pulling.

    (*h*) *Incorrect horizontal phasing (picture moves to the sides with a vertical bar between the two pictures)*—Check for:

      (*i*) leaky wave-shaping capacitor,

      (*ii*) defective AFC diodes.

6. Video Amplifier Circuit

It is a wide-band amplifier which increases the amplitude of the composite video signal before applying it to the picture tube grid or cathode. The video amplifier could be dc coupled, ac coupled, or partially dc coupled with upper 3 db down frequency extending up to about 5 MHz.

    (*a*) *No picture, raster normal, sound satisfactory*—Check for:

      (*i*) short between grid and cathode of picture tube,

      (*ii*) no B_+ supply to video amplifier,

      (*iii*) defective video amplifier tube/tansistor/IC,

      (*iv*) open compensating coils,

      (*v*) open load resistor,

(*vi*) open coupling capacitor

(*vii*) open bias resistor,

(*viii*) open contrast control potentiometer.

(*b*) *Smeared picture* (Fig. 30.19)—Check for:

(*i*) open peaking (compensating) coils,

(*ii*) low value of coupling capacitor,

(*iii*) open cathode/emitter or other bypass capacitors,

(*iv*) misalignment of IF amplifier states.

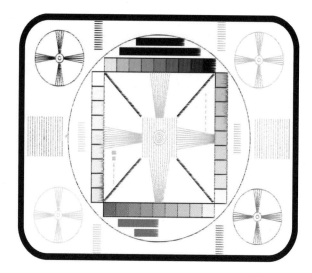

Fig. 30.19. *Smeared picture caused by excessive low frequency response with phase distortion in the video amplifier.*

Note. If blacks in the picture are not fully black the reason could be a shorted video coupling capacitor.

(*c*) *Sound bars on picture* (Fig. 30.20)—Check for:

(*i*) defective sound trap circuit.

(*ii*) misalignment of IF stages (sound IF receiving excessive gain).

(*d*) *Ringing (oscillations in picture detail)*—Check for:

(*i*) open shunt resistance across peaking coils,

(*ii*) improper shielding.

(*e*) *Weak (dim) picture*—Check for:

(*i*) weak tube/transistor/defective IC,

(*ii*) defective picture tube (rare),

(*iii*) incorrect bias,

(*iv*) faulty contrast control,

(*v*) excessive AGC bias.

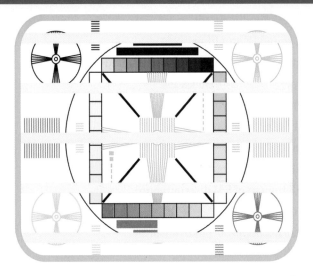

Fig. 30.20. White bars on the picture due to presence of sound signal at the output of video amplifier.

(*f*) *Intermittent picture*—Check for:

(*i*) defective tube/transistor/IC,

(*ii*) poor soldering at joints,

(*iii*) dirty or worn out brightness and contrast controls,

(*iv*) intermittent capacitors or resistors in circuit due to rise in temperature.

7. Video Detector

It demodulates the picture (IF) signal to recover composite video signal. It also heterodynes picture and sound IFs to produce intercarrier sound signal.

No sound, no picture, raster normal—Check for:

(*i*) open diode,

(*ii*) open bandwidth compensation coil,

(*iii*) defective filter circuit components,

(*iv*) open coupling capacitor,

(*v*) short in the last IF transformer secondary,

(*vi*) incorrect biasing circuit (if provided).

8. Sound Section

It amplifies the intercarrier sound signal picked up from the video detector or first video amplifier. An FM detector demodulates it and the resulting audio signal is amplified by an audio amplifier before it feeds the speaker.

(*a*) *No sound, picture and raster normal*—Check for:

(*i*) faulty tube/transistor/IC,

(*ii*) low dc supply,

    (*iii*) defective discriminator circuit,

    (*iv*) misaligned sound IF amplifier,

    (*v*) open circuit from sound take-off point to sound IF input.

  (*b*) *Weak or distorted sound*—Check for:

    (*i*) weak tube/transistor/faulty IC,

    (*ii*) low sound IF amplifier gain,

    (*iii*) low dc supply,

    (*iv*) improper alignment of sound IF,

    (*v*) improper alignment of FM discriminator,

    (*vi*) defective loudspeaker,

    (*vii*) partially shorted audio output transformer.

  (*c*) *Hum in sound, picture and raster normal*—Check for:

    (*i*) open filter capacitor in L.V. supply,

    (*ii*) open ground lead of volume control,

    (*iii*) poor shielding,

    (*iv*) poor alignment of FM detector,

    (*v*) heater cathode short in tube circuits.

  (*d*) *Buzz in sound, picture and raster normal*—Check for:

    (*i*) improper alignment of the secondary of FM detector,

    (*ii*) improper alignment of sound IF and limiter stages,

    (*iii*) open capacitor in ratio detector,

    (*iv*) improper picture IF alignment.

9. AGC Circuit

The AGC circuit develops a voltage proportional to the channel signal strength which is used for controlling gain of IF and RF stages. RF amplifier is normally supplied with delayed AGC.

  (*a*) *Weak picture*—This is due to excessive bias voltage to IF and RF amplifier stages. Check for:

    (*i*) incorrect adjustment of AGC control,

    (*ii*) faulty tube/transistor/IC,

    (*iii*) shorted AGC filter,

    (*iv*) faulty keying circuit.

  (*b*) *Loss of sync*—If AGC is low, one of the IF stages may become over-loaded and cause distortion of sync pulses. Check for:

    (*i*) weak AGC tube/transistor/faulty IC,

    (*ii*) low value of delay resistor,

    (*iii*) inadequate AGC filtering.

(*c*) *Dark picture and buzz in sound*—Check for:

   (*i*) defective tube/transistor/faulty IC,

   (*ii*) leaky AGC filter capacitor,

   (*iii*) low AGC on strong signals.

(*d*) *Negative picture* (Fig. 30.21)—The negative or reversed black and white in the picture is caused by reversal of the polarity of video signal. This occurs as a result of rectification in an overloaded amplifier which is usually the last IF stage. Check for:

   (*i*) absence of composite video signal from the video amplifier section to the AGC circuit.

   (*ii*) absence of reference flyback pulses from the H.O.T.,

   (*iii*) faulty tube/transistor/IC in the AGC circuit,

   (*iv*) leaky or shorted capacitor in AGC circuit.

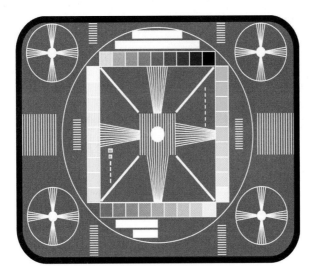

Fig. 30.21. *Negative picture.*

10. Video (picture) IF Section

The IF section provides most of the gain and is AGC controlled. It needs precise tuning for vestigial sideband correction and successful working of intercarrier sound system.

(*a*) *No picture, no sound, but raster normal*—Check for:

   (*i*) no L.V. supply.

   (*ii*) open tube/transistor/inoperative IC,

   (*iii*) excessive AGC bias,

   (*iv*) open/shorted IF transformer winding,

   (*v*) break in printed board (module),

   (*vi*) open cathode/emitter resistance.

(*b*) *Poor contrast or weak picture* (Fig. 30.22)—Check for:

Chapter 30

(*i*) low B_+ supply,

(*ii*) weak tube/tansistor/faulty IC,

(*iii*) incorrect AGC,

(*iv*) misaligned IF stages.

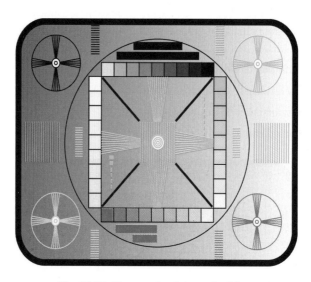

Fig. 30.22. *Poor contrast or weak picture.*

11. **Tuner Section**

(*a*) *No picture, no sound, raster normal*—Check for:

(*i*) defective tube/transistor,

(*ii*) faulty contacts on selector switch,

(*iii*) dry joints

(*iv*) break in printed circuit board,

(*v*) low dc voltage supply.

(*b*) *Weak picture, weak sound, raster normal*—Check for:

(*i*) weak or gassy tube/defective transistor.

(*ii*) incorrect AGC bias,

(*iii*) poor antenna connections,

(*iv*) open balun.

(*c*) *Intermittent picture and sound*—Check for:

(*i*) loose antenna and lead-in cable connections,

(*ii*) faulty selector switch contacts,

(*iii*) dry and loose joints.

(*d*) *Snow on picture and/or noisy sound* (Fig. 30.23)—Check for:

(*i*) incorrect dc supply voltage,

(*ii*) dry and loose joints,

(*iii*) defective RF amplifier,

(*iv*) weak local oscillator,

(*v*) open balun.

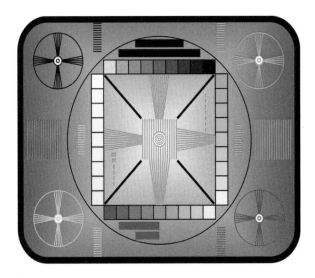

Fig. 30.23. White speckled background of snow due to poor signal to noise ratio.

12. Miscellaneous Faults

(*a*) *Ghost picture* (Fig. 30.24)—Check for:

(*i*) wrong balun connections,

(*ii*) improper shielding of tuner,

(*iii*) wrong orientation of antenna.

Fig. 30.24. Ghost images due to reflections of arriving signals.

Chapter 30

(b) *Picture not centered, sound normal*—Check for:

   (i) misadjusted centering controls,

   (ii) defective centering controls.

(c) *Only a portion of the picture area in focus*—Check for:

   (i) incorrect voltage in the focus electrode circuit.

(d) *Brightness control partially or completely inoperative*—Check for:

   (i) defective brightness control

   (ii) low B_+ voltage to brightness control,

   (iii) defective picture tube (rare).

(c) *Contrast control ineffective*—Check for:

   (i) excessive input signal at antenna,

   (ii) defective contrast control,

   (iii) shorted AGC bypass capacitor.

(f) *Retrace lines visible*—Check for:

   (i) open input capacitor or resistor in blanking circuit,

   (ii) open bypass capcitor to blanking circuit.

30.4 TROUBLE SHOOTING COLOUR RECEIVERS

The general procedure for trouble shooting and servicing colour receivers is the same as for monochrome receivers. The chart given in section 30.1 can also be followed for localizing defective sections of the receiver. This is so, because, except for the chroma section of the receiver, the remaining circuitry performs identical functions in both B and W and colour receivers. However, the chart given on the next page can serve as a good guide for identifying defective section(s) in a colour receiver. It may be mentioned that most auxiliary circuit blocks in the chroma section affect the colour killer circuit and disable it when they become inoperative. Thus, it may be necessary to bypass the colour killer block while locating a faulty circuit in the chroma section.

30.5 SERVICING CIRCUIT MODULES

(a) Printed-Circuit Servicing Techniques

Printed-circuit wiring is used in television receiver manufacture to ensure uniformity of product, speed of assembly, and ease in servicing. When working on printed-circuit boards, extreme care should be exercised while soldering and desoldering components. Due to the nature of laminating process, the printed circuit board should not be exposed to prolonged high temperatures. In manufacturing, excessive heating is prevented by a special technique called wave soldering. To avoid excessive heating a small soldering iron (25 to 40 watts) should be used. Whenever possible the soldering iron should be kept next to or below the printed-circuit board to permit the solder to flow onto the iron, thus preventing the spread of solder over the board. The soldering iron should always be wiped clean before desoldering. Special desoldering tools are available where a suction device is used. The solder on melting is sucked into a tube clearing the point of solder. While working on a printed board, care must be taken not to smear solder across other connections or leave splinters adhering to the board.

(b) Repairing Cracks in Printed-Circuit Boards

Cracks in printed-boards may make good contact while cold and open up as the receiver heats up. To repair this type of crack a copper wire should be soldered across the crack. The wire will expand and contract along with the copper pattern and thus form a permanent joint.

Fault Localization in a Colour Receiver

No.	Defective stage :	Symptoms				Remarks
		Raster	Picture	Colour	Sound	
1.	Video IF/ video detector	Clean raster at maximum contrast setting	No picture	No colour	No sound/ normal	
2.	Luminance channel amplifier	No visible raster even at maximum brightness setting	No picture	No colour	Normal	
3.	Sound channel	Normal	Normal	Normal	No sound	
4.	Vertical deflection	Bright horizontal line	No picture	No colour	Normal	
5.	Horizontal deflection	No raster	No picture	No colour	No sound	B_+ supply for most sections of the receiver is obtained from the line (horz) output stage
6.	Colour killer circuit	Normal	Picture with colour snow at maximum setting of saturation control	—	Normal	
7.	Chroma bandpass amplifier	Normal	Black and white picture	No colour	Normal	
8.	Burst gate amplifier	Normal	Normal	No definite colours (colours roll)	Normal	Colour killer circuit assumed operative

(Contd.)...

Chapter 30

9.	Automatic frequency and phase control circuit	Normal	Normal	Loss of colour synchronization, no proper colours in the picture	Normal	
10.	Reference subcarrier oscillator	Normal	Normal	No colour	Normal	
11.	Colour demodulator (U. V.)	Normal	Normal	Loss of blue or red colour in the picture depending on which demodulator is inoperative	Normal	The green colour reproduction will also be affected.
12.	Colour difference amplifiers $(R–Y), (G–Y), (B–Y)$	Normal	Normal	Loss of corresponding colour and other colours produced by mixing this colour	Normal	Picture will not have proper colours

(c) Integrated Circuits (ICs)

It is now feasible to assemble into a single silicon chip dozens of units including diodes, transistors, FETs, resistors and associated wiring with overall size limited only by the necessity of bringing out connecting leads. The tiny integrated circuit has number identified leads brought out for external connections. The ICs have many packages, like circular square or rectangular. Some ICs have prongs that fit into holes in the printed circuit board for soldering to the metallic-conducting strips.

Schematically ICs are often shown in triangular form. Only those terminals are usually numbered that are actively used for the particular circuit or electronic device formed. Because of their tiny structure and the necessity to avoid overheating, great care must be exercised while soldering and removing ICs.

Testing of ICs. Testing of ICs is a complex procedure due to large number or components inside the package. Various testers are now available for testing different ICs. If the fault is isolated to the IC and its external circuit, it is usually easier to check the external components of the circuit first. If they are good, the IC is suspected and should be replaced with a new one.

In most schematics, the detail of the internal construction of the IC is not shown. Testing in such cases is restricted to checking of dc and signal voltages at the various pins as shown on the schematic.

(d) Servicing Plug-in Modules

The trend in present day receivers is towards use of plug-in modules. Suitable inter-circuitry is often used to unify transistors and components needed for performing certain television into one sectional module. Thus a module may contain all the circuitry for the vertical sweep system; sound section; video detector, video amplifier and so on. Such modulus with plug-in facilities, simplify servicing to some extent and reduce component replacement time. While the use of replaceable modules often minimizes the number of steps required for trouble shooting, it is still necessary to check connecting circuits outside the module. Frontpanel operating controls can be used for checking circuit operation because they are not located in the sealed unit but are connected to the terminals of the plug-in module. Disadvantages of the use of modules include the need for stocking various replacement modules of different manufactures, and their higher cost compared to replacement of a single transistor, diode, capacitor etc.

30.6   SAFETY PRECAUTIONS IN TELEVISION SERVICING

The following general safety precautions should be observed during operation of test equipment and servicing of television receivers:

(*i*) A contact with ac line can be fatal. Line connected receivers must have insulation so that no chassis point is available to the user. After servicing all insulators, bushes, knobs etc. must be replaced in their original position. The technician must use an isolation transformer whenever a line connected receiver is serviced.

(*ii*) Voltage in the receiver, such as B_+ and EHT can also be dangerous. The service man should stand or sit on an insulated surface and use only one hand when probing a receiver. The interlock and the back cover of the receiver must always be replaced properly to ensure that the receiver's high voltage points are not accessible to the user.

(*iii*) Fire hazard is another major problem and must get the attention it deserves. Technicians should be very careful not to introduce a fire hazard in the process of repairing TV receivers. The parts replaced must have correct or higher power rating to avoid overheating. This is particularly important in high power circuits.

(*iv*) The picture tube is another source of danger by implosion. If the envelope is damaged, the glass may shatter violently and its pieces fly great distance with force. It can cause serious injury on hitting any part of the body. Though in modern picture tubes internal protection is provided but it is necessary to handle it carefully and not to strike it with any hard object.

(*v*) Many service instruments are housed in metal cases. For proper operation, the ground terminal of the instrument is always connected to the ground of the receiver being serviced. It should be made certain that both the instrument box and the receiver chassis are not connected to the hot side of the ac line or any point above ground potential.

(*vi*) All connections with test leads or otherwise to the high-voltage points must be made after disconnecting the receiver from ac mains.

(*vii*) High voltage capacitors may store charge large enough to be hazardous. Such capacitors must be discharged before connecting test leads.

(*viii*) Only shielded wires and probes should be used. Fingers should never be allowed to slip down to the meter probe tip when the probe is in contact with a high voltage circuit.

Chapter 30

(ix) The receiver should not be connected to a power source which does not have a suitable fuse to interrupt supply in case of a short circuit in the line cord or at any other point in the receiver.

(x) Another hazard is that the receiver may produce X-radiation from the picture tube screen and high voltage rectifier. It is of utmost importance that the voltage in these circuits are maintained at the designed values and are not exceeded.

REVIEW QUESTIONS

1. Describe briefly the basic trouble shooting procedure a technician must employ to quickly localize the fault in a TV receiver. Justify how a fault can be localized by observing the picture and listening to the sound.

2. Tabulate likely fault(s) and faulty section(s) of the receiver for the following visual indications:
    (i) no sound, no picture but raster normal,
    (ii) sound normal but no raster,
    (iii) poor interlacing,
    (iv) keystone effect in the raster,
    (v) negative picture,
    (vi) poor resolution,
    (vii) audio bars on picture screen.
    (viii) weak picture and sound but of good quality.

3. Indicate the effect of following faults on the raster, picture and sound.
    (i) open filter choke,
    (ii) defective (gassy) damper diode,
    (iii) high resistance in the EHT lead,
    (iv) horizontal oscillator not working,
    (v) Excessive AGC voltage,
    (vi) vertical oscillator drifts,
    (vii) fine tuning not proper,
    (viii) shunt resistance across the peaking coil open,
    (ix) brightness control open,
    (x) open balun.

4. Enumerate the visual and aural (sound) indications that would result in a colour receiver when the following circuits of the chroma section become defective/imperative:
    (i) Colour killer circuit,
    (ii) Automatic frequency and phase control circuit,
    (iii) Reference subcarrier oscillator,
    (iv) (R–Y) video amplifier,
    (v) 'U' demodulator.

5. Write short notes on:
    (i) printed circuit board servicing,
    (ii) testing of ICs,
    (iii) servicing plug-in modules.

6. Enumerate safety precautions which must be observed while servicing a television receiver.

31

Satellite Television

The main drawback of terrestrial broadcasting is in its limiting range because of the earth's curative which eventually breaks the signal path thus preventing reception over long distances. This problem is solved with geo-stationery satellites that orbit the globe at the same speed as the rotation of earth. The curvature of earth thus no longer presents any problem and communication over long distance is carried out through satellites.

At the receiving end, the signals received are very weak and special dish antennas and equipments are needed to process them. The distribution of received signals to the subscribers is either through cables or retransmission over the existing terrestrial network. Direct-To-Home reception is also possible for which each installation needs its own receiving dish antenna and a decoder-receiver. All aspects of satellite transmission, reception and signal distribution are briefly described in this chapter.

31.1 SATELLITE COMMUNICATION SYSTEM

Conceptually satellite television is very much similar to broadcast television. It is a wireless system for delivering television programmes directly to the viewer's house or through a Cable network. Both broadcast *i.e.,* terrestrial television and satellite television transmit programmes via a radio signal. However, the main drawback of broadcast television is its limited range. The radio signals from the broadcasting antenna propagate in a straight line. In order to receive these signals, the receiving antenna has to be in the direct 'LINE-OF-SIGHT" of the transmitting antenna. The main obstacle to this is the earth's curvature which eventually breaks the signal's line of sight thus preventing reception over long distances.

Satellite television solves the problem of range by transmitting broadcast signals from satellites orbiting the earth. Since satellites are high in the sky, signals can be received over long distances by using special antennas called Dish-Antennas. The television satellites are in a GEOSYNCHRONOUS orbit meaning that they stay in one place in the sky relative to the earth. Each satellite is launched into space at about 7000 miles per hour reaching approximately 22300 miles (*i.e.,* 35887 kms–usually referred to as 36000 kms) above the earth. At this speed and altitude, the satellite revolves around the planet (earth) once every 24 hours–the same period of time it takes the earth to make one full revolution. In other words, the satellite keeps exact pace with the earth's rotation. This way, the receiving dish antenna has to be directed only once at the satellite and it will continue to pick up signals without any adjustment.

Satellite Signal Path

In satellite communication, the transmitting station sends information to the satellite as RF waves which in turn retransmits it to the receiving ground station. The frequencies used for this purpose are in the microwave band from 3 to 30 GHz (one GHz $=10^9$ Hz) because at these ultra high frequencies, the atmosphere no longer acts as barrier and signals are able to travel out into space and back without any obstruction, absorption or deflection. Figure 33.1 shows the basic technique of communication by a satellite. The signals from the studio or elsewhere are sent to the transmission earth station which converts them to the assigned UP-LINK microwave frequency band. The transmitting dish antenna then beams them in the direction of unseen satellite. The satellite is equipped with its own antenna which receives the up-link signals and feeds them to a receiver. The receiver is designed to amplify and change the signals to another base frequency called DOWN-LINK frequency. This is done primarily to prevent interference between the up-link and down-link frequencies. The frequency converted wideband signal is then fed to a transmitter located in the satellite which provides additional amplification and retransmits it back to earth through another antenna. The receiving earth station which is located far-away from the transmitting earth station, suitably orients its dish antenna to receive the down link signal. The picked up signal is given enough amplification and converted back to its original form as at the sending end source. The band-width allotted for up-link and hence down-link transmission is 500 MHz, wide enough to accommodate about 12 TV channels each originating from a different studio location but combined as a band at the earth transmitting station.

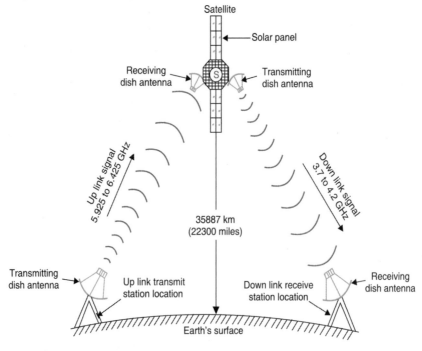

Fig. 33.1. Signal paths in a satellite communication system.

31.2 TELEVISION SIGNAL PROCESSING AND TRANSMISSION

Electronic signals can be analog or digital. An analog signal like the output of a microphone can have any amplitude within a range, for different frequency components. In contrast, in a digital system, information is represented in discrete or digitized form rather than continuous as in analog. The most common is the binary form, which means that for any signal only two discrete states are possible which are normally denoted as 'O' and '1'. In digital circuits, the 'O' state is usually represented by 0 volt or a very low voltage and the '1' state is represented by a voltage around 2 to 5 volts. The usage of only two discrete signal level results in a very significant noise immunity advantage for digital circuits. If the noise level at the input is very small, it is sharply attenuated between the input and output while the logic signals are transmitted at full amplitude. This is one of the main merits of signal processing by digital means.

While it is very advantageous to carry out signal processing in digital form, the real world is analog. Therefore, analog signals must first be converted to digital form before processing, and the digital results must be converted back to analog form for human consumption. For this, special high speed converters have been designed. An A/D (analog to digital) converter converts analog signal to digital form as a binary pulse train depending on the amplitude of analog signal at any given instant. Similarly, a D/A (digital to analog) converter does the reverse to convert the processed digital signal back to analog form.

Digitization of signal path has gained importance due to many advantages of the digital system. These include noise suppression and improved quality of reproduction. Its impact on television has been almost a complete transition from analog to digital form of signal encoding. Such a transformation has resulted in improved picture quality and better sound output. For A/D convention of video and audio signals, their analog from as obtained from TV camera output and microphone respectively are sampled at a very fast rate but at fixed discrete levels to obtain corresponding data streams in terms of 0's and 1's which are actually low and high voltage levels causing together a square wave nature continuous output stream. These are FM modulated on a suitable carrier and then up-converted to the allotted micro frequency band before up-linking to the satellite.

Data Multiplexing

The entire signal data is sent in the form of packets separately for video and audio components. These packets are time multiplexed meaning, data of each packet is transmitted in a sequential manner. This is controlled by a multiplexing control system operated electronically. Multiplexing takes place at a very fast rate to ensure steady stream of the contents of each packet at the receiving end.

Modulation and Transmission

Frequency modulation (FM) is preferred in all satellite communication systems because of its greater noise immunity as compared to amplitude modulation (AM). The sequence of operations is shown in Fig. 31.2. The base band video, audio and associated signals are frequency modulated around a centre carrier frequency of 70 MHz and the resulting output is converted to the 6 GHz range before high power amplification and transmission.

Chapter 31

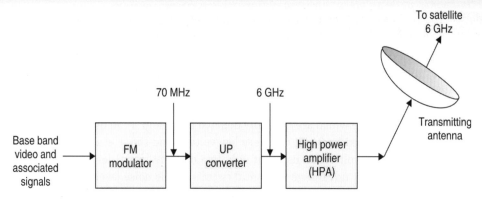

Fig. 31.2. Simplified block diagram of a transmitting earth station

Transmitting Dish Antenna. Attenuation of microwave signals, both during up-linking and down-linking, is very large. Therefore; special parabolic reflector antennas called dish antennas are used to obtain high gain and directivity. This is achieved by using a horn in conjunction with a parabolic reflector which is a large dish shaped structure made of either screen mesh or metal. The energy radiated by the horn is pointed at the reflector which focuses it into a narrow beam and directs it towards its destination. The same dish antenna works in the opposite way at the receiving end for collecting down-link signals.

The transmitting dish antennas are very big in size because these have to handle large up-link signal power. They are located close to the transmitting site to keep losses to a minimum in coaxial cables that link them. Such dish antennas are very large in diameter and mounted on strong foundations to prevent any shaking by high speed winds. Also there is provision to move the dish horizontally and tilt it up and down for precisely directing it towards the receiving antenna of the assigned satellite.

31.3 DOWN-LINK SIGNALS - RECEPTION AND DECODING

For beaming signals from a satellite down to earth, its down linking antenna is inclined at a particular angle which depends on its parking location in the sky and 'Foot Print' (signal receiving) area on the earth. This angle is called 'Look Angle' of the satellite. Therefore, to acquire any satellite, the earth station's dish antenna must point at the 'look angle' of the satellite, meaning orient itself for capturing maximum signal strength. For this, two angles called Elevation and Azimuth are specified. The angle of elevation is the angle which appears between the line from the earth station antenna to the satellite and the line between the antenna and earth's horizon. The azimuth angle is the angle of direction and is measured clockwise from north pole taken as zero angle point and the plane containing earth station and satellite antennas.

Orientation of Dish Antenna

As explained above AZIMUTH and ELEVATION angles are required for correct orientation of dish antenna. These are calculated by a set procedure, Then, for acquiring any satellite for a given location the dish antenna is first set in the horizontal plane at the corresponding azimuth

Chapter 31

angle with the help of a common magnetic compass. Similarly, for setting the dish at known elevation angle, an ANGLE METER' is used. This is illustrated in Fig. 31.3.

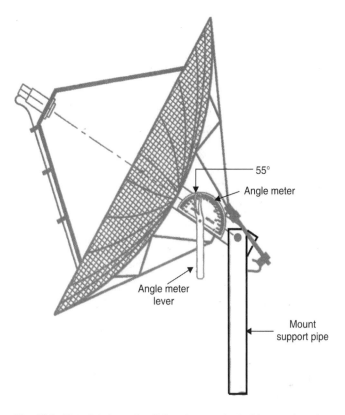

55°

Angle meter

Angle meter lever

Mount support pipe

Fig. 31.3. Pictorial view of a dish antenna oriented for an elevation angle (EL) = 55 degree and azimuth angle (AZ) = 27 degree.

RECEPTION AND DECODING. The radiated microwave signal on reception by the earth dish antenna is converted by the LNBC (Low Noise Block Converter) to RF output at 950 to 1450 MHz. It is then processed by the Receiver-Decoder to obtain composite video and audio outputs. These are then amplitude modulated in the usual way on assigned channel carries for distribution to subscribers in different ways. A simplified functional block diagram of the receiver-decoder is shown in Fig. 31.4.

Its various sections are:

(*i*) MICRO-CONTROLLER (µC). It controls all the signal processing blocks and selects the desired channel on receiving key command from a remote controller.

(*ii*) TRANSPORT IC. The signal after receiving demultiplexing and error correction is processed by the transport IC to isolate video and audio data packets. The separated video and audio payloads are then sent to receptive video and audio decoders. A decoder is a circuit block built around a digital IC to perform desired functions.

Chapter 31

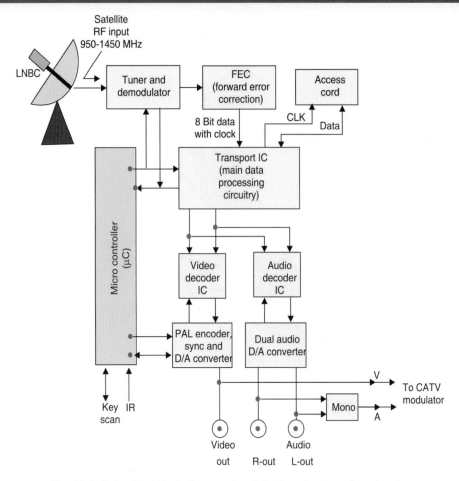

Fig. 31.4. A simplified block diagram of a digital broadcast receiver decoder.

(*iii*) **VIDEO AND AUDIO DECODERS.** These two ICs separately decode the corresponding video and audio data and send these to corresponding PAL video and audio encoders, which in turn provide desired outputs as shown in the block diagram.

31.4 SIGNAL DISTRIBUTION TO SUBSCRIBERS

There are three ways by which television TV channel signals can be made available to subscribers. These are:

    (1) Cable Television (CATV)

    (2) Domestic Broadcast System

    (3) Direct To Home (DTH) Satellite Television.

All the three models are separately explained:

    1. *Cable Television.* This is the most common system because it does not have the restriction of channel allocations as is necessary in terrestrial broadcasts. Thus it can offer a variety of programmes on a large number of channels. In big cities, cable

TV stations are located in different locations, each independent of each other and deliver programmes on around 60 to 100 channels to their subscribers. For collecting signals from different satellites, cable operators install 20 to 30 dish antennas each with its own LNBC which in turn feed into a large number of tuner-receiver decoders. As explained in the previous section, each decoder gives video and audio signal as its output which become input to the channel modulators.

2. *Domestic Broadcast System.* Domestic broadcast from geo-stationery Satellite is another way that is used for TV signal distribution all over the country. In this, signals from the chosen satellite are received by a large number of fairly small and simple earth receiving stations located all over the country. The digital signals thus received are processed to obtain video and audio signals as per PAL TV standards. These are then processed as in conventional analog AM transmitters and rebroadcast terrestrially by low power transmitters (LPT) operating at different channels in the VHF band. In India, there are over 800 such small stations that cover practically the entire populated area of the country.

3. *Direct-To-Home (DTH) Satellite Television.* The Direct–To–Home (DTH) or Direct Satellite Broadcast (DSB) system enables viewers to access directly many channels of high quality digital video programming over a vast area from one or more high powered KU-Band satellites. There is no need for complex cable networks for signal distribution. KU-Band frequencies are preferred because these are not prone to interference from ground point-to-point communication and also need much smaller dish antennas. These are positioned on top of the building and directed at the Look-Angle of the wanted satellite. Since C-band and KU-band signals do not interfere with each other thus allowing a single satellite to relay both types of transmission.

REVIEW QUESTIONS

1. Explain how the limiting range of terrestrial broadcasting is solved with geostationery satellite that orbit the globe at the same speed as the rotation of earth.
2. What is look-angle of a satellite? Explain how the ground station dish antenna is correctly oriented to obtain signals from a particular satellite.
3. With the help of block schematic of receiver decoder shown in Fig. 31.4 explain how the video and audio signals received from RF signals are obtained with video and audio decoders and converters.
4. Explain briefly the three main systems that are used to distribute TV channel signals to subscribers. What is the main merit of DIRECT-TO-HOME (DTH) satellite television system?

Chapter 31

32

Liquid Crystal and
Plasma Screen Television

From the time television became available, TV receivers have been built around a cathode-ray tube (CRT) for display of pictures. In a colour receiver, three guns (R, G & B) in the CRT fire beams of electrons towards the screen to excite corresponding colour phosphors coded on its inner surface. The phosphor atoms on excitation emit red, green and blue colour lights which on scanning enable a picture on the CRT screen. Though the pictures are quite bright and crisp, such receivers are bulky, mainly due to the weight of cathode-ray tube.

However, over the years, demand for bigger screen receivers has been growing for better viewing of television programmes. While efforts were made to increase the CRT screen size, but because of limitations in glass tube technology it has not been possible to go beyond around 40″ screens. As such, research work had been going on to develop screens larger than possible with CRTs. Liquid Crystal Display (LCD) was the first to emerge as an alternative and initially found applications in computer monitors, laptops and as indicators in watches and various equipments. In a LCD display, the screen consists of a liquid crystal solution in-between two clear glass panels. An electric current passed through the solution causes the crystals to act like a shutter, either blocking the incident light or allowing it to pass through. This phenomena is used to cause light and dark areas on the LCD screen which when regulated result in pictures.

Another alternative that become available is the Plasma display, a better choice for TV screens. In it, a solution is coated on inner side of the glass panels. The solution has millions of phosphor coated miniature glass bubbles containing plasma, which is a gas made up of free flowing ions and electrons. An electric current flow through the solution causes certain plasma containing bubbles to emit ultra-violet (UV) rays which trigger the phosphors to produce colour lights. These when combined and controlled, result in colour pictures on the screen.

This chapter aims at introducing the functioning of LCD and PLASMA panels and their applications as television receiver screens.

32.1 LCD TECHNOLOGY

Liquid Crystals. In solids, molecules always maintain their orientation and stay in the same position with respect to each other. In liquids, molecules change their orientation and move anywhere in the liquid. However, there are some substances where the molecules tend to maintain their orientation like in solids but move around to different locations as in liquids. These are called liquid crystals.

Liquid Crystal Types. Most liquid crystals have rod shaped molecules and are classified as Thermotropic or Lyotropic. Thermotropic liquid crystals can be either Isotropic or Nematic. While molecules in Istropic Liquid crystals are random in their arrangement but in Nematic type these have a definite order or pattern. Their pattern can be changed on application of magnetic or electric charge across them and this forms the basis of LCD displays.

Operation. Basically LCD screens (liquid crystal displays) consist of a very large number of liquid crystals pushed in the space between two glass plates. If an electric charge is applied to liquid crystal molecules they untwist. On straightening out they change the angle of light passing through them so that it no longer matches the angle of the top polarized filter. Consequently, no light can pass through that area of the LCD, which makes it darker than the surrounding areas. If the glass panel is divided into a large number of sections insulated from each other, the nature of applied charge applied to them will produce either dark or light areas. Insulating electrodes are added to the panel for making connections to various sections. In practice all LCD displays consist of millions of tiny sub-areas called pixels as in CRT screens. Each receives the charge through columns and rows of conducting material. The combined action of untwisting the liquid crystals on all the pixels is to cover the entire screen area, with black, grey and white mini-areas. The sequence and amplitude of charge voltage supplied to all the pixels in accordance with the input video signal result in a compact pattern or picture on the display screen.

Liquid crystal materials emit no light of their own. The type of LCD panels used in laptops are lit with built-in fluorescent tubes. A white diffusion panel put behind the LCD screen redirects and scatters the light evenly to ensure uniform display. In LCD type TV screens the backlight is light blue in colour as is seen on conventional TV receivers.

32.2 LCD SCREENS FOR TELEVISION

If voltage applied to a liquid crystal is carefully controlled, it can be made to partially untwist to allow some light to pass through it. By doing this in exact and very small increments, LCDs can create a grey scale that varies from full brightness to darkness. Most displays as also in television create 256 levels of brightness per pixel. An LCD that can show colours must have three sub-pixels each with red, green or blue filter to create one colour pixel. By careful control of the intensity of applied voltage to each sub-pixel, a range of 256 colour shades can be obtained which is enough to display full spectrum of visible light. This will need separate addresses for each sub-pixel with corresponding increase in columns rows and an enormous number of TFTs (thin film transistors) etched onto the glass. Though this appears to be an impossible task but present day advances in micro-electronics technology enable to build such LCD panels for different applications including television screens. LCD screens enable good colour reproduction and high contrast. These can be built in larger sizes than possible with CRT tubes. Though large, they are light in weight and can be installed easily, even hung from a wall. In general, LCD displays have fixed resolution and colour display but this varies with the viewing angle.

Chapter 32

32.3 | PLASMA AND CONDUCTION OF CHARGE

The basic idea of a plasma display is to illuminate tiny coloured fluorescent lights to form an image. Each pixel is made up of three fluorescent lights - a red, a green and a blue light. The variation of intensity of these colour lights results in a full range of colours on the display panel.

Plasma

The central element in a fluorescent light is plasma, a gas made up of free flowing ions and electrons. Under normal conditions a gas is mainly made up of uncharged particles where individual gas atoms have equal number of protons and electrons which balance each other so that the atom has a net zero charge. If many free electrons are introduced into the gas by establishing an electronic voltage across it, the situation changes very quickly. The free electrons collide with atoms knocking loose other electrons. With a missing electron, an atom loses its balance to become an ion with a net positive charge. In the plasma thus created with electric current flowing through it, negatively charged particles rush towards its positively charged area and positively charged particles move towards the negatively charged area. In this mad rush, particles are constantly bumping into each other. These collisions excite gas atoms in the plasma, causing them to release photos of energy in the form of ultraviolet light. This sequence is illustrated in Fig. 32.1. The ultraviolet photons cause the release of visible light photons that illuminate the display.

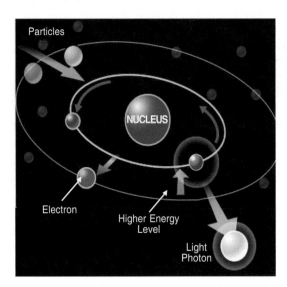

Fig. 32.1. Illustration of how light photons are released.

32.4 | PLASMA TELEVISION SCREENS

Xenon or Neon gas atoms are used in plasma screen televisions. The gas is contained in hundreds of thousands of tiny cells positioned between two plates of glass. Long electrodes are also sandwiched between the glass plates on both sides of the cells. The address electrodes are behind the cells along the rear glass plate. The transport display electrodes which are surrounded by an insulating dielectric material and covered in a magnesium oxide protective layer are mounted above the cell along the front glass plate. Both sets of electrodes extend across the entire screen. The display electrodes are arranged in horizontal rows along the screen and the address electrodes in vertical columns, thus forming the basic grid. As obvious, the idea is the same as explained for LCD displays to enable the charge to reach each cell in the screen.

To ionize the gas in a particular cell, the plasma display's computer charges the electrodes that intersect at the cell. The computer does so thousands of times in a small fraction of a second charging each cell in turn. When the intersecting electrodes are charged on application of a small voltage, an electric-current flows through the gas in the cell. As explained earlier, the current creates a rapid flow of charged particles which simulate the gas atoms to release ultraviolet photons. The released ultraviolet photons interact with phosphor material coated on the inside wall of the cell. When an ultraviolet photon hits a phosphor atom in the cell, one of its electrons jumps to a higher energy level and the atom heats up. When the electron falls back to its normal level, it releases energy in the form of visible light photon which illuminate the screen.

The phosphors in a plasma display give-off colour light when they are excited. Each pixel is made up of three separate subpixel cells with different colour (R, G, B) phosphors. Their colours blend to create the overall colour of the panel. By varying the pulses of current flowing through the cells, the control system can increase or decrease the intensity of each subpixel colour to create hundreds of different combinations of red, green and blue. In this way, the control system can produce colours across the entire visible spectrum. In television, the control system sends synchronized current pulses in accordance with the R, G, and B video signals obtained on demodulating and processing the received channel signal. This results in colour pictures similar to those being televised at the transmitting station.

There is no flicker as all the phosphor excited pixels react at the same time during one frame of scanning. There is also no backlight and no projection of any kind. As such, the light emitting phosphors result in bright pictures with rich colours and wide viewing angle. Though plasma screens are thin, they are heavy and consume lot of power. These are also fragile and often need professional help to install them.

32.5   SIGNAL PROCESSING IN PLASMA AND LCD TV RECEIVERS

To begin with RF input is processed as in digital receivers and satellite decoders. Later, the output data is fed to panels through a computer to produce pictures. The sequence of signal processing is as follows:

RF Input

The receivers are designed to receive both digital satellite and terrestrial analog and digital transmissions. These are also HD ready, meaning can receive HDTV transmissions. In both 625 line PAL and 525 NTSC due to interlaced scanning, there is no picture information for

Chapter 32

quite a few lines due to vertical retrace. There is a gap of about 49 lines in 625 and 45 in the 525 line systems. So the active lines per frame are 576 for PAL and 480 for the American system. These are usually referred as 5762 and 4801 line TV systems.

Tuner and TV Section

The tuner which is microprocessor controlled is designed to accept all the above stated TV signals. Remote control enables selection of desired transmission and channel. The tuner otherwise is as in modem colour receivers. The selected RF signal is demodulated and amplified in the tuner in the conventional way to obtain IF signal. From this onwards, it is all digital. The demodulated and selected IF signal is digitized and then fed to digital signal processing circuits with the aid of a micro-computer. Before further processing video and audio IF signals are separated.

Audio Signal Processing

The Audio Codec processes the input signal and passes output to the audio processor. This enables audio output which is then converted to analog form by a D/A converter and on amplification fed to the loudspeakers.

Video Signal Processing

The video codec produced on a VLS chip along with video processor processes the IF signal to provide luminance and chrominance outputs. The luminance signal on due processing sets the brightness and contrast of the picture in accordance with the users settings. The chromance signal on processing provides R, G, G video components which on matrixing result in R, G, B video outputs.

Sync Signals

The IF output also feeds into a sync-processor. It separates horizontal and vertical sync pulses which are used to keep the reproduced picture in step with the transmitting picture details.

Central Control Unit (CCU)

The central control unit is in fact a control computer that controls other chips and translates the user's instructions from a remote controlled transmitter.

Creating the Picture

The most complex job is of interfacing of R, G, B outputs and sync details with the inputs on the Plasma and LCD screens. It is done by a computer that is programmed to receive, process and deliver proper outputs at exact intervals and for predetermined periods to the address lines (columns and rows) of all the liquid crystals in LCD screens and subpixels in Plasma screens. All inputs to the computer are digital. In case of interlaced data, it is first converted to progressive data. The computer also controls the size of picture produced and image scanning rate in conformity with the chosen transmission. The manner in which the display screens produce pictures is explained in previous sections.

It is now seen that a Plasma/LCD receiver is a sophisticated equipment having two distinct sections. The first is a high quality digital receiver that on processing provides audio, video

Chapter 32

and sync signals. The second section is an inbuilt computer which is programmed to interpret these signals to create pictures on the Plasma/LCD panel as per user's input through a remote control unit.

32.6   PERFORMANCE OF LCD TV RECEIVERS

Early LCD sets were widely discredited for their poor overall image quality, most notably the ghosting on fast-moving images, poor contrast ratio, and muddy colours. In spite of many predictions that other technologies would always beat LCDs, massive investment in LCD production and manufacturing has addressed many of these concerns. Present day LCD receivers apart from becoming increasingly price competitive have the edge over plasma in several other key areas. LCDs tend to have higher native resolution than plasmas of similar size, which means more pixels on a screen. LCDs also tend to consume less power than plasma screens. With some newer 'Eco' panels, LCD sets use almost half power then equivalent plasmas, with the trade-off being lower brightness. In terms of bulk, LCDs are also generally lighter than similar-sized plasmas, making it easier to move around or wall mount. This is because LCDs use plastic in their screen make-up whereas plasmas tend to use glass.

LCDs sets can in theory be built at any size, with production yields being the primary constraint. As yields increased common LCD screen size grew from 14 to 30 inch, to 42 inch then 52″ and 65″ sets are widely available. LG Electronics has recently announced its 55″ OLED (Organic LED) TV claiming much superior performance.

32.7   PERFORMANCE OF PLASMA COLOUR TV RECEIVER

The development of both Plasma and LCD televisions began almost at the same time. The only major problem Plasma TV designers faced was screen "burn-in". This occurs when an image is left too long on the screen resulting in the ghost of the image "burned-in". Newer sets are less susceptible to this because of improved technology and introduction of features like screen savers and pixel orbiting.

In general both Plasma and LCD sets produce good quality pictures but Plasma produce higher levels of brightness and contrast levels than the LCDs because the pixels of their screen structure are either 'ON' or 'OFF' at any given instant. In terms of bulk, LCDs are generally lighter with depth around 2 inches than similar sie Plasmas which are about 3 inches deep. LCD supporters believe that LCDs have a longer life span than Plasma screens. This may be true for earlier receivers but the present Plasmas have a life span of more than 20,000 hours of viewing which is nearly the same as of LCDs. Plasmas are heavier, use more power and run hotter than LCD televisions. Therefore, more planning is required for mounting them on the wall. Plasmas are generally best installed by professionals. For LCDs, end users can easily do so themselves and can even fix the receiver on a suitable stand. To be competitive 'PANASONIC' has also marketed relatively light weight PLASMA. TV receivers. Equally bright pictures are received on these receivers. fig 32.2 shows such a receiver.

Chapter 32

Fig. 32.2. Plasma receiver's side view showing its light weight and slim screen

Nevertheless, if the need is for a big screen television– 50 inches and above, PLASMA continues to be a preference. Big Plasma TV sets with screen sizes 103 and 150 inches are available. Though bigger, and enable excellent pictures, these are very expensive.

32.8 PROJECTION TELEVISION

There has always been a quest for viewing movies and other entertaining programmes on bigger screen television receivers. Cathode-ray tube screens top out at 40″ (= 100cm) or so.

Chapter 32

This led to the development of projection TV receivers where small CRTs form the picture which on magnification is projected by a reflective mirror onto the screen located on the upper part of TV receiver. However, pictures thus formed are not very bright and lack details. To circumvent these drawbacks a new technology called Micro Elector-Mechanical Systems (MEMS) was developed for use in projection receivers. In this system, chips like Digital Micromirror Devices (DMD) are used to develop the picture. The pictures thus obtained on large screens are very bright with excellent details. Such projection receivers-REAR and FRONT display type are now used in Home Entertainment Theatres. This, along with the installation of a Dolby surround sound stereo speaker system for the accompanying audio in the hall brings the Home Theatre close to a commercial cinema complex. A typical setup is show below in Fig. 32.3.

Fig. 32.3. Screen side view of a modern Home Theatre set-up.

REVIEW QUESTIONS

1. Explain briefly the construction of an LCD screen panel for use in television.
2. What is Plasma? Describe how it enables production of colour pictures on the TV screen.
3. Explain the distinguishing features of LCD receivers and compare these with the performance of other Plasma cousins.

Chapter 32

33

Three Dimensional (3D) Television

The two-dimensional (length and breadth) pictures as we see on the television receiver screen look flat because these lack depth. However, in three-dimensional (3-D) viewing depth is also depicted on the screen and the picture appears to have all the qualities of a live scene as viewed with natural vision. The 3-D picture seems to extend beyond the screen at its back and also in the front. As an illustration, the road in any picture would appear stretching back behind the screen. Similarly, bullets fired in any scene will look as if shooting out of the screen and a cricket ball during a huge sixer would appear to fall into the viewer's lap. We human beings are able to see in three-dimensions since our eyes (left and right) see a slightly different image of the same subject while viewing and the brain interprets these as a single composite image in 3-D instead of two overlapping pictures. This is the clue to 3-D television.

While many systems to obtain 3D pictures were developed earlier but later another technique which looked superior came into use. In this system, the TV had an LCD screen, on which both the left and right images were shown in alternate columns of picture elements. Covering the screen were corresponding columns of double convex lenses for directing the images to respective eyes. However, the 3D pictures thus obtained, though superior still lacked the desired resolution and display of colours.

The real breakthrough came around the year 1990 when signal processing techniques like data compression codecs, high speed A/D and D/A digital converters and fast operating computers became available for precise manipulation of the two video signals. Also, in the meantime superior LCD display screens became available that enable better picture formation, thus dispensing with the earlier CRT screens.

While research work on LCD, PLASMA and 3D TVs continued side by side the success in obtaining correct stereoscopic effect to create 3-dimensional pictures eluded perfection for many years. However, now 3D ready TVs have been demonstrated which can operate in 3D mode with special glasses in addition to the regular 2D mode. In these TVs LCD shutter viewing glasses are used where the TV through sensers tells the glasses which eye should see the image being exhibited at the moment thus creating the illusion of a 3D image. For further enhancing the picture quality, scanning *i.e.*, refresh rate has been increased. This was not possible earlier with CRT screen.

33.1   BASICS OF 3D TECHNOLOGY

With the arrival of 3D TV receivers interest sprung-up all-around to know the basics of TV technology and its operating features. This is best answered in question-answers form as follows:

1.   **What is 3D TV Technology.** The 3D system relies on a visual process called STEROSPIS which enables 3D perception. This comes out of the fact that the eyes of an adult human lie about 2.5 inches ($\approx$ 6.5 cm) apart which lets each eye see objects from slightly different angles. The combined effect of this on the viewer's mind is that of a three dimensional picture. This ability of our eyes is used in 3D TVs by showing the same object from different angles on the screen to cause 3D perception.

2.   **How is 3D effect obtained on a 2D screen.** A 2D TV screen showing 3D content displays two separate images simultaneously of the same object, one intended for the viewer's right eye and the other for the left eye. These appear intermixed with one another if seen without special glasses. To prevent this, special tinted glasses are used which enable the eye to see the two images separately thus creating the same illusion as obtained on seeing the object directly. This amounts to stereoscopic effect to cause a third dimension *i.e.*, depth to current display which is typically limited to only length and breadth (2D).

3.   **What is the difference between the OLD and NEW 3D Technology.** The earlier 3D methods were based on the use of two colour signals in different ways. The pictures though seen with special glasses, were usually discoloured. The resolution of pictures was also not good due to the limitations of 50/60 Hz scanning rate with CRT screens.

In the new 3D technology this problem has been solved by using special active liquid crystal shutter glasses which work by very quickly blocking the left and then the right eye in sequence at a higher rate. To ensure this sequencing, electronics in the form of an IC chip and battery is provided in the glasses to obtain exact synchronism between the glasses and pictures appearing on the screen of TV receiver. The Sync is achieved through sensers fixed on the TV panel and glasses by an infrared signal.

4.   **Is it Necessary for Everyone Watching 3D TV to Use Special Glasses.** Yes, everyone sitting around 3D TV must wear special glasses to see the 3D effect. Without glasses the image on the TV screen will appear doubled, distorted and for most practical purposes, unwatchable. Currently there is no technology that enables watching 2D and 3D content simultaneously without glasses.

5.   **Is New TV Necessary for 3D.** As of now the current H.D. TVs cannot be upgraded to support the new 3D formats. Thus a new 3D TV will be needed to watch 3D programmes.

6.   **Is 3D TV any Good or Just a Gimmick to Sell a New Product.** Watching 3D TV on a 2D screen is making eyes to believe that it is 3D presentation. As such, 3D viewing may in a simple way be called a gimmick or a deception to the viewer's eyes. To begin with, watching 3D content can cause headache after extended period. For children it can be more strenuous as their eyes are closer together than of an adult. But when evaluating 3D TV if it is "any good" it will be worth recalling all that has been described, while answering other questions. It can be argued that we human beings adopt that appears to be good and in due course of time

Chapter 33

get used to new concepts. Thus, after viewing 3D content regularly it will look natural and thus acceptable.

33.2 PRESENT STATUS OF 3D RECEIVERS

Broadcasters all over the world have started 3D transmissions but it is not growing, may be due to the lack of available software and high cost of quality 3D receivers. Such sets are called 3D.-Ready, because these can operate in 3D mode in addition to the regular 2D mode in conjunction with LCD shutter glasses where the TV tells the glasses which eye should see the image being exhibited at the moment creating stereoscopic image.

While major TV receiver makers like LG, Panasonic, Samsung, Sony, Toshiba, Vizio and Philips are advertising their 3D receivers but their sale appear to be limited. Hopefully, when improved versions of 3D receivers become available, not needing special glasses, and regular broadcasts of 3D programmes become common will be the right time to buy a 3D TV receiver.

REVIEW QUESTIONS

1. What is 3D TV technology and how is it difference from the 2D presentation?
2. Why is it necessary to use special glasses for watching 3D programmes?
3. What is the present status of 3D broadcasting and availability of 3D ready receivers?

Appendices

Appendix A

Conversion Factors and Prefixes

1 ampere (A)	= 1 C/sec	1 mile	= 5,280 ft
1 angstrom unit (Å)	= 10^{-10}m		= 1.609 Km
1 coulomb (C)	= 1 A-sec	milli (m)	= × 10^{-3}
1 electron volt (eV)	= 1.60×10^{-19} J	nano (n)	= × 10^{-9}
1 farad (F)	= 1 C/V	1 newton	= 1 Kg-m/sec^2
1 foot (ft)	= 0.305 m	Permeability	
1 giga (G)	= × 10^9	of free space	
1 henry (H)	= 1 V-sec/A	(μ_0)	= $4\pi \times 10^{-7}$H/m
1 hertz (Hz)	= 1 cycle/sec	Permittivity	
1 inch (in.)	= 2.54 cm	of free space	
1 joule (J)	= 10^7 ergs	(e_0)	= $(36\pi \times 10^9)^{-1}$
kilo (K)	= × 10^3		F/m
1 kilogram (Kg)	= 2.205 lb	pico (p)	= × 10^{-12}
1 kilometer (Km)	= 0.622 mile	1 pound (lb)	= 453.6 g
1 lumen	= 0.0016 W	1 tesla (T)	= 1 Wb/m^2
	(at 0.55 µm)	1 ton	= 2,000 lb
1 lumen per square		1 volt (V)	= 1 W/A
foot	= 1 ft-candle	1 watt (W)	= 1 J/sec
mega (M)	= × 10^6	1 weber (Wb)	= 1 V-sec
1 meter (m)	= 39.37 in.	1 weber per	
micro (µ)	= × 10^{-6}	square meter	
1 micron	= 10^{-6}m	(Wb/m)^2	= 10^4 gauss
1 mil	= 10^{-3} in.		

Note 1: Since nano and kilo-pico mean the same multiplier *i.e.*, 10^{-9}, nano-farads (nF) or (n) and kilo-pico farads (KpF) are used interchangeably to denote capacitance values.

Note 2: To minimise congestion of information on the diagrams, the symbol Ω is often omitted from resistor values *i.e.*, 50 Ω, 100 Ω, 10 KΩ, 1 MΩ etc. appear as 50, 100, 10 K, 1 M etc.

Appendix B
Transient Response and Wave Shaping

Deflection coils are essentially reactive-resistive circuits. In order to obtain linear rise of current for uniform deflection of the electron beam, non-sinusoidal voltages are applied to the yoke coils. This involves transient response in LCR circuits. Similarly RC and RL circuits are used for defferentiating and integrating synchronizing pulses for control of scanning oscillators. In order to fully understand the generation of such waveshapes it is necessary to have a good grasp of the phenomena of transient response and time constant in such circuits.

RC circuit—In the circuit of waveshapes B.1 (a), a step pulse of magnitude E and of relatively long period as compared to $RC*$ of the circuit is applied to the series RC combination. Assuming that the capacitor has no charge on it initially i.e., $v_C = 0$ at $t = 0$, the voltage across C rises exponentially when current flows through it and approaches E as limit according to the equation $v_C = E\,(1 - e^{-t/RC})$. This equation can be easily derived by writing the general equation of voltage drops around the circuit:

$$E - IR - \frac{1}{C}\int_0^t i\,dt = 0 \tag{B.1}$$

On differentiating this and solving for i we obtain

$$i = I_0\,e^{-t/RC} \tag{B.2}$$

where
$$I_0 = E/R \tag{B.3}$$

Substituting this in equation (B.2) we obtain

$$i = E/R\,e^{-t/RC} \tag{B.4}$$

$$\therefore \qquad v_C = \frac{1}{C}\int_0^t I_0\,e^{t/RC}\,dt = \frac{1}{C}\int_0^t E/R\,e^{-t/RC}\,dt$$

$$= E\,(1 - e^{-t/RC}) \tag{B.5}$$

Capacitor charge—The equation (B.4) implies that on application of the step voltage (at $t = 0$) the amount of current flow is maximum $(= E/R)$ since there is no voltage across the capacitor to oppose the applied voltage. The capacitor charges most rapidly at the beginning of the charging period. However, as C charges, the capacitor voltage increases and the net drive voltage decreases thereby reducing the circuit current. The fall in current is exponential (equation B.4) and so is the rise of voltage across the capacitor (equation B.5). The voltage

---

$*R \times C = V/I \times Q/V = \dfrac{I \times t}{I} = t$. Thus RC product has the dimension of time.

across the resistor (v_R) is always equal to $i \times R$. This voltage has the same waveshape as that of the current because the resistance is constant. The waveshapes shown in Fig. B.1 (a) illustrate the exponential behaviour of the circuit. Note that when C is fully charged the current is zero and voltage across R also drops to zero. At any instant the voltage drops around the circuit must equal the applied voltage. Thus when $E = 10$ V, then at any time when C is charged to 6.3 V, $v_R = 3.7$ V.

Capacitor discharge—As soon as the input voltage become zero, current starts flowing in the opposite direction and the capacitor discharges exponentially till v_C becomes zero. This is also illustrated in Fig. B.1 (a). Such a behaviour can be verified by writing voltage loop equation at $t = t_1$. Thus

$$0 = \frac{1}{C} \int_{t_1}^{t} i\,dt + Ri \qquad (B.6)$$

Equation (B.6) on solving yields the following results

$$i = -E/R\ e^{-(t-t_1)/RC} \qquad (B.7)$$

and

$$v_C = E\ e^{-(t-t_1)/RC} \qquad (B.8)$$

These equations confirm the statements made above. The voltage across the capacitor decreases most rapidly at the beginning of the discharge because v_C, now acting as the applied voltage has the highest value and can drive maximum discharge current around the circuit. The magnitude of the discharge current has the peak value $= E/R$ at $t = t_1$ and then decreases because of the declining value of v_C. After C is completely discharged the current is zero. The voltage v_R has the same wave shape as that of the current.

RC time constant—The series resistance R limits the amount of current. A larger value of C requires a longer time to charge to the final value. Thus both R and C determine the rate at which the circuit current rises and falls. A convenient measure of the charge and discharge time of the circuit is the RC product. Since this has the dimension of time it is called time constant (τ) of the circuit. With R in ohms and C in farads, τ has the dimension of seconds. From equation (B.5) it is obvious that v_C/E after one time constant (τ) is 63 percent. Thus the capacitor charges to 63 percent of the maximum applied voltage in a time equal to one RC. Similarly equation (B.4) yields the value of $i = I_0\ e^{-1} = 0.37\ I_0$ for $\tau = 1$. Thus the current drops by 63 percent of the maximum value to become 37 percent of its original value. In the same manner during discharge the time required for C to discharge 63 percent or down to 37 percent of its original voltage, is one time constant (RC).

Note that the capacitor is practically fully charged to the applied voltage after a time equal to four* time constants. On discharge, the capacitor voltage is reduced to almost zero voltage after four time constants.

In conclusion note the following facts about the charge and discharge of the capacitor. When the applied voltage is more that v_C the capacitor will charge. The capacitor will continue to charge as long as the applied voltage is maintained and is greater that v_C. The rate of charging is determined by the RC time constant. When v_C equals the applied voltage, however, the capacitor cannot take any more charge regardless of the time constant. Furthermore, if the

Appendix B

*In a period = 4 RC, the voltage across C is 98 percent of the applied voltage.

Appendix B

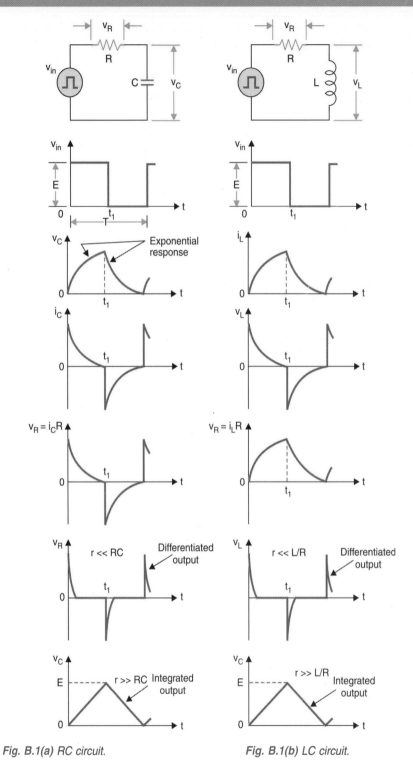

Fig. B.1(a) RC circuit.   Fig. B.1(b) LC circuit.

applied voltage decreases the capacitor will discharge. The capacitor will discharge as long as its voltage is greater than the applied voltage. Once the capacitor voltage drops to zero, it cannot discharge any more. Thus the capacitor charges as long as the applied voltage is greater than v_C. Similarly the capacitor discharges as long as v_C is able to produce discharge current. If C is discharging and the applied voltage changes to become greater than v_C, the capacitor will stop discharging and start charging, In any case both charge and discharge paths are exponential and obey the general equations derived earlier. Also it should be noted that 63 percent change of voltage in one RC time refers to 63 percent of the net voltage available for producing charging and discharging currents. As an example if 10 V is applied for charging a capacitor that already has 2 V across it, the capacitor voltage will increase by 63 percent of 8 V in one RC time, adding 5.04 V. This when added to the original 2 V produces 7.04 V across the capacitor.

RL circuit—In B.1 (*b*), a series RL circuit and associated voltage and current waveshapes are shown. Note that when voltage is applied to such a circuit, the current cannot attain its steady-state value instantaneously but builds up exponentially because of the self-induced voltage across the inductance. The general equation of the voltage drop around the circuit is

$$E = iR + L\,\frac{di}{dt} \tag{B.9}$$

On solving this equation we get $i = E/R\,(1 - e^{-Rt/L})$ $\hfill$ (B.10)

and $\qquad\qquad\qquad\qquad v_L = Ee^{-Rt/L}$ $\hfill$ (B.11)

Similarly when the step voltage E drops to zero at $t = t_1$, Equation (B.9) yields

$$i = E/Re^{-(R/L)(t-t_1)} \tag{B.12}$$

and $\qquad\qquad\qquad\qquad v_L = -\,Ee^{-(R/L)(t-t_1)}$ $\hfill$ (B.13)

Thus it is obvious from the above results that the rise of inductive current corresponds to rise of voltage across the capacitor (see Fig. in B.1). Similarly the fall of voltage across the inductor corresponds to fall of current in the RC circuit. The transient waveshapes are exactly the same, with current in the inductive circuit substituted for capacitive voltage. The voltage across R follows the current variations in either case.

The time constant for an inductive circuit is given by $\tau = R/L^{\dagger}$, where τ is in seconds, with L in henries and R in ohms. The time constant indicates similar 63 percent change as in the RC circuits. The transient response is completed in four time constants. However, note that higher resistance provides a shorter time constant for the current rise or decay in an inductive circuit.

Universal time constant curves—Fig. B.2 shows universal time constant curves and these can be used for determining voltage and current values for any amount of time in RC and RL circuits. Note that the horizontal axis is in units of time constants rather than the absolute values. The vertical axis denotes percentage change in full voltage or current.

Differentiated output—The time constant which is at least one-fifth (preferrably one-tenth) of the period of the applied voltage is called a short time constant. Such a combination permits

---

$\dagger L/R = \dfrac{V \times t}{I} \times I/V = t$. Thus L/R has the dimension of time.

the capacitor in an RC circuit to fully charge and discharge much before the next cycle of the input voltage commences. In such a circuit (both R and C small) the voltage across R is called differentiated output because its amplitude can change instantaneously in either polarity. In the circuit of Fig. B.1 (a), $iR + \dfrac{1}{C}\displaystyle\int_0^t i\,dt = v_{\text{in}}$. With a short time constant (both R and C small)

i.e.

$$R \ll 1/2\pi fC, \frac{1}{C}\int_0^t i\,dt \approx C \times v_{\text{in}}$$

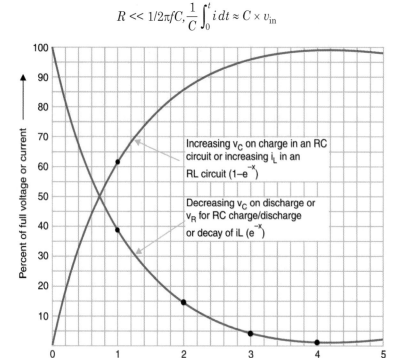

Fig. B.2. Universal time constant curves.

This on differentiating yields

$$i \approx C\,\frac{dv_{\text{in}}}{dt}$$

$$\therefore \qquad v_R = iR \approx RC\frac{dv_{\text{in}}}{dt} \tag{B.14}$$

Thus as illustrated in Fig. B.1 (a) the voltage across R, when RC is quite small, is the differentiated version of the input votage. As shown, it consists of positive and negative spikes corresponding to the leading and trailing edges of the square wave input. Such pulses are used for timing applications.

In an RL circuit having a short time constant, v_L provides (see Fig. B.1 (b)) differentiated output of the input signal. The instantaneous value of v_L can be calculated as $L\,di/dt$ where di/dt is the rate of change of current.

Intergrated output—A long time constant is at least five times (preferably ten times) the period of the applied voltage. Then the capacitor in an RC circuit cannot take any appreciable charge before the applied voltage drops to start the discharge. Also, there is little discharge before the charging voltage is applied again. In such a circuit (Fig. B.1 (*a*)) where both R and C are large ($R \gg 1/2\pi fC$)

$$v_i \approx i \times R$$

$$\therefore \qquad i = \frac{v}{R}$$

and

$$v_C = \frac{1}{C} \int_0^t idt \approx \frac{1}{RC} \int_0^t v_i dt \qquad\qquad\qquad \text{(B.15)}$$

Thus the output voltage (v_C) in an RC circuit having a large time constant, is the integrated output of the input voltage. In an RL circuit v_R can provide integrated output of the input signal since the current cannot change instantaneously in such a circuit.

Waveshaping—Based on the behaviour of RC and RL circuits to non-sinusoidal inputs, such circuits are used for waveshaping. Fig. B.3 (*a*) shows consequences of differentiating and integrating two typical non-sinusoidal voltage sources. In television receivers sync pulses are passed through such circuits to obtain trigger pulses for the deflection oscillators. Similarly, as shown in Fig. B.3 (*b*) a rectangular voltage provides a sawtooth current in an inductive circuit. A non-sinusoidal voltage of nearly this shape is applied to the deflection coils of a picture tube to obtain linear deflection on the electron beam.

Appendix B

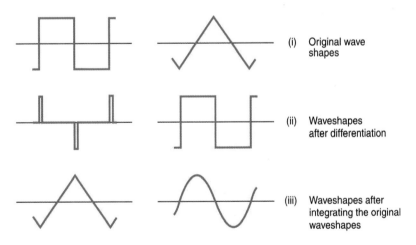

(i) Original wave shapes

(ii) Waveshapes after differentiation

(iii) Waveshapes after integrating the original waveshapes

Fig. B.3(a). Consequences of differentiating and integrating typical non sinusoidal waveshapes

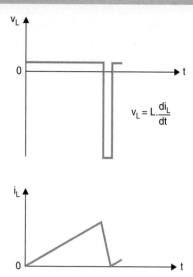

Fig. B.3(b). *Application of a rectangular voltage to an inductive circuit results in a sawtooth current through it.*

Appendix C

Television Broadcast Channels

Band I	Lower VHF range	41 to 68 MHz
Band III	Upper VHF range	174 to 230 MHz
Band IV	UHF range	470 to 598 MHz
Band V	UHF range	606 to 870 MHz

(Band II (88 to 108 MHz) is used for FM broadcasting)

Television Channel Allocation (CCIR) in Bands I and III)

Band	Channel No.	Frequency range	Picture carrier Frequency (MHz)	Sound carrier Frequency (MHz)
I	1	41–47 (not used)		
(41-68 MHz)	2	47–54	48.25	53.75
	3	54–61	55.25	60.75
	4	61–68	62.25	67.75
III	5	174–181	175.25	180.75
(174–230 MHz)	6	181–188	182.25	187.75
	7	188–195	189.25	194.75
	8	195–202	196.25	201.75
	9	202–209	203.25	208.75
	10	209–216	210.25	215.75
	11	216–223	217.25	222.75
	12	223–230	224.25	229.75

In the UHF bands while the channel width remains the same at 7 MHz, a band gap of 1 MHz is allowed between adjacent channels to prevent any mutual interference. The frequency allocations are as given on the next page.

649

UHF BAND IV (470–598 MHz) Channels 21–36

Band (UHF)	Channel No.	Frequency range (MHz)	Picture Carrier frequency (MHz)	Sound Carrier frequency (MHz)
IV	21	470 to 478	471.25	476.75
Channel	22	478 to 486	479.25	484.75
21–36	23	486 to 494	487.25	492.75
	24	494 to 502	495.25	500.75
	25	502 to 510	503.25	508.75
	26	510 to 518	511.25	516.75
	27	518 to 526	519.25	524.75
	28	526 to 534	527.25	532.75
	29	534 to 542	535.25	540.75
	30	542 to 550	543.25	548.75
	31	550 to 558	551.25	556.75
	32	558 to 566	559.25	564.75
	33	566 to 574	567.25	572.75
	34	574 to 582	575.25	580.75
	35	582 to 590	583.25	588.75
	36	590 to 598	591.25	596.75

UHF BAND V (606–870 MHz) Channels 37–69

Channel No.	Frequency band (MHz)	Channel No.	Frequency band (MHz)
37	606–614	54	742–750
38	614–622	55	750–758
39	622–630	56	758–766
40	630–638	57	766–774
41	638–646	58	774–782
42	646–654	59	782–790
43	654–662	60	790–798
44	662–670	61	798–806
45	670–678	62	806–814
46	678–686	63	814–822
47	686–694	64	822–830
48	694–702	65	830–838
49	702–710	66	838–846
50	710–718	67	846–854
51	718–726	68	854–862
52	726–734	69	862–870
53	734–742		

Note : With channel range from a to b MHz, the picture carrier is $(a + 1.25)$ MHz, and sound carrier $(b - 1.25)$ MHz.

Index

Index

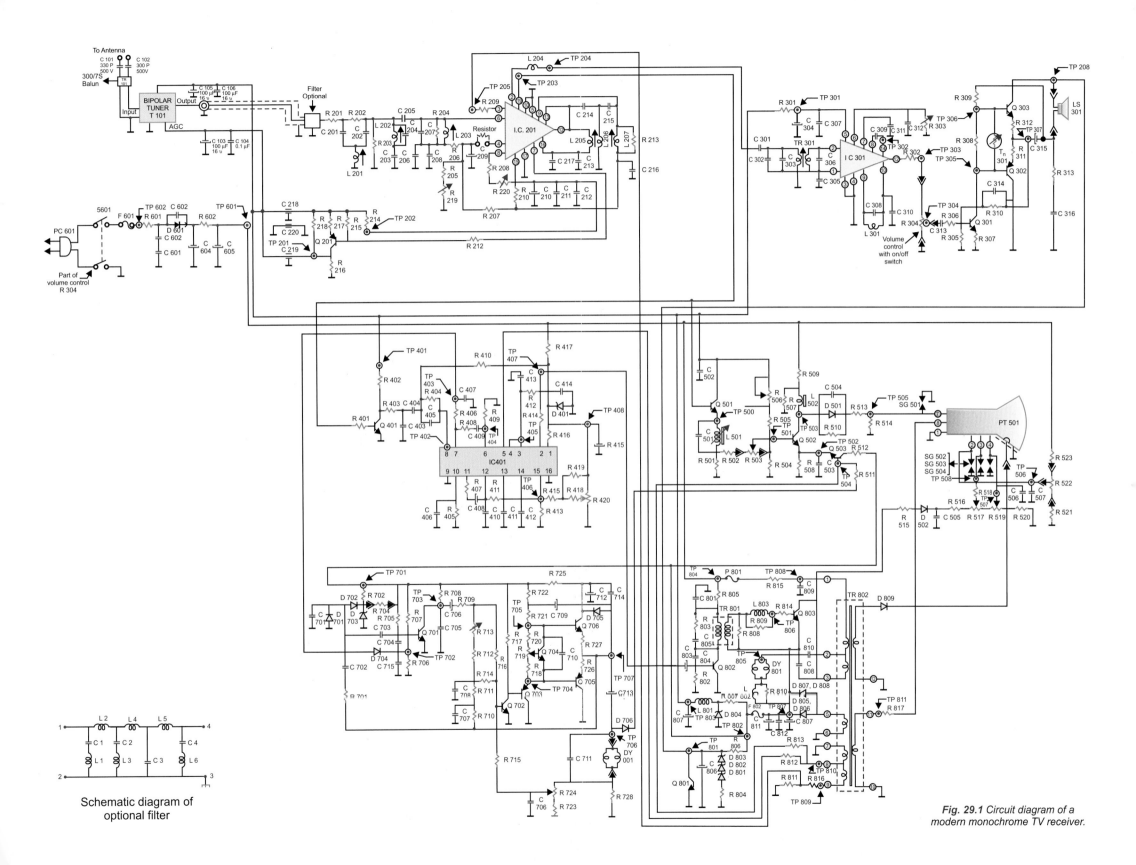

Schematic diagram of optional filter

Fig. 29.1 Circuit diagram of a modern monochrome TV receiver.